PLASMA
ASTROPHYSICS

PLASMA ASTROPHYSICS

T. Tajima
The University of Texas at Austin

K. Shibata
National Astronomical Observatory of Japan

The Advanced Book Program

PERSEUS PUBLISHING
Cambridge, Massachusetts

Copyright © 2002 by T. Tajima and K. Shibata

Westview Press books are available at special discounts for bulk purchases in the United States by corporations, institutions, and other organizations. For more information, please contact the Special Markets Department at The Perseus Books Group, 11 Cambridge Center, Cambridge MA 02142, or call (617) 252-5298.

Published in 2002 in the United States of America by Westview Press, 5500 Central Avenue, Boulder, Colorado 80301–2877, and in the United Kingdom by Westview Press, 12 Hid's Copse Road, Cumnor Hill, Oxford OX2 9JJ

Find us on the World Wide Web at www.westviewpress.com

A Cataloging-in-Publication data record is available from the Library of Congress.
ISBN 0-8133-3996-0

The paper used in this publication meets the requirements of the American National Standard for Permanence of Paper for Printed Library Materials Z39.48–1984.

10   9   8   7   6   5   4   3   2   1

Frontiers in Physics

David Pines, Editor

Volumes of the Series published from 1961 to 1973 are not officially numbered. The parenthetical numbers shown are designed to aid librarians and bibliographers to check the completeness of their holdings.

Titles published in this series prior to 1987 appear under either the W. A. Benjamin or the Benjamin/Cummings imprint; titles published since 1986 appear under the Addison-Wesley imprint.

Volumes published from 1974 onward are being numbered as an integral part of the bibliography.

Editor's Foreword

The problem of communicating in a coherent fashion recent developments in the most exciting and active fields of physics continues to be with us. The enormous growth in the number of physicists has tended to make the familiar channels of communication considerably less effective. It has become increasingly difficult for experts in a given field to keep up with the current literature; the novice can only be confused. What is needed is both a consistent account of a field and the presentation of a definite "point of view" concerning it. Formal monographs cannot meet such a need in a rapidly developing field, while the review article seems to have fallen into disfavor. Indeed, it would seem that the people who are most actively engaged in developing a given field are the people least likely to write at length about it.

Frontiers in Physics was conceived in 1961 in an effort to improve the situation in several ways. Leading physicists frequently give a series of lectures, a graduate seminar, or a graduate course in their special fields of interest. Such lectures serve to summarize the present status of a rapidly developing field and may well constitute the only coherent account available at the time. One of the principal purposes of the *Frontiers in Physics* series is to make notes on such lectures available to the wider physics community.

As *Frontiers in Physics* has evolved, a second category of book, the informal text/monographs, has played an increasingly important role in the series. In an informal text or monograph an author has reworked his or her lecture notes to the point at which the manuscript represents a coherent summation of a newly-developed field, complete with references and problems, suitable for either classroom teaching or individual study.

Plasma Astrophysics is just such a volume. The authors are distinguished plasma theorists and theoretical astrophysicists who have taught a graduate course on this topic since 1988. They provide a lucid account of the nonlinear processes and instabilities which are the source of so many of the violent phenomena observed in the universe. Intended for the graduate student, their book introduces the reader to the remarkably broad range of astrophysical phenomena (including the early universe) which involve plasma processes in an essential way. With its focus on astronomical phenomena, it complements two earlier volumes in this eries (Setsuo Ichimaru's *Statistical Plasma Physics* and Tajima 's earlier Frontiers volume *Computational Plasma Physics*), and should prove of interest to both observational and theoretical astrophysicists. It gives me great pleasure to welcome Dr. Shibata and welcome back Dr. Tajima, to the ranks of authors in the *Frontiers in Physics* series.

David Pines
Urbana, IL
August, 1997

Contents

Preface

The 20th Century has witnessed the transformation of astronomy from celestial mechanics to astrophysics. We discovered not only various stars, but also many kinds of congregates of stars: globular clusters, galaxies, then clusters of galaxies, superclusters, and so on. In a sense this parallels (in a reverse way in scale and in hierarchy) the discovery in physics that starts from atoms, then goes down in scale to electrons and nuclei, and then subnuclear particles, etc. In the latter part of this century, however, astronomy has not only advanced with optical telescopes but has also been galvanized by other instruments such as radio telescopes, X-ray and γ-ray telescopes, and even non-photon instruments (such as neutrino detectors).

With a broad brush, we might paint the current picture of astronomy so as to say that while optical astronomy may have presented a peek into the structure of the constituents of the Universe, such as stars and galaxies – often in a magnificently stable configuration – these new windows of observation have revealed that there are far more amorphous objects such as nebulae, sheets, filaments, and voids; and constituent astrophysical objects themselves often present violent processes such as flares, shocks, accretion disks, and jets. Our picture of the Universe has changed from the quiet Universe to the violent Universe. We will find in this book that this tinder box of the violent Universe is often filled with a combination of gravitational and electromagnetic interactions, both of which have long-range interactions. Further, we will learn that these processes underlie many different astrophysical objects from stars to interstellar media to galaxies to clusters. In these processes the constituent matter is often a plasma and it acts not only as the constituent matter but also as a medium through which the astrophysical setting becomes so volatile. Therefore, the science and the language of science to describe this situation must not be of a static nature, but has to be a dynamical one.

Physics (or science in general) comes in with two fundamental approaches. One is by deduction (analysis) and the other by induction (synthesis). The first approach is that of (a broad sense of) particle physics, which includes "elementary" particle physics, nuclear physics, and atomic and molecular physics. The second approach is that of statistical physics (or matter physics), which includes statistical mechanics, solid-state (or condensed matter) physics, hydrodynamics, plasma physics, and astrophysics. We note that plasma astrophysics is the confluence of two of the synthetic physics approaches, as both plasma physics and astrophysics handle volatile unstable "matter." We consider the main intellectual challenge of plasma astrophysics to lie in understanding and describing this very matter and its processes. Contemplating its nature, we decided that it is not sufficient for this book to describe a classification of various astrophysical objects (or phenomena) and their properties, as these objects of our interest are often too violent to be discussed as a stationary subject. In contrast to this traditional vertical approach (with warp), perhaps we may need a *horizontal approach weaving various processes (with woof)*, as we discuss below. We observe that the underlying processes are the ones that manifest these phenomena of signature events in astrophysics. Moreover, we observe that these processes may recur in astrophysical objects in disparate spatial (and temporal) scales, as plasma and astrophysical phenomena are both hierarchical in nature due to their fundamental interaction being a long-range one. (The hierarchical nature of plasma process has been emphasized in Tajima's book "Computational Plasma Physics" in this series.)

Thus we decided to describe the plasma astrophysics in this book through fundamental processes that govern various plasma astrophysical phenomena. In this way we believe that the current book may be complementary to Parker's "Cosmical Magnetic Fields," and perhaps it may be said that the present one presents a modern viewpoint of plasma astrophysics, or plasma physics itself. With this philosophy in mind, it should be noted, we do not reproduce the full view of introductory plasma physics. We emphasize the description of the above "modern" viewpoints of plasma physics that may not be found in conventional plasma physics books. An excellent book to fill in the basic plasma physics knowledge may be found, e.g., in S. Ichimaru's "Statistical Plasma Physics I" in this series.

In Chapter 1 we introduce our view of plasma astrophysics through fundamental processes of quasi-magnetostatic equilibria, quasi-hydrostatic equilibria, and non-equilibria. In Chapter 2 after a *very* brief survey of plasma physics, we present unique approaches to the astrophysical plasmas and their processes. The conventional plasma physics approach should be to establish equilibria, and then to study their stability, and to go on to study its nonlinear behavior. In astrophysics, we believe, this conventional approach often may not be sufficient (nor adequate). This is because in many astrophysical settings the phenomena we observe are the end results of the nonlinear evolution of instabilities fully played out or entering into their own self-consistent regimes far removed from these "initial equilibria." Instead, we discuss fluctuations, structures, nonlinear dynamics, relaxation, self-organization, and general relativistic processes. These are, we believe, much more fitting to the violent astrophysical plasmas, while the conventional approach may be to the more quiescent, benign, contained laboratory plasmas.

In Chapter 3 we describe the fundamental processes in hydrostatic and magnetostatic objects. Primary processes are: dynamo, buoyancy due to magnetic fields, and magnetic reconnection. Each of these processes, as commented above, can manifest itself in various astrophysical objects such as the Sun, galaxies, and disks. In Chapter 4, we discuss the fundamental processes in gravitational objects (i.e., non-equilibria). We focus on the gravitational contraction process, flows with shear, and jets. The criterion of choice of these processes has been that each process for discussion recurs in many astrophysical settings and is a key to the current outstanding problems. There may be many classic processes that have been solved for specific astrophysical problems. However, they may not fit such a criterion if they are self contained and do not have any impact on contemporary problems of astrophysical interest. Conversely, even if the problem is a classic one, if it has relevance in contemporary problems, we try to present such. An example of the latter is Parker's solar wind solution, which is not only a solution to the solar wind, but may have importance for jets and other phenomena. Thus our choice of subjects hopefully makes the book vibrant with contemporary subjects and topics interleaved each other and necessarily open-ended to future developments. Therefore, we regard the book's natural home as this Frontiers in Physics series, as this monograph is a snapshot at our current knowledge.

Chapter 5 is dedicated to the applications of plasma astrophysics to the furthest arena, cosmology. Since cosmology has been discussed mostly in the context of general relativity and sometimes of particle physics, but seldom in the plasma astrophysics context, this chapter is a first attempt to lay down a few cornerstones of the foundation. In this sense this chapter serves as an anticipation or a design of the future exciting developments in this field.

This book stems from a series of lectures given by the authors in the past decade. The first lecture series, a graduate physics course entitled "Plasma Astrophysics," taught by one of the authors (TT), was opened in the Spring Semester of 1988 at the Physics Department of The University of Texas at Austin. It was during this period that the other author (KS) was on sabbatical at Austin and we jointly charted out the foundation of the course. (A similar course was once again given in the Spring of 1993 by TT.) Later KS gave a similar lecture series at Osaka University (1990 and 1994), at the University of Tokyo (1994), and at Kyoto University (1995).

The present book is intended primarily for graduate students, though most of it is digestible by advanced undergraduate students. At the same time the materials herein and the fashion in which we cover them should act as stimuli for researchers working at the frontiers of plasma astrophysics. In order for the reader to get a glimpse of some of the current theoretical techniques, sometimes we dare to leave in relatively heavy mathematics, particularly when they illuminate "modern" (or hard to find in other textbooks) approaches. In order to guide the readers navigation of the materials in this book, we put asterisks in the table of contents to indicate those sections that are more advanced and appropriate for research purposes, and may be skipped on the first round of reading.

The initiation of this lecture course was motivated by our realization of the confluence of two significant developments. The last few decades have seen many rapid and often revolutionary discoveries in astrophysics, such as pulsars, jets, cosmic microwave and X-ray backgrounds, galactic and extragalactic magnetic fields, γ-ray bursters, and a supernova. Also in the last few decades, a deep and sophisticated branch of physics, plasma physics, has matured, often incubated in applications to laboratory (fusion) plasmas. On the one hand, the rich discoveries of new phenomena often cry out for solutions that are difficult to understand without a deep grasp of plasma physics theory have been too closely tied to the specifics of fusion developments, and it seems to be losing its earlier basis in astrophysics in the tradition of Lyman Spitzer, H. Alfvén, S. Chandrasekhar, and others. We felt that a serious plasma physical study of such astrophysical problems should not only be extremely productive to astrophysics, but should also pose a vigorous stimulus to grow a new branch of plasma physics that attempts to lay the groundwork for the understanding of astrophys-

be skipped on the first round of reading.

We benefited immensely from discussion, camaraderie, and support from our teachers, colleagues, and students: S.I. Akasofu, S. Cable, W. Chou, B. Coppi, J. Daniel, J.M. Dawson, D. Gilden, T. Hanawa, T. Hirayama, M.R. Hayashi, S. Hirose, K. Holcomb, W. Horton, S. Ichimaru, C. Kennel, R. Kinney, Y. Kishimoto, T. Kosugi, T. Kudoh, R. Kulsrud, J.N. Leboeuf, R. Lovelace, K. Makishima, R. Matsumoto, M. Mori, M. Ohyama, S. Oliveira, V. Petviashvili (late), R. Rosner, N. Rostoker, T. Sawa, M. Shimojo, T. Taniuti (late), S. Tanuma, S. Tsuneta, Y. Uchida, F. Waelbroeck, T. Watanabe, S. Yashiro, and T. Yokoyama. Without the ceaseless dedication and incomparable skills of Mrs. Suzy Mitchell, this endeavor would never have been realized. Also, the tireless efforts of Ms. Darla Bouse in literature search and other vital assistance are noted. Professor J. Hawley kindly reviewed this book in draft form and gave us valuable comments. Professor E.N. Parker made a thorough, conscientious, and penetrating review of this from the first page to the end, and made innumerable constructive suggestions, which contributed much improved text. For these we would like to thank with sincerest gratitude. Ever since 1980 when TT moved to Austin, National Science Foundation has been supporting this plasma astrophysics work and we are glad to respond to the generosity with the completion of this book. The U.S. Department of Energy also supported the development of basic plasma science. The Institute for Fusion Studies as well as the University Cooperative Society Award of The University of Texas at Austin has supported this effort.

T. Tajima
K. Shibata
Austin, TX
March, 1997

Chapter 1

Introduction

1.1 Overall Structure of Plasma Physics and Astrophysics

Plasma astrophysics, as we cover it here, includes space plasma physics, solar physics, and plasma astrophysics in a narrower sense. Namely, it is a branch of plasma physics that studies celestial objects, phenomena, and their evolution; it is at the same time a branch of astrophysics whose perspective and techniques are derived from those of plasma physics.

Plasma astrophysics is among the youngest disciplines in astrophysics. Plasma physics, itself, as we generally identify it today, began merely three or four decades ago. In our opinion, plasma astrophysics will become an increasingly important discipline of physics and astronomy and eventually an essential component central to both plasma physics and astrophysics. This is based on the following four observations on modern astrophysics:

(i) Overwhelmingly major constituents of the universe are made of *plasmas*. This is true not only of the Universe we presently observe, but also of the past. The Universe before the recombination epoch ($t \sim 10^{13}$ sec after the *Big Bang*) was all plasma. Certainly between $t \sim 10^{-2}$ sec and $t \sim 10^{13}$ sec after the Big Bang the main interactive force was electromagnetic and the main constituent matter was plasma (see Fig. 1.1). Immediately after the *recombination*, most matter became neutral atoms, while perhaps one part out of a million or so remained to be ionized, thus forming weakly ionized plasmas. Even a minute fraction of ionization could be important, as this portion of matter could strongly interact with magnetic fields. Although we don't know exactly when, we do know that later the Universe reheated mainly due to the gravitational energy to form the Universe that is primarily made up of plasmas. One guess is somewhere around $t \sim 10^{16} - 10^{17}$ sec certainly not much later than the birth of quasars. Thus the traditional approach of plasma physics finds itself in a curious situation. In terms of equilibrium theory, although most laboratory plasma configurations are in or near equilibrium (otherwise, we would likely have lost it), many astrophysical plasmas may be out of equilibrium. In terms of stability theory, plasma physicists have discovered thousands of instabilities (e.g. Hasegawa, 1975), and in terms of transport theory, most hot plasmas have been found to exhibit so-called 'anomalous'

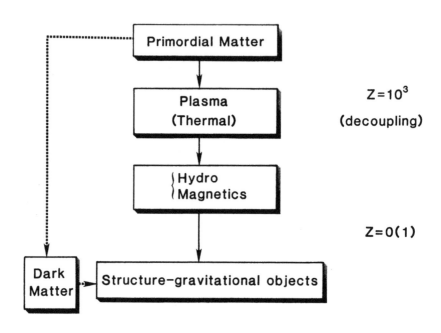

FIGURE 1.1 Evolution of universe, plasmas, and magnetic fields.

transport, much faster transport than the classical collisional transport. This theoretical situation might prompt one to ask if the framework of the traditional plasma physics is an appropriate paradigm, particularly for plasma astrophysics.

(ii) The traditional astronomy after Galileo is based primarily on optical telescopes for observation. Our view of the Universe has been that of the quiet Universe. Stars are stable, their relative motions, also. The interaction is gravitation alone. However, recent progress in observations beyond optical telescopes (but also including progress in optical telescopes) such as radio waves and x-rays, has shown a new emerging view of the Universe. Unlike the *quiet Universe* primarily gathered from optical data and based on the interaction of gravitation alone, an *active* and violent *Universe* gathered from a wide spectrum of observations such as radio, infrared (optical included), x-ray, gamma ray and other data has taken shape, which is based on the interplay of both gravitational interaction and plasma (electromagnetic) interaction. These observations show much more than stars as basic astronomical objects, such as clusters, nebulae, jets, disks, filaments, and cellular structures. The combination of these two interactions proves to be quite unstable, volatile, and often violent.

FIGURE 1.2 Hierarchical structure of our universe.

(iii) The theoretical description of astronomical phenomena has been quite limited until recently. It was mostly local theory such as local thermodynamics and local nuclear reaction theory (zero dimension theory). Saha'a law for interstellar media and Schwarzschild (1958)'s stellar structure are two of these examples in the former (0D thermodynamics), while Hayashi's model of stellar evolution (Hayashi *et al.*, 1962) is an example of the latter. In astrophysics, however, morphology and global structures play important roles. To explain such, we need to develop a global dynamic theory that applies to systems even in nonequilibrium, and nonlinear physics that can describe volatile, violent dynamics processes. Because of these the field has been left a relatively untouched, almost virgin area. These characteristics of recent astrophysics have much in common with plasma physics which often deals with higher than 0 and is after three-dimensional dynamics, global dynamics and morphology, nonequilibrium plasma behavior and nonlinear dynamics. Because astrophysics provides such difficult theoretical challenge, it can offer a unique opportunity to develop or test theory, a testing ground, that may be impossible in a terrestrial laboratory.

As the spatial scale of astrophysical plasmas is nearly infinite, a study of plasma turbulence in which energy may be cascaded to modes of greater scale ad infinitum (see Fig. 1.2). We may be able to observe morphologies of fully developed processes with their development time amply guaranteed, as the temporal scale is again nearly infinite. Recently physicists (plasma physicists certainly among them) have come to realize the need to learn or develop the language (and mathematical treatment) of describing such phenomena. The development of *nonequilibrium statistical mechanics* (for example, Bogoliubov, 1962), that of plasma physics in particular, the narrow sense of *nonlinear dynamics* (for example, Lichtenberg and Lieberman, 1983), *knot theory* (for example, Kaufman, 1987), and other fledgling fields may be tapped in this endeavor. Among these newly emerging languages and mathematics, perhaps most important (it seems to the author) is the progress of *computational physics*, in particular, that of computational plasma physics (Tajima, 1989). The progress in this area has been spurred on both by the intrinsic need of understanding complex nonlinear plasma behaviors and by the explosive progress of computers and associated innovative softwares/algorithms.

(iv) Astrophysical phenomena range from small to extremely large. For example, the basic astrophysical constituent objects range from, say, dusts, boulders, stars, clusters of stars, galaxies, superclusters, and large structures. Important characteristics in astrophysical phenomena are that these disparate scales and phenomena appear in a *hierarchical* manner and yet they are in some ways interconnected. That is, each hierarchy of structure and interaction does not close by itself but interrelates to other hierarchies. This is probably due to the nature of the long-range interaction of gravitation.

The traditional theoretical structure of plasma physics investigated in fusion research or laboratory plasma studies is as follows: the equilibrium is first established and studied, followed by study of (linear) stability and, if stable (perhaps nonlinearly), followed by study of transport processes of plasmas. The transport study tracks evolution of the plasma in the slowest time scale of interest. On the other hand, if the plasma is unstable, it becomes imperative to determine nonlinear evolution of the instability and possibly its saturation. However, this last step is a notoriously difficult one for theorists and remains the least-developed branch of plasma physics. One possible reason for this, other than the inherent difficulty involved in nonlinear physics, is that as far as global *magnetohydrodynamics (MHD) instabilities* are concerned, once they become unstable, they are difficult to get stabilized (nonlinearly), and the plasma suffers severe distortion, disruption, and perhaps losses. Thus, the linear stability analysis, as it often turns out, gives an appropriate margin of plasma confinement and presence. However, this is not necessarily the case when instabilities are not global MHD (due to such agents as magnetic shear) or are resistive or kinetic in origin. In these instances modes saturate after some growth due to a variety of mechanisms such as turbulence, soliton formation, modification of the unstable profile, etc. If this happens, the transport study has to incorporate this new plasma state, not the original state.

In astrophysics, on the other hand, the system is open, subject to gravitational fields but not subject to specific boundaries. Most often, all we see is the morphology of the evolved plasma state not necessarily tied with the linear theoretical prediction. For certain classes of phenomena that are not very violent, the above described traditional path of analysis may

6

still be applicable. For certain classes of phenomena that are rapid or violent, however, such an approach may be not only difficult but also inapplicable. In either case, the major interest and research thrust in plasma astrophysics is to understand the evolved state of plasma and thus most often nonlinear physics of the evolved plasma. Nonlinear physics, therefore, plays the major and central role in astrophysics.

Here we wish to characterize more modern (perhaps not universally accepted, but certainly our view of) plasma physics that will be immersed in our analysis of astrophysical phenomena. We discern five such points.

(i) We classify and approach the problems of plasma astrophysics in terms of the *fundamental processes* rather than the traditional categorization of equilibrium *states*, their stability properties and their resultant transport properties. At the same time, we try to avoid getting into a "zoology" of various minute processes and morphologies in plasma astrophysics by counting tit-for-tat.

(ii) We recognized that the processes of plasma are rich and varied, but they are *hierarchical* and *interrelated* at various levels (Tajima, 1989). We have said similarly on features in astrophysics. Here the reason for such structure is due to plasma's long-range Coulombic interaction.

(iii) The plasma dynamics will be looked at from *evolutionary* point of view, as opposed to or in addition to the approach of "determining the state of plasma." This characteristic, along with the difficulty (or impossibility) of any experimentation on astrophysical objects attaches importance to the approach of computer simulation, a third new method of Gedanken experiment via fast modern digital computers. We will naturally emphasize these aspects of plasma astrophysics in what follows.

(iv) We recognize that the plasma dynamics fairly easily and quite often evolve unstably and become *nonlinear*.

(v) Such plasma evolution is, by and large, *structured but complex*. Such a plasma may exhibit self-similar, fractal, or highly convoluted patterns (Mandelbrot, 1983). The plasma is likely to be far from equilibrium, far from homogeneous and uniform, i.e., it can be highly intermittent, filamentary, turbulent, or chaotic. Characterization by topology and/or morphology (Moffatt *et al.*, 1992) can be important, as higher dimensional structures in addition to zero or one-dimensional local properties matter.

By listing these characteristics one realizes that these attributes have much in common with those we listed for astrophysics. We believe that the coincidence is not accidental but springs from the deep common roots of these two disciplines. We will review plasma physics in Chap. 1 from this 'astrophysical perspective' based on these five points. Figure 1.3 depicts the relationships among the gravity, magnetics, and plasmas that play each role and interact with each other.

Those subjects within the field of plasma astrophysics that concern high frequency phenomena are omitted altogether in this book. For example, of immense interest is the coupling between radiation and plasma in such an environment as the pulsar magnetosphere. Nevertheless, we omit such a subject here. This is partly because many previous textbooks (for example, Melrose, 1980) and review articles cover such topics excellently. This is also because we want to sharpen our focus primarily on evolutionary plasma astrophysical process of the Universe. A major goal of astrophysics may be said to be the understanding of the "present" Universe (as we "see" it) and with that to understand the origin of the Universe and thus its evolution. In this book we make an attempt to understand the evolution of the Universe from the plasma astrophysics point of view. There seems to have been relatively few textbooks that tackle such a problem, although related topics have been covered by individual articles or books such as Zeldovich et al. (1983), Parker (1979, 1994), and Alfvén (1963) etc. This relative paucity of literature certainly is due in part to the young age of this field. Because of its infancy, perhaps, plasma astrophysics has so far failed to present a full scope of the world view it is eventually expected to cover.

For example, the classic topics of MHD dynamo of stars and planets and the evolution of these objects are an important area of plasma astrophysics. The dynamo is believed to play an important (if not crucial) role in planetary and stellar evolution. It may be argued that the terrestrial dynamo has played an important role in shaping the history of earth's atmosphere and thus the evolution of life on earth. Furthermore, it is very likely that the solar dynamo and its related activities such as convection, sunspots, flares etc. have played an equally important role in evolution. In fact, Eddy (1977) found from his radioactive carbon identification that up to 1 million years ago peaks of sunspot activities coincided with times of warm climate. In this sense, plasma astrophysics is ultimately connected with paleontology and the other related sciences of life. Together with evolution/or pulsations of the Sun the terrestrial dynamo may have had some bearing on the shaping of various episodes and the pulsations of extinction and bifurcation of species through the variation of radiation and climatological changes on earth. However, in spite of the potentially tremendous impact on related fields, developments in specific subject fields such as dynamo are still too primitive and remain largely speculative domains at this time. We hope within a decade or two that the horizon will change markedly.

Another example of a potentially important area of contribution by plasma astrophysics is the physics of star formation in molecular clouds (Shu et al., 1987). Obviously, a star is the most fundamental object in astrophysics. It has long been thought that the formation of a star is purely a result of gravitation. In recent years, however, it has become increasingly evident that the influence of plasma and magnetic fields plays a pivotal role in star formation. In fact, Shu et al. say (1987) "... magnetic fields are probably the crucial ingredient because without them the observed levels of rotation and turbulence in molecular clouds would be very difficult to understand."

1.2 Introductory Survey

1.2.1 Hierarchical Structure of the Universe

The aim of the plasma astrophysics is to clarify the physics of plasmas in various celestial objects, including the physics of celestial objects themselves. Thus, we must have fundamental information about celestial objects; their size, mass, and so on.

8

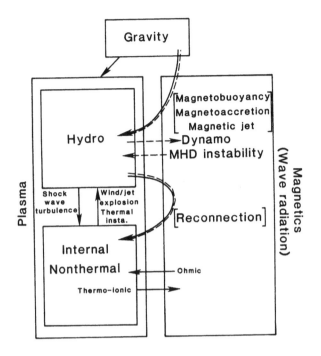

FIGURE 1.3 Relation between gravity, magnetics, and plasmas.

The Universe includes many kinds of celestial objects. One important empirical rule for celestial objects is that the Universe has a hierarchical structure. That is, many stars gather to form a galaxy, and many galaxies gather to form a cluster of galaxies, and many clusters gather to form a supercluster, etc. Actually, our Universe is a super-supercluster itself, and it is the largest celestial object we know of. The size of our universe is about 10^{28} cm, or ten billion light years, and the mass is about 10^{22} solar masses, or 10^{55} g.

Table 1.1 shows typical celestial objects in the hierarchical order. "Our objects" is the one we live inside. "General" means the general term for celestial objects in those levels. Sun–stars and Earth–planets are written separately from other objects by the horizontal line. That is, these objects (Sun-stars, Earth-planets) are different from planetary systems, galaxies and so on, in their essential nature. This will be discussed in the next section.

our object	general	size (cm)	mass($M_\odot$)
our Universe	(Universes ?)	10^{28}	10^{22}
(local group)	cluster of galaxies	10^{25}	10^{14}
Galaxy	galaxies	10^{22}	10^{11}
Solar system	(planetary systems)	10^{15}	1
Sun	stars	10^{11}	1
Earth	planets	10^{9}	10^{-6}

TABLE 1.1 Hierarchy of the Universe. $1M_\odot = 2 \times 10^{33}$ g

1.2.2 Structure of Celestial Objects

Recent developments in observational astronomy have revealed that the some celestial objects are shaped in the so-called *core-halo structure*. The largest object having such a structure is a cluster of galaxies. A giant elliptical galaxy, called cD galaxy, is situated in the center of the cluster of galaxies, and many normal galaxies form a halo configuration around the cD galaxy (Fig. 1.4).

Another example of the core-halo structure is found in a galaxy. In this case, the core is a nucleus which probably consists of one supermassive black hole with an accretion disk surrounding it. In the case of spiral galaxies, the main structure is the disk rather than the halo, because of the rotation in these galaxies. (It should be noted that in a spiral galaxy there is a tenuous [gaseous] halo outside the disk, but we will neglect this aspect for the moment.)

Finally, a well-known example of a core-halo structure is a planetary system. In this case, the core is the star, and the halo (or disk) consists of planets. The stars and planets are only a part of the core-halo structure and are therefore essentially different from galaxies and clusters which are entire parts of a core-halo structure.

It should be noted that these celestial objects having core-halo structures contain diffuse components. These diffuse components consist of plasmas, molecules, dusts, magnetic fields, and cosmic rays, and play an important role in the evolution and activity of celestial objects.

The similar structures of celestial objects such as the "core-halo" structure is due to the fundamental nature of gravitational fields, known as gravo-thermodynamics (Lynden-Bell and Wood, 1968, Saslaw, 1986). The "core-halo" structure emerges in different scales because there is no inherent length scale in the gravitational force. This is the reason why the universe has a hierarchical structure.

These objects are not in a static state, partly because of the fundamental nature of the gravitational force. The detailed structures of the objects depends on their age (or evolutional stage), and hence we will consider the evolutionary effect on celestial objects in the next section.

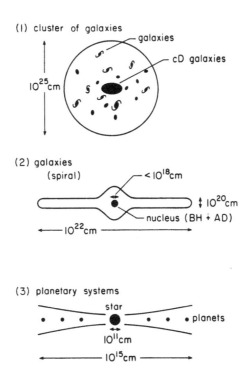

FIGURE 1.4 Core-halo structure of astronomical objects.

1.2.3 Evolution and Activity of Celestial Objects

Celestial objects are formed by the gravitational contraction of the parent cloud. The angular momentum (rotation of the cloud) and the magnetic field play an important role in the gravitational contraction of the cloud. Thus the process is classified into three regimes according to the relative importance of gravitational (f_g), centrifugal (f_c), and magnetic forces (f_m); (1) nonrotating, nonmagnetic cloud ($f_c \ll f_g$, $f_m \ll f_g$), (2) rotating, nonmagnetic cloud, ($f_c \sim f_g$, $f_m \ll f_g$), (3) rotating, magnetic cloud ($f_c \sim f_g$, $f_m \sim f_g$). The results of contraction are (see also Fig. 1.5);

1. If there is no rotation, the final result is the spherical 'core + halo' structure. (example: cluster of galaxies, globular clusters, red giants)

2. If there is rotation, the final result is a 'core + disk' structure.

   (example: galaxies, planetary systems)

11

Formation of Celestial Objects

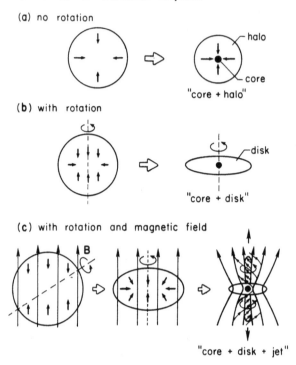

(a) no rotation

"core + halo"

(b) with rotation

"core + disk"

(c) with rotation and magnetic field

"core + disk + jet"

FIGURE 1.5   (a) Core-halo structure; (b) core-disk structure; and (c) core-disk-jet structure.

3. If there is rotation and a magnetic field, the final result is a 'core + disk + jet' structure.
(example: active galactic nuclei, star forming regions)

The cores evolve to hydrostatic stars (or black holes), and the disks evolve to steady disks, where the gas is in hydrostatic equilibrium in the vertical direction of the disks. Various magnetic activities occur on the surface of these hydrostatic objects, because the magnetic energy density can be larger than the thermal energy density (i.e., low plasma β, the ratio of the gas pressure to the magnetic pressure) due to the low plasma density on the surface of these objects, even if the plasma β is much larger than unity in their interior.

The development of radio astronomy in the 1960s revealed that radio jets (lobes) with enormous energy (10^{60} ergs) are ejected from the nuclei of active galaxies. Recent (from 1980 on) developments in radio (mm) and infrared astronomy have revealed jet-like outflows similar to extragalactic radio jets in star forming regions. Furthermore, research in X-ray and UV astronomy in the 1970s and 1980s has revealed exotic activity around the compact objects (neutron stars, white dwarfs), and has revealed a lot of magnetic activity even around

12

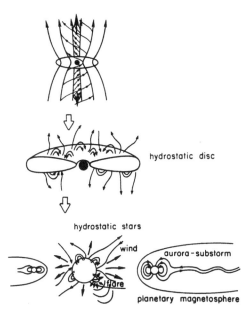

Evolution of Cores and Disks

hydrostatic disc

hydrostatic stars

wind

aurora-substorm

flare

planetary magnetosphere

FIGURE 1.6 Evolutionary sequence of non-equilibrium objects, hydrostatic objects, and magnetostatic objects.

the normal 'quiet stars' (Vaiana, 1983; Rosner *et al.*, 1985).

The central problem in plasma astrophysics is clarifying the origin and physics of such newly-discovered activity in various celestial objects.

1.3   Classification of Celestial Objects

1.3.1   Three Categories of Celestial Objects

As we discussed in the previous section, the structure of celestial objects depends largely on their age. Thus, we will attempt here to classify celestial objects in three categories, according to both the age of the objects and the role of magnetic field.

In old objects, there is rarely an energy source in the interior, so the main body is nearly static. If there is vigorous activity, the energy source is in the outer region. The time scale of the change in magnetic configuration (usually magnetic diffusion time or dynamo time) is much longer than the dynamical time scale of active phenomena around the objects. Examples are planetary magnetospheres, and accreting magnetic neutron stars. We call these objects *celestial objects in quasi-magnetostatic equilibrium.*

On the other hand, young objects are not static. They are usually undergoing gravitational contraction. Thus, their main source of energy is the gravitational potential energy released in the accretion of mass or the contraction of the objects themselves. Magnetic fields are highly deformed by the gravitational motion, although in the very early stages magnetic fields sometimes prevent the objects from contracting gravitationally. Examples are star forming regions, and active galactic nuclei. We call these objects *celestial objects in nonequilibrium.*

The final category in our classification scheme is objects with medium ages which are in quasi-static equilibrium but have energy sources in their interiors. These are called *celestial objects in quasi-hydrostatic equilibrium.* Because of the interior energy source, these objects have convective motion which couples with differential rotation to generate magnetic fields by dynamo action. Magnetic fields are slowly deformed by convective motion and differential rotation so that various magnetic activities similar to solar activity occur at the surface of these objects. Examples are stars, galactic disks, and accretion disks.

The evolutionary sequences of these three objects are illustrated in Fig. 1.68 (see also Table 1.2).

Equilibrium	age	main source of energy	examples
magnetostatic	old	external	planetary magnetosphere magnetic neutron stars
hydrostatic	medium	internal	stars galactic disks accretion disks
nonequilibrium	young	gravitational	star forming regions active galactic nuclei

TABLE 1.2 Classification of Celestial Objects

1.3.2 Fundamental Processes in Objects in Quasi-Magnetostatic Equilibrium

A prototype of these objects is the planetary, in particular terrestrial, *magnetosphere* (Fig. 1.7a). Its energy source is solar wind kinetic energy. A bow shock (collisionless shock) is created just in front of the magnetosphere because the solar wind collides with it. Kelvin-Helmholtz instability occurs in the contact surface between high speed solar wind plasma and low speed magnetospheric plasma. Magnetic reconnection occurs in the tail of the magnetospheric neutral sheet, and high energy particles are accelerated associated with magnetic reconnection.

14

Celestial Objects in Quasi-Magnetostatic Equilibrium

(planetary magnetosphere)

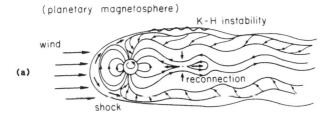

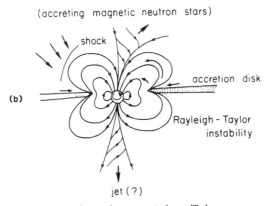

FIGURE 1.7 The structure of objects in quasi-magnetostatic equilibrium.

The open magnetic flux tube extending from the polar part of the planet is twisted by the planet's rotation, and the twist propagates as the tortional Alfvén wave, which may produce polar flow by the nonlinear effect.

In the case of accreting magnetic neutron stars, the situation is somewhat different (Fig. 1.7b), although many of the basic physics are common. When the accretion is spherical, the shock is created just outside of the magnetosphere. On the other hand, if the accretion occurs in disk geometry, the accretion is stopped at the Alfvén surface, where the mass is supported by the magnetic field against the gravity so that the Rayleigh-Taylor instability may occur. The magnetic flux tube extending from the poles is twisted by the rotation of the star, and the nonlinear Alfvén wave, including the centrifugal effect, may drive the bipolar jets along the polar flux tube.

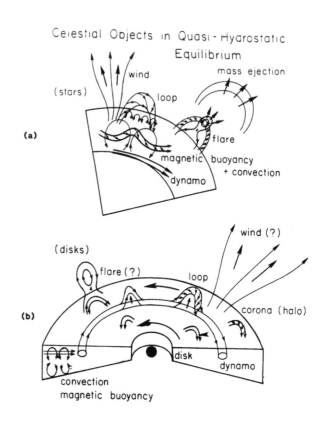

Celestial Objects in Quasi-Hydrostatic Equilibrium

(a)

(stars)

wind

mass ejection

loop

flare

magnetic buoyancy
+ convection

dynamo

(disks)

(b)

flare (?)

loop

wind (?)

corona (halo)

disk

dynamo

convection
magnetic buoyancy

FIGURE 1.8 The structure of objects in quasi-hydrostatic equilibrium.

	ideal or nonideal MHD	steady	nonsteady
low β region	nonideal MHD	corona	flare
	ideal MHD	wind	mass ejection (jet)
		magnetic structure	
high β region		magnetoconvection	
		magnetic buoyancy	
		$\uparrow$	
		dynamo	
		$\uparrow$	
		differential rotation	
		convection	

TABLE 1.3 Interrelation of Fundamental Processes in Hydrostatic Objects

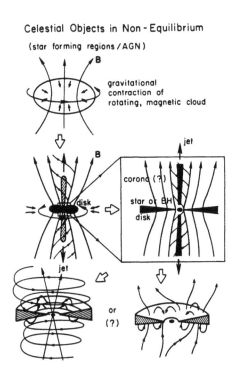

Celestial Objects in Non-Equilibrium

(star forming regions / AGN)

FIGURE 1.9 The structure of objects in non-equilibrium.

1.3.3 Fundamental Processes in Objects in Quasi-Hydrostatic Equilibrium

Apparently, a prototype for these objects is the Sun (Fig. 1.8a, Table 1.3). Magnetic fields are generated in (or just below) the convection zone by dynamo action which is a result of coupling between convection and differential rotation. The magnetic flux tube thus created rises upward due to the magnetic buoyancy, being also controlled by the convective turbulent motion. The emerging magnetic flux produces various quasi-static magnetic structure in the solar atmosphere, such as a flux tube (sunspots, faculae), prominence (filament), and coronal loop. Since the plasma β is much larger than unity below the photosphere, magnetic flux tubes are markedly deformed, twisted, and entangled by the turbulent convective motion. Thus, part of kinetic energy of turbulent motion is stored as magnetic energy in the tube. The magnetic energy stored in the magnetic flux tube is released after the tube emerges to the solar atmosphere where the plasma β is less than unity. Flares, prominence eruptions, and coronal mass ejection are examples of the explosive energy release of stored magnetic

17

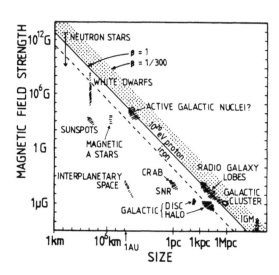

FIGURE 1.10 Magnetic field strength of various celestial objects as a function of their sizes (from Hillas, 1984).

energy. On the other hand, an example of quasi-steady release of magnetic energy may be the coronal loop. If the loop is open to interplanetary space, a wind is created.

Although there are many uncertainties in the disks (especially in the case of accretion disks), one possibility is that the disk has a structure similar to the solar interior-corona structure as illustrated in Fig. 1.8b.

1.3.4 Fundamental Processes in Objects in Non-Equilibrium

Since a prototype of these objects is the *star forming region*, we will summarize here the basic physics of star forming regions (Fig. 1.9). (The same basic physics may be applied to the formation of galaxies.) In the first stage of gravitational contraction of cloud, magnetic force sometimes exceeds gravitational force. Thus, the cloud contraction is nearly quasi-static. The important processes in this first stage are angular momentum and magnetic flux loss, by which the cloud can start dynamical contraction.

In the next dynamical contraction stage, the mass loss (jet or outflow) occurs because

18

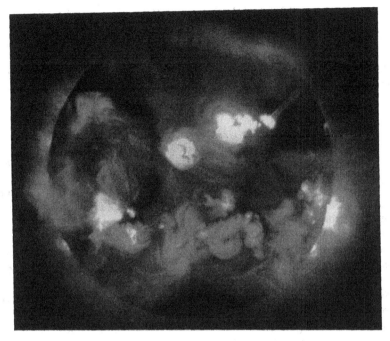

1991 November 12   11:30:28 UT

FIGURE 1.11   The soft X-ray (∼ 1 keV) image of the solar corona on Nov. 12, 1991 taken with the soft X-ray telescope aboard Yohkoh. It is seen that the corona shows many interesting phenomena; i.e., large scale arcade near the north pole, x-ray jets, helical coronal loops, dark coronal holes, bright coronal loops, and so on (courtesy of ISAS).

of the release of the gravitational potential energy. The extraction of angular momentum and the magnetic flux are still important. The quasi-hydrostatic core and accretion disk surrounding it start forming in this stage. An accretion shock is generated on the surface of the core.

The final stage of gravitational contraction is characterized by the core-disk structure and associated activities. The object in this stage is the intermediate one between objects of the quasi-static equilibrium and the nonequilibrium. Thus, we can expect both basic physics seen in both categories; mass loss (wind or jet), angular momentum loss, magnetic flux loss, accretion shock, flares, corona, and so on. Since the luminosity of the core is huge, the radiation pressure may also play an important role in producing mass loss in the case of AGN.

1.4 Typical Examples of Celestial Objects

In this section, we give an overview of actual observations of typical celestial objects, by taking examples over a broad range of astrophysical objects: the solar corona, galactic disks, active galactic nuclei, and cluster of galaxies. We will gain insight into physical parameters of these celestial objects. (See also Fig. 1.10 and Tables 1.4 and 1.5).

1.4.1 Solar Corona

The solar corona has been observed with our naked eye during a total eclipse for more than a few thousand years. It is only 50 years, however, since the solar corona was found to consist of hot plasmas with a temperature of a few million K (Edlen, 1942), far higher than that of the solar photosphere. The heating mechanism of the corona still remains to be one of the challenging puzzles in astrophysics. Since a few million K plasmas emit strong soft X-rays ($\sim$ 1keV), the corona is now best observed in soft X-rays.

Figure 1.11 shows the solar corona observed with the soft X-ray telescope aboard the solar X-ray observing satellite *Yohkoh* (Ogawara *et al.*, 1991). It is seen that the corona is not a uniform gas layer but consists of many loops. Many of these bright coronal loops are situated in the vicinity of sunspots, called *active regions*, and the morphology of the loops suggest the presence of strong magnetic fields along these loops. In fact, it has been established that sunspots have strong magnetic fields of order of a few thousand Gauss (e.g., Bray and Loughhead, 1964), and the comparison of soft X-ray coronal loops with theoretically calculated magnetic field lines (assuming potential field) show excellent agreement to a first approximation (Altschuler *et al.*, 1977, Sakurai and Uchida, 1977).

Such strong magnetic fields are the ultimate origin of nearly all solar active phenomena, e.g., flares, jets, and even the corona itself. It is also seen in Fig. 1.11 that there are dark areas, called *coronal holes*, which are now known as the source region of the fast solar wind. The speed of the fast solar wind is about 800 km/s, much greater than that ($\sim$ 400 km/s) of the normal solar wind emanating from *quiet regions*, the intermediate brightness coronal regions in Fig. 1.11. Although the speed of the normal solar wind is explained by Parker's theory of thermally driven wind (see Sec. 4.3), the origin of the fast solar wind still remains a puzzle.

It is worthwhile here to briefly discuss typical numbers of physical quantities in the corona (see also Tables 1.6–1.9, where fundamental parameters in plasmas and useful formulae to calculate them are listed using typical coronal values as examples). The typical field strengths and (particle number) density in the solar corona is 1–100 G (see Fig. 1.12), and $10^8 - 10^{10}$ cm^{-3}.

The time scale of active phenomena such as flares is 10–1000 sec (of order of 10–100 t_A), which is longer than the electron-ion collision time $\tau_{ei} \sim 0.01 - 1$ sec. Here, $t_A = L/V_A$ is the *Alfvén time*, the time for an Alfvén wave to travel a distance L with Alfvén velocity V_A. On the other hand, the typical size of coronal phenomena ranges from $\sim 10^9$ cm to 7×10^{10} cm, marginally greater than the electron mean free path $\lambda_{\mathrm{mfp}} \sim 10^7 - 10^9$ cm but

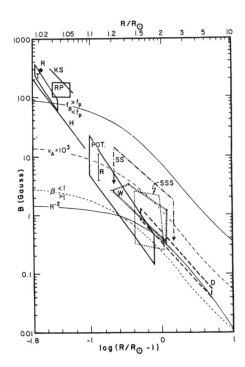

FIGURE 1.12 Distribution of magnetic field strengths above active regions in the solar corona as a function of height, based on the radio observations (from Dulk and McLean, 1978).

much greater than the ion gyro radius $\sim 100\,\text{cm}$. Hence the MHD approximation holds well in the direction perpendicular to the magnetic field. As for the parallel direction, however, the MHD approximation sometimes breaks down (such as in the flare impulsive phase). The magnetic Reynolds number ($R_m = LV_A/\eta_m = t_d/t_A$, where t_d is the magnetic diffusion time) is extremely high, of order of 10^{13}, so that the dissipation process must occur in a small scale, which is, of course, governed by a collisionless process, yet a subject of great debate. It should be emphasized that these properties of coronal plasmas (magnetic Reynolds number $\gg 1$, collisionless dissipation, etc.) are all applicable to astrophysical diffuse plasmas such as hot plasmas in galactic disks/halos and in cluster of galaxies.

The *plasma beta* (β = gas pressure/magnetic pressure) in the corona is much smaller than unity, i.e., the corona is made up of a *low β plasma*;

$$\beta_{\text{corona}} = \frac{p_{\text{gas}}}{p_{\text{mag}}} = \frac{8\pi p}{B^2} \simeq 0.01 \left(\frac{n}{10^9\ \text{cm}^{-3}}\right) \left(\frac{T}{2\times 10^6\ \text{K}}\right) \left(\frac{B}{30\ \text{G}}\right)^{-2}.$$

21

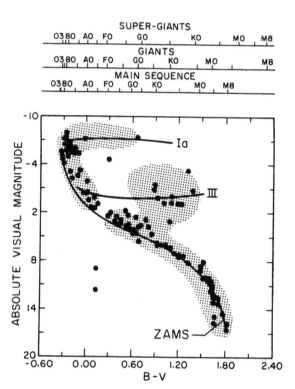

FIGURE 1.13 Stars showing strong X-rays in HR diagram (Vaiana, 1983; Rosner *et al.*, 1985).

Hence in the corona the magnetic force dominates the gas pressure force. The magnetic field morphology is essentially that of sheets (Parker, 1994). This is in contrast to the situation in underlying gas layers, e.g., in the photosphere, where the plasma β is of order unity in magnetic regions (sunspots and small scale flux tubes) or much larger than unity in non-magnetic regions; i.e., the plasma in these regions is a *high β plasma*. Hence the magnetic field in the photosphere and the convection zone is easily compressed by the gas pressure force and takes the form of *isolated flux tubes*. A penchant for such morphology of plasmas is surveyed in Chap. 2. Since the convection velocity in the photosphere is of the order of a few km/s, the dynamic pressure $\rho v^2 \simeq 10^4 - 10^5$ dyne/cm^2 is much greater than the magnetic pressure for average field strength $(B \sim 10\,\mathrm{G})$, $B^2/8\pi \simeq 4$ dyne/cm^2, so that the plasma motion twists and stretches magnetic fields freely, storing magnetic energy in the tube. (The plasma β based on the average field strength in the photosphere is $\sim 10^4$.) On the other hand, in the corona the magnetic field controls the plasma motion and heating, by releasing the stored magnetic energy. Note that coronal plasmas are cataclysmically heated once the magnetic energy is released, because $\beta \ll 1$. The dynamical coupling between the

22

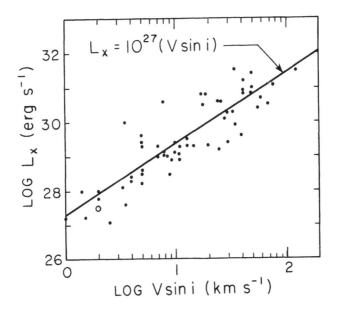

$$L_x = 10^{27}(V \sin i)$$

FIGURE 1.14 The relation of stellar X-ray luminosity to stellar rotational velocity (Vaiana, 1983; Rosner *et al.*, 1985).

low β corona and the high β photosphere through magnetic fields is a key physical process underlying the coronal activity such as flares and coronal heating.

It is not only the Sun that has a hot corona; it has been revealed by the Einstein observatory that almost all stars in HR diagram emit X-rays, implying ubiquitous existence of coronae in these stars (see Fig. 1.13; Vaiana, 1983; Rosner *et al.*, 1985). Remarkably, X-ray luminosities from low mass stars are correlated well with their rotational velocities (Fig. 1.14), which suggests that dynamo-generated magnetic fields are the ultimate origin of stellar coronae.

In summary, coronal active phenomena such as flares, jets, and even the corona itself are produced by magnetic fields. How and why the magnetic energy is released in these phenomena are central issues in solar physics and plasma astrophysics. We will discuss the mechanism of solar flares in detail in Chap. 3.

1.4.2 Galactic Disks

Our Galaxy (the Milky Way galaxy) is a typical spiral galaxy rotating at 200 km/s, and has a flat gas disk with radius of more than 10 kpc ($\sim 3 \times 10^{22}$ cm) and thickness of about 100 pc ($\sim 3 \times 10^{20}$ cm). The total amount of gas is of the order of 1 percent of the total stellar mass ($\sim 10^{11}$ solar mass). The galactic gas disk consists of a plasma with three phases, *cold* (~ 30 K), *warm* ($\sim 8 \times 10^3$ K), and *hot* ($\sim 10^6$ K) components (McKee and Ostriker 1977). The cold component corresponds to molecular clouds (dark clouds) observed with radio molecular lines, while the warm component is observed with the neutral hydrogen 21 cm line, and the hot component are observed with soft X-rays. Typical densities of the three phases are $n_{\text{cool}} \sim 10^2$ cm^{-3}, $n_{\text{warm}} \sim 0.4$ cm^{-3}, and $n_{\text{warm}} \sim 10^{-2}$ cm^{-3}, so that the gas pressure is approximately balanced among the three phases.

Figure 1.15 shows the distribution of magnetic field vectors found from the observation of optical polarization due to dust grains. It can be seen that magnetic fields are nearly parallel to the galactic plane with some wavy patterns or loop structures, suggesting the inflation of magnetic field via supernova remnants or magnetic buoyancy (i.e., the Parker instability, see Sec. 3.2). The observed field strengths (e.g., Sofue *et al.*, 1986) are of the order of a few μG, though the absolute value increases with increasing density up to 30–100 μG in molecular clouds (see Fig. 1.16).

In the galactic disk the energy densities of thermal plasmas, magnetic fields, and cosmic rays are all comparable, of order of 10^{-12} erg/cm^3 ~ 1 eV/cm^3. In other words, the plasma β is about unity in the galactic disk. Hence a rough estimate based on equipartition is sometimes used to derive physical quantities. Recall, however, that what the equilibrium statistical physics teaches is the equipartition among each degree of freedom of modes, not that among the total energies of different elements (such as magnetic, thermal etc.). The fact that the equipartition among the latter is found in the galactic disk plasma is a testimony that there exists a strong coupling among these three elements such as turbulence. This will be further examined in Chap. 2 and Chap. 4. Of course, the energy of the galactic rotation is dominant, and there is no doubt that the ultimate source of energy of the galactic dynamo and associated activity is the energy of the galactic rotation, as will be treated in Chap. 3.

In addition to the above three phases, it has been revealed through recent X-ray astronomy satellites (e.g., EXOSAT, ASCA, etc.) that there is one more component, superhot plasmas with temperature of $10^7 - 10^8$ K, called the *galactic ridge X-ray emission* or *GRXE* (Warwick *et al.*, 1985, Koyama *et al.*, 1986, Yamauchi *et al.*, 1996). The thermal pressure of GRXE is larger than the average pressure in the galactic disk, and must be confined by some way, possibly by the magnetic force. Heating of GRXE plasmas could also be a result of magnetic heating via reconnection (e.g., Makishima, 1996; Tanuma *et al.*, 1997).

The galactic disk is probably similar to accretion disks, though the galactic disk contains much more complicated physical processes than accretion disks, including star formation, supernovae, superbubbles, cosmic rays, and so on. Stars are born in molecular clouds with activities, e.g., bipolar flows. Massive stars produce massive stellar winds and evolve into supernovae, creating hot plasmas with $\sim 10^6$ K in supernova remnants or in superbubbles. Such violent activities of the gas disk are also observed in external galaxies. Figure 1.17

24

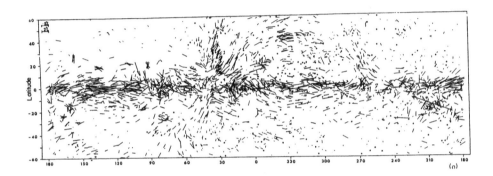

FIGURE 1.15 Distribution of polarization vectors (parallel to magnetic field lines) in the Galactic disk (Mathewson and Ford, 1970). The length of each line is proportional to the polarization degree. This shows rough sketch of magnetic field configuration in the Galactic disk.

shows a sketch of an optical image of the spiral galaxy NGC253 (Sofue *et al.*, 1994) which shows many interesting features such as loops, filaments, shells, etc., many of which are related to a vigorous star formation activity with possible influence of magnetic fields.

Although there is not much observational evidence (hard to observe), hot plasmas (with a few milion K) exist in halos of galaxies, which are similar to the solar corona. The properties of these *galactic coronae* are not well known, though it is possible that they are a result of outflows of hot or superhot components or the consequence of direct magnetic heating (e.g., Parker, 1992).

Some discussion on possible magnetic processes in galacic disks, such as dynamo, magnetic buoyancy, and reconnection, will be discussed in Chap. 3. (Some of the discussions on accretion disks in Chap. 4 will also be applied to galactic disks.)

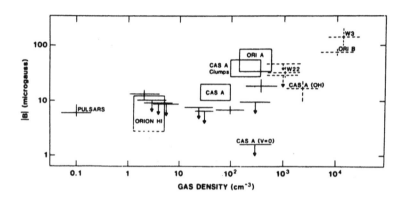

FIGURE 1.16 The relationship between observed field strength and mass density (from Troland and Heiles, 1986). The magnetic field strengths are measured using Zeeman effect.

1.4.3 Active Galactic Nuclei

Active galactic nuclei (AGN), i.e., nuclei of active galaxies (*quasars, radio galaxies,* and *Seyfert galaxies*), are one of the most enigmatic objects in our universe. They are far from us (at millions of light years away from us) and show an early stage of galaxy evolution.

Some of them produce remarkable bipolar radio jets/lobes extending to 100 kpc–1 Mpc from the nucleus (see Fig. 1.18), and their luminosity amounts to $\sim 10^{46}$ erg/s. They are the most energetic phenomenon in our universe. AGNs are now observed with almost all electromagnetic wave frequencies, ranging from radio to gamma rays.

From the condition that the luminosity must be smaller than the *Eddington luminosity*,

$$L_{\text{Edd}} = 4\pi G M m_p c/\sigma_T \simeq 1.3 \times 10^{38} M/M_\odot \quad \text{erg/s},$$

which is the luminosity when the radiation pressure is equal to the gravitational force, where σ_T is the Thomson scattering cross-section and m_p is the proton mass. It has been

26

FIGURE 1.17 A sketch of dark filaments found in an optical image of spiral galaxy NGC253 (from Sofue *et al.*, 1994). Dark filaments comprises dark arcs, loops and/or bubbles, vertical filaments, and short dust filaments.

postulated that there is a supermassive black hole with $\sim 10^8 \, \mathrm{M_\odot}$ in AGN (e.g., Rees, 1984). The ultimate source of the radiant energy is probably the gravitational energy released during the accretion of plasmas on to the supermassive black hole from its accretion disk (see Chap. 4).

The ion temperature of plasmas near black holes is estimated to be $\sim 10^{12} \, \mathrm{K}$ (= virial temperature), whereas the electron temperature is thought to be $\sim 10^9 \, \mathrm{K}$ due to the short cooling time. In such relativistically high temperature plasmas (we call *relativistic plasmas*), the electron-positron pair creation/annihilation is important.

The radio jets are probably ejected from the close vicinity of *black holes* (i.e., $r < 10^{14} \, \mathrm{cm} \sim 10 r_S \sim 10^{-4} \, \mathrm{pc}$, where $r_S = 2 \, GM/c^2$ is the *Schwarzchild radius*), because relativistic motion of jets is observed as *superluminal motion*. The Lorentz factor is estimated to be 2–10 for superluminal jets. However, the velocity of jets has not yet been directly measured with the Doppler shift measurement. It is not even known whether the jets are normal electron-ion plasmas or *electron-positron plasmas*. On the other hand, pulsar winds

27

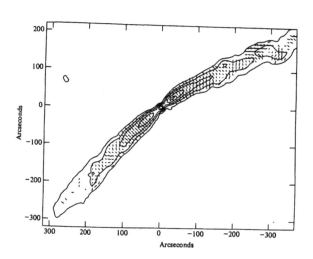

FIGURE 1.18 Radio jet ejected from IC4296 taken with VLA at 1465 MHz (from Killeen, Bicknell, and Ekers). Vectors represent projected magnetic field with lengths proportional to fractional polarization.

are believed to consist of electron-positron plasmas.

There is no doubt about the existence of magnetic fields and energetic electrons in *radio jets*, since the radio emission has been found to be synchrotron emission. Recently, from X-ray observations by ASCA, the magnetic field strength in the lobe of Fornax A was estimated to be 2–4 μG (Kaneda *et al.*, 1995). In the case of a radio jet shown in Figure 1.18, magnetic field vectors seem to be perpendicular to the axis of a jet, implying a helically twisted magnetic field configuration along the jet, though general conclusions have not yet been reached on the magnetic field configuration in radio jets. The formation mechanism of jets from accretion disks via a magnetic force will be discussed in detail in Chap. 4.

The Galactic center radio lobe (Sofue and Handa, 1984) and radio arc (Yusef-Zadeh *et al.*, 1984) seen in the center of our Galaxy may be a prototype of AGN and/or jets, though their size is rather large, $\sim 10 - 100$ pc. The beautiful *filamentary structure* of the radio arc (Fig. 1.19) is suggestive of the influence of strong magnetic fields, and indeed the strong magnetic fields of order of $10 - 100\mu$G have been discovered in the radio arc (Inoue *et al.*,

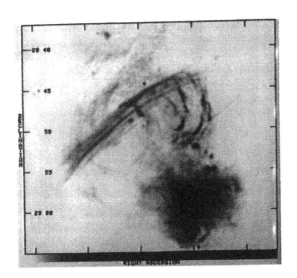

FIGURE 1.19 Radio arc in our Galactic center (Yusef-Zadeh, *et al.*, 1984).

1984, Tsuboi *et al.*, 1985).

1.4.4 Clusters of Galaxies

Intergalactic matter in *clusters of galaxies* has been found to be very hot, $\sim 10^7 - 10^8$ K, emitting X-rays of about $10^{43} - 10^{45}$ erg/s from the region around the cluster core (~ 1 Mpc). Since this temperature is comparable to the virial temperature of the gravitational motion of galaxies ($\sim 400 - 1200$ km/s), it is believed that the source of energy is the gravitational energy due to the cluster potential. Figure 1.20 shows the X-ray image of the Virgo Cluster superposed on an optical image. It is seen that the distribution of hot plasmas is nearly spherical.

The electron density of cluster plasmas is about $10^{-4} - 10^{-2}$ cm^{-3} and magnetic field strengths are observed to be $1 - 2\mu$ G which extend to ~ 0.5 Mpc from the cluster centers (Kim *et al.*, 1990, 1991). By the interstellar medium standard, this field is surprisingly strong, comparable with the disk interstellar field in our Galaxy ($\sim 3 - 4\mu$ G) (Kronberg,

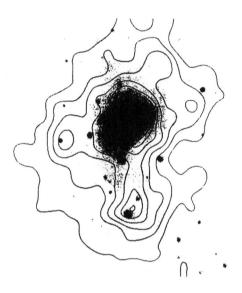

FIGURE 1.20 Soft X-ray image (contours) of cluster of galaxy, the Virgo cluster, superposed on its optical image (from Boehringer 1996). Original X-ray image is from Schindler *et al.* (1995), and the optical image is from Binggeli *et al.* (1987).

1994). Since the dynamical time for the typical length scale of this field (the sound crossing time over $\sim 0.5\,\mathrm{Mpc}$) is comparable to 10^9 yr, a question arises whether there is enough time to create such large scale magnetic fields in the early evolution of clusters.

As emphasized in the discussion of the solar corona, the magnetic field plays an important role in justifying the hydrodynamic treatment (i.e., the MHD approximation). In cluster plasmas, the mean free path of electrons is $\lambda_{\mathrm{mfp}} \sim 10^{23}\,\mathrm{cm} \sim 30\,\mathrm{kpc}$, exceeding the size of galaxies, while even $0.1\ \mu\mathrm{G}$ magnetic fields make the effective electron's 'mean free path' (i.e., gyro radius) as small as 10^9 cm, justifying the MHD approximation in the direction perpendicular to magnetic fields. The presence of magnetic fields results in the so-called 'insulation' of the fluid motion of the plasma from the thermal individual motion of plasma particles. More discussion is found in Chap. 2.

Although the energy density of $1 - 2\mu\mathrm{G}$ fields is much less than that in the hot gas (the plasma beta $\beta \sim 100$), their influence on thermal conductivity is sufficient to indirectly

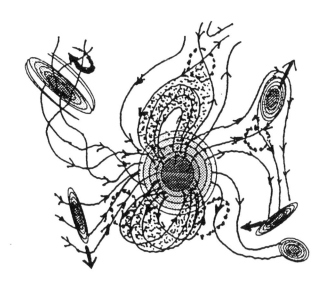

FIGURE 1.21 Hypothetical magnetic field configuration and associated activity in cluster of galaxies (from Makishima, 1997).

incluence the dynamics of the intergalactic matter, and the long term evolution of galaxies in clusters (Kronberg, 1994). We should recall here again that the plasma β based on average field strength in the solar photosphere is $\sim 10^4$, though we observe various magnetic structures (sunspots and intense flux tubes) in the photosphere, and more active magnetic phenomena in upper rarefied atmospheres. Hence it is possible that magnetic fields (in particular, locally enhanced strong fields similar to sunspots) play a certain role in heating and confining *hot cluster plasmas*, and even in the formation and evolution of galaxies (Makishima, 1996).

In this regard, we note that in the centers of dense, 'cooling flow' clusters, stronger field strengths $\sim 30\mu$ G have been observed (Taylor and Perley, 1993). In this case, the magnetic pressure is comparable to the gas pressure, and would affect not only the dynamics of the *cooling flow* itself but also the evolution of cluster plasmas.

The role of magnetic fields in cluster plasmas, including their effects on the formation and evolution of galaxies, and the origin of *cosmological magnetic fields* are very new, challenging fields in plasma astrophysics. Related these problems are the problem of the *cosmic X-ray background*, which may be of the cosmological origin. These problems will be discussed in detail in Chap. 5.

Problem 1–1: In the *HR diagram* the optical spectrum is related to the stellar mass, which in turn is related to the equilibrium properties of stellar interior (particularly that of the thermonuclear burning). In Fig. 1.14 the X-ray luminosity is related to the stellar rotation. What physics relates these quantities? Why the advent of X-ray and radio astronomy reveals violent actions of the Universe (in which plasmas play important roles), while the optical astronomy often has shown quiet ones?

Problem 1–2: It was found (Gray, 1982) that the stellar rotation is related to the stellar spectrum types. Also there is a relation between the stellar angular momentum and stellar mass. What physics can you think of to relate these quantities (or properties)?

Problem 1–3: Calculate the fundamental length scales (Table 1.6) and time scales (Table 1.7) for plasmas in galactic disks, active galactic nuclei, and cluster of galaxies. Discuss the applicability of the magnetohydrodynamic approximation to these objects on the basis of this calculation.

Problem 1–4: Discuss the methods of observations of magnetic fields in celestial objects.

Problem 1-5: If there is no magnetic fields in our universe, what would be changed in the evolution and structure of celestial objects?

	objects	B(G)	L(cm)	$n(cm^{-3})$	T(K)
(A)	Earth	0.31	6.4×10^8		
	Solar Wind at 1AU	6×10^{-5}	1.5×10^{13}	1	10^6
	Jupiter	4	7×10^9		
	Pulsar (Neutron Star)	$10^8 - 10^{12}$	10^6		
(B)	Solar center	?		10^{26}	10^7
	– conv. zone	$10^4(?)$	10^{10}	10^{23}	10^6
	– photosphere (spots)	2×10^3	10^9	10^{17}	10^4
	– corona	10	10^{10}	10^8	10^6
	Magnetic stars	$10^3 - 10^4$	10^{11}		d
	Galactic disks	5×10^{-6}	$10^{20} - 10^{22}$	1	10^4
	– halos	10^{-6}	10^{23}	10^{-3}	10^6
(C)	Molecular cloud	$10^{-5} - 10^{-4}$	10^{19}	10^3	10
	– core	$10^{-4} - 10^{-3}$	10^{18}	10^7	10
	Bipolar flows	10^{-4}	10^{18}	10^3	10
	AGN nucleus	$10^{-1} - 10^{-3}$	10^{18}		
	– jets	$10^{-3} - 10^{-5}$	$10^{18} - 10^{24}$		
	– lobes	$10^{-6} - 10^{-7}$	10^{23}		
	Intergalactic (in cluster)	$10^{-6} - 10^{-8}$	10^{25}	$10^{-3} - 10^{-6}$	$(10^6 - 10^7)$
	– (intercluster)	$< 10^{-9}$		$(10^{-6} - 10^{-9})$	

TABLE 1.4 Typical Fundamental Quantities in some Celestial Objects

B: magnetic field strengths

L: characteristic scale

n: particle number density

T: temperature

ref: Asseo and Sol (1987).

objects	magnetic	thermal	kinetic	rotation	grav.
(A) Solar Wind at 1AU	10^{-10}	10^{-10}	10^{-9}		10^{-11}
(B) Solar convection zone	10^7	10^{13}	10^{6*}	10^9	10^{14}
– photosphere	10^5	10^5	10^3		10^8
– corona	4	0.01	10^{-4}		0.1
Galactic disks	10^{-12}	10^{-12}	10^{-12}	10^{-9}	10^{-9}
(C) Molecular Cloud	4×10^{-10}	10^{-12}	10^{-11}	10^{-11}	10^{-11}

TABLE 1.5 Various Energies in Some Celestial Objects in erg cm^{-3}
*turbulent velocity $v = 0.03$ km/s: Parker (1979) p. 145
magnetic energy $= B^2/8\pi$
thermal energy $= \frac{3}{2}nkT$
kinetic energy $= \frac{1}{2}\rho v^2$
rotational energy $= \frac{1}{2}\rho v_{\rm rot}^2$
gravitational energy $= GM\rho/r$

Debye length	$\lambda_D \equiv \left(\frac{kT}{4\pi n e^2}\right)^{1/2} \simeq 2\left(\frac{T}{10^6\,\mathrm{K}}\right)^{1/2}\left(\frac{n}{10^9\,\mathrm{cm}^{-3}}\right)^{-1/2}$ cm
electron Larmor radius	$r_{Le} \equiv \frac{v_{th,e}}{\Omega_e} = \frac{c}{eB}(m_e kT)^{1/2} \simeq 2\left(\frac{B}{10\,\mathrm{G}}\right)^{-1}\left(\frac{T}{10^6\,\mathrm{K}}\right)^{1/2}$ cm
ion Larmor radius	$r_{Li} \equiv \frac{v_{th,i}}{\Omega_i} = \frac{c}{eB}(m_i kT)^{1/2} \simeq 10^2\left(\frac{B}{10\,\mathrm{G}}\right)^{-1}\left(\frac{T}{10^6\,\mathrm{K}}\right)^{1/2}$ cm
(electron) collisionless skin depth $\lambda_e \equiv \frac{c}{\omega_{pe}} \simeq 30\left(\frac{n}{10^9\,\mathrm{cm}^{-3}}\right)^{-1/2}$ cm
electron mean free path	$\lambda_{\rm mfp} = \frac{v_{th,e}}{\nu_{ei}} = \frac{m_e^2 v_{th,e}^4}{ne^4 \ln\Lambda} \simeq 10^7\left(\frac{n}{10^9\,\mathrm{cm}^{-3}}\right)^{-1}\left(\frac{T}{10^6\,\mathrm{K}}\right)^2$ cm
pressure scale height	$H \equiv \frac{C_s^2}{\gamma g} = \frac{R_g T}{\mu g} \simeq 6 \times 10^9 \left(\frac{T}{10^6\,\mathrm{K}}\right)\left(\frac{g}{g_\odot}\right)^{-1}\left(\frac{\mu}{0.5}\right)^{-1}$ cm
Schwarzschild radius	$r_g \equiv \frac{2GM}{c^2} \simeq 3 \times 10^5 \left(\frac{M}{M_\odot}\right)$ cm

TABLE 1.6 Fundamental Length Scales in Plasmas
$g_\odot = GM_\odot/R_\odot^2 \simeq 2.74 \times 10^4$ cm^2/s, $R_\odot \simeq 7 \times 10^{10}$ cm

electron plasma frequency	$\omega_{pe} \equiv \left(\frac{4\pi n e^2}{m_e}\right)^{1/2} \simeq 10^9 \left(\frac{n}{10^9\,\mathrm{cm}^{-3}}\right)^{1/2} \; H_z$
electron gyrofrequency	$\Omega_e \equiv \frac{eB}{m_e c} \simeq 2 \times 10^8 \left(\frac{B}{10\,\mathrm{G}}\right) \; H_z$
ion gyrofrequency	$\Omega_i \equiv \frac{eB}{m_i c} \simeq 10^5 \left(\frac{B}{10\,\mathrm{G}}\right) \; H_z$
collision frequency*	$\nu_{ee} \cong \nu_{ei} = \frac{n e^4 \ell n \Lambda}{m_e^2 v_{th,e}^3} \simeq 10^2 \left(\frac{n}{10^9\,\mathrm{cm}^{-3}}\right) \left(\frac{T}{10^6\,\mathrm{K}}\right)^{-3/2} \; H_z$
energy relaxation time	$\tau_{ie} \simeq \frac{1}{\nu_{ie}} \simeq \left(\frac{m_i}{m_e}\right) \frac{1}{\nu_{ei}} \simeq 20 \left(\frac{n}{10^9\,\mathrm{cm}^{-3}}\right)^{-1} \left(\frac{T}{10^6\,\mathrm{K}}\right)^{3/2} \; s$
Alfvén time	$t_A = \frac{L}{V_A} = \frac{L(4\pi\rho)^{1/2}}{B} \simeq 10 \left(\frac{L}{10^9\,\mathrm{cm}}\right) \left(\frac{n}{10^9\,\mathrm{cm}^{-3}}\right)^{1/2} \left(\frac{B}{10\,\mathrm{G}}\right)^{-1} \; s$
sound crossing time	$t_s = \frac{L}{C_s} = \frac{L\mu^{1/2}}{(\gamma R_g T)^{1/2}} \simeq 50 \left(\frac{L}{10^9\,\mathrm{cm}}\right) \left(\frac{T}{10^6\,\mathrm{K}}\right)^{-1/2} \left(\frac{\mu}{0.5}\right)^{1/2} \; s$
conductive cooling time[†]	$t_{\mathrm{cond}} = \frac{3nkL^2}{\kappa_0 T^{5/2}} \simeq 4 \times 10^2 \left(\frac{L}{10^9\,\mathrm{cm}}\right)^2 \left(\frac{n}{10^9\,\mathrm{cm}^{-3}}\right) \left(\frac{T}{10^6\,\mathrm{K}}\right)^{-5/2} \; s$
radiative cooling time[††]	$t_{\mathrm{rad}} = \frac{3kT}{nQ(T)} \simeq \frac{3kT}{10^{-16}T^{-1}n} \simeq 4 \times 10^3 \left(\frac{n}{10^9\,\mathrm{cm}^{-3}}\right)^{-1} \left(\frac{T}{10^6\,\mathrm{K}}\right)^2 \; s$
magnetic diffusion time	$t_d = \frac{L^2}{\eta_m} = \frac{4\pi\sigma L^2}{c^2} = \frac{4\pi e^2 n L^2}{\nu_{ei} m_e c^2} \simeq 10^{14} \left(\frac{L}{10^9\,\mathrm{cm}}\right)^2 \left(\frac{T}{10^6\,\mathrm{K}}\right)^{3/2} \; s$
free fall time	$t_{\mathrm{ff}} = \frac{R}{V_k} = \frac{R^{3/2}}{(GM)^{1/2}} \simeq 10^3 \left(\frac{R}{R_\odot}\right)^{3/2} \left(\frac{M}{M_\odot}\right)^{-1/2} \; s$

TABLE 1.7 Fundamental Time Scales in Plasmas

* Λ is the Coulomb logarithm ($\simeq 20$ for $T \simeq 10^6$ K) (Λ^{-1} is the plasma parameter. See Table 1.8.)

[†] $\kappa_0 \simeq 10^{-6}$ cgs (Spitzer conductivity coefficient)

[††] $Q(T)$ is the radiative loss function and is assumed to be $\sim 10^{-16}\,T^{-1}$ (erg cm^{3}s^{-1}) for 10^5K $<$ T $< 10^7$K

(Craig, 1981)

electron thermal velocity $v_{th,e} \equiv \left(\frac{kT}{m_e}\right)^{1/2} \simeq 3 \times 10^8 \left(\frac{T}{10^6\,\mathrm{K}}\right)^{1/2}$ cm/s

ion thermal velocity $\quad v_{th,i} \equiv \left(\frac{kT}{m_i}\right)^{1/2} \simeq 10^7 \left(\frac{T}{10^6\,\mathrm{K}}\right)^{1/2}$ cm/s

sound velocity $\quad C_s \equiv \left(\frac{\gamma p}{\rho}\right)^{1/2} = \left(\frac{\gamma R_g T}{\mu}\right)^{1/2} \simeq 10^7 \left(\frac{T}{10^6\,\mathrm{K}}\right)^{1/2} \left(\frac{\mu}{0.5}\right)^{-1/2}$ cm/s

Alfvén velocity $\quad V_A \equiv \frac{B}{(4\pi\rho)^{1/2}} = 10^8 \left(\frac{B}{10\,\mathrm{G}}\right)\left(\frac{n}{10^9\,\mathrm{cm}^{-3}}\right)^{-1/2}$ cm/s

Kepler velocity $\quad V_k \equiv \left(\frac{GM}{R}\right)^{1/2} \simeq 5 \times 10^7 \left(\frac{R}{R_\odot}\right)^{-1/2} \left(\frac{M}{M_\odot}\right)^{1/2}$ cm/s

TABLE 1.8 Fundamental Velocities in Plasmas

plasma parameter $\quad \Lambda^{-1} \equiv (g \equiv)(n\lambda_D^3)^{-1} \simeq 10^{-8} \left(\frac{T}{10^6\,\mathrm{K}}\right)^{-3/2} \left(\frac{n}{10^9\,\mathrm{cm}^{-3}}\right)^{1/2}$
(Coulomb coupling parameter $\Gamma \equiv \Lambda^{-2/3}$)

electrical conductivity $\quad \sigma \equiv \frac{e^2 n}{\nu_{ei} m_e} = \frac{m_e v_{th,e}^3}{e^2 \ell n \Lambda} \simeq 10^{16} \left(\frac{T}{10^6\,\mathrm{K}}\right)^{3/2}$ s^{-1}

magnetic diffusivity $\quad \eta_m \equiv \frac{c^2}{4\pi\sigma} \simeq 10^4 \left(\frac{T}{10^6\,\mathrm{K}}\right)^{-3/2}$ cm^2 s^{-1}

viscosity $\quad \mu \equiv \frac{m_i^{1/2}(kT)^{5/2}}{e^4 \ell n \Lambda} \simeq 0.1 \left(\frac{T}{10^6\,\mathrm{K}}\right)^{5/2}$ g cm^{-1} s^{-1}

magnetic Reynolds number $R_m \equiv \frac{LV_A}{\eta_m} \simeq \frac{t_d}{t_A} \simeq 10^{13} \left(\frac{L}{10^9\,\mathrm{cm}}\right)\left(\frac{T}{10^6\,\mathrm{K}}\right)^{3/2}\left(\frac{B}{10\,\mathrm{G}}\right)\left(\frac{n}{10^9\,\mathrm{cm}^{-3}}\right)^{-1/2}$
(Lundquist number)

Reynolds number $\quad R \equiv \frac{Lv}{\mu/\rho} \simeq 10^2 \left(\frac{L}{10^9\,\mathrm{cm}}\right)\left(\frac{v}{10^7\,\mathrm{cm/s}}\right)\left(\frac{T}{10^6\,\mathrm{K}}\right)^{-5/2}\left(\frac{n}{10^9\,\mathrm{cm}^{-3}}\right)$

TABLE 1.9 Other Fundamental Parameters in Plasmas

References

Alfvén, H. and Fälthammer, C.G., *Cosmical Electrodynamics* (Oxford University Press, Oxford, 1963) p. 80.

Altschuler, M.D., Levine, R.J., Stix, M., and Harvey, J.W., Solar Phys. **51**, 345 (1977).

Arnold, V.I., *Mathematical Methods of Classical Mechanics*, (Springer-Verlag, Berlin, 1978).

Asseo, E., and Sol, H., Phys. Rep. **148**, 307 (1987).

Boehringer, H., in MPE Report **263**, 537 (1996).

Bray, R.J. and Loughhead, R.E., *Sunspots*, (Dover, 1964) Chap. 1.

Craig, I.J.D., in *Solar Flar Magnetohydrodynamics*, ed. Priest, E.R. (Gordon and Breach Science Pub., New York, 1980).

Dulk, G.A., and McLean, D.J., Solar Phys. **57**, 279 (1978).

Eddy, J.A., *Climatic Change*, Vol. 1 (D. Reidel Publishing Co., Dordrecht-Holland) p. 173.

Edlen, B., Z. Ap. **22**, 30 (1942).

Gray, D.F., Astrophys. J. **261**, 259 (1982).

Hayashi, C., Hoshi, R., and Sugimoto, D., Prog. Theor. Phys. Suppl. No. 22 (1962).

Hasegawa, A. *Plasma Instabilities and Nonlinear Effects*, (Springer-Verlag, Berlin, 1975).

Hillas, A.M., Ann. Rev. Astron. Astrophys. **22**, 425 (1984).

Ichimaru, S., *Statistical Plasma Physics*, (Addison-Wesley, Reading, MA, 1992).

Inoue, M. *et al.*, Publ. Astr. Soc. Jpn. **36**, 633 (1984).

Kaneda, H. *et al.*, Astrophys. J. Lett. **453**, L13 (1995).

Kaufman, L., *Knots*, (Princeton U. Press, Princeton, 1987).

Killeen, N.E.B., Bicknell, G.V., and Ekers, R.D., Astrophys. J. **302**, 306 (1986).

Kim, K.T., Kronberg, P.P., Dewdney, P.E., and Landecker, T.L., Astrophys. J. **355**, 29 (1990).

Kim, K.T., Tribble, P.C., and Kronberg, P.P., Astrophys. J. **379**, 80 (1991).

Koyama, K., Makishima, K., Tanaka, Y., and Tsunemi, H., Publ. Astr. Soc. Jpn. **38**, 121 (1986).

Kronberg, P.P., Rep. Prog. Phys. **57**, 325 (1994).

Lichtenberg, A.J., and Lieberman, M.A., *Regular and Stochastic Motion*, (Springer-Verlag, Berlin, 1983).

Lynden-Bell, D., and Wood, R., Mon. Not. R. Astr. Soc. **138**, 395 (1968).

Makishima, K., *Proc. ASCA Symposium on the X-ray Imaging and Spectroscopy of Cosmic Hot Plasmas*, eds. Makino, F., and Ohashi, T. (ISAS, 1996) p. 171.

Makishima, K., (1997).

Mandelbrot, B.B., *The Fractal Geometry of Nature*, (Freeman, NY, 1983).

Mathewson, D.S., and Ford, V.L., Mon. Not. R. Astr. Soc. **74**, 139 (1970).

McKee, C.F., and Ostriker, J.P., Astrophys. J. **218**, 148 (1977).

Melrose, D.B., *Plasma Astrophysics*, (Gordon and Breach, NY, 1980).

Moffatt, H.K., G.M. Zaslavsky, P. Comte, and M. Tabor, eds., *Topological Aspects of the Dynamics of Fluids and Plasmas*, (Kluwer, Dordrecht, 1992).

Ogawara, Y. *et al.*, Solar Phys. **136**, 1 (1991).

Parker, E.N., *Cosmical Magnetic Fields*, (Clarendon Press, Oxford, 1979), p. 314.

Parker, E.N., Astrophys. J. **401**, 137 (1992).

Parker, E.N., *Spontaneous Current Sheet in Magnetic Fields*, (Oxford U. Press, 1994).

Rosner, R., Golub, L., and Vaiana, G.S., Ann. Rev. Astron. Astrophys. **23**, 413 (1985).

Rees, M., Ann. Rev. Astron. Astrophys. **22**, 471 (1984).

Sakurai, T., and Uchida, Y., Solar Phys. **52**, 397 (1975).

Saslaw, W.C., *Gravitational Physics of Stellar and Galactic Systems*, (Cambridge Univ. Press, 1985).

Schwartzshild, M., *Structure and Evolution of Stars*, (Princeton Univ. Press, 1958).

Sofue, Y., and Handa, T., Nature **310**, 568 (1984).

Sofue, Y., Wakamatsu, K., and Malin, D.F., Astron. J. **108**, 2102 (1994).

Sofue, Y., Fujimoto, M., and Wielebinski, R., Ann. Rev. Astron. Astrophys. **24**, 459 (1985).

Shu, F.H., Adams, F.C., and Lizano, S., Ann. Rev. Astron. Astrophys. **25**, 23 (1987).

Tajima, T., *Computational Plasma Physics*, (Addison-Wesley, Reading, Mass., 1989).

Tanuma, S., *et al.*, submitted to Astrophys. J. (1997).

Taylor, G.B., Perley, R.A., Astrophys. J. **416**, 554 (1993).

Troland and Heiles, C., Astrophys. J. **301**, 339 (1986).

Tsuboi, M., *et al.*, 1986, Astrophys. J. **92**, 818 (1986).

Vaiana, G.S., *Proc. IAU Symp. on Solar and Stellar Magnetic Fields*, (Reidel, 1983), p. 165.

Warwick, R.S., Turner, M.J.L., Watson, M.G., and Willingale, R., Nature, **317**, 218 (1985).

Yamauchi, S., *et al.*, Publ. Astr. Soc. Jpn. **48**, 15 (1996).

Yusef-Zadeh, F., Morris, M., Chance, O., Nature **310**, 557 (1984).

Zeldovich, Ya.B., Ruzmaikin, A.A., and Sokoloff, D.D., *Magnetic Fields in Astrophysics* (Gordon and Breach Science Pub., 1983).

Chapter 2

Plasma Physics—Modern Astrophysical Perspectives

2.1 General Survey

In this chapter we briefly review plasma physics from the point of view of plasma astrophysics. We first note the long-range nature of the Coulombic or electromagnetic interaction and the gravitational interaction. Special to these two kinds of force fields is the long-range potential of $1/r$; these forces can reach infinitely far (so-called infrared divergence) and yet become infinitely strong at near distances (so-called ultraviolet divergence). This long-rangedness of forces gives rise to long-ranged orderedness, manifesting itself in structures with various spatial (and temporal) scales. Because of this nature of the forces a *hierarchy* of structures and phenomena of plasmas and gravitational systems emerges. A well-known hierarchical structure of galaxies that interact gravitationally with each other shows clusters of galaxies, superclusters of clusters, and super-superclusters of clusters (Devoucleurs, 1980). Similarly small particles form a hierarchical structure (Hoyle, 1953; Scalo, 1985) such as nano- to micro-particles, dusts, boulders/meteors, and planets that ultimately form such a structure as a star. And, of course, stars form a galaxy.

Similarly plasmas due to long-ranged electromagnetic interaction exhibit hierarchical structures (Tajima, 1989). Typically, the shortest time scale (or largest frequency) in a plasma comes with radiation. The next shortest time scale comes with the plasma oscillation which is associated with the electron plasma frequency $\omega_{pe} = \sqrt{4\pi n e^2/m}$. An ionic time scale in a magnetized plasma comes with the lower hybrid oscillation whose frequency is in the range of the ion plasma frequency ω_{pi}. Longer than these are the time scales of Alfvén waves, ion acoustic waves, drift waves etc., that is, low frequency waves of plasmas. Still larger than these time scales are the time scales of reconnection, resistive instabilities, and finally those of collisions and transport. This hierarchy is shown in Table 2.1. Corresponding to this is the spatial hierarchy of scales. Generally speaking, we discern that the longer the time scale of a phenomenon, the larger its spatial scale.

This hierarchical nature of plasma manifests itself as a kind of structure made up of many layers of different levels of physical phenomena. But these are yet ultimately interrelated to

Cosmic Time	Epoch (Interaction)	
	Particle Physics (superstring etc.) (GUT etc.)	$>10^{13}$ GeV
10^{-2} sec		
10^{0}	(weak interaction) Nuclear Physics (strong interaction)	GeV MeV
10^{1}		
	Plasma Physics	keV
10^{13} sec		
	Atomic Physics	eV
	(Molecular Physics)	meV
10^{18} sec	Gravitational Physics	$<10^{-8}$ eV

	Electron	RF Heating			Resistive	
	Cyclotron Plasma Wave	Lower Hybrid Wave	MHD	Driftwaves	MHD Trapped Particles	Transport Confinement Collisions
Timescale	$\lesssim$ psec	$\sim$ nsec	$\sim \mu$ sec	$\sim 10\mu$ sec	msec	sec
Frequency	$\Omega_e = \frac{eB}{m_e c}$ $\omega_{pe} = \left(\frac{4\pi n e^2}{m_e}\right)^{1/2}$	$\Omega_i = \frac{eB}{m_i c}$ $\omega_{pi} = \left(\frac{4\pi n e^2}{m_i}\right)^{1/2}$	$\omega_A = k v_A$	$\omega_0 = \frac{cT_e n k_y}{eB}$	$\omega_A^{\alpha} \nu^{1-\alpha}$	ν

TABLE 2.1 Hierarchy of time scales in plasma physics. While we show the cosmological hierarchy in times (a), the table in (b) shows various characteristic timescales ranging from electronic nature to ionic nature to much slower ones in the case of typical laboratory plasma.

each other. Let us examine a case of *magnetic reconnection*, for example.

Collisions give rise to non-vanishing resistivity, which in turn, brings in fundamental changes in (ideal) MHD. In the ideal (or non-collisional) MHD the fluid is stuck to the magnetic field line, whereas with resistivity (or other "non-ideal" effects) the fluid slips away from the field line. One of the most important consequences of this is the possibility of the reconnection of magnetic field lines. An instability which grows due to the finite resistivity, tearing the field lines (the tearing instability), was discussed by Furth, Killeen, and Rosenbluth (1963). They found the growth rate of the instability between the Alfvén frequency ω_A (the ideal MHD frequency) and the collision frequency ν. (Note that $\omega_A \gg \nu$ for a typical case.) The resistive MHD phenomena are much faster than the collisional phenomena such as the collisional transport process. The disruption of the tokamak confinement magnetic fields may take place as a nonlinear evolution of these resistive MHD processes. An example of subtlety of interrelationship between different hierarchies of plasma interaction may be well exemplified in the case of collisionless (and sometimes semi-collisional) reconnection (Drake and Lee, 1977). In collisionless reconnection a microscopic electron motion along a magnetic field line resonates with the growing reconnecting magnetic island dynamics and thus acts as a dissipative mechanism, just as a collision acts similarly in collisional reconnection. Thus the microscopic ballistic motion of electrons couple to the time scale much slower than the Alfvén time scale and could give rise to an agent for a global change of magnetic topology. In the present book we are primarily interested in the higher levels of this hierarchy, namely those of the physics with long time scale-large spatial scale. In other words we put emphasis on plasma astrophysics in the evolutionary point of view and barely touch upon the physics with fast time scale and microscopic spatial.

Another important feature in plasma astrophysics is the *interplay* between the gravitational and electromagnetic interactions. As is well known, there are four fundamental forces: the strong interaction, the electromagnetic force, the weak interaction, and the gravitation, in descending order of strength. In these four, the strong and weak interactions are short-ranged, while the electromagnetic and gravitational forces are long-ranged. Except for the weak interaction which is quite local in nature (by heavy W and Z particle exchanged), the influence of the other forces is inversely proportional to their strength. This is intuitively clear because it is difficult that the interaction potential energy far exceeds the rest mass energy. This leads us to a very important conclusion that *the weaker the force of interaction is, the farther it influences in the Universe*. This principle not only applies among different kinds of forces but with a particular force.

Consider a particle (say, an electron or an ion) under the influence of gravitoelectromagnetic forces

$$\frac{d\mathbf{p}}{dt} = q\mathbf{E} + q\frac{\mathbf{v}}{c} \times \mathbf{B} + \mathbf{F}_g, \tag{2.1.1}$$

where $\mathbf{F}_g$ is the gravitational force. The strong interaction (or the 'color' force) is so strong that it causes 'pionization' beyond the distance of the size of a nucleus ($\sim$ one fermi $= 10^{-13}$cm) so that it can be neglected in the present consideration. Again the weak interaction is so local that we neglect for the present purpose. Within the electromagnetic interaction, in general, we discern the electric and magnetic interactions. [It is possible within certain limiting (weakly relativistic) conditions that the general relativistic equation (the Einstein equation) can be cast into a similar split of 'electric-type' and 'magnetic-type' interaction of gravitation] (Thorne and MacDonald, 1982). From Eq. (2.1.1) it is clear that for nonrelativistic motions ($v/c \ll 1$) the electric term (the first term on the right-hand side) is much larger than the second term (and the third). The lack of magnetic monopoles (and thus the lack of the magnetic current) brings in this lopsidedness or superiority of electric over magnetic fields

$$\frac{\partial \mathbf{B}}{\partial t} + (4\pi \mathbf{J}_m) = -c\nabla \times \mathbf{E}, \tag{2.1.2}$$

$$\frac{\partial \mathbf{E}}{\partial t} + 4\pi \mathbf{J}_e = c\mathbf{\nabla} \times \mathbf{B}, \tag{2.1.3}$$

where $\mathbf{J}_e$ is the electric current while $\mathbf{J}_m$ is the (nonexistent) magnetic current with a corresponding equation of motion being

$$\frac{d\mathbf{p}}{dt} = q\left(\mathbf{E} + \frac{\mathbf{v}}{c} \times \mathbf{B}\right) + \left[q_m\left(\mathbf{B} \pm \frac{\mathbf{v}}{c} \times \mathbf{E}\right)\right] + \mathbf{F}_g, \tag{2.1.4}$$

where q_m is (nonexistent) magnetic charge. Just as the pionization took place in the strong interaction, when the potential energy of the electric interaction becomes large enough (i.e. of the order of the kinetic energy of the particles), a particular electric charge will strongly attract the opposite charge so that the net electric field is much reduced. This effect is commonly called the *Debye screening* (Ichimaru, 1973). On the other hand, the magnetic force was weaker to begin with and there exists no magnetic parallel to the Debye screening. As a net result, the weaker of the two forces of electromagnetic interaction, the magnetic interaction, can influence the plasma state far more globally. These far-reaching characteristics of magnetic field, in contrast to electric fields, are buttressed by another characteristic of electromagnetic fields. That is, the predominance of magnetic fields in low-frequency phenomena. From Faraday's law

$$\frac{\partial \mathbf{B}}{\partial t} = -c\mathbf{\nabla} \times \mathbf{E}, \tag{2.1.5}$$

the ratio of the electric field to magnetic field is given by

$$\frac{E}{B} = \frac{\omega}{kc} = \frac{v_{ph}}{c}, \tag{2.1.6}$$

where ω and k are the frequency and wavenumber of a particular phenomenon and $v_{ph} = \omega/k$ is its phase velocity. The slower the phenomenon (and thus the phase velocity) is, the more predominant the effects of magnetic fields are over those of electric fields. As we set out in the Introduction to emphasize the evolutionary aspect of plasma astrophysics, we are primarily concerned with slow time scales that refer to evolution. This underscores the prominent role of magnetic fields in plasma astrophysics of our interest.

In spite of (or because of) the fact that the gravitational interaction is much weaker than the electromagnetic interaction, ultimately it can reach farther. It sounds a little paradoxical but we saw such is the case for the electric and magnetic interactions. In the case of the gravitational interaction we see no repulsive interaction and thus no parallel to the electric Debye screening effect arises. As a consequence, the more global and massive the system is, the more important the gravitational interaction becomes. Thus many fundamental astrophysical phenomena and objects are shaped by the gravitational interaction on the zeroth order approximation. This is certainly one of the main reasons why the traditional astrophysics has focused on the gravitational interaction. On the other hand, it becomes clear from the above discussion that the magnetic interaction has also a *global reach*. This is the fundamental reason why the magnetic field effects (and thus plasma effects) are as fundamental to astrophysics as the gravitational effects.

There is, however, one fundamental difference in nature between the magnetic and gravitational interactions. The source of the gravitational interaction, of course, is the mass. The mass of matter_is cumulative; i.e. if more matter is added, the mass increases accordingly. The governing equation for this is a Poisson-type equation

$$\nabla^2 \phi_g = 4\pi\rho = 4\pi \int d\rho, \qquad (2.1.7)$$

where ϕ_g is the gravitational potential and ρ and $d\rho$ are the mass density and differential mass density of matter ($d\rho \geqq 0$). The source of the magnetic interaction is the current. Within the magnetoinductive approximation (Darwin, 1942; Nielsen and Lewis, 1976) which neglects the displacement current in favor of the electric current and is relevant to slow time scale phenomena the governing equation may be written as

$$\nabla^2 \mathbf{A} = \frac{4\pi}{c} \mathbf{J}, \qquad (2.1.8)$$

where the vector potential is related to $\mathbf{B}$ as $\mathbf{B} = \nabla \times \mathbf{A}$ and $\mathbf{J}$ is the current density. Note that Eq. (2.1.8) takes the same structure as Eq. (2.1.7) except for the vectorial form. However, the current density is *not* cumulative unlike the mass density. Positively and negatively directed currents can annihilate each other and only a remnant current survives.

Here it now becomes necessary to introduce a third and very important nature of astrophysical plasmas and objects. Astrophysical systems are not bounded by some walls etc. and not closed or isolated. They are unbounded, forming a hierarchical structure (as we already discussed), and they are *open*, interacting with other systems. They are ever evolving, not staying in a static equilibrium. Because the astrophysical system is open, it is most likely in non-equilibrium, be it near equilibrium or far from it, and possesses nonvanishing current densities. Thus the *nonequilibrium* nature of the Universe (sometimes people call turbulence, fluctuations, noise, etc.) gives rise to nonvanishing fluctuations of sources of currents and thus magnetic fields, which reach far into various corners of the cosmos.

The evolution and nonequilibrium nature of the Universe has been mathematically formulated using a generalized concept of entropy (Lynden-Bell, 1969; Sugimoto, 1981). We are, however, unaware of an equivalent development that includes magnetic effects.

It is believed that the Universe evolved very rapidly after the *Big Bang* and experienced various epochs. Most phenomena or structures we presently see in our Universe are controlled by gravitational physics and plasma physics, as we discussed in the previous section. As we look more carefully or farther, however, we see remnants or 'live actions' of earlier epochs of the Universe. The most famous of them all is the *cosmic background radiation* (Wilson et al., 1965). We may be able to classify these various epochs in terms of the predominant governing interaction. For example, around 10^{13} seconds after the Big Bang, the temperature of the Universe became low enough (~ 3000 K) that the photon opaqueness disappeared and the *recombination* of hydrogen atoms began to take place. This is the transition between the electromagnetic dominated epoch (the radiation epoch or more precisely, the plasma epoch)

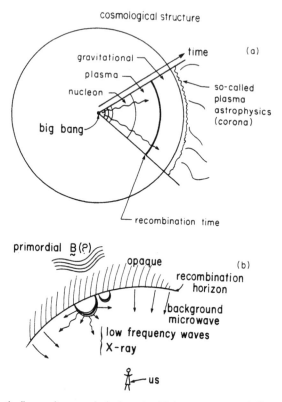

cosmological structure

FIGURE 2.1 Schematic diagram for cosmological epochs. High temperature early Universe in the middle and lower temperature later Universe outwards are described. (a) When the temperature dips below the hydrogen recombination temperature (around 3000 K), the opaque plasma yields its place to the transparent neutral gas, while a small component of it remains as hot plasmas (either due to reheating or a remnant of the early epoch or sustenance of it through perhaps magnetic activities). This point is schematically shown in (b).

and the matter-(or gravitation-) dominated epoch (or *gravitational epoch*). There may be a brief transition period of an atomic physics-dominated era. In general, the higher the characteristic energy level of a particular interaction, the earlier that interaction becomes important to shape the evolution of the Universe. A schematic description of this suggestion is shown in Fig. 2.1.

Prior to $\sim 10^{-2}$ second from the Big Bang interactions other than the electromagnetic and gravitational interactions dominate; i.e. the strong interaction-dominated epoch, the weak interaction (and some perhaps more fundamental than all these forces). Around 10^0 second the fog of strong interaction had cleared up but there have been many electron-positron pairs and these began to annihilate each other. This brief period ($\lesssim 10^0$ sec) is thus characterized as a part of the plasma epoch with the electron-positron plasma. Although this plasma was opaque, it is an important subject of plasma astrophysics. The long period after the electron-positron annihilation ceased and till the recombination of hydrogen atoms began, the main phase of the *plasma epoch*, is certainly a subject of plasma astrophysics. Although many people (Rees, 1988; Weinberg, 1977) suggest that this epoch is an uneventful and thus uninteresting period of the evolution of the Universe, it is uncertain that this is the case. For example, under the current views the formation of galaxies has not been satisfactorily explainable. Partially due to such a view on the role of plasmas and partly due to oversight there has been little work on this topic to date. Some attempt to reconsider this situation will be described in Chap 5.

After the recombination (the gravitational epoch) the main agent to drive the evolution of the Universe is the gravitational interaction. On the other hand, the electromagnetic interaction and plasma effects play indispensable roles in this process in more subtle ways than during the plasma epoch. This can be expected, considering our discussion in Sec. 1.1 on the role of plasmas in the (present) Universe. In what follows we briefly survey plasma physics in its very essence that is important from the point of view of astrophysics. There are many excellent plasma physics textbooks, including Ichimaru (1973) and Sagdeev and Galeev (1969). In the present book we have no intention of repackaging these. For example, Ichimaru's book has a clear and first-principle discussion of kinetic statistical aspects of plasmas, while Sagdeev and Galeev's book presents a concise description of representative approaches in nonlinear plasma physics. In these textbooks a traditional statistical physics description of plasma physics (Ichimaru, 1995), or a traditional nonlinear dynamical plasma physics (Sagdeev and Galeev, 1969; Horton and Ichikawa, 1996) have made an excellent survey. However, the emergence of the recent and modern awareness on *structures, complexity* etc. is on the rise (see for example, Moffatt *et al.*, 1992). These new insights are slow to influence science in general and analytical science such as physical sciences in particular. These new branches of physical science methods which may include *chaos, fractal* theory, complexity theory, structure formation, relaxation theory etc. But these remain, still, in the background of science and at best on the back of the scientist's consciousness. It has not fully integrated into a whole paradigm. We will not try to load all these developments here, but rather pick a few that we believe are very important to the modern plasma astrophysics understanding as well as those that may not be found systematically elsewhere. (If materials of a certain modern view are already concisely available, we refer the reader to such references, instead). Among the many important subjects we concentrate on *fluctuations* (and their potential for structure formation), structure formation and the filamentary constructed plasmas, nonlinear coupling of slow and fast plasma motions, *relaxation* and dissipation of plasma structures, *self-organization* of plasma through the instability settling near criticality (the phenomenon called the self-organized critical transport), formulations of general relativistic plasma physics and new plasma equilibria. We merely try to point out a few important cornerstones of plasma physics with focus on such topics as open systems, driven processes, hierarchical structure, and nonlinear physics.

2.1.1 Nonmagnetized Plasma

In Table 2.1 we showed typical hierarchy of time (frequency) scales of plasma physics. In the second row time scales are shown that are present in a plasma whether it is nonmagnetized or magnetized. In the first row time scales are listed that appear when the plasma is immersed in a magnetic field. Herewith we discuss linear properties of the plasma first and then go on to nonlinear properties later. In this section we begin with a nonmagnetized plasma.

A uniform nonmagnetized plasma in a thermal equilibrium is isotropic. In such a plasma there are two distinct phenomena, electrostatic and electromagnetic. The electrostatic (electric) field $\mathbf{E}$ is parallel ($\mathbf{E}\|\mathbf{k}$) to the wavevector of propagation $\mathbf{k}$ and is called *longitudinal*. There are two time (and correspondingly spatial) scales for this plasma associated with oscillatory motions of electrons and ions: the plasma oscillation period $2\pi\omega_{pe}^{-1}$ in which electrons oscillate overshooting their charge excesses and the ion-acoustic oscillation period $2\pi(kc_s)^{-1}$ in which ions oscillate overshooting their charge excesses but electrons follow ions to screen them and thus reduce the strength of the overshooting restoring force and the frequency of the oscillation from what would be $(2\pi\omega_{pi}^{-1})$ if there were no electron screening. The typical wavelength of plasma oscillations $(2\pi/k)$ is the Debye length $\lambda_D = (T/4\pi ne^2)^{1/2}$ or longer, while that of ion-acoustic oscillations is similar to the plasma wavelength $(2\pi/k)$ or longer. These modes are eigenmodes of the homogeneous plasma given by the condition

$$\epsilon\mathbf{E} = 0 \longrightarrow \epsilon = 0, \tag{2.1.9}$$

where ϵ is the scalar dielectric function (Ichimaru, 1973).

In addition to these oscillatory time scales there are dissipative time scales: the collision time and the Landau damping (Ichimaru, 1973) time. For typical astrophysical (or for that matter most) plasmas the plasma parameter $g \equiv \Lambda^{-1}(\Lambda = n\lambda_D^3)$ is much smaller than unity. The fundamental equations of plasmas, the BBGKY hierarchy (Bogoliubov, 1962; Ichimaru, 1973), start with the first BBGKY equation

$$\frac{\partial f_1}{\partial t} + \mathbf{v}\cdot\frac{\partial}{\partial\mathbf{x}}f_1 + \frac{e\mathbf{E}}{m}\cdot\frac{\partial f_1}{\partial\mathbf{v}} = Cg\frac{\partial}{\partial\mathbf{v}}\int f_2 dx_2, \tag{2.1.10}$$

where C is a constant of order unity and f_1 and f_2 are the one-body and two-body distribution functions of plasma particles in phase space of one and two particles, respectively. The right-hand side of Eq. (2.1.10) is g times smaller than terms on the left-hand side. When the right-hand side is ignored or ignorably small, Eq. (2.1.10) reduces to the collisionless plasma equation, the *Vlasov equation*. The small right-hand side of Eq. (2.1.10) represents correlation effects or more simply collisional effects. The collision time scale is much longer than the above-mentioned plasma oscillation time scales in most plasmas by the ratio of g^{-1} ($g \ll 1$). In addition to collisions the Landau damping (Ichimaru, 1973) is present in (nearly) collisionless plasmas as a result of interaction between waves and particles. For waves that are excited prominently in such a plasma the Landau damping time scale is typically much longer again than the oscillation time scales (depending on the wavelength of oscillations and plasma temperature). In a dusty plasma it is possible to have much larger effective plasma parameter (Ikezi, 1986), as dust particles carry much larger charge per particle.

The electromagnetic (electric) field $\mathbf{E}$ is perpendicular ($\mathbf{E} \perp \mathbf{k}$) to the wavevector of propagation $\mathbf{k}$ and is called *transverse*. The plasma dispersion relation for electromagnetic waves in Fourier space may be written as

$$\mathbf{k} \times (\mathbf{k} \times \mathbf{E}) + \frac{\omega^2}{c^2} \mathbf{E} = \frac{\omega_p^2}{c^2} \mathbf{E}. \qquad (2.1.11)$$

Since the dispersion relation is written as $\omega^2 = \omega_p^2 + k^2 c^2$, $\omega > \omega_p$ and $\omega/k > c$, the time scale of EM waves is shorter than the plasma period and the phase velocity of electromagnetic waves in a plasma is greater than the speed of light. Because of this there is no wave-particle interaction for electromagnetic waves in contrast with electrostatic waves. Thus no Landau damping for the transverse waves. As for the fluid description and its validity for collisionless (or hot) plasma, see Problem 2-2.

2.1.2 Magnetized Plasma

When a plasma contains a magnetic field, it becomes anisotropic. The plasma behavior along the direction of the magnetic field and that perpendicular to it are now distinct. To specify these directions with respect to the magnetic field, we use the words parallel ($\|\mathbf{B}$) and perpendicular ($\perp \mathbf{B}$). Note that we still use the words longitudinal ($\|\mathbf{k}$) and transverse ($\perp \mathbf{k}$) with respect to the wavevector. Let us consider a homogeneous plasma. Eigenmodes in this system are determined by

$$\boldsymbol{\epsilon} \cdot \mathbf{E} = 0 \longrightarrow \det|\boldsymbol{\epsilon}| = 0, \qquad (2.1.12)$$

where the dielectric function $\boldsymbol{\epsilon}$ is a tensor, reflecting anisotropy of the plasma. Equation (2.1.12) yields a number of new modes (Ichimaru, 1973; Krall and Trivelpiece, 1973).

New electrostatic modes that are primarily propagating in perpendicular directions are: the *electron cyclotron mode* (Bernstein, 1960) at or near the frequency of $\omega \sim n\Omega_e$ where the electron cyclotron frequency is $\Omega_e = eB/mc$; the *ion cyclotron mode* (*ion Bernstein wave*) at or near $\omega \sim n\Omega_i$ with $\Omega_i = eB/Mc$ (M is the ion mass); the *upper hybrid wave* with $\omega \sim (\omega^2 + \Omega_e^2)^{1/2}$; and the *lower hybrid wave* with the frequency in the vicinity of $\omega = \omega_{pi}$, where ω_{pi} is the ion plasma frequency $\omega_{pi} = (4\pi n e^2/M)^{1/2}$. The latter two modes are relatives of the plasma (Langmuir) wave and the ion-acoustic wave in the sense that when the magnetic field strength is gradually reduced they tend to become the plasma wave and the ion-acoustic wave, respectively. As the propagation direction ($\mathbf{k}$) is gradually turned from the perpendicular to parallel direction with respect to $\mathbf{B}$, the frequencies (and other properties) of these two waves in a magnetized plasma approach those of the corresponding waves in an unmagnetized plasma. The upper hybrid frequency and lower hybrid frequency are higher than these respective mode frequencies. In general the frequency represents the strength of the restoring force and is proportional to its square roots. In the directions perpendicular to the magnetic field the charge accumulation exerts a repulsive force on charged

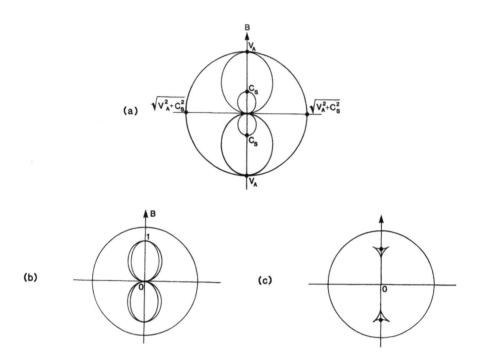

FIGURE 2.2 Three branches of magnetohydrodynamic waves. The compressional Alfvén (the middle lobes), and the magnetized sound waves (the innermost) for a low $\beta(< 1)$ case. The external magnetic field direction is the axis of the lobes.

particles just as in an unmagnetized plasma, but in addition the rigidity of the magnetic field effectively increases the repulsive force and thus increases the eigenfrequencies of modes perpendicular to the magnetic field. (This newly acquired rigidity due to the magnetic field will be revisited in a problem of nontrivial "vacuum" of the Yang-Mills fields in Sec. 5.5.). The two cyclotron modes (electron and ion) are completely new and no corresponding modes exist in an unmagnetized plasma. As already indicated implicitly in the above, there enters no new physics for electrostatic waves in the direction parallel to the magnetic field.

For electromagnetic modes there appear a set of new waves, the Alfvén (1942) waves. There are three branches (see Fig. 2.2). These waves are very often handled by fluid theory and thus can be called *magnetohydrodynamic (MHD) waves*. The reason why these waves can be handled by fluid theory is that electrons and ions move (nearly) identically with the drift $(c\mathbf{E} \times \mathbf{B}/B^2)$ velocity, thus forming a fluid of a plasma (see below). The *shear Alfvén wave* primarily propagates along the magnetic field with the phase velocity $\omega/k = v_A \cos\theta$, where

θ is the angle between $\mathbf{B}$ and $\mathbf{k}$. This wave propagates as if a rubber string (representing a magnetic field line) vibrates and its vibration propagates along the string. The frequency of this wave $\omega = k_\parallel v_A$ is between those of the *compressional Alfvén wave* and the *magnetized sound wave*. In low $\beta(\beta < 1)$ plasmas where $\beta = 8\pi n T/B^2$ the ratio of the plasma kinetic energy density and the magnetic energy density, the compressional Alfvén wave has highest frequency and phase velocity, while the magnetized sound wave has the lowest among the three branches. In high $\beta(\beta > 1)$ plasmas the order between the compressional Alfvén wave and the magnetized sound wave reverses. Sometimes the mode with faster phase speed is called the *fast mode MHD wave*, and the slower mode is called the *slow mode MHD wave*. See Fig. 2.2(b). For low $\beta(\beta < 1)$ plasmas the compressional Alfvén wave (fast mode) has the frequency of $\omega = kv_A[\frac{1}{2}(1+\beta)+\frac{1}{2}\sqrt{(1 + \beta^2 - 4\beta\cos^2\theta}]^{1/2}$, while the magnetized sound wave (slow mode) has $\omega = kv_A[\frac{1}{2}(1+\beta)-\frac{1}{2}\sqrt{(1 + \beta)^2 - 4\beta\cos^2\theta}]^{1/2}$. In the parallel or near parallel propagation the shear Alfvén branch has the superposition of the left-circular polarization and the right-circular one. The whistler wave (an electron effect dominated wave) has the right-circular polarization. The shear Alfvén wave becomes a left-polarized electromagnetic ion cyclotron wave, as the wavenumber increases much beyond Ω_i/v_A. The *whistler wave* becomes an electromagnetic electron cyclotron wave, as the wavenumber exceeds Ω_e/v_A. In the near-perpendicular propagation the compressional Alfvén wave becomes a lower hybrid wave, as the wavenumber increases. See Fig. 2.3.

An important characteristic of a magnetized plasma is that the plasma easily becomes inhomogeneous where the magnetic field varies in space even under a local thermodynamical equilibrium. Such inhomogeneities in plasmas are known to excite new classes of waves and further bring in numerous *plasma instabilities* (Mikhailovskii, 1974). These instabilities take place in configuration space. Configuration space instabilities in general tend (albeit not always) to convert inhomogeneity into homogeneity in plasma. In equilibrium the density gradient ∇n, pressure gradient ∇P, magnetic gradient ∇B, and the gravitational acceleration $\mathbf{g}$ may be present. In a magnetized plasma a charged particle gyrates on a Larmor orbit around a guiding center. These gradients bring in a new effect, the drift of particle's *guiding center*. The *drift velocity* is perpendicular to the vector of the inhomogeneity ($\nabla B, \nabla n, \nabla P$, or $\mathbf{g}$) and to the direction of $\mathbf{B}$. For example, the value of the drift velocity in inhomogeneous density is given by

$$v_d = \kappa \frac{cT_e}{eB},\qquad (2.1.13)$$

where $\kappa = (\partial n/\partial x)/n$, the inverse of the density scale length. A wave associated with the drift, the *drift wave*, has the frequency of the order of $\omega^* \equiv kv_d$. [The physical mechanism of the wave may be found in Tajima (1989), for example.] The wave propagation direction $\mathbf{k}$ is typically in the direction of the drift. The frequency of the drift wave in general falls much below the acoustic wave frequency:

$$\omega_* \sim (k\rho_i)(\kappa c_s) < \kappa c_s < kc_s = \omega_{ia},\qquad (2.1.14)$$

where ρ_i is the ion Larmor radius and we assumed that the ion and electron temperature are similar. This entails that this slow time scale phenomenon plays an important role in slow

51

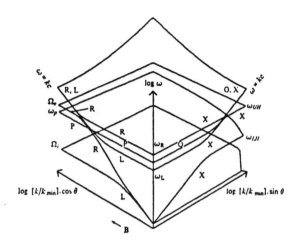

FIGURE 2.3 Polar log plots for the five dispersion surfaces for the underdense case. [log ω vs. log$(k/k_{\min})$ and θ.] Oakes *et al.*, 1979.

processes such as the transport process and evolution.

The density gradient drift that was discussed, often called the *diamagnetic drift*, as the sense of the drift is to counter the plasma current direction that sustains the magnetic inhomogeneity and thus the density inhomogeneity. Besides this diamagnetic drift there are various kinds depending on the origin of the drift. The $\mathbf{E} \times \mathbf{B}$ *drift* has the velocity $\mathbf{v}_E = c\mathbf{E} \times \mathbf{B}/B^2$ and the polarization drift $\mathbf{v}_p = cd\mathbf{E}/dt/\Omega_\sigma B$ ($\sigma = i$ or e). The electron $\mathbf{E} \times \mathbf{B}$ drift $\mathbf{v}_E^{(e)}$ is identical to the ion $\mathbf{E} \times \mathbf{B}$ drift $\mathbf{v}_E^{(i)}$, including the sense. This means that the electric field emersed perpendicular to the magnetic field produce no charge separation due to the $\mathbf{E} \times \mathbf{B}$ drift.

It is important to note that this $\mathbf{E} \times \mathbf{B}$ drift velocity is coherent, i.e. is independent not only of species (ions and electrons), but also of the thermal velocity of any particle. Further, it is important to notice that the presence of the magnetic field eliminates the ballistic motion perpendicular to the field and thus the *Landau pole* $(\omega - \mathbf{k} \cdot \mathbf{v})^{-1}$, which leads to the disappearance of the *Landau damping* (i.e. the kinetic interaction between the wave and particles) (Bernstein, 1960). [Only at higher frequencies *cyclotron resonance* can enter]. In another word, the introduction of magnetic fields into a plasma makes a certain (not perfect but important) insulation of macroscopic field (or wave, or fluid) motion from microscopic particle (or kinetic) motion. This partial detachment of the fluid motion from the thermal energy is one of the main reasons why so many interesting phenomena in magnetized plasmas emerge and will be discussed in much more detail in the following sections.

If the electric field is time dependent, the *polarization drift* appears and does cause charge separation, as the ion polarization drift is the mass ratio (M/m) times larger than the electron polarization drift (and in the opposite sense). Because this drift can give rise to polarization of charge in the plasma, it is called the polarization drift. The size of the polarization with respect to the $\mathbf{E} \times \mathbf{B}$ drift is typically very small in low frequency waves such as MHD waves:

$$\frac{|\mathbf{v}_p^{(\sigma)}|}{|\mathbf{v}_E|} = \frac{\omega}{\Omega_\sigma} \ll 1, \tag{2.1.15}$$

where ω is the typical frequency of the temporal change of $\mathbf{E}$. In spite of its smallness this drift plays an important role in the drift wave physics (Horton, 1985). The drifts due to inhomogeneity of a plasma include the ∇B *drift* ($\mathbf{v}_{\nabla B} = v_\perp^2 \nabla B \times \hat{b}/2B\Omega$), the *curvature drift* ($\mathbf{v}_{\mathrm{cur}} = -v_\parallel^2 [(\hat{b} \cdot \nabla)\hat{b}] \times \hat{b}/\Omega$), and the *gravitational drift* ($\mathbf{v}_g = c(\mathbf{g} \times \hat{b})/\Omega$), where $\hat{b} = \mathbf{B}/B$. If there is a gradient in the plasma flow v (∇v), i.e. shear flow, including the shear in the drift velocity, the Kelvin-Helmholtz instability (Chandrasekhar, 1961) may develop.

The MHD modes in an inhomogeneous (as well as in a homogeneous) plasma in the absence of dissipation such as resistivity are either oscillatory or purely growing (or damping), but not both simultaneously. This is because the MHD equations can be cast in a self-adjoint form and the eigenvalue ω^2 is real, i.e. either $\omega^2 \geqq 0$ (oscillatory) or $\omega^2 < 0$ (purely growing or damping). The theory is based on the variational principle and is called the *energy principle* of MHD (Bernstein *et al.*, 1958). This is of course in stark contrast to kinetic modes, which in general have ω^2 as complex.

2.2 Nonlinear Physics

As plasmas are quite volatile and unstable matter as mentioned in the Introduction in part due to the long-range Coulombic interaction and in part due to the often driven and open nature of astrophysical plasmas in particular, modes in plasma can easily grow in a large amplitude. Therefore, in great many cases plasmas exhibit nonlinear stages of evolution and interaction. In fact, plasmas more often than not are in these nonlinear regimes. In many laboratory plasma experiments, including fusion experiments the experimentalist (and the theorist) are interested first how to control this potentially very volatile matter. Once the plasma becomes unstable, most global instabilities keep growing until the perturbation hits the vessel that contains the plasma and the plasma would most likely get disrupted or disappear. This along with much easier tractability of linear theoretical approaches has imprint on the character of the traditional (laboratory) plasma physics, i.e. domination of linear theories and paucity of nonlinear ones. Even in laboratory plasma physics, however, the nonlinear physics attracts increasing attention these days. In astrophysics in contrast we see "the end result" of instabilities and nonlinear evolution of instabilities and all nonlinearities are "played out." We are interested in the ultimate fate or stage of these nonlinear plasma interactions and their morphology. Thus nonlinear plasma physics plays a central and all-important role in plasma astrophysics. In what follows we briefly survey some of the essences of nonlinear plasma physics.

Consider an example of a hydrodynamic equation of motion

$$\frac{\partial v}{\partial t} + v \cdot \frac{\partial v}{\partial x} = \nu \nabla^2 v. \tag{2.2.1}$$

The evolution of the velocity, the first term on the left-hand side, is governed by the nonlinear advection, the second on the left-hand side, and by the viscous diffusion, the term on the right-hand side of Eq. (2.2.1). The relative importance of the nonlinear advective term with respect to the linear term on the right-hand side determines the strength of the nonlinearity of the particular situation. This ratio of the second term on the left-hand side of Eq. (2.2.1) to the term on the right-hand side is called the *Reynolds number R*. The greater R is, the stronger the nonlinearity of the system is. When $R \gg 1$, the system is highly nonlinear and perhaps exhibits *strong turbulence*.

In the limit of $1/R = 0(R \to \infty)$, i.e. in the absence of the dissipative term on the right-hand side of Eq. (2.2.1), there exists an exact implicit solution of Eq. (2.2.1)

$$v = f(x - vt), \tag{2.2.2}$$

where f is an arbitrary function. Given a form of f, say a sinusoidal function, this solution exhibits well-known evolution of the *wave steepening* because where $|v|$ is large the propagation of the wave, the phase velocity is large, so that the portions with larger amplitude $|v|$ overtake those with less amplitude. When this process continues, the crest of the wave can hang out and the velocity v becomes multivalued with respect to x. In a real physical world this is not permitted. In fact, the dissipation is rarely exactly zero, no matter how small it may be. In this more realistic case in which R is very much larger than unity but not infinite, however, an entirely different scope of physics appears. In the other limit with R being not large the physics is again relatively simple and well known, i.e. the laminar flow physics (Bachelor, 1967; Landau and Lifshitz, 1975). It is notoriously difficult, however, for $\infty > R \gg 1$ and not much is known in spite of recent rapid advancement of knowledge in this regime. The reason for difficult is, unlike for R not being very much larger than unity, in case of $R \gg 1$ there appear all kinds of spatial scales appear and we can specify no specific scale for the phenomenon. A wave appears within a wave and that wave appears within a wave.....*ad infinitum*. A vortex within vortex within vortex....*ad infinitum*. A picture is shown in Fig. 2.4 to illustrate this situation a bit similar to the Indian mythic elephant structure in Fig. 1.2.

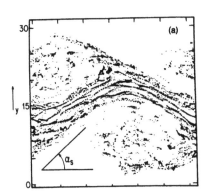

FIGURE 2.4 Kelvin-Helmholtz instability generated vortex and vortex within it. (Tajima & Leboeuf, 1980).

In magnetohydrodynamics we have another important evolutionary equation in addition to Eq. (2.2.2)

$$\frac{\partial \mathbf{B}}{\partial t} = \nabla \times (\mathbf{v} \times \mathbf{B}) + \eta \nabla^2 \mathbf{B}, \tag{2.2.3}$$

where η is the resistivity. The first term on the right-hand side of Eq. (2.2.3) is the induction term and nonlinear, while the second term is the resistive diffusion term and linear. The *Lundquist* (or magnetic Reynolds) *number* $L(\equiv R_m)$ is the ratio of the nonlinear term to the linear term. When $R_m \gg 1$, the system is highly nonlinear and renders such phenomena as magnetic turbulence, plasma disruption, etc. On the other hand, when $1/R_m = 0$ ($R_m \to 0$), the plasma fluid element is frozen in the field line. Since fluid elements do not cross each other, the correspondingly field lines do not either, i.e. no field line reconnection. In this case, therefore, there is no possibility for change of *magnetic topology*. In contrast when $\infty > R_m \gg 1$, topology not only can change but can render a very complex structure. For example, one may expect *magnetic islands* are formed due to reconnection, but within islands one may find an island structure within which an island structure....*ad infinitum*. For another example, we consider twisting of a magnetic flux tube (Steinolfson and Tajima, 1987). As we twist the tube, an axial current is induced first with a diffuse profile [see Fig. 2.5(a)] and then with a sharper profile [Fig. 2.5(b)]. As we further twist, the current column becomes unstable against the twist-kink instability (Zaidman and Tajima, 1988) [Fig. 2.5(c)]. When we further twist the tube, an individual kink current would suffer further kink within itself

55

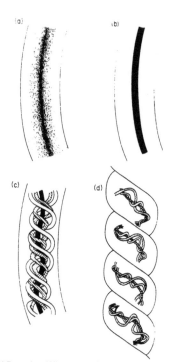

FIGURE 2.5 Morphology of a twisted flux tube. When we twist the extremes of a magnetic flux tube, the current along the flux tube (and thus the original magnetic field) is generated (diffusely) in the center of the tube (a). As the twist continues, the induced current increases and its distribution is peaked near the center (b), which leads to a strongly helical magnetic field. As the twist further proceeds, this current column becomes unstable against the kink instability and forms a helical current structure (c). As further twists are added, there may be a transition from the helical to the superhelical structure (d).

[Fig. 2.5(d)]. This process could continue *ad infinitum*, if the magnetic Reynolds number is extremely large.

We now discuss properties of nonlinear stages of plasmas. Sagdeev and Galeev's textbook (1969) eloquently made exposition of some of the nonlinear processes i.e. the quasilinear stage, saturation, and weak turbulence. Transitions from one kind of nonlinear state to another have been studied extensively recently (Feigenbaum, 1978; Swinney and Golub, 1978; Ginet and Sudan, 1987). As a result of these transitions the system tends to shift from more orderly laminar flows to more chaotic turbulent flows. We will discuss this further in Sec. 2.3. When the system is sufficiently nonlinear, it develops into a strongly turbulent state. An alternative development (and sometimes the simultaneous consequence) to the above is the commencement of explosive processes (Tajima and Sakai, 1986). In order for the system to develop such highly nonlinear behaviors, it is necessary to have a drive such as the gravitational energy, the thermal or fusion energy, the mechanical energy (like winds) etc. If there is no drive, the system tends to the thermodynamical equilibrium, a thermal death: everything is equally distributed and no specific structure or evolution of such exists.

In the present day astrophysical objects or Universe the gravitational interaction and the expansion of the Universe determine the overall drive of various processes. Although the overall entropy of the Universe may increase, a local entropy may decrease. For example, a galaxy roughly keeps its size and shape in spite of the Hubble expansion of the Universe. Thus the distance between two galaxies may become larger and larger, as time passes. Thus any creation or increment of entropy within a galaxy can be thrown away to the intergalactic space which is ever more abundant to absorb excess entropy. In this way the galaxy or its subsystem can dispense entropy and go into a more organized state and further away from a thermal equilibrium. Emersed into this cosmic background of volatile entropy reduction is the ever unstable and mercurial plasma world.

In astrophysics, after all, morphology is important, because that is what we observe. We can list some of the more probable morphologies that plasmas in nonlinear stages want to take. Filamentation and the vortex structure of current or flow are two of them. Dual to this is filamentation of magnetic flux. Sunspots (Babcock, 1959), solar magnetic flux tube "quantization" in granules and others, are examples of this. Jupiter red spots, terrestrial typhoons are examples of this. Density depression is another kind of morphology that a plasma likes to take in nonlinear stages. Either electrostatic waves (Zakharov, 1972) or electromagnetic waves (Kurki-Suonio and Tajima, 1988) causes such depression. Such a structure often takes a solitary (or soliton) structure (Kruskal *et al.*, 1964). Auroral cavities (Wagner, 1981) are an example. Knots, twist (kink), and braiding are another kind of morphology we often witness. A pictorial representation of knotty loop is shown in Fig. 2.9. Astrophysical jets often develop a knotty flow pattern, for example. The magnetic field structure in the solar corona may well be a braiding structure (Parker, 1992). Cellular structures of current, magnetic and electric fields, and density are another example of the astrophysical plasma morphology. We see collisional and collisionless shocks, rotational discontinuity, double layers (Babcock, 1959; Joyce and Montgomery, 1973; Sato and Okuda, 1980; Wagner *et al.*, 1981), and galactic cells as some of the examples of this morphology. Blobs, clumps, and holes are another example of the form that plasmas in cosmos tend to take. The phase space holes and clumps have been discussed by Lynden-Bell (1969) and Berk, *et al.* (1969). Condensation and radiation instabilities have been a subject of investigation of late (Steinolfson and Van Hoven, 1987; Drake, 1987). Similar structures may be created by other mechanisms such as thermal instabilities, drift waves (Terry and Diamond, 1985), the Rayleigh-Taylor (or interchange) instability (Sharp, 1985), and the Parker (ballooning) instability (Shibata *et al.*, 1989). Turbulence is a very common phenomenon and morphology of plasma astrophysics. Weak turbulence (Kadomtsev, 1965), strong turbulence, and fully developed turbulence (Kolmogorov, 1941) are different varieties of turbulence developments. The fully developed turbulence possesses no specific spatial scales other than the overall system size and the molecular dissipation scale. Sometimes intermittency (McWilliams *et al.*, 1983) comes into play in strong turbulence. Intermittent turbulence often exhibit blobby or clumpy structures.

We take an example of the density depression by nonlinear waves for more detailed analysis. In 1972 Zakharov analyzed the density depression process by nonlinear electrostatic equation. He started out with the two-fluid equations (electron and ion) of motion and related continuity equations and Poisson's equation. This set of equations has the high frequency Langmuir waves and the low-frequency acoustic waves. The problem Zakharov considered was that larger amplitude Langmuir waves interact with the plasma. These high-frequency waves in a slower time scale give rise to a new effective force. Consider the equation of motion for electrons:

$$\frac{\partial \mathbf{v}_e}{\partial t} + \mathbf{v}_e \cdot \nabla \mathbf{v}_e = -\frac{e}{m} \mathbf{E} - \frac{1}{n_e m} \nabla P_e. \tag{2.2.4}$$

$$\frac{\partial n_e}{\partial t} + \nabla \cdot (n_e v_e) = 0 \tag{2.2.5}$$

and Poisson's equation

$$\nabla \cdot E = -4\pi e(n_e - n_i), \tag{2.2.6}$$

where the ion density n_i may be regarded as constant in the high frequency. As we know (for example Chen, 1984), Eqs. (2.2.4)–(2.2.6) yield the Langmuir waves. If we coarse-grain these equations over time, i.e. time-average these over the electron time scale, we end up with equations for longer time scales. We write down the electron velocity as

$$\mathbf{v}_e = \mathbf{v}_e^{(0)} + \mathbf{v}_e^{(1)} + \cdots, \tag{2.2.7}$$

where $\mathbf{v}_e^{(0)}$ is the high-frequency component and $\mathbf{v}_e^{(1)}$ the lower frequency one. The Langmuir wave is described by $\mathbf{v}_e^{(0)}$. In order to obtain the equation for the slower component we substitute $\mathbf{v}_e^{(0)}$ into the nonlinear term, the second term on the left-hand side of Eq. (2.2.4), where the zeroth order velocity $\mathbf{v}_e^{(0)}$ is given as

$$\mathbf{v}_e^{(0)} = -\frac{e}{m\omega_p} \mathbf{E}. \tag{2.2.8}$$

Then from Eq. (2.2.4) we obtain

$$\frac{\partial \mathbf{v}_e^{(1)}}{\partial t} = -\frac{1}{m} \nabla \left(\frac{e^2 |\mathbf{E}|^2}{2m\omega_p^2} \right) - \frac{1}{n_e m} \nabla P_e. \tag{2.2.9}$$

The first term on the right-hand side of Eq. (2.2.9) takes a form of the force potential Φ

$$\Phi = \frac{e^2 |\mathbf{E}|^2}{2m\omega_p^2}, \tag{2.2.10}$$

which is called the *ponderomotive potential*. Zakharov (1972) derived the following set of equations utilizing Eq. (2.2.9)

$$\frac{\partial^2 n_i}{\partial t^2} - c_s^2 \nabla^2 n_i = \frac{1}{8\pi M} \nabla^2 |E|^2, \tag{2.2.11}$$

$$i\frac{\partial \mathbf{E}}{\partial t} + \frac{3}{2} \omega_p \lambda_D^2 \nabla(\nabla \cdot \mathbf{E}) = \omega_p \left(\frac{n_i}{n_0} \right) \mathbf{E}. \tag{2.2.12}$$

58

Equation (2.2.11) was derived from the ion equation and describes the low-frequency ion acoustic phenomena. Equation (2.2.12) was derived from the electron equation and describes the nonlinear evolution of the high-frequency field $\mathbf{E}$. Equation (2.2.12) is often called the nonlinear Schrödinger equation, as the "potential" $\omega_p(n_i/n_0)$ in Eq. (2.2.12) is nonlinearly determined by Eq. (2.2.11). The term on the right-hand side of Eq. (2.2.12) may be considered as a "cubic term," because n_i according to Eq. (2.2.12) gives rise to the third harmonics, a ubiquitous effect of nonlinearity of plasmas (and fluids). We will see a similar point in Sec. 2.3.

Zakharov has obtained exact solutions to the system of Eqs. (2.2.11) and (2.2.12). In one dimension there is a steady profiled *soliton* solution which produces sech$(x - v_s t)$-like function for E (and corresponding one for n_i), where v_s is the soliton propagation speed. Such a solution contains a large set of wavenumbers and frequencies, not restricted by the third harmonic. This is due to high nonlinearity. The steady profile of soliton is a result of the balance of the spatial dispersion [the second term on the left of Eq. (2.2.12)] with the cubic nonlinear term on the right of Eq. (2.2.12). Some solitons or structures in electron-positron plasmas have been discussed in Tajima-Taniuti (1990). In dimensions higher than one there are solutions which can *"collapse"*, i.e. the structure is spatially peaking as time progresses and higher mode numbers appear. The collapse can happen in finite time in the absence of additional nonlinearities, which have been neglected in the current derivation. This collapse phenomena are of very important class of nonlinear physics and appear in numerous occasions deemed important in astrophysics (see e.g. Sec. 3.3). The reverse of collapse is expansion or *explosion*. The main nonlinearity in the system of Eqs. (2.2.11) and (2.2.12) is the cubic nonlinearity, as the factor n_i/n_0 in the right-hand side of Eq. (2.2.12) is proportional to $|E|^2$ according to Eq. (2.2.11). Sometimes the nonlinearity is quadratic such as $\mathbf{J} \times \mathbf{B}$ (or $\nabla \times \mathbf{B} \times \mathbf{B}$) and sometimes more complicated. Examples of this may be found in Sec. 3.2.

2.3 Turbulence

It is well acknowledged that plasmas are quite unstable. Two main reasons can be thought of for this phenomenon—plasma's propensity for instability—one is because it is amorphous as contrasted to a solid which has a rigid structure with symmetry; the second is because it has the electromagnetic long-range interaction as opposed to, for example, a glass which is amorphous but not as volatile as a plasma. The first reason makes the plasma at least as unstable as any fluid. In addition, the long-rangedness of interaction makes the plasma often unstable because any particle in the plasma can affect the rest of the medium and thus even how the boundary is given can affect the entire plasma. This global interaction in combination with the amorphous fluidity of the plasma makes the plasma quite volatile.

When a plasma instability develops, the amplitude of the most unstable mode or a set of unstable modes grows. When the amplitude is sufficiently small, the linearization of the original nonlinear equations can be carried out, as nonlinear terms are smaller than linear terms in this regime. This stability theory, however, is left for the thousands of original

59

articles and textbooks on the subject (e.g. Mikhailovskii, 1974) and no further discussion is given here.

In a thermal equilibrium plasma with no instability, it is known that the ratio of the potential energy to the kinetic energy is of the order of the plasma parameter g (see, e.g. Ichimaru, 1973) and this is usually much less than unity ($g \ll 1$). As is well known, *equipartition* of energy among independent degrees of freedom is realized in thermal equilibrium and thus $\langle E_k^2 \rangle / 8\pi = T/2$ and $m\langle v^2 \rangle / 2 = T/2$, where the angular brackets are for statistical average and T is the temperature of the plasma (in the energy unit). The smallness of the ratio mentioned above may be easily understood by noting that the modes of electrostatic fluctuations E_k^2 are present typically over the Debye wavenumber ($|\mathbf{k}| < k_D$), beyond which fluctuations are severely suppressed, while the degrees of particle thermal motions are over the total number of particles. Thus

$$\frac{\sum_{\mathbf{k}} \langle E_\mathbf{k}^2 \rangle / 8\pi}{\sum_i \frac{m\langle v_i^2 \rangle}{2}} \sim \frac{V k_D^3 (T/2)}{N(T/2)} = \frac{1}{n\lambda_D^3} \equiv g. \tag{2.3.1}$$

In another word the number of relevant *collective modes* is far smaller than the number of particles. We will discuss the level of electromagnetic thermal fluctuations in much detail with rigorous theory of the *fluctuation-dissipation theorem* in Sec. 5.3.

As we have commented in Sec. 2.1, however, in a magnetized plasma a partial detachment of fluid motion from the thermal energy is realized. In Sec. 1.4 we have seen that in many astrophysical plasmas such as that in a galactic disk the energies of magnetic field and fluid motion are of the same level as the thermal energy. This means that since the number of collective modes in magnetic fields and fluid motion are a lot smaller than that of individual thermal motion of particles, the individual energy per mode $\langle B_\mathbf{k}^2 \rangle / 8\pi$ and $\rho\langle v_\mathbf{k}^2 \rangle / 2$ is a lot greater (approximately by a factor of g^{-1}) than the thermal energy per mode, $T/2$, where T is the tempeature of individual particles. (Note that in a magnetized plasma $v_\mathbf{k} = c\mathbf{E}_\mathbf{k} \times \mathbf{B} / B^2$). See Problem 2-3. In a limit we can regard this detachment from thermal energy we can isolate the MHD degrees of freedom, the magnetic energy $\langle B_\mathbf{k}^2 \rangle / 8\pi$ and fluid kinetic energy $\rho\langle v_\mathbf{k}^2 \rangle / 2$. Such a theory that regards the interaction only among these modes is discussed in Sec. 2.3.2, where we discuss the equipartition of energies among these MHD modes. Even in this situation under the circumstance if there is an invariant additional to the energy of the MHD modes, there emerges nonequal partition among modes. We then consider the case with coupling between the MHD modes and thermal (individual kinetic) motions (i.e. the dissipative case) in Sec. 2.3.3. When the dissipation of macromode's energy into heat and thus the thermal particle motion, the allowed domain of wavenumbers expands toward a greater wavenumber cutoff, which is usually now set by the viscous wavelength. However, typically this cutoff wavenumber for viscous damping is still far less than $n^{1/3}$, the inverse of the interatomic distance. Further discussion in Sec. 2.3.3. It is important to realize that either little coupling or a small amount of coupling of MHD modes with the rest of thermal plasma allows the magnetized plasma to often easily acquire *structure formation*, as a substantially large portion of system's energy is distributed among a relatively small (compared to N, the number of particles) amount of macroscopic degrees of freedom. This problem is so important a consequence for astrophysical plasmas that we treat the topic not only in this section, but also in Secs. 2.4 and 2.5. The self-organizing tendency of magnetized plasmas, thus arises both from the plasma's penchant for instabilities that provide energy into the macroscopic collective modes and from the macro modes (partial) detachment from the thermal energy. In Sec. 2.4 we primarily discuss *self-organization* of plasmas even in the absence of (or little) dissipation, that of the *filamentary* tendency of magnetized plasmas (but of high β). In Sec. 2.5 we discuss self-organization of plasmas with the presence of dissipation, which allows a whole set of new nontrivial equilibria. In astrophysical observation of plasma phenomena, it is often the case to observe super-brilliant ('super-thermal') phenomena in long wavelength (such as microwave) domains. This is because in many cases the 'effective temperature' of the long wavelength collective modes (which emit such radiation) represents such an enhanced energy distribution much beyond the temperature associated with the thermal motion of the plasma. On the other hand, it is often the case that on optical wavelengths the temperature represents that of the thermal motion, while in X-rays the temperature corresponds to the energetic components of thermal motions.

2.3.1 Onset of Turbulence

In the following we keep a hydrodynamical (type of) instability in mind, as in contrast to a kinetic one. Other cases will be briefly discussed later. When an order parameter, which controls the system's 'order' or stability/instability, becomes sufficiently large, such as the Reynolds number for the flow of an ordinary fluid, the Rayleigh number for the heat convective fluid, the Grashoff number for the vertical slot, etc., nonlinearity of the system becomes important and turbulence ensues.

An example may be illustrated by the *Navier-Stokes equation*

$$\frac{\partial \mathbf{v}}{\partial t} + \mathbf{v} \cdot \nabla \mathbf{v} = \frac{1}{R_e} \nabla^2 \mathbf{v}, \tag{2.3.2}$$

where the velocity, spatial and temporal scales are normalized nondimensionally so that the viscosity ν becomes the inverse Reynolds number. When there is a flow $\mathbf{v}_0$ in the fluid (actually we normalize the velocity with respect to $\mathbf{v}_0$), we can linearize the velocity $\mathbf{v} = \mathbf{v}_0 + \delta\mathbf{v}$. Such linearization, in general, gives rise to analysis of the stability of the flow and the onset of instability and the transition to turbulence. To simplify the matter, we express the perturbation amplitude by A (such as $\delta\mathbf{v}$ in the above problem). Since the hydrodynamic problem does not have the intrinsic (real) frequency within the system itself, the generic equation in the linear approximation may be written as

$$\frac{\partial A}{\partial t} = \gamma A + \alpha A^3. \tag{2.3.3}$$

where γ is the linear growth rate and α is constant (Landau and Lifshitz, 1978). Note that the right-hand side of Eq. (2.3.3) starts with the linear term followed by a cubic nonlinear term. The reason why the quadratic nonlinear term is not kept is due to the parity of the mode. More specifically, suppose that the predominant mode has a wavenumber $\mathbf{k}$ [or $\cos(\mathbf{k} \cdot \mathbf{r})$]. The quadratic term yields $\cos^2(\mathbf{k} \cdot \mathbf{r})$, which contains wavenumbers of $2\mathbf{k}$ and 0. As Eq. (2.3.3) looks at the mode at $\mathbf{k}$, these mode numbers cannot contribute to Eq. (2.3.3). The cubic terms contain modes of wavenumbers $3\mathbf{k}$ and $\mathbf{k}$ (and $-\mathbf{k}$). This arises from the second iteration of the nonlinearity in Eq. (2.3.2). The latter $\mathbf{k}$, of course, is the mode under investigation. This is the second term appearing on the right-hand side of Eq. (2.3.3). This effect of having the cubic nonlinearity from the original quadratic nonlinear term is rather general, often found in hydrodynamics, plasma physics, and nonlinear optics (Bloembergen, 1965).

When we multiply the complex conjugate A^* on Eq. (2.3.3), we obtain

$$\frac{\partial |A|^2}{\partial t} = 2\gamma |A|^2 + \alpha |A|^4. \tag{2.3.4}$$

If we are interested in a steady-state ($\partial_t = 0$), we set

$$0 = 2\gamma |A|^2 + \alpha |A|^4. \tag{2.3.5}$$

For Eq. (2.3.5) to have a finite solution, α must be negative, as γ is positive in the unstable regime. The steady-state amplitude of Eq. (2.3.4) is

$$|A|^2 = -\frac{2\gamma}{\alpha}. \tag{2.3.6}$$

Landau suggested that near the threshold of the instability the (linear) growth rate γ can be expanded around the *critical order parameter* R_c:

$$\gamma \sim \gamma'(R - R_c). \tag{2.3.7}$$

Then the saturation amplitude of the most unstable mode can be expressed as

$$|A|_{\text{sat.}} \sim \sqrt{2\gamma'/|\alpha|} \, \sqrt{R - R_c}. \tag{2.3.8}$$

The *order parameter* (Reynolds number) dependence of the saturated amplitude Eq. (2.3.8) is shown in Fig. 2.6(a) (near $R \gtrsim R_c$). This square root dependence of the saturated amplitude on the order parameter turned out to be found in many hydrodynamical instabilities. In some instances as the amplitude increases during the instability, nonlinearities and thus higher order terms beyond A^3 should become important:

$$\frac{\partial A}{\partial t} = \bar{\gamma} A + \bar{\alpha} A^3 + \bar{\beta} A^5 + \bar{\delta} A^7 + \cdots . \tag{2.3.9}$$

Again the appearance of only odd powers is due to the fact that the response at the original predominant mode takes odd times interactions. The intensity ($|A|^2$) equation now becomes

$$\frac{\partial |A|^2}{\partial t} = f(|A|^2), \tag{2.3.10}$$

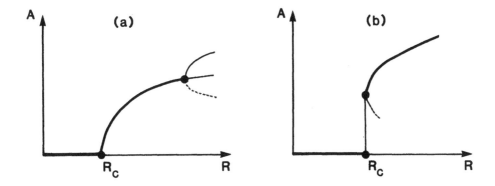

FIGURE 2.6 The threshold Reynolds number (or structure parameter) in the instability and the saturation amplitude. (a) The soft transition and the bifurcation, (b) the hard transition (and the possibility of hysteresis).

where

$$f(a) = \gamma a + \alpha a^2 + \beta a^3 + \delta a^4. \tag{2.3.11}$$

Equations (2.3.11) [or Eqs. (2.3.10) and (2.3.11)] may be called the *Landau-Ginzburg equation*. Excluding some trivial cases such as the signs of α and δ are the same, [which produces no qualitative difference beyond Eq. (2.3.4)] for example, we can sketch a typical behavior of Eq. (2.3.11). If there are four intersections for $f(a) = 0$, which correspond to steady-state (saturation) amplitudes, the first one, $a = 0$, is the trivial pre-instability solution, the second and the fourth are stable solutions, while the third one is an unstable solution. The second solution has already appeared in Eq. (2.3.6); the fourth is a new interesting solution. Such a phenomenon may be called *bifurcation*. In fact, experimentally and/or computationally in many different instabilities one finds such bifurcation. It is known also that often more than once bifurcation happens as the order parameter is raised further and further. See Fig. 2.6(a) at large R.

Often this series of bifurcation is followed by (a) chaotic state(s). The Feigenbaum (1978) series and limit are the most striking example.

In some systems in contrast to the Landau type of transition (the 'soft' transition) the transition can be more sudden (the 'hard' transition) as shown in Fig. 2.6(b). In a system of many degrees of freedom many different modes may have different transition critical numbers. Such a case as the previous chaotic transition can eventually lead to turbulence.

When many modes are excited instead of a predominance of a mode, it is no longer possible to follow each mode's dynamics exactly. We can only construct a statistical theory. Most well-developed theory of plasma turbulence in this stage may be the *quasilinear theory* (see, for example, Sagdeev and Galeev, 1969) and the *weak turbulence* (for more turbulent cases, see, for example, Kadomtsev, 1965). We refer the reader for discussion of these subjects.

When turbulence is fully developed (*fully developed turbulence*) or strong (i.e. the amplitude expansion becomes invalid), each mode couples to many other modes and they in turn couple with many other modes and *vice versa*. The most well-known theory of spectra of hydrodynamic turbulence is that of Kolmogorov (see, e.g. Landau-Lifschitz, 1975) in a dissipative (open) system.

Some of the statistical dynamical treatments of interacting magnetohydrodynamic turbulence modes in a closed (energy conserving) system may be found in Montgomery *et al.* (1993).

2.3.2   Nondissipative Fully Developed Turbulence in MHD

When a system of (M)HD is energetically closed (such as in an opaque local plasma in a disk, or a cosmological plasma before recombination), many modes develop through the nonlinear coupling. Eventually (nearly) all modes of availability are excited. In a limit where dissipation is neglected these modes lead to a statistical equilibrium only among themselves, nearly detached from their thermal particle motions.

When all available modes are excited, it is thought that all the space phase points of MHD modes are visited with an equal probability (the Boltzmann's equal probability ansatz). Usually a system of complete energetical insulation is treated as a microcanonical ensemble. It is well known in statistical mechanics that with a large number of degrees of freedom a canonical ensemble statistics with a given "temperature" has so little fluctuations that the most probable energy of the system is so sharply determined. Thus the treatment derives identical results derived from the microcanonical one. Here we thus take the approach of the canonical ensemble of MHD modes that are "contact" with the heat reservoir. We can think of the heat reservoir as the real thermal energy bath which can extract its own energy via an instability to the MHD collective modes and also can gain energy by absorbing the collective modes' energy. In this sense the system of MHD modes is not totally insulated, but is considered to be dissipationless because the dissipation mechanism does not determine the energy spectrum and does not cause energy flows among wavenumbers in a steady state. The resultant canonical distribution is characterized by the invariants. The energy is the best known invariant of the MHD system. In a 2-D MHD system there are three quadratic invariants of motion (Fyfe *et al.*, 1977) which are the total energy, the *cross helicity*, and the mean square potential

$$E = \frac{1}{2}\int(\mathbf{v}^2 + \mathbf{B}^2) = \frac{1}{2}\sum_k \frac{1}{k^2}(|\omega(\mathbf{k})|^2 + |j(\mathbf{k})|^2),$$

$$P = \frac{1}{2}\int \mathbf{v}\cdot\mathbf{B} = \frac{1}{2}\sum_k \frac{1}{k^2}\omega(\mathbf{k})j(-\mathbf{k}),$$

$$A = \frac{1}{2}\int A_z^2 = \frac{1}{2}\sum_k \frac{1}{k^4}|j(\mathbf{k})|^2. \tag{2.3.12}$$

The *partition function* for the *canonical distribution* with these three invariants is given as

$$Z = \int \exp(-\alpha E - \beta P - \gamma A). \tag{2.3.13}$$

Here α, β, γ and the inverse temperatures associated for the distribution of energy, cross-helicity and the square potential. As commented in the intorduction of this section, the temperature of this distribution may be greatly enhanced because of a large amount of energy per mode may be sustained. Using the standard statistical mechanical technique to calculate expectation values using the partition function, we can derive a predicted spectral distribution of the current and vorticity densities j and ω as

$$\left\langle|j|^2\right\rangle = \frac{1}{2}\frac{k^2}{\alpha - \beta^2/4\alpha + \gamma/k^2},$$

$$\left\langle|\omega|^2\right\rangle = \frac{k^2}{2}\left(\frac{1}{\alpha} + \frac{\beta^2/4\alpha^2}{\alpha - \beta^2/4\alpha + \gamma/k^2}\right). \tag{2.3.14}$$

In the particular case in which $\langle\mathbf{v}\cdot\mathbf{B}\rangle = 0$, β is zero, and the spectra for $\mathbf{v}$ and $\mathbf{B}$ are

$$\left\langle|\omega|^2\right\rangle = \frac{k^2}{\alpha}$$

$$\left\langle|j|^2\right\rangle = \frac{k^2}{\alpha + \gamma/k^2}. \tag{2.3.15}$$

The spectra of statistical averages of energy, cross-helicity and squared potential are shown in Fig. 2.7. The current spectrum of Eqs. (2.3.14) allows for a spectrum peaked at k_{min} (which is determined by system's size) when γ is negative. It can also be shown that

$$\sum_k \left\langle|\mathbf{v}|^2\right\rangle - \sum_k \left\langle|\mathbf{B}|^2\right\rangle = \gamma\frac{2A}{\alpha}. \tag{2.3.16}$$

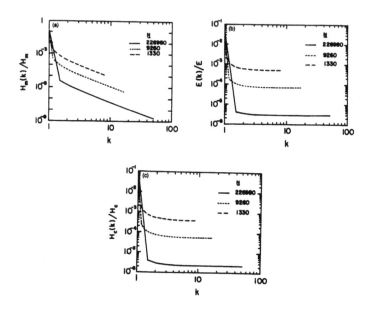

FIGURE 2.7 The Fourier statistics with invariants additional to the energy (such as helicity or cross-helicity) shown possibilities of non-equipartition and condensation of Fourier energy at the lowest wavenumber, leading to structure formation (after Stribling and Matthaeus (1990)).

Whenever the magnetic energy is greater than the kinetic energy, γ will be negative, and the magnetic field will exhibit large-scale structure.

There is, however, no possibility in Eq. (**??**) for vorticity ω to exhibit large-scale structure; indeed, $\langle|\omega|^2\rangle$ is independent of γ. An interesting point is that the quantity $\int v^2 - B^2$ is not an invariant of the ideal MHD equations, but the sign of the expected value of this quantity still is the determining parameter for whether there will be large-scale magnetic structure or not. In the filamentary picture, this quantity is exactly conserved (see Sec. 2.3.1).

The vorticity 'energy' ω^2 spectra from Eq. (2.3.15) gives a kinetic energy spectrum $k^{-2}\langle|\omega|^2\rangle$ partitioned equally between the modes regardless of the parameters of the system, and even when the magnetic field is identically zero. The spectrum is the same as if one had ignored the effects of *enstrophy* (vorticity integral) in the derivation of Eqs. (2.3.12); indeed one has ignored these effects, since enstrophy is not an invariant of MHD. Still, as

66

the magnetic field vanishes, the invariants of the system change. The total energy becomes the kinetic energy normally, but the mean square magnetic potential (A) goes to zero. The enstrophy, meanwhile, changes at a rate proportional to the magnetic field strength. As the magnetic energy vanishes, the enstrophy should be included as an invariant in the calculations, but it is not clear how to do this in a truncated Fourier theory.

We have seen an example of structure formation even for ideal MHD closed systems under certain conditions [particularly the third invariant A in Eq. (2.3.12) was important]. If $A = 0$ or $\gamma = 0$ i.e. the temperature for A is infinite) on top of $\beta = 0$ in Eq. (2.3.13), however, the distribution of Eq. (2.3.14) reduces to the more familiar canonical distribution, which of course yields the well-known equipartition law for v^2 and B^2 spectra, i.e., no structure formation. A related subject (and similar technique) will be treated in Sec. 2.4 with the emphasis on structure formation.

2.3.3 Fully Developed Dissipative Turbulence

When the system is considered to be energetically open (dissipative), that is, there is an input of energy (through some kind(s) of instabilities or other mechanism) and an output of energy (through dissipation), the spectrum of modes (i.e. the mode distribution) takes quite a different outlook from the closed system (or open, but energy input and output happen in the same wavenumber domain) we considered in Sec. 2.3.2. If the input energy is not matched with the dissipation at a given time, the system takes an episodic oscillation in energy. When the input and output match, a steady state is realized. In this steady state typically energy is input from a large scale nature (at a small wavenumber k_0) and dissipation of mode energy takes place in a small scale nature such as molecular nature (at a large wavenumber). Thus there is energy flow in wavenumber space in contrast to the thermal equilibrium solutions in the system considered in Sec. 2.3.2. Kolmogorov (1942) analyzed this situation. He noted that away from the input wavenumber (k_0) region and from the output wavenumber $(1/a)$, (where a is the size at which molecular viscosity becomes significant), little energy input and output of the mode energy to the external system takes place. See Fig. 2.8.

For a given wavenumber region (say, k), the energy flows in from the smaller wavenumber region adjacent to k and flows out to the larger wavenumber region. In this region k, since no energy is dissipated or input, all energy input and output have to be balanced. Kolmogorov called such a region the *inertial domain*. Within the inertial domain, he noted, the same physics happens even if the wavenumber may vary $(k, k', k'', \cdots)$ and thus self-similarity holds. Consider a particular k (which can be typically $k_n \equiv 2^n k_0$). Let $\pi(k)$ be the energy flux in k space, which is invariant with respect to k. The energy flux can be expressed in terms of the energy of the mode $\varepsilon_n = v_n^2$ (we write $v_{k_n} \equiv v_n$) and the timescale is called the eddy turnover time $\tau_n = 2\pi/k_n v_n = \ell_n/v_n$.

$$\pi_n = \frac{v_n}{\ell_n} v_n^2 = \text{const} \equiv \epsilon. \qquad (2.3.17)$$

From Eq. (2.3.17) using the expression of ε_n we obtain

$$\varepsilon_n \propto \epsilon^{2/3} \ell_n^{2/3}. \qquad (2.3.18)$$

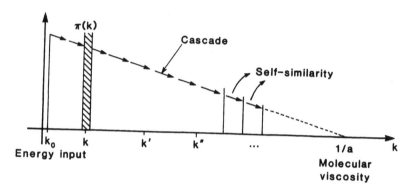

FIGURE 2.8 Kolmogorov's inertial domain and turbulent energy flow.

Since ε_n is defined as

$$\varepsilon_n = \int_{k_{n-1}}^{k_n} E(k)\,dk, \qquad (2.3.19)$$

we arrive from (2.3.17) at the expression

$$E(k) = c\varepsilon^{2/3} k^{-5/3}. \qquad (2.3.20)$$

Equation (2.3.20) is called the *Kolmogorov spectrum* of fully developed (or strong) turbulence and $k^{-5/3}$ dependence is now well known, observed in a variety of natural turbulence spectra such as in oceanographic turbulence, interplanetary turbulence, and some astrophysical turbulence (see Fig. 3.12). The number of degrees of freedom available in the turbulence is dependent of the size of the inertial domain. The largest $k_{max} = 1/a$ is related to $(\varepsilon \nu^3)^{1/4}$ and the number of modes up to this k_{max} for a given volume V of the plasma is

$$N = \left(\frac{k_{max}}{k_{min}}\right)^3 = \mathrm{Re}^{9/4}. \qquad (2.3.21)$$

68

Note that this number is typically far grater than that treated in Sec. 2.3.4, but still typically far smaller than the number of particles and thus the total degrees of freedom.

2.4 Structure Formation and Plasma Filaments

Astrophysical plasmas often exhibit striking morphologies, that can be variously characterized by such words as *filamentary, knotty, cellular......* . By and large in high β (here β is the ratio of the plasma pressure to the magnetic energy density) plasmas, the main morphology is filamentary, while in low β plasmas the main structure is cellular. See, for example, Parker (1994) on cellular or sheet-like plasma-dynamics in low β. Here we consider the former filamentary case in detail. See Fig. 2.9 for a variety of filamentary astrophysical plasmas. For the latter some discussion will be found in Sec. 3.3. A recent snapshot survey on structure formation may be found in Moffatt *et al.* (1992).

In thermodynamics, as we discussed in Sec. 2.3, we learn that a thermal equilibrium state is where chaos is maximized, no structure is permitted, and all modes are equiparticipated (the equipartition law). The kinetic motions and their fluctuations can give rise to a seed for structure formation. When we consider more coarse-grained and larger scale structure, we need to consider the fluid description of plasmas. The statistical theory of magnetohydrodynamic fluctuations based on the Gibbs statistics has been investigated [see, for example, Montgomery *et al.* (1993)]. These studies show that unless there is an additional constraint (or conserved) quantity other than energy (Stribling and Matthaeus, 1990) as we have seen in Sec. 2.3.2, the wavenumber spectrum is still the equipartition distribution and gives rise to no structure formation. It is also interesting to note that, similar to plasma's equipartition (in wavenumber for thermal systems), the gravitational system in its linear phase the system does maintain the equipartition and of course gives rise to the well known gravitational structure formation in its nonlinear phases. However, in plasmas it has been thought that since the uniform plasma equilibrium is nonlinearly stable, there would be no structure formation. In general, or more precisely if the system under study is small in scale or is (infinitely) long after its disturbance, the above notion of plasma equilibria without structure holds.

However, in recent studies [see, for example, Kinney *et al.* (1993) and reference therein] we began to learn that in a system that is spatially large or the time since its disturbance is not great, the equilibrium property can be quite different from the Gibbs statistics and the notion of faceless structureless state does not hold. High Reynolds and/or Lundquist number plasmas that often satisfy these conditions may be found abundantly in astrophysical plasmas. In many of these plasmas, studies have found that vortices and filaments in the fluid persist for a long time, and in fact they are the main characteristics. It is also the case that when we look at cosmical plasma structures they tend to form vortical or filamentary plasmas. Thus it may be a more convenient (and even correct) representation if we represent the fluid by filaments (i.e. by ensemble of delta functions) instead of by waves (i.e. by Fourier sums of sinusoidal functions). Though the delta function representation and the Fourier one are equally valid mathematical basis functions, very subtle statistical mechanical deviations do occur. (This is believed to be due to nonuniform convergence) (Tasso, 1987; Davidson,

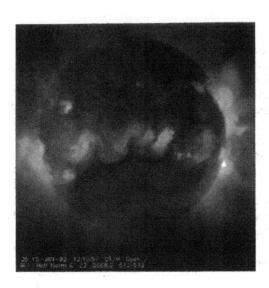 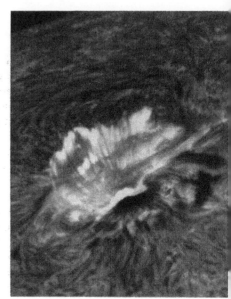

FIGURE 2.9 Variety of filamentary astrophysical plasmas. (a) Solar active regions (bright spots) and coronal holes (dark regions) taken by the Yohkoh; (b) H_α photographs of solar fluxes (the lower left) by the courtesy of H. Zirin; (c) Solar emerging magnetic flux (courtesy of H. Zirin). The Cirrus nebula by the Hubble telescope (J. Hestre and after D. Fischer and H. Duerbeck, 1996); (d) Radio jets ejected from AGN (after Parma *et al.*, 1987).

1954). In fact the filamentary construct of hydrodynamics and magnetohydrodynamics will be shown to possess thermal equilibria (spectra) different from those in the Fourier picture. Since they are not equipartitioned, the plasma emerges with structure. It should be noted that these structures that emerged, however, are structures in a closed system, as opposed to those which may emerge in an open system that has energy input and/or dissipation. The latter structure emergence is often called the *dissipative structure formation* pioneered by Prigogine and will be discussed in Sec. 2.5 in more detail.

An additional element that aids structure formation in plasmas is the coexistence of gas and plasma phases. This problem was pioneered by Field (1965). The admixture of gas and plasma makes the medium thermodynamically unstable, thus lending this matter to form structures. This important subject deserves more research. A recent review by Meerson (1996) is referred to as a latest survey.

 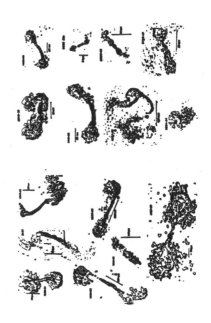

FIGURE 2.9 continued

2.4.1 Filaments vs. Continua in Plasmas

A *filamentary construct* of magnetohydrodynamical plasma dynamics is conveniently expressed in terms of the *Elsasser* (1950) *variables* (see Sec. 2.3.2 in detail). This approach is similar to discrete vortex models of hydrodynamical turbulence (Onsager, 1949; Edwards and Taylor, 1974), which cannot be expected in general to produce results identical to those based on a Fourier decomposition of the fields, as we mentioned above. In a highly intermittent plasma, the induction force is small compared to the convective motion, and when this force is neglected, the plasma vortex system is described by a Hamiltonian. Canonical and microcanonical statistical calculations of such filaments of a plasma show that both the vorticity and the current spectra are peaked at long wavelengths, a deviation from the equipartition, and thus leading to structure formation. [These results differ from previous Fourier-based statistical theories (Sec. 2.3.2), but it is found that when the filament calculation is expanded to include the inductive force, the results approach the Fourier equilibria in the low-temperature limit, and the previous Hamiltonian plasma vortex results in the high-temperature limit.] A statistical calculation in the canonical ensemble and numerical simulations in 3D show that a nonzero large-scale magnetic field is statistically favored, and that the preferred shape of this field is a long, thin tube of flux (Kinney *et al.*, 1994).

Because of the engineering problem of maintaining a confined ionized gas, the traditional theoretical approach to describing plasmas has been to identify stable and unstable configurations. While such knowledge may tell the experimentalist what to avoid building, it does not help the theorist understand what the naturally-occurring states of a plasma are. For such knowledge, a probabilistic approach is necessary (Montgomery *et al.*, 1978; Kinney *et al.*, 1994). Since at a fundamental level any fluid is no more than a collection of particles, one might naively expect a fluid's equilibrium states to conform to the Gaussian of "true" thermodynamic equilibrium. Presumably after sufficient time, and in the absence of external influences, any fluid should eventually reach such an equilibrium, but how long must one wait, and what transpires meanwhile? The time required for the particles to reach their ultimate distribution may be extremely long, and they may occupy some quite long-lived states in the interim. Indeed, if a fluid description is appropriate for a given system, then any statistical analysis of that system should be based on the fluid equations, which will produce states quite different from a particle model. To perform statistical mechanics on a fluid system, some discrete representation must be chosen. That the results of the discrete system will approach the continuous system in some limit must be taken on faith. A point of great subtlety is that discrete representations of continuous objects are in no sense unique, and results obtained via one representation may differ from those obtained through another (Royer, 1984).

Two discrete representations which have been shown to be effective in modeling ideal hydrodynamical turbulence in two dimensions are a Fourier-mode representation and a discrete-vortex representation (Kraichnan and Montgomery, 1980). The two approaches do not yield identical results, but both display interesting thermodynamical features in that the system exhibits self-organizational tendencies in the discrete vortex representation. The question remains of which gives a more realistic description of a physical fluid. This question may be only phenomenologically resolvable.

There is a well-known tendency for a fluid to form intermittent structures when dissipation is small. In addition to casual observations of naturally-occurring vortices in day-to-day life, experimental observations of thin films of super fluid Helium (Schwartz, 1990) have been rich in vortex structure. Very high-resolution, spectral-based numerical simulations of high Reynolds-number (McWilliams, 1984; Brachet *et al.*, 1988) fluids have consistently shown a tendency towards intermittency, specifically a formation of sharply peaked axisymmetric vortex filaments which are quite persistent in time. While such structures can be represented by Fourier modes, the phase correlations represented by their coherence are very difficult to treat in a statistical theory. Statistical theories based on a truncated Fourier representation are unable to predict the formation of these filaments, the presence of which has significant effects on dynamical quantities such as cascade rates (McWilliams, 1984).

In a recent high-accuracy Fourier (spectral) represented 2D MHD simulation Kinney *et al.* (1995) have shown that there emerges a certain class of structure formation. Shown

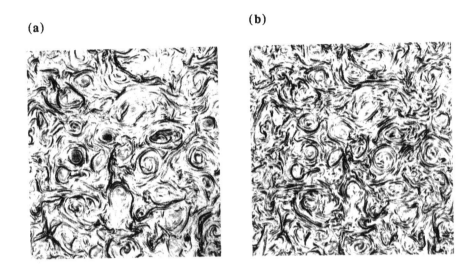

(a)

(b)

FIGURE 2.10 Long-living structures in 2D MHD (a later stage) (a) J_z and (b) ω_z (Kinney *et al.* 1995).

in Fig. 2.10(a) and (b) are the structures of current ($\mathbf{j} = \nabla \times \mathbf{B}$) and vorticity ($\boldsymbol{\omega} = \nabla \times \mathbf{v}$). These figures imply a largest scale vortical structure emerges after a random Fourier noise start. This vortical structure will not decay away for a long time (in a time compared with the eddy turn-over time). it is also noted that the structure of current is similar to that of vorticity.

Sec. 5.2 (cosmological dynamo discussion) this property will be noted again and exploited. In Fig. 2.10(c) we show several cases of runs starting with various different values of plasma β (this particular β is the ratio of the enstrophy to the squared current integral) as a function of time. Even though we started with various β's, we tend to end up with only the classes of β's; two attractors are the one with high β and the other with $\beta \lesssim 1$. This behavior of splitting into two attractors in terms of β will be again found in the context of dynamo β in Sec. 3.1 and in that of accretion disk β in Sec. 4.2

The case for intermittent structures in plasmas is equally strong, if not stronger. An analysis of two-dimensional magnetohydrodynamics (MHD) using a closure method (Pouquet, 1978) shows that a cascade of current to small wavelengths is to be expected, with

73

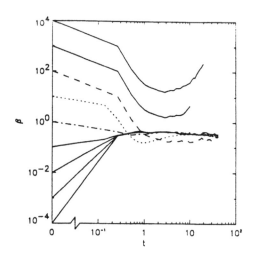

FIGURE 2.10 (continued) (c) Plasma β (here the ratio of entropy to the current energy) is a function of time, showing two attractors, one with a large β and the other with $\beta \lesssim 1$.

singularities formed in finite time from smooth initial conditions. Since strongly intermittent magnetic fields were first observed in the solar atmosphere (Stenflo, 1973), much evidence of intermittent, filamentary plasma structures has been found in many astrophysical plasmas (Yusef-Zadeh *et al.*, 1988). Figure 2.9 is a strong testimony for this. Laboratory plasmas have also displayed key features indicating intermittency (Jha *et al.*, 1992), and two-dimensional spectral fluid simulations at high resolution (Orszag and Tang, 1979; Biskamp *et al.*, 1990) have shown strongly peaked structures in both current and vorticity.

In the following we describe a representation of ideal MHD that employs singular structures as its fundamental objects (Kinney *et al.*, 1994). A two-dimensional, discrete-vortex formalism for MHD is developed for filaments with fixed strengths, though filaments whose strength is allowed to yield similar results. Results are compared with truncated Fourier results of both MHD and neutral fluid theories.

2.4.2 Filamentary Construct of Two-Dimensional MHD*

An MHD fluid is described in terms of two vector fields, the fluid velocity $\mathbf{v}(\mathbf{x}, t)$ and the magnetic field $\mathbf{B}(\mathbf{x}, t)$, in addition to two scalar fields, the fluid density ρ and pressure p. The pressure and density are usually related by an equation of state (we take $\rho = $ const), and in a dissipationless system with incompressible flow, the vector fields obey

$$\frac{\partial \mathbf{v}}{\partial t} + (\mathbf{v} \cdot \nabla)\mathbf{v} = -\frac{\nabla p}{\rho} + \frac{(\nabla \mathbf{B}) \times \mathbf{B}}{4\pi\rho},$$

$$\frac{\partial \mathbf{B}}{\partial t} = \nabla \times (\mathbf{v} \times \mathbf{B}),$$

$$\nabla \cdot \mathbf{v} = \nabla \cdot \mathbf{B} = 0. \tag{2.4.1}$$

We will first treat the two-dimensional case, i.e. all quantities are independent of some coordinate (say z), and $\mathbf{v}_z = \mathbf{B}_z = 0$. In this case, the vorticity, $\boldsymbol{\omega} \equiv \nabla \times \mathbf{v} = \omega\hat{\mathbf{z}}$, and current $\mathbf{j} \equiv \nabla \times \mathbf{B} = j\hat{\mathbf{z}}$ are scalars. Elsasser (1950) pointed out that the incompressible MHD equations, Eq. (2.4.1), can be written in the form

$$\frac{\partial \mathbf{u}}{\partial t} + (\mathbf{w} \cdot \nabla)\mathbf{u} = -\nabla\eta,$$

$$\frac{\partial \mathbf{w}}{\partial t} + (\mathbf{u} \cdot \nabla)\mathbf{w} = -\nabla\eta,$$

$$\nabla \cdot \mathbf{u} = \nabla \cdot \mathbf{w} = 0, \tag{2.4.2}$$

where the new variables are

$$\mathbf{u} = \mathbf{v} + \mathbf{B}, \qquad \mathbf{w} = \mathbf{v} - \mathbf{B}, \tag{2.4.3}$$

$$\eta = p + \frac{1}{2}\mathbf{B}^2: \tag{2.4.4}$$

The equations have been cast in dimensionless form, with the velocity measured in units of an arbitrary constant v_0, magnetic field measured in units of $B_0 = \sqrt{4\pi\rho v_0^2}$, and p in units of ρv_0^2. The quantity η is the fluid pressure plus the magnetic pressure. The ratio $2p/\mathbf{B}^2$, the "plasma beta," is a measure of the strength of the magnetic field relative to the pressure, and has important effects on the dynamical behavior of the system. Other parameters exist which define similar behavior regimes, such as the ratio of fluid kinetic energy to magnetic field energy. In the statistical theory (Kinney et al., 1994), the difference between the kinetic and magnetic energies will emerge as a natural parameter upon which the expected states will depend.

Let us define functions Ω^s and A^s by

$$\Omega^u = \nabla \times \mathbf{u}, \mathbf{u} = \nabla \times \mathbf{A}^u,$$

$$\Omega^w = \nabla \times \mathbf{w}, \mathbf{w} = \nabla \times \mathbf{A}^w. \tag{2.4.5}$$

We will use a general species superscript s to indicate one of either u or w. In a neutral fluid, the vorticity is conservatively advected through the fluid. We seek an analogous result for our Ω^s's. We define the differential operators

$$D^u\mathbf{V} \equiv \partial_t\mathbf{V} - \nabla \times (\mathbf{u} \times \mathbf{V}),$$

$$D^w\mathbf{V} \equiv \partial_t\mathbf{V} - \nabla \times (\mathbf{w} \times \mathbf{V}),$$ (2.4.6)

and a source term

$$\mathbf{S} = \mathbf{\Omega}^w \times \mathbf{\Omega}^u + \left[\mathbf{\Omega}^w \cdot \nabla\mathbf{u} - \mathbf{\Omega}^2 \cdot \nabla\mathbf{w}\right] + \frac{1}{2}\left[\nabla^2(\mathbf{u} \times \mathbf{w}) - (\nabla^2\mathbf{u}) \times \mathbf{w} - \mathbf{u} \times (\nabla^2\mathbf{w})\right].$$ (2.4.7)

In the two-dimensional case, $\mathbf{\Omega} = \Omega\hat{\mathbf{z}}$, $\mathbf{A} = A\hat{\mathbf{z}}$, and $\mathbf{S} = S\hat{\mathbf{z}}$. We choose a gauge in which $\nabla \cdot \mathbf{A} = 0$, so that $\Omega^s = -\nabla^2 A^s$. The equations of motion are then written as

$$\partial_t\Omega^u + \mathbf{w} \cdot \nabla\Omega^u = S,$$ (2.4.8)

$$\partial_t\Omega^w + \mathbf{u} \cdot \nabla\Omega^w = -S.$$ (2.4.9)

At this point, no departure from the original MHD equations has been made. The form of Eqs. (2.4.8) and (2.4.9), however, suggests a useful representational picture. Let Ω^s take the form

$$\Omega^s = \sum_i \alpha_i^s \delta(\mathbf{x} - \mathbf{x}_i^s).$$ (2.4.10)

Equations (2.4.8) and (2.4.7) are solved by the motion of the filaments if

$$\frac{d\mathbf{x}_i^u}{dt} = \mathbf{w}(\mathbf{x}_i^u) \; , \; \frac{d\mathbf{x}_i^w}{dt} = \mathbf{u}(\mathbf{x}_i^w),$$ (2.4.11)

$$\sum_i \frac{d\alpha_i^u}{dt}\delta(\mathbf{x} - \mathbf{x}_i^u) = -\sum_i \frac{d\alpha_i^w}{dt}\delta(\mathbf{x} - \mathbf{x}_i^u) = S(\mathbf{x}).$$ (2.4.12)

The induction of current represented by S is manifested as a simultaneous increase in the strength of u-filaments and a decrease in the strength of w-filaments.

As an illustration for the essence in the statistics of u-w filaments, we consider the case where the source S effect is neglected, so that individual filament strengths do not change in time. See Kinney et al. (1994) for more general cases where this assumption need not be made. Let us have N_u filaments of type u and N_w of type w, $N_u + N_w = N$. The u's will have strength $\pm\alpha^u$ and w's $\pm\alpha^w$, and the domain will be a periodic box in (x, y) with volume V. The potentials are given by

$$A^s(\mathbf{x}) = -\int G(\mathbf{x}|\mathbf{x}')\Omega^s(\mathbf{x}')d\mathbf{x}' = -\sum_{i \in s} \alpha^s G(\mathbf{x}|\mathbf{x}_i^s),$$ (2.4.13)

where G is the Green's function for Poisson's equation, $\nabla^2 G = \delta(\mathbf{x} - \mathbf{x}')$, with appropriate boundary conditions. In an infinite domain, $G(\mathbf{x}|\mathbf{x}') \propto \ln|\mathbf{x} - \mathbf{x}'|$.

The filaments move by

$$\frac{d\mathbf{x}_i^u}{dt} = \mathbf{w}(\mathbf{x}_i^u) = \nabla \times \hat{\mathbf{z}} A^w(\mathbf{x}_i^u) \tag{2.4.14}$$

$$\frac{d\mathbf{x}_i^w}{dt} = \mathbf{u}(\mathbf{x}_i^w) = \nabla \times \hat{\mathbf{z}} A^u(\mathbf{x}_i^w), \tag{2.4.15}$$

so this system of filaments is Hamiltonian with conjugate variables that are the usual Cartesian coordinates $(\sqrt{|\alpha|}x_i, \sqrt{|\alpha|}y_i)$. The Hamiltonian is

$$H(\mathbf{x}_1, \ldots, \mathbf{x}_N) = -\sum_{i \in u}\sum_{j \in w} \alpha_i^u \alpha_j^w G(\mathbf{x}_i|\mathbf{x}_j), \tag{2.4.16}$$

which may also be written

$$H = \int A^u \Omega^w = \int \mathbf{u}\dot{\mathbf{w}} = \int \mathbf{v}^2 - \mathbf{B}^2. \tag{2.4.17}$$

An isolated u or w filament has $\mathbf{v}$ and $\mathbf{B}$ fields of equal magnitude, and therefore makes no contribution to H. A key difference between discrete-vortex and Fourier-mode models of neutral fluids is that a singular vortex filament has infinite self-energy. This energy is usually subtracted as a constant from the Hamiltonian, although the scaling of this constant with the system size must be treated carefully (Seyler, 1976; Montgomery and Joyce, 1974). In the Fourier description, the self-energy of structures do not appear explicitly, and cannot be subtracted out. The filamentary MHD representation differs from the Fourier description in a similar way, except that the self-energy of filaments, being exactly zero when S is neglected, need not be subtracted away. In addition to being an interaction energy, H is a parameter whose sign determines whether the fluid is kinetically or magnetically dominated.

2.4.2.1 Statistics of Filaments*

Given a statistical system with Hamiltonian H, a canonical distribution of $u - w$ filaments can be described with the Boltzmann factor $e^{-\beta H}$ as we have done in Sec. 2.3.2. In our fixed-strength u-w filament system, the partition function is

$$Z = \int e^{-H/T} \prod_{i,s} d\mathbf{x}_i^s. \tag{2.4.18}$$

The choice of coordinates in the above integral is important. Only coordinates $\{q_i\}$ for which the volume element $\prod_i dq_i$ is constant during the evolution of the system give meaningful results. This requirement on the coordinates is known as *Liouville's theorem*, and it is automatically satisfied if the coordinates are the conjugate variables of a Hamiltonian system.

Following the derivation of Kinney *et al.* (1994), the partition function turns out to be

$$Z = \prod_k \frac{k^2}{k^2 - \gamma^2/k^2}, \tag{2.4.19}$$

77

in which γ is the (negative) normalized inverse temperature

$$\gamma = -\frac{\sqrt{N_u N_w}\alpha^u \alpha^w}{VT}.$$ (2.4.20)

The expected spectra for the filament density

$$\rho_s(\mathbf{k}) = \frac{1}{V}\sum \alpha_i^s e^{ikx_i^s}$$

are found to be:

$$\left\langle |\rho_s|^2 \right\rangle = \frac{N_s \alpha^{s^2}}{V^2}\frac{k^2}{k^2 - \gamma^2/k^2},$$

$$\left\langle \mathrm{Re}(\rho_u \rho_w^*) \right\rangle = \frac{\sqrt{N_u N_w}\alpha^u \alpha^w}{V^2}\frac{\gamma}{k^2 - \gamma^2/k^2}.$$ (2.4.21)

Ensemble-average spectra of usual fluid quantities can be calculated from these density spectra by the relations

$$\left\langle |\omega|^2 \right\rangle = \frac{1}{4}\left\langle |\rho_u + \rho_w|^2 \right\rangle$$

$$\left\langle |j|^2 \right\rangle = \frac{1}{4}\left\langle |\rho_u - \rho_w|^2 \right\rangle$$

$$\left\langle \mathbf{v}\cdot\mathbf{B} \right\rangle = k^{-2}\frac{1}{2}\left(\left\langle |\rho_u|^2 \right\rangle - \left\langle |\rho_w|^2 \right\rangle\right).$$ (2.4.22)

In particular, when $N_u = N_w$ and $\alpha^u = \alpha^w$, $\langle \mathbf{v}\cdot\mathbf{B} \rangle = 0$, and the $|\omega|^2$ and $|j|^2$ spectra are

$$\left\langle |\omega|^2 \right\rangle = \frac{N_s \alpha^{s^2}}{2V^2}\frac{k^2}{k^2 - \gamma},$$

$$\left\langle |j|^2 \right\rangle = \frac{N_s \alpha^{s^2}}{2V^2}\frac{k^2}{k^2 + \gamma}.$$ (2.4.23)

The Hamiltonian takes the expected value

$$\langle H \rangle = \sum_k \left\langle |\mathbf{v}|^2 \right\rangle - \sum_k \left\langle |\mathbf{B}|^2 \right\rangle = \gamma\frac{2V}{N_s \alpha_s^2}\sum_k \frac{1}{k^4}|\omega|^2 |j|^2,$$ (2.4.24)

and is of the same sign as γ, which is constrained to $|\gamma| < k_{\min}^2$ because $\langle |\omega|^2 \rangle$ and $\langle |j|^2 \rangle$ must be positive.

In comparison with Eq. (2.4.22), Kraichnan's truncated Fourier analysis for neutral fluids (Kraichnan and Montgomery, 1980), based on the invariance of the kinetic energy and the enstrophy, gives an equilibrium spectrum for the vorticity

$$\left\langle |\omega|^2 \right\rangle = \frac{k^2}{\alpha k^2 + \beta}.$$ (2.4.25)

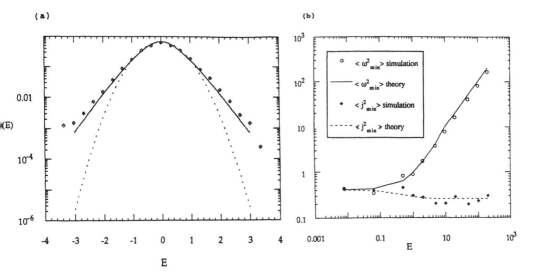

FIGURE 2.11 Statistics of 2D filaments as a function of energy. Theory of filamentary construct (real line) and dots (2D simulation) and broken line for Gaussian statistics are shown (a). A deviation from Gaussian statistics is evident. (b) Energy distribution in velocity and magnetic fields as a function of energy. The higher the energy, the more energy unbalance between the kinetic and magnetic energies occurs (Kinney *et al.*, 1994).

In the neutral fluid spectra, either α or β may be negative, within appropriate restrictions that keep $\langle |\omega|^2 \rangle$ positive. Obviously, $\gamma > 0$ MHD states correspond to $\alpha > 0$, $\beta < 0$ neutral fluid states. By choosing appropriate temperatures, the denominator may be made arbitrarily small, and any total energy may be attained. The energy will be contained predominantly at k_{min}, which is an indication of large-scale structure in ω. If a system contains equal numbers of filaments of both signs, such a state must take the form of one large cluster of each sign of filament. When $\gamma < 0$ in the MHD system, the magnetic field shows structure of the same form, with large-scale current distributions.

One of the important consequences of this theory was that the spectrum of filaments in terms of energy E does not show the Gaussian, but rather a broader non-Gaussian statistics. See. Fig. 2.11.

2.4.3 Filamentary Objects in Three-Dimensional MHD Turbulence*

The equation describing the dynamics of filaments Eq. (2.4.11) can be integrated. The solution can be written in terms of the Biot-Savart-like expression for u and w fields as

$$\mathbf{u}(\mathbf{x},t) = \sum_i \alpha_i^u \int \frac{\mathbf{x} - \mathbf{r}_i^u(s,t)}{|\mathbf{x} - \mathbf{r}_i^u(s,t)|^3} \times \mathbf{r}_i^{u\prime}\, ds,$$

$$\mathbf{w}(\mathbf{x},t) = \sum_j \alpha_j^w \int \frac{\mathbf{x} - \mathbf{r}_j^w(s,t)}{|\mathbf{x} - \mathbf{r}_j^w(s,t)|^3} \times \mathbf{r}^{w\prime}_{\;j}\, ds, \tag{2.4.26}$$

where $\mathbf{r}_i^s(s,t)$ is a curve in terms of a parameter s that determines the spatial location of the filaments, and $\mathbf{r}^{s_i} = \partial \mathbf{r}_i^s / \partial s$. This approach has already been used for hydrodynamical vortex filament simulations (Pumir and Siggia, 1987). Since this decomposition of the filament coordinates in 3D is still rather intractable, for actual calculations, however, here we describe the vortices in terms of a multipole expansion, i.e. as a sum over derivatives of the Green's function (Kinney et al., 1994)

$$\mathbf{A}^s(\mathbf{x},t) = \sum_{n=0}^{\infty} \frac{1}{n!} \mathbf{M}^{(n)}_{i_1,\ldots,i_n}(t) \frac{\partial^n}{\partial x_{i_1} \ldots \partial x_{i_n}} G(\mathbf{x}|\mathbf{x}^s(t)). \tag{2.4.27}$$

A filament, here, may be looked upon as a sewed-up string of dipoled and higher order poles. [We will revisit this picture in Sec. 5.3 and Fig. 5.13.] An arbitrary field can be represented by suitably chosen distributions of point particles possessing internal degrees of freedom (the multipole moments **M**). The equations of motion of a u-vortex are solved by

$$\dot{\mathbf{x}}^u = \mathbf{w}(\mathbf{x}^u)\dot{\mathbf{M}}^{(n)}_{i_1,\ldots,i_n} - (\mathbf{M}^{(n)}_{i_1,\ldots,i_n} \nabla)\mathbf{w}(\mathbf{x}^u) = \mathbf{S}, \tag{2.4.28}$$

and similarly for w-filaments. The source term is unique only up to an additive gradient, because such a gradient added to the potentials does not actually change the fields. The source term again represents the induction of current in the advecting vortices.

In three dimensions, the lowest nontrivial multipole moment for a divergenceless field is a point dipole or vortex ring. In hydrodynamics, one must cope with the problem that a vortex loop moves under its own influence at a speed that scales inversely with the loop size. The problem of infinitely speedy rings is not present in our system, however, since a loop does not self-interact. The ith component of the dipole moment is given in terms of the jth component of $\mathbf{M}_k^{(1)}$ by $D_i = \epsilon_{ijk} M_{j;k}^{(1)}$.

The turbulent fluid in this representation is given by the positions and dipole strengths of the vortices, i.e. by the $6N$ variables $\mathbf{D}_i^s$ and $\mathbf{x}_i^s$. Defining the dipole field as

$$\mathbf{D}^s(\mathbf{x}) \equiv \sum_i \mathbf{D}_i^s \delta(\mathbf{x} - \mathbf{x}_i^s), \tag{2.4.29}$$

the field arising from the dipoles is given by

$$\mathbf{u}(\mathbf{x}) = \mathbf{D}^u(\mathbf{x}) - \nabla \int \nabla' \cdot \mathbf{D}^u(\mathbf{x}') G(\mathbf{x}|\mathbf{x}') d\mathbf{x}', \tag{2.4.30}$$

and the dipole moments change according to

$$\dot{\mathbf{D}}_i^u = -\nabla(\mathbf{D}_i^u \cdot \mathbf{w})|_{\mathbf{x}_i^u}, \quad \dot{\mathbf{D}}_i^w = -\nabla(\mathbf{D}_i^w \cdot \mathbf{u})|_{\mathbf{x}_i^w}. \tag{2.4.31}$$

The evolution of the dipoles can thus be described by

$$\dot{\mathbf{x}}_i = \frac{\partial H}{\partial \mathbf{D}_i}, \quad \dot{\mathbf{D}}_i = -\frac{\partial H}{\partial \mathbf{x}_i}, \tag{2.4.32}$$

and $\mathbf{u}(\mathbf{x})$ and $\mathbf{w}(\mathbf{x})$ are given by Eq. (2.4.26) and H by Eq. (2.4.17).

Kinney *et al.* (1994) have studied once again the canonical distribution for the Hamiltonian for dipole filaments in 3D. Their result is similar to the expected orientation of a system of magnetic dipoles. When the 'temperature' drops below a critical temperature (the Curie point), the dipoles align themselves. The critical temperature T_c is a function of the dipole strengths and the geometric mean of u and w filament densities n_u and n_w:

$$T_c = \frac{1}{3} D^u D^w \sqrt{N_u N_w}/V. \tag{2.4.33}$$

For $|T| < T_c$, nontrivial structure formation emerges; i.e. u dipoles form a u string (or filament) and w dipoles form a w string. For T positive, the dipoles align themselves with u's anti-parallel to w's. This is the minimum energy state, because it aligns the magnetic moments, but cancels the kinetic energy that the dipoles carry. In this simplified model with the magnitude of D^s fixed, the phase-space is bounded, and negative T states also exist, with all dipoles aligned parallel. However, in the full description, the magnitude of D^s is unconstrained and the phase-space is unbounded. In this case, only positive T is allowable, so we expect that a physical system will only admit states in which u and w dipoles are anti-aligned, i.e. states with a mean magnetic field.

Figure 2.12 shows the computational volume of these fixed-magnitude runs. The initial magnetic field is generally vertical. Plotted in this volume are surfaces on which $B^2 = 0.7B_{max}^2$. Stronger fields are present within the volume defined by this surface. Initially, B^2 is within 30% of the maximum throughout nearly the entire computational volume. The general attraction between oppositely-aligned dipoles brings u and w dipoles together, and the magnetic field becomes more localized. Finally the magnetic field becomes dominantly concentrated into a vertical column. These structures live for a time and then decay away, although the exact mechanism for this decay is still unclear.

The presence of coherent structures has come to be recognized as one of the most important features of neutral turbulence. Structures have profound effects on the dynamics of turbulence, but are not well represented by truncated Fourier statistics. While point-vortex theories are also limited in how they can describe extended structures in real fluids, they can provide a useful starting point. The search for structures in MHD turbulence is only just beginning, but evidence of strong intermittency in plasmas has long been observed, and there is every reason to believe structures will play important roles in magnetic turbulence as well. The present model assumes structure at the smallest scale, but the theory predicts structure on a large scale. This "self-organization" of the system takes place in the absence of both external forcing and dissipation.

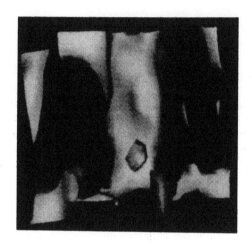

FIGURE 2.12 A three-dimensional structure in 3D MHD. An equicontour of magnetic field energy, based on the filamentary construct. (Kinney *et al.*, 1994).

2.5 Relaxation and Dissipation

In the presence of dissipation, plasma can undergo a relaxation from the original state toward a "relaxed" state. An ultimate *relaxed state* may be that of thermal equilibrium of the plasma. However, in most of the situations of plasma physics, in particular that of plasma astrophysics, the lifetime of the plasma under consideration is finite in view of the dissipative time scales. In this situation it is necessary to consider the relaxation process of the plasma in detail and we often find surprisingly rich structures and processes (Taylor, 1986). Taylor introduced the *helicity* as invariant, as opposed to the energy in the relaxing plasma. This is because the helicity is a *topological invariant* and relaxes, in general, slowly as compared with energy in a dissipative plasma. Equilibria with helicity as invariant often give rise to nontrivial structures. In the following, using the relaxation theory, we look for nontrivial equilibria that are spatially localized. There are interests to spontaneous structure formation such as in cosmological plasmas (see Sec. 5.3).

2.5.1 Dissipative Vorticity Equation*

We survey the relaxation theory (Oliveira *et al.* 1995, a generalization of Taylor's) based on the canonical momentum of each fluid species in a multicomponent plasma. The generalized helicity, as a topological quantity, has a lifetime greater than that of the energy. The variational principle suggests vortices structures. We are interested in localized solutions, assuming the existence of a separatrix. Two-dimensional and three-dimensional solutions are considered for an electron-positron-proton plasma as well as an ideal MHD plasma.

We seek localized solutions of electromagnetic fields that do not cause charge separation. The localizability is required in virtue of the no external field boundary condition. Mathematically, at spatial infinity the fields must vanish. We may require the existence of a separatrix beyond which the fields decrease fast enough to have the total field energy finite. These solutions must be stable. The relaxation theory is appropriate to investigate this problem because of its self-organization feature which is in the spirit of self-generated and/or self-maintained configurations in a cosmological plasma, as we expect no 'external' energy source or sink. The macroscopic equations of a plasma with N species are:

$$\nabla \cdot \mathbf{B} = 0, \tag{2.5.1}$$

$$\nabla \cdot \mathbf{E} = \frac{4\pi}{c} \sum_{a=1}^{N} q_a n_a, \tag{2.5.2}$$

$$\nabla \times \mathbf{B} = \frac{4\pi}{c} \sum_{a=1}^{N} q_a n_a \mathbf{v_a} + \frac{\partial}{c \partial t} \mathbf{E}, \tag{2.5.3}$$

$$\nabla \times \mathbf{E} = -\frac{\partial}{c \partial t} \mathbf{B}, \tag{2.5.4}$$

$$\frac{\partial}{\partial t} n_a + \nabla \cdot (n_a \mathbf{v_a}) = 0, \tag{2.5.5}$$

$$\nabla \cdot \mathbf{v_a} = 0, \tag{2.5.6}$$

$$n_a m_a \left(\frac{d}{dt}\right)_a \mathbf{v_a} = q_a n_a \left(\mathbf{E} + \frac{\mathbf{v_a}}{c} \times \mathbf{B}\right) - n_a m_a \nabla \phi_G - \nabla P_a + \mathbf{R_a}, \tag{2.5.7}$$

where $(d/dt)_a = \partial/\partial t + \mathbf{v_a} \cdot \nabla$ and

$$\mathbf{R_a} = \mu_a \nabla^2 \mathbf{v_a} - m_a n_a \sum_b \nu_{ab}^c (\mathbf{v_a} - \mathbf{v_b}), \tag{2.5.8}$$

with the viscosity μ_a and the collision frequency for different fluids ν_{ab}^c. We assume an equation of state $P_a = P_a(n_a)$.

Let the electric and magnetic fields be given by their potentials $\mathbf{E} = -\nabla\phi - \partial/c\partial t \, \mathbf{A}$ and $\mathbf{B} = \nabla \times \mathbf{A}$. If we use the canonical momentum

$$\mathbf{p_a} = m_a \mathbf{v_a} + \frac{q_a}{c} \mathbf{A} \tag{2.5.9}$$

of each of the species of the plasma and eliminate $\mathbf{A}$ in favor of $\mathbf{p_a}$ and $\mathbf{v_a}$, we obtain from the equation of motion (2.5.7)

$$\frac{\partial}{\partial t} \mathbf{P_a} = -\mathbf{v_a} \times \boldsymbol{\Omega_a} - \nabla \epsilon_a + \mathbf{r_a}, \tag{2.5.10}$$

where the *generalized vorticity* is

$$\boldsymbol{\Omega_a} = -\nabla \times \mathbf{P_a} = -m_a \nabla \times \mathbf{v_a} - \frac{q_a}{c} \mathbf{B}, \tag{2.5.11}$$

ϵ_a is the energy of the component a:

$$\epsilon_a = \tfrac{1}{2} m_a \mathbf{v_a^2} + q_a \phi + m_a \phi_G + \frac{P_a}{n_a}, \tag{2.5.12}$$

and

$$\mathbf{r_a} = \frac{\mathbf{R_a}}{n_a} - \frac{P_a}{n_a} \nabla (\log n_a). \tag{2.5.13}$$

Applying the curl on (2.5.10) we obtain an equation for the vorticity:

$$\frac{\partial}{\partial t} \boldsymbol{\Omega_a} = \nabla \times [\mathbf{v_a} \times \boldsymbol{\Omega_a}] - \nabla \times \left(\frac{\mathbf{R_a}}{n_a} \right). \tag{2.5.14}$$

In the limit of low-viscosity-high-density and low interfluid collision frequency

$$\frac{\mu_a}{n_a} \ll \min \left[L m_a v_a; L^2 \frac{q_a}{c} B \right], \tag{2.5.15}$$

$$\nu_{ab}^c \ll \min \left[\frac{v_a}{L}; \frac{v_b}{L}; \frac{q_a B}{m_a c} \right], \tag{2.5.16}$$

where L is a typical length in the problem, we can neglect $\mathbf{r_a}$ and $\mathbf{R_a}$. In this limit the equilibrium configuration will be the one in which ϵ_a is the level surface function for field lines of $\mathbf{v_a}$ and $\boldsymbol{\Omega_a}$:

$$\mathbf{v_a} \times \boldsymbol{\Omega_a} = \nabla \epsilon_a, \tag{2.5.17}$$

$$\mathbf{v_a} \times \nabla \epsilon_a = 0, \tag{2.5.18}$$

$$\boldsymbol{\Omega_a} \cdot \nabla \epsilon_a = 0. \tag{2.5.19}$$

The energy is constant along the streamlines of both the velocity and the vorticity fields.

Except possibly in subdomains where $\nabla \epsilon_a = 0$, the streamlines of $\mathbf{v_a}$ and $\boldsymbol{\Omega_a}$ lie on surfaces $\epsilon_a = $ const. The topology of these surfaces is determined by the topology of the sets of points at which $\nabla \epsilon_a = 0$: these points may be isolated, or they may fill three-dimensional subdomains (Moffatt, 1985).

Let us consider a stationary solution propagating with some velocity $\mathbf{u}$, for which $\partial/\partial t = -\mathbf{u} \cdot \nabla$ and we get from (2.5.14)

$$[(\mathbf{v_a} - \mathbf{u}) \times \boldsymbol{\Omega_a}] = -\nabla(\epsilon_a - \mathbf{u} \cdot \mathbf{p_a}) \tag{2.5.20}$$

instead of (2.5.17). Now the divergenceless fields $\boldsymbol{\Omega}$ and $\mathbf{v_a} - \mathbf{u}$ lie on the level surfaces of $\epsilon_a - \mathbf{u} \cdot \mathbf{p_a}$. For localized solutions $\mathbf{v_a} \to 0$ and $\mathbf{B} \to 0$ as $\mathbf{r} \to \infty$. We conclude that outside the *separatrix*

$$\epsilon_a - \mathbf{u} \cdot \mathbf{p_a} = 0, \tag{2.5.21}$$

and so

$$\boldsymbol{\Omega_a} = \alpha_0(\mathbf{r})(\mathbf{v_a} - \mathbf{u}), \tag{2.5.22}$$

such that $\alpha_0(\mathbf{r})\mathbf{u}$ approaches zero at the spatial infinity. Note that $\alpha_a(\mathbf{r})$ is constant along $(\mathbf{v_a} - \mathbf{u})$. So $\mathbf{u} \cdot \nabla \alpha = 0$ and $\mathbf{u} \cdot \nabla \alpha = 0$. Of course a sufficient condition for existence of vortices is $\alpha = 0$ outside the separatrix.

2.5.2   Generalized Relaxation Theory*

Now we relate the equations obtained above with a general relaxation theory. The evolution of the fields, determined by (2.5.9) in the limit (2.5.15)–(2.5.16) and spatially constant density, preserves the *generalized helicity*:

$$I_a^h = \int \mathbf{p_a} \cdot \boldsymbol{\Omega_a} d^3x - \oint \mathbf{p_a} \cdot dl_1 \oint \mathbf{p_a} \cdot dl_2, \tag{2.5.23}$$

where the integration is over the whole spatial volume and the line integrals appear for multiply-connected spaces. This definition is gauge independent (Horiuchi and Sato, 1988). Indeed the time derivative of (2.5.23) is

$$\frac{\partial}{\partial t} I_a^h = \oint [-\epsilon_a \boldsymbol{\Omega_a} + (\mathbf{p_a} \cdot \boldsymbol{\Omega_a})v_a - (\mathbf{p_a} \cdot \mathbf{v_a})\boldsymbol{\Omega_a}] \cdot dS.$$

So, for the boundary conditions $\boldsymbol{\Omega} \cdot \mathbf{n} = 0$ and $\mathbf{v_a} \cdot \mathbf{n} = 0$, we have $\partial/\partial t\, I_a^h = 0$.

If we include the dissipation terms we find, assuming the above boundary conditions,

$$\frac{\partial}{\partial t} K_a^h = 2 \int \mathbf{r_a} \cdot \boldsymbol{\Omega_a} d^3x. \tag{2.5.24}$$

For the special case $\mathbf{r_a} \cdot \boldsymbol{\Omega_a} = \lambda_a \mathbf{p_a} \cdot \boldsymbol{\Omega_a}$, we obtain the evolution of the helicity as $I_a^h(t) = I_a^h(0) \exp[2 \int^t \lambda_a(t')dt']$. It can increase, decrease, or be constant, depending on the behavior of $\lambda_a(t)$ in time. The total energy

$$E_{total} = \int \frac{1}{2} \left[\sum_a n_a m_a v_a^2 + \frac{1}{4\pi} (E^2 + B^2) \right] d^3x \tag{2.5.25}$$

decreases with time as

$$\frac{\partial}{\partial t} E_{\text{total}} = \int \sum_a \mathbf{R_a} \cdot \mathbf{v_a} d^3 x. \tag{2.5.26}$$

In the ideal limit neglecting $\mathbf{R_a}$ the total energy is, of course, conserved. But in general both helicity and total energy can decay in time. From the expressions (2.5.26) and (2.5.24) the rates of decay of helicity and total energy are different. I_a^h are *topological quantities* and we have some reason (Horiuchi and Sato, 1988) that the helicity does not decay as fast as the total energy does (Isichenko and Marnachev, 1987). Indeed, since changing in helicity involves changing in the topology of the lines, breaking and reconnecting them, it takes some time to happen, while the dissipation of energy does not have such a constraint.

We estimate phenomenologically the lifetime for decreasing the total energy and the helicity as

$$\tau_{\text{energy}} = \min \left[1/\nu_a^c; n_a m_a L^2/\mu_a \right] \tag{2.5.27}$$

$$\tau_{\text{helicity}} = \max \left[1/\nu_a^c; n_a m_a L^2/\mu_a \right]. \tag{2.5.28}$$

The case $1/\nu_d^c \ll n_a m_a L^2/\mu_a$ is one in which the dissipation of the energy is through the interfluid collisions, usually at small scales compared to the other case $1/\nu_a^c \gg n_a m_a L^2/\mu_a$ in which the energy is dissipated through the viscosity of each fluid species.

The *variational principle* is: minimize E_{total}, subject to the constraint that $\sum_a I_a^h = $ const. Let $\delta\phi, \delta\mathbf{A}, \delta\mathbf{p_a}$ be the variations of the electrostatic potential, the vector potential and the canonical momentum, respectively. Then the variational 'stationarity' condition

$$\delta \left(E_{\text{total}} - \lambda \sum_a I_a^h \right) = 0 \tag{2.5.29}$$

leads to

$$\nabla \cdot \mathbf{E} = 0, \tag{2.5.30}$$

$$\nabla \times \mathbf{B} = \frac{4\pi}{c} \sum_{a=1}^N q_a n_a \mathbf{v_a}, \tag{2.5.31}$$

$$\mathbf{\Omega_a} = -\frac{n_a}{2\lambda} \mathbf{v_a}. \tag{2.5.32}$$

The first two equations above are of no surprise. The last equation is a special case solution for Eq. (2.5.17). In this case the generalized vorticity field lines are frozen in the fluid.

To check the stability of these configurations, we make a second variation on (2.5.29), use Eqs. (2.5.30)–(2.5.32), and integrate by parts. We get

$$\delta^2 \left(E_{\text{total}} - \lambda \sum_a I_a^h \right) \Bigg|_{\text{extreme}} = \int d^3 x \left[\sum_a \frac{n_a m_a}{2} (\delta v_a)^2 \right.$$

86

$$+ \frac{1}{4\pi} \left((\nabla\delta\phi)^2 + \sum_{ij}(\partial_j\delta A_i)^2 \right) - 2\lambda\sum_a \delta(\nabla \times \mathbf{p_a}) \cdot \delta\mathbf{p_a} \right]. \tag{2.5.33}$$

If the last term does not change sign, we can make the configuration stable by appropriate choice of λ. This term is called average perturbation *spirality* in connection to amplifications of vortex disturbances in planetary atmospheres (Petviashvili and Yankov, 1989).

Combining (2.5.31), (2.5.32), and (2.5.11), we get

$$\nabla \times \mathbf{B} = \frac{4\pi}{c} \left[\left(2\lambda\sum_a \frac{q_a^2}{c} \right) \mathbf{B} + 2\lambda\sum_a (m_a q_a \nabla \times \mathbf{v_a}) \right]. \tag{2.5.34}$$

In the next subsection we study the more general equilibrium equation (2.5.17) with planar and axial symmetry for an electron-positron-proton plasma as well as ideal MHD plasmas. We also discuss the *Beltrami property* and various special solutions such as magnetic vortices, parallel vortices, and dynamic vortices.

2.5.3   Electron-Positron Proton Plasma*

Electron and positron fluids are likely to have a steady state with the same velocity $\mathbf{v}$ due to the strong coupling with the isothermal photon pressure. Therefore the canonical momentum (2.5.9) and generalized vorticity (2.5.11) are: for the electrons $(-)$ and positrons $(+)$: $(\mathbf{v}_+ = \mathbf{v}_- = \mathbf{v})$

$$\mathbf{p}_\pm = m\mathbf{v} \pm \frac{e}{c}\mathbf{A} \tag{2.5.35}$$

$$\mathbf{\Omega}_\pm = -\nabla \times \mathbf{p}_\pm = -m\nabla \times \mathbf{v} \mp \frac{e}{c}\mathbf{B} \tag{2.5.36}$$

and similarly for the protons (with a subscript i)

$$\mathbf{p}_i = m_i\mathbf{v}_i + \frac{e}{c}\mathbf{A} \tag{2.5.37}$$

$$\mathbf{\Omega}_i = -\nabla \times \mathbf{p}_i = -m_i\nabla \times \mathbf{v}_i - \frac{e}{c}\mathbf{B}. \tag{2.5.38}$$

The displacement current is neglected: $\partial/c\partial t\, \mathbf{E} \ll \nabla \times \mathbf{B}$ and assume the quasineutrality condition: $n_i = n_- - n_+ \equiv \delta n^*$. We consider δn^* constant and the ions velocity much smaller than the electron-positron velocity: $\mathbf{v}_i \ll \mathbf{v}$. From Ampére's law (2.5.3)

$$\mathbf{v} = -\frac{c}{4\pi e\delta n^*}\nabla \times \mathbf{B}. \tag{2.5.39}$$

Configurations in which this relationship between $\mathbf{v}$ and $\mathbf{B}$ holds are called "*magneti* vortices" (Petviashvili and Yankov, 1989). Let us use some appropriate units: spatial coor dinates $= c/\omega_p^*$, time $= mc/eB_0$, $\omega_p^* \equiv \sqrt{4\pi\delta n^* e^2/m}$, and $\mathbf{H} \equiv \mathbf{B}/B_0$. Therefore the electro and positron vorticities (2.5.32) are:

$$\mathbf{\Omega}_\pm = \mp \mathbf{H} - \nabla^2 \mathbf{H}, \qquad (2.5.40)$$

$$\frac{\partial}{\partial t} \mathbf{\Omega}_\pm = \nabla \times [\mathbf{\Omega}_\pm \times (\nabla \times \mathbf{H})]. \qquad (2.5.41)$$

Now let us solve Eq. (2.5.41) for the case in which one spatial coordinate, say z, is ignorable. Physically it means that the typical length in the z-direction is much larger than the typical length in the x-y-plane. ($\partial A_z/\partial t \ll \partial A_z/\partial y$ and $\partial A_y/\partial z \ll \partial A_y/\partial x$). Therefore we can write the magnetic field as:

$$\mathbf{H} = [\nabla a(x,y,t) \times \hat{\mathbf{z}}] + h(x,y,t)\hat{\mathbf{z}}, \qquad (2.5.42)$$

The scalar fields a and h from Eq. (2.5.42) with Eq. (2.5.41) satisfy

$$\frac{\partial}{\partial t} h = 0, \qquad (2.5.43)$$

$$\frac{\partial}{\partial t} \nabla^2 h + (h, \nabla^2 h) = 0, \qquad (2.5.44)$$

$$\frac{\partial}{\partial t} a + (h, a) = 0, \qquad (2.5.45)$$

$$\frac{\partial}{\partial t} \nabla^2 a + (h, \nabla^2 a) = 0, \qquad (2.5.46)$$

where $(f,g) \equiv \hat{\mathbf{z}} \cdot [\nabla f \times \nabla g]$. Also we get $\nabla^2 h = P[p]$.

For a vortex propagating in the x-y-plane with velocity u Eqs. (2.5.44)–(2.5.47) give us

$$\nabla^2 a = Q[h + \hat{\mathbf{z}} \cdot (\mathbf{u} \times \mathbf{r})], \qquad (2.5.47)$$

$$a = R[h + \hat{\mathbf{z}} \cdot (\mathbf{u} \times \mathbf{r})]. \qquad (2.5.48)$$

P, Q and R are arbitrary functions of the arguments in the brackets.

Let us solve a "linear" case $a = h + uy$, in which $\mathbf{u}$ is in the x-direction and Q is given by Oliveira *et al.* (1995):

$$Q[a] = \begin{cases} -c^2(h + uy) & \text{for } r < r_0 \\ +d^2(h + uy) & \text{for } r > r_0. \end{cases} \qquad (2.5.49)$$

The general solution for h (continuous up to the first derivative) is given by:
for $r < r_0$

$$h = u \left[\left(\frac{dr_0 K_2(dr_0)}{K_1(dr_0)} + B_1 d K_1'(dr_0) \right) \frac{J_1(cr)}{c J_1'(cr_0)} - r \right] \sin\phi, \qquad (2.5.50)$$

for $r > r_0$

$$h = u \left(\frac{r_0}{K_1(dr_0)} + B_1 \right) K_1(dr) \sin\phi, \qquad (2.5.51)$$

where $J_l(cr_0) = 0$.

Another possible choice of the arbitrary $Q[a]$ is:

$$Q[a] = \begin{cases} -c^2(h + uy) & \text{for } r < r_0, \\ 0 & \text{for } r > r_0, \end{cases} \qquad (2.5.52)$$

which gives us the solution:

$$h = \begin{cases} u \left(\frac{2}{cJ_1'(cr_0)} J_1(cr) - r \right) \sin\phi, & \text{for } r < r_0 \\ -ur_0^2 \sin\phi/r, & \text{for } r > r_0 \end{cases} \qquad (2.5.53)$$

with $J_1(cr_0) = 0$.

Both solutions above are dipole-like solutions. The first solution has finite total energy while the second has a logarithmic divergence. We emphasize that these solutions represent physically filamentary *vortex structures*. Thus these solutions are again related to the discussion in Sec. 2.4. At a large enough scales this solutions are thin (r_0 very small) "strings" that may eventually close itself. A good ensemble of this filaments may form more complex structures in this large scale.

Relaxed state

The relaxed state configuration obeys Eqs. (2.5.30)–(2.5.32). Oliveira *et al.* (1995) concluded that for the electron-positron-proton plasma case

$$\mathbf{H} = -\frac{\delta n^*}{4\lambda} \nabla \times \mathbf{H}, \qquad (2.5.54)$$

$$\nabla^2 \mathbf{H} = \frac{n_+ + n_-}{4\lambda} \nabla \times \mathbf{H}. \qquad (2.5.55)$$

Compatibility of these equations gives $(4\lambda)^2 = \delta n^*(n_+ + n_-)$. Then we get a Helmholtz-like equation for $\mathbf{H}$:

$$\nabla^2 \mathbf{H} = -\frac{n_- + n_+}{n_- - n_+} \mathbf{H}. \qquad (2.5.56)$$

The scales of these solutions are $\sqrt{\frac{n_- - n_+}{n_- + n_+}} \frac{c}{\omega_p^*} = c/\omega_p$ where ω_p^* is the plasma frequency for the density $n_- + n_+$, which means the size is $\sqrt{\frac{n_- + n_+}{n_- - n_+}}$ times the collisionless skin depth of ee^+ plasma. It is well known that the general solution for the divergenceless fields satisfying (2.5.57) is

$$\mathbf{H} = \nabla \times (\widetilde{m}u) + \frac{1}{\alpha} \nabla \times (\nabla \times (\widetilde{m}u)), \qquad (2.5.57)$$

89

where $\widetilde{m}$ is a unitary vector, and u satisfies the scalar Helmholtz equation:

$$\nabla^2 u = -\alpha^2 u,$$

where $\alpha \equiv \sqrt{\frac{n_- + n_+}{n_- - n_+}}$. For the cosmological phasing e$^+$eP, $\alpha \approx 10^4$. This estimate is based on the observed limits for the asymmetry of matter over anti-matter.

2.5.4 Ideal MHD Structures*

The e$^+$eP plasma structures can combine to form larger scales structures in MHD. Therefore it is appropriate to investigate localized solutions represented by MHD. The set of equations used are

$$\mathbf{B} \cdot \nabla n = \mathbf{v} \cdot \nabla n = 0, \tag{2.5.58}$$

$$\nabla \cdot \mathbf{v} = 0, \tag{2.5.59}$$

$$\nabla \cdot \mathbf{B} = 0, \tag{2.5.60}$$

$$nm \left(\frac{d}{dt} \right) \mathbf{v} = \frac{\mathbf{J}}{c} \times \mathbf{B} - \nabla P - mn \nabla \phi_G, \tag{2.5.61}$$

$$\nabla \times \mathbf{B} = \frac{4\pi}{c} \mathbf{J}, \tag{2.5.62}$$

$$\frac{\partial}{\partial t} \mathbf{B} = \nabla \times (\mathbf{v} \times \mathbf{B}). \tag{2.5.63}$$

where

$$\frac{d}{dt} = \frac{\partial}{\partial t} + \mathbf{v} \cdot \nabla. \tag{2.5.64}$$

The equation of motion (2.5.61) can be rewritten as

$$\frac{\partial}{\partial t} \mathbf{v} - [\mathbf{v} \times (\nabla \times \mathbf{v})] + \frac{1}{4\pi mn} [\mathbf{B} \times (\nabla \times \mathbf{B})] = -\nabla \left(\frac{P}{mn} + \tfrac{1}{2} v^2 + \phi_G \right). \tag{2.5.65}$$

We assume the density is constant. Then the time evolution equations have the following three conserved integrals

$$I_\pm = \int \left(\mathbf{v} \pm \frac{\mathbf{B}}{\sqrt{4\pi mn}} \right)^2 d^3 \mathbf{r}, \tag{2.5.66}$$

$$I_h = \int \mathbf{A} \cdot \mathbf{B} d^3 \mathbf{r}. \tag{2.5.67}$$

Let us look for static solutions:

$$[\mathbf{v} \times (\nabla \times \mathbf{v})] - \frac{1}{4\pi mn}[\mathbf{B} \times (\nabla \times \mathbf{B})] = \nabla \left(\frac{P}{mn} + \tfrac{1}{2}v^2 + \phi + G \right), \qquad (2.5.68)$$

$$\nabla \times [\mathbf{v} \times \mathbf{B}] = 0. \qquad (2.5.69)$$

There are three possible vortices, depending on how $\mathbf{v}$ and $\mathbf{B}$ are related. They are called parallel, magnetic and dynamic vortices (Petviashvili and Yankov, 1989; Oliveira et al., 1995). For *parallel vortices*

$$\mathbf{v} = \pm \frac{M}{\sqrt{4\pi mn}} \mathbf{B}, \qquad (2.5.70)$$

$$\frac{M^2 - 1}{4\pi mn}[\mathbf{B} \times (\nabla \times \mathbf{B})] = \nabla \left(\frac{P}{mn} + \tfrac{1}{2}v^2 + \phi_G \right). \qquad (2.5.71)$$

There is a degeneracy when $M = 1$ and $P/mn + \tfrac{1}{2}v^2 + \phi_G = $ const, which corresponds to *Alfvén vortices*. Notice that

$$(M^2 - 1)\epsilon \equiv 4\pi \left(P + mn\tfrac{1}{2}v^2 + mn\phi_G \right) \qquad (2.5.72)$$

is constant along the streamlines of the magnetic field.

Let it be a 3D axially symmetric field:

$$r\mathbf{B} = \hat{\phi} \times \nabla\psi + \hat{\phi}f[\psi] \qquad (2.5.73)$$

and $\epsilon = \epsilon[\psi]$. So we obtain the *Grad-Shafranov equations*:

$$\Delta^*\psi = -ff' - r^2\epsilon', \qquad (2.5.74)$$

where prime means derivatives with respect to the argument ψ and the *Grad-Shafranov operator* $\Delta^* \equiv r\,\partial/\partial r\,1/r\,\partial/\partial r + \partial^2/\partial z^2$. This operator often appears in plasma physics and in particular in Sec. 2.8 and in Sec. 4.3 again. Let the separatrix be a sphere of radius a. Then we take $\epsilon[\psi]$ and $f[\psi]$ to be linear inside the sphere and zero outside of it. It turns out that ψ vanishes outside the separatrix. The inside equation and general solution (Berk et al., 1986) are as follows:

$$\Delta^*\psi = -k^2\psi + cr^2, \qquad (2.5.75)$$

$$\psi = \frac{c}{k^2}r^2 + \sum_{n=2}^{\infty} A_n C_n^{-1/2}\left(\frac{z}{R}\right)\sqrt{R}\,J_{n-1/2}(kR), \qquad (2.5.76)$$

where $J_{n-1/2}$ is the Bessel function and $C_n^{-1/2}$ is the Gegenbauer function. We impose continuity for ψ and its first derivative. This procedure leads us to:

$$A_n = -\delta_{n,2}\frac{c}{k^2}\,1/j(ka), \qquad (2.5.77)$$

$$\tan ka = \frac{3ka}{3 - (ka)^2}. \qquad (2.5.78)$$

Therefore

$$\psi = \frac{c}{k^2} \left[1 - \frac{j(kR)}{j(ka)} \right] f^2, \qquad (2.5.79)$$

where

$$j(\xi) = \frac{(\sin \xi - \xi \cos \xi)}{\xi^3}. \qquad (2.5.80)$$

Other localized numerical solutions were found (Petviashvili and Yankov, 1989) for $f = \sqrt{2/(n+1)}\, \psi^{n+1/2}$ and $\epsilon' = -\psi$ for $n = 2.3$. These solutions have a preferred direction of strong interaction along, say, the z-axis. Therefore this solitary vortices have a tendency to form linear polymer-like structure. In turn these "polymers" may form even larger structures and so on (Tajima *et al.*, 1992b).

Localized solutions for e^+eP plasmas were found in the form of long strings (mathematically 2D) solutions. Filamentary e^+eP may form localized 3D solution in MHD. The localized solutions in MHD may also form larger scale structures in a polymer-like shape. These solutions may represent more natural structures than ones found in earlier work (Tajima and Taniuti, 1990) in one dimension in electron-positron plasmas. These may be of great importance to formation of *isothermal perturbations* during the radiation epoch of the universe. This scenario provides one possible way for formation of structures of later epochs that is consistent with the observed uniformity and isotropy of the Microwave Background Radiation (Tajima *et al.*, 1992c) (see Sec. 5.3).

2.6 Stochastic Processes

In Secs. 2.3 and 2.4 we review spectra of turbulence for both closed and open systems. In a wider range of astrophysical spectra (distributions) we often encounter nonthermal spectra (i.e. *non-Gaussian distribution*). This is both surprising and not surprising. Not surprising, as astrophysical phenomena are often out of equilibrium. It is nonetheless surprising that these nonthermal spectra often take a form of power law, a surprisingly coinciding universality.

In Fig. 2.13 we show several examples of distribution of astrophysical spectra that show a clear power law tendency. We wonder why such wide range emergence of power laws where phenomena share little physical common ground. Perhaps a general stochastic nature can underline these occurrences. This astrophysical ubiquity motivates us to formulate this problem and to study the power law origin (seemingly) regardless of specific physical mechanisms except for a certain common stochastic nature.

Here we briefly survey (non-gaussian) *stochastic processes*. If the stochastic process is gaussian, a standard statistics (the canonical distribution) applies, as we did in Sec. 2.3.2. Chandrasekhar (1943)'s classic review covers the standard technique in stochastic processes and *gaussian statistics*. Here we introduce a general result by Mima *et al.* (1990) for the *Markov process* (i.e. stochastic process which has the character that what happens at a given instant of time t depends only on the state of the system at time t) that leads to *non-gaussian* spectra. Let $f(\varepsilon, t)$ be the probability of this process at energy ε at time t. Using

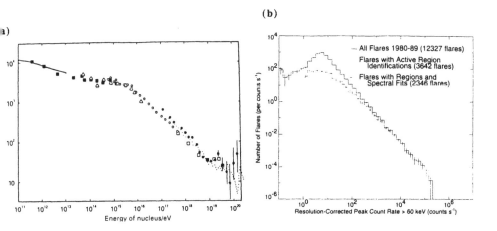

(b)

FIGURE 2.13   Various phenomena in astrophysics that exhibit the power-law spectrum. (a) cosmic ray (Hillas, 1984); (b) solar flare (Kucera *et al.*, 1997); (c) meteorites (Dohnanyi, 1978); (d) interstellar clouds (Scalo, 1985).

the *transition probability* $W(\varepsilon; \Delta\varepsilon)\Delta t$ that energy ε changes by $\Delta\varepsilon$ during time interval Δt (Chandrasekhar, 1943), we can write as

$$f(\varepsilon; t + \Delta t) = \int f(\varepsilon - \Delta\varepsilon; t) W(\varepsilon - \Delta\varepsilon; \Delta\varepsilon) \Delta t d(\Delta\varepsilon), \qquad (2.6.1)$$

where the normalization is $\int W_\tau(\varepsilon; \Delta\varepsilon) d(\Delta\varepsilon) = 1$. From Eq. (2.6.1) we obtain

$$\frac{\partial f(\varepsilon, t)}{\partial t} = \int f(\varepsilon - \Delta\varepsilon) W(\varepsilon - \Delta\varepsilon; \Delta\varepsilon) d(\Delta\varepsilon) - f(\varepsilon) W(\varepsilon; \Delta\varepsilon) d(\Delta\varepsilon) + C[f(\varepsilon, t)], \quad (2.6.2)$$

where the first term on the right-hand side of Eq. (2.6.2) is the increasing term, the second the decreasing, and the last the collision term. By Taylor expanding $|\Delta\varepsilon| \ll \varepsilon$,

$$\frac{\partial}{\partial\tau} f = \frac{\partial}{\partial\varepsilon} \left\{ \int_{-\infty}^{\infty} d(\Delta\varepsilon) \Delta\varepsilon \, W(\varepsilon, \Delta\varepsilon) f(\varepsilon) + \frac{\partial}{\partial\varepsilon} \left(\int_{-\infty}^{\infty} d(\Delta\varepsilon) \frac{\Delta\varepsilon^2}{2} W(\varepsilon, \Delta\varepsilon) f(\varepsilon) \right) \right\}$$
$$+ \frac{\partial}{\partial\varepsilon} \left(R(\varepsilon) f(\varepsilon) \right), \qquad (2.6.3)$$

where the last term is the Coulomb drag term $R(\varepsilon) = R_0(\varepsilon/\varepsilon_0)^{-1/2}$.

93

(c)

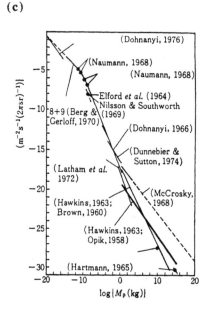

(d)

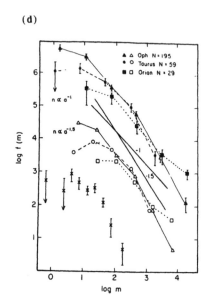

FIGURE 2.13 continued

Let us look for a steady-state spectrum of Eq. (2.6.3), $\partial f/\partial t = 0$. We have

$$B(\varepsilon)f(\varepsilon) + \frac{\partial}{\partial \varepsilon}\left[A(\varepsilon)f(\varepsilon)\right] = C, \qquad (2.6.4)$$

where

$$B(\varepsilon) = \int d(\Delta\varepsilon)\Delta\varepsilon\, W(\varepsilon, \Delta\varepsilon),$$

$$A(\varepsilon) = \int d(\Delta\varepsilon)\frac{\Delta\varepsilon^2}{2}\, W(\varepsilon, \Delta\varepsilon).$$

From Eq. (2.6.4) with $C = 0$, after integration we obtain the Gaussian (Maxwellian) distribution, $f(\varepsilon) \propto e^{-C'\varepsilon}$. This is the standard statistics.

However, in nature (including astrophysics) often power-law spectra are frequently observed as witnessed in Fig. 2.13 (although representing a very small subsection of such)..

94

When

$$B(\varepsilon) = \int_{-\infty}^{\infty} d(\Delta\varepsilon)\Delta\varepsilon\, W(\varepsilon, \Delta\varepsilon) = 0, \tag{2.6.5}$$

we obtain a spectrum different from the conventional Gaussian. When does the first moment B of W vanish? If W is sharply peaked, $B \neq 0$, while if W is *self-similar*, then B (nearly) vanishes. In the nonrelativistic case $A(\varepsilon) \sim A_0(\varepsilon/\varepsilon_0)^{2n+1}$. With the boundary condition $f(\varepsilon) \to 0$ as $\varepsilon \to \infty$, we obtain from Eq. (2.6.4) a *power-law spectrum* for $f(\varepsilon)$

$$f(\varepsilon) = C_0 \left(\frac{\varepsilon}{\varepsilon_0}\right)^{-2n+1} \exp\left[-\frac{R_0}{A_0} \int^{\varepsilon} d\varepsilon \left(\frac{\varepsilon}{\varepsilon_0}\right)^{-2n-3/2}\right]. \tag{2.6.6}$$

The first factor is the power-law spectrum, while the second factor can show residual exponential dependence. In summary, when the physical process through W is self-similar, there is no characteristic scale and we tend to have the power-law physics, a general result. Because in astrophysics phenomena many processes tend to be self-similar, we may conclude that ubiquity of self-similarity in astrophysics gives rise to the ubiquity of distribution of a power law.

2.7 Anomalous Transport

The presence of a magnetic field, even a weak one, can change the properties of transport in a plasma drastically, particularly that across the magnetic field. In high temperature plasmas while electrons (and ions) can rapidly propagate along the field line to conduct themselves and their thermal energy, they are trapped by the magnetic field and largely insulated from each layer. The length over which these particles can conduct heat (and other physical qualities) changes from the *mean free path* to the *Larmor radius*, while the time over which the exchange of energy takes place is the collision time for cases with or without magnetic field. When the magnetic field is closed, particles may be confined and the transport coefficients may be drastically reduced. A stark contrast between the bright soft X-ray emitted from hot plasma electrons from the *coronal loop* (a closed configuration) regions (often called active regions) and the relatively darker regions with open field lines (often called *coronal holes*) of the Sun may be a good example to demonstrate such transport contrast with or without the presence of closed magnetic field lines and thus the presence of the *cross-field transport* (i.e. *magnetic insulation*). It is evident, however, if the magnetic field lines are not nested and some of them meander out of the inner region toward the outer one, the transport is once again enhanced. In astrophysics magnetic fields take often patchy, knotty, and turbulent forms. In these magnetic fields particles are much less confined than in a laminar, *nested magnetic field* configuration. The transport in the former situation is often called anomalous, (as compared to collisional transport over the Larmor radius scale), indicating that some chaotic (or patchy) field lines and/or excitation of collective modes conspires to induce enhanced transport over the (normal) collisional transport.

Anomalous transport is the general term to designate the transport process that is beyond the *classical transport* due to collisions and their induced processes. The presence of plasma turbulence means *enhanced fluctuations* (see Sec. 2.3). The *diffusion coefficient* is determined by (Chandrasekhar, 1961)

$$D = \frac{\Delta r^2}{\Delta t}, \tag{2.7.1}$$

where Δr is the typical spatial step size and Δt the temporal size. In collisional magnetized plasma the transport (diffusion) coefficient is determined by the Larmor radius ρ for Δr and the collision time ν^{-1} for Δt. Thus the diffusion coefficient goes like

$$D \sim \rho^2 \nu, \tag{2.7.2}$$

which is inversely proportional to B^2, where B is the strength of magnetic field. This scaling on B sometimes is called classical or somewhat more broadly *gyroBohm*-like. This scaling may be called the standard transport, or in some (or many) occasions a naive result. However, even in early plasma experiments the transport coefficient often far deviated from the classical value. Bohm *et al.*(1949) postulated this by his superb intuition or fact that D should be

$$D \sim \frac{1}{16} \frac{cT}{eB} \propto \rho^2 \Omega, \tag{2.7.3}$$

which is inversely proportional to B and Ω is the cyclotron frequency. This *Bohm scaling* may be interpreted as the time scale for diffusive step Δt is much reduced from ν^{-1} to Ω^{-1} (in most relevant plasmas) with the spatial step size being ρ in both cases; alternatively with $\Delta t \sim \nu^{-1}$, the spatial scale is enhanced. The latter possibility is often more realistic. When there occur collective modes (E_k) due to some instabilities such as the drift wave instability, electrons and ions of the plasma can move together as a fluid motion $v_k = cE_k \times B/B^2$ (Sec. 2.1). If the size of this migration is $\lambda = 2\pi/k (> \rho_i)$, then the transport is enhanced over the value $\rho_i^2 \nu$. In magnetically confined plasmas such as *tokamaks* the enhancement of transport is sometimes up to a few orders of magnitude. In astrophysical plasmas such as the convective zone of the Sun this coherent motion is huge compared with collisional process. We first briefly survey the convective instability and its transport. This instability involves a fluid motion. We then review a recent theoretical development in kinetic instability and its associated heat transport. Even though the latter instability is of more microscopic kinetic nature, we see common properties in the transport theory.

2.7.1 Convective Transport

Under the gravitation g, the plasma (such as the stellar convection zone) is stratified. For the equilibrium $\nabla P = -\rho g$, an *adiabatic perturbation* yields the temperature gradient in the vertical (r) direction

$$\left(\frac{T}{L_T}\right)_{ad} = \left(\frac{dT}{dr}\right)_{ad} = \frac{T}{P}\left(1 - \frac{1}{\gamma}\right)\frac{dP}{dr} = -\frac{g}{c_p}, \tag{2.7.4}$$

where γ is the adiabatic gas index and c_p the isobaric specific heat. Schwarzschild (1906) pointed out that convective instability occurs if

$$\frac{1}{L_T} - \left(\frac{1}{L_T}\right)_{ad} > 0 \qquad (2.7.5)$$

where the first term is the actual temperature gradient (of the star) dT/Tdr. When convection occurs, the tendency is to reduce the temperature (sometimes called the *structural temperature*) gradient until Eq. (2.7.5) becomes an equality. The convective motion is described by the continuity equation, the equation of motion, and the heat conduction equation. When the convection ensues, bubbles (or convective vortices) are formed with a characteristic length ℓ, velocity v, and a temperature difference ΔT between the bubble and the surrounding, which is

$$\Delta T = \ell \left[\frac{dT}{dr} - \left(\frac{dT}{dr}\right)_{ad}\right]. \qquad (2.7.6)$$

When the convective bubble merges with its surrounding after traveling the distance ℓ, the length is called a *mixing-length*, and it is typically related to the *scale height* $H \equiv T/mg$. The convective instability can be characterized by the Rayleigh number R

$$R \equiv \frac{g\alpha}{\chi\nu}\frac{\Delta T}{\ell}\ell^4, \qquad (2.7.7)$$

where α, χ, ν are the coefficient of volume expansion, the heat conductivity, and the kinematic viscosity, respectively (see also Sec. 3.1). When the *Rayleigh number* exceeds a critical number R_c, instability occurs. For a certain $R > R_c$, a certain set of modes of convection (low wavenumber modes) are destabilized and their growth rates become positive. See. Fig. 2.14. In terms of the mass density ρ of the plasma (or gas)

$$\Delta\rho = \frac{\rho\Delta T}{T} = \ell \left[\frac{d\rho}{dr} - \left(\frac{d\rho}{dr}\right)_{ad}\right], \qquad (2.7.8)$$

which will yield a buoyancy force (see Sec. 3.1)

$$f = Vg\Delta\rho = V\rho g\ell \left[\frac{1}{L_T} - \left(\frac{1}{L_T}\right)_{ad}\right], \qquad (2.7.9)$$

where V is the volume of the bubble. The work done on the bubble over the convective distance ℓ gives rise to the bubble convection velocity as

$$v = \frac{\ell}{2}(g)^{1/2}\left[\frac{1}{L_T} - \left(\frac{1}{L_T}\right)_{ad}\right]^{1/2}. \qquad (2.7.10)$$

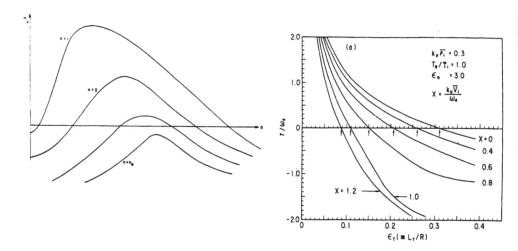

FIGURE 2.14 (a) Growth rates of convective modes at a given Rayleigh number $R > R_c$ (Spiegel, 1966) as a function of horizontal wavenumber k and n is the vertical mode number; (b) the growth rate and the heat flux as a function of the order parameter ϵ_T for the ion temperature gradient instability (Kishimoto et al., 1996).

The *heat flux* due to the convective flow is

$$Q = \frac{1}{2}\, c_p\, \rho v \ell^T \left[\frac{1}{L_T} - \left(\frac{1}{L_T} \right)_{ad} \right],$$

(2.7.11)

which with Eq. (2.7.10) becomes

$$Q = \frac{1}{4}\, c_p\, \rho \ell^2 g^{1/2} T \left[\frac{1}{L_T} - \left(\frac{1}{L_T} \right)_{ad} \right]^{3/2}$$

(2.7.12)

The heat transport, enhanced much beyond the collisional process, due to the convection is proportional to the three halves power of the deviated amount away from the *critical*

98

temperature gradient (the adiabatic gradient). Because of this, sometimes this is called the *superadiabatic* instability. If the deviation is too large, the heat flux Eq. (2.7.12) becomes large so that the interior heat of a star is rapidly convected out and the temperature gradient settles until the deviation becomes tiny. In nearly all actual cases thus the deviation stays small and thus the interior temperature of the star can be obtaind by simply integrating $(dT/dr)_{ad}$ from the surface of the star inward. As we shall see in Sec. 2.7.2, even for kinetic (drift) instabilities in magnetically confined plasmas a physical process of transport is now found to be similar to the stellar convection. Perhaps a detailed theoretical and experimental study of heat transport in magnetically confined plasmas may shed light on the invisible (opaque) stellar interior processes.

2.7.2 Self-Organized Critical Transport

Ever since Bohm's heuristic suggestion some four decades ago intense debates have raged among the scientists of magnetically confined plasmas. Meanwhile a large amount of computer simulation, first in 2D slab and 3D slab, have been carried out to investigate this complex phenomenon. In many situations where they tried to incorporate the magnetic field shearing due to the toroidal current, these simulations see the gyroBohm-like behavior. This is because the *magnetic shear* localizes the mode at each rational surface where the resonance condition is satisfied: away from the rational surface the radial (i.e. in the direction across the field line) structure of the mode is "shredded" by the magnetic shear. Plasmas are characterized by the long-ranged electromagnetic force, i.e. all charged particles are affected by all the other particles no matter how far they may be. And thus the boundary conditions severely affect the transport and so do the global configurations and structures of the plasma, such as the magnetic topology. In recent experimental and theoretical understanding and thus convergence of predictions in toroidal plasmas revealed that in fact the global topology of the equilibrium toroidal plasma, such as that of a tokamak has a definitive impact on the stability and transport of the plasma. In fact recent theory (Kishimoto *et al.*, 1996) indicates that the stability and the transport of toroidal plasmas are intricately coupled to each other and cannot separately be described. The time scale hierarchy is dictated by this interplay. As an illustration, let us look into this situation a bit further. In toroidal plasmas the *toroidicity* (the geometrical effect) induces to couple modes with each other at adjacent resonant surfaces. This leads to a chain of modes successively tied together, so much so to form a (even) macroscopic radial mode. If this large mode structure manifests itself, the spatial scale for transport across the magnetic field is characterized not by the Larmor radius, but by a larger patch or the macroscopic scale such as the temperature (or density) scale length. This leads to a relaxation of the plasma profile over this macroscopic length. If the plasma is driven by external or internal sources of heat etc., depending upon the strength of sources and the size of scale length, different equilibrium temperature (or density) profiles and temporary features (such as relaxation oscillations) will arise. In the following we look at this physics in some detail. As a result the anomalous heat conduction in a toroidal plasma results and holds the following remarkable properties distinct from a cylindrical plasma: (i) the development of radially extended potential streamers localized to the outside of the torus, (ii) more robust ion temperature gradient instability, (iii) radially constant eigenfrequency, (iv) global temperature relaxation, (v) radially increasing heat conductivity, and (vi) the Bohm-like behavior.

The evolution of the background distribution function f_0 may be described by a *quasi-linear theory*: $\partial f_0/\partial t = \mathcal{L}[E(\hat{f}), \hat{f}]$, where $\hat{f}$ is the perturbed distribution, E the electric field generated by $\hat{f}$, and $\mathcal{L}$ is the usual Vlasov operator (see Kishimoto *et al.*, 1996 for more detail). When we make a second moment of velocity of this quasilinear equation, we obtain an equation describing the background temperature evolution:

$$\frac{3}{2}n\frac{\partial T}{\partial t} + \nabla \cdot \mathbf{Q} = 0, \tag{2.7.13}$$

where the source term is neglected (consistent with the simulation situation), T refers to the background temperature and the heat flux $\mathbf{Q}$ is the second moment of velocity of the term $\mathcal{L}[E(\hat{f}), \hat{f}]$ and it takes the form

$$r\chi\frac{\partial T}{\partial r} = rQ = \mathrm{Re}\int d\mathbf{v}\,\frac{1}{2}mv^2\,\frac{c\mathbf{E}^* \times \mathbf{B} \cdot \hat{\mathbf{e}}_r}{B^2} \cdot g\,\exp(i\mathcal{L}),$$

$$= \frac{n_0\,r\,k_\theta ce}{B}|\phi(r)|^2 G(\Omega, \eta_i, L_T, r), \tag{2.7.14}$$

where χ is the *heat conductivity*, $\mathbf{Q}$ is the *heat flux* due to the fluctuating electric fields E (or potential ϕ) and G is the normalized heat flux [of the order unity and a function of macroscopic plasma variables Ω, η_i, L_T, r etc. (see Kishimoto *et al.*, 1996 including the definition and detail of this equation.)]. Although the kinetic processes make the form of function G a complicated one, a general tendency of G is such that we may be able to approximate as

$$G \propto \left[\frac{1}{L_T} - \left(\frac{1}{L_T}\right)_{cr}\right]. \tag{2.7.15}$$

Thus the form of heat flux Q here is similar to Eq. (2.7.11).

In the early stage of an initial value problem before saturation the temperature profile rapidly relaxes toward global profile. In the later stage the temperature profile relaxes slowly, maintaining the functional form and its temperature gradient only gradually decreasing toward the *marginal stability*. Since there exist these two distinct time scales in relaxation, we introduce a multiple time expansion in our quasilinear equation (2.7.14). A systematic expansion may be carried out by the well-known *reductive perturbation method* (Washimi and Taniuti, 1966), introducing a smallness parameter ε (which will be determined below). We expand the time and space scales, temperature, and heat flux and conductivity as follows:

$$\frac{\partial}{\partial t} = \frac{\partial}{\partial t_0} + \varepsilon^2\frac{\partial}{\partial t_1} + \varepsilon^4\frac{\partial}{\partial t_2} + \cdots,$$

100

$$\frac{\partial}{\partial r} = \frac{\partial}{\partial r_0} + \varepsilon \frac{\partial}{\partial r_1} + \varepsilon^2 \frac{\partial}{\partial r_2} + \cdots,$$

$$T = T_0 + \varepsilon T_1 + \cdots,$$

$$Q = \varepsilon Q_0 + \varepsilon^2 Q_1 + \cdots,$$

$$\chi_0 = \varepsilon \chi_i^0 + \varepsilon^2 \chi_i^2 + \cdots, \tag{2.7.16}$$

where T_0 is the global temperature profile. From Eq. (2.7.16) the present reductive perturbation theory is crucially based on the fundamental properties of relaxation of the (tokamak-like) toroidal plasma we described. The important point is that the instability is radially global and vigorous in toroidal plasmas so that the plasma parameters have to relax rapidly until they approach sufficiently near the stable profile. However, as long as there is enough heat reservoir or there is energy input, the sustained finite fluctuations will cause a certain amount of heat flux and thus dissipation, which has to be compensated by the weak but still unstable wave activities. The amount of sustained fluctuations thus is a function of the energy (or power) input (or the availability of the heat reservoir). Then, the heat flux $(\mathbf{Q}_0, \mathbf{Q}_1)$ is expanded in terms of (T_0, T_1) and $\chi_i^{(0)}, \chi_i^{(1)}$ as

$$\mathbf{Q}_0 = -\chi_i^{(0)} \nabla_0 T_0,$$
$$\mathbf{Q}_1 = -\chi_i^{(0)} \nabla_0 T_1 - \chi_i^{(1)} \nabla_0 T_0 - \chi_i^{(0)} \nabla_1 T_0. \tag{2.7.17}$$

In general each quantity is a function of both t_0 and t_1 and r_0 and r_1 (and higher order variables). We assume $\partial T_0 / \partial t_0 = 0$, where t_0 represents the fast time scale. From this expansion, we obtain a series of transport equations:

$$\mathcal{O}(\varepsilon) : \frac{3}{2} n_i \frac{\partial T_1}{\partial t_0} + \nabla_0 \cdot \mathbf{Q}_0 = 0, \tag{2.7.18}$$

$$\mathcal{O}(\varepsilon^2) : \frac{3}{2} n_i \frac{\partial T_0}{\partial t_1} + \nabla_0 \cdot \mathbf{Q}_1 + \nabla_1 \cdot \mathbf{Q}_0 = S, \tag{2.7.19}$$

by assuming that the density n_0 is maintained constant, where S is a heat source.

In Eq. (2.7.18), if there is finite divergence for $\mathbf{Q}_0$, i.e. $\nabla \cdot \mathbf{Q}_0(r) \neq 0$, it will force T_1 to rapidly relax such that $T_1 \rightarrow 0$. Thus, on average, we obtain $\partial T_1 / \partial t_0 \rightarrow 0$ and therefore $\nabla \cdot \mathbf{Q}_0 \rightarrow 0$, which leads to the condition

$$r Q_0(r) = r \chi_i^{(0)}(r) \frac{\partial T_0}{\partial r} \simeq \text{const}, \tag{2.7.20}$$

over a range of r such that r_0 belongs to a region with a given value of r_1. We find that the constraint that the plasma fluctuating temperature T_i deviation away from the relaxed T_0 profile rapidly washes away leads to a plasma *self-organized critical* state. This is a bit similar to a sandbox (Bak *et al.*, 1988): when we tilt a sandbox beyond the marginal angle, the sand catastrophically cascades down the slope to relax the gradient to tend to the marginal. Equation (2.7.20) dictates that critical state. The global T_0 profile is self-organized to be $T_0 \sim \exp(-r/L_T)$ for a certain mode whose stability is determined by the temperature gradient L_T^{-1}; the constraint in $\mathcal{O}(\varepsilon)$ given by Eq. (2.7.20) requires that the thermal conductivity varies as

$$\chi_i^{(0)} \sim \frac{\text{const.}}{r} \exp\left(\frac{r}{L_T}\right). \tag{2.7.21}$$

The thermal conductivity $\chi_i^{(0)}$ given by Eqs. (2.7.21) found to show radially increasing profiles. In the next order in ε, we obtain the slow variation of the background temperature $T_0(t_1)$ through divergence of Q_1 in Eq. (2.7.19). Note that the slow variation is induced as a variation of the scale length which is given by $\partial T_0/\partial t_1$. Sometimes this self-similarity in transport is called the "profile consistency" (Coppi, 1983).

In order to evaluate heat flux Q, here we employ the quasilinear expression given by

$$r_1 Q_0(r_1) = \frac{n_i c e}{B} n q(r_1) |\phi(r_1)|^2 \text{ Re } G, \tag{2.7.22}$$

where n is a given toroidal mode number and $k_\theta = nq(r_1)/r_1$ is used in Eq. (2.7.14). We emphasized here the subscript 1 to indicate the spatial dependence of parameters over a larger scale (r_1). In order to have the constraint given by Eq. (2.7.19) i.e. $rQ_0(r) = \psi(= \text{ const.})$ in $\mathcal{O}(\varepsilon)$, the radial profile of the electrostatic potential is self-consistently determined as follows

$$|\phi_0(r)|^2 = \frac{\psi B}{n_i c e n \, q(r) \text{ Re } G}, \tag{2.7.23}$$

where $|\phi_0(r)|^2$ is also an order ε and compensates the radial dependence of q and G. The finite divergence over a small scale (r_0) of Q_1 as well as the divergence of Q_0 over a larger spatial scale (r_1) become the driving forces of the self-similar relaxation of the temperature T_0 which takes place in slower time scale t_1.

2.7.3   Landau-Ginzburg Transport Equations*

In order to solve the transport equations, Eqs. (2.7.18) and (2.7.19), the heat flux $\mathbf{Q}$ has to be determined self-consistently by the plasma properties, such as fluctuation level, plasma temperature gradient [see Eq. (2.7.14)]. We have to therefore determine an equation that describes the evolution of $|\phi|^2$. The linear growth rate of the instability is linearly proportional to the difference of the (normalized) gradient of temperature μ ($\propto L_T^{-1}$) from the critical gradient μ_c [Kishimoto et al., 1996] around marginal stability by

$$\gamma_L = \gamma_0(\mu - \mu_c). \tag{2.7.24}$$

Once again this is similar to the convective instability [Eq. (2.7.9) and the deviation from the critical Rayleigh number R_c]. It is in fact possible to write this instability criterion in terms of the Rayleigh number vs. the critical Rayleigh number (Horton and Hu, 1997). We define a nonlinear total growth rate by

$$\gamma(W) = \gamma_L - \gamma_{NL}(W),$$

where γ_{NL} represents the rate of nonlinear damping that saturates the linear mode, and we describe the damping rate as $\gamma_{NL} = \gamma_N W$ [where normalized fluctuations $W \equiv |\phi/\langle T_i \rangle|^2$]. We determine γ_N by considering $\mathbf{E} \times \mathbf{B}$ (MHD fluid motion) trapping of a particle by a radially extended nonlocal mode. From an analysis of the saturation level, we obtain a scaling of thermal diffusion near the critical gradient.

It should be emphasized that since a plasma relaxes and achieves the critical gradient state, a temperature gradient of the plasma is close to that of marginal stability:

$$\mu - \mu_c \ll \mu_c. \tag{2.7.25}$$

If the temperature gradient (μ) were too large and much away from the critical value (μ_c), a large amount of convection of heat would happen, which would cool the central temperature down, leading back to the condition Eq. (2.7.25). Once again the situation is similar to the stellar temperature profile seen in Sec. 2.7.1. As discussed in Eq. (2.7.18), if Eq. (2.7.25) is not fulfilled, a rapid temperature evolution enforces the plasma to quickly attain the condition (2.7.25) in a short time scale (t_0). Thus the temporal smallness expansion parameter ε in Eq. (2.7.16) is related to Eq. (2.7.25) by

$$\varepsilon^2 = \frac{\gamma_0}{\omega_d} (\mu - \mu_c), \tag{2.7.26}$$

where ω_d is the normalizing frequency (such as linear eigenmode frequency). The turbulent energy equation thus derived takes the following form:

$$\frac{\partial W}{\partial t_1} = 2[\gamma_0(\mu - \mu_c) - \gamma_N W]W. \tag{2.7.27}$$

This equation resembles the *Landau-Ginzburg equation* in the critical phenomena as in phase transition, where the critical slowing-down takes a very long growth time, just like in our present case of the critical gradient of temperature. We have seen this equation already in Eq. (2.3.4). Various phases and transitions among them in the Landau-Ginzburg equation for plasma transport have been reviewed by Itoh *et al.* (1996). The slowness of this evolution is explicitly written by Eq. (2.7.27) through a slow time scale (t_1). As is evident, the right-hand side ordering is consistent with the left-hand side, as Eqs. (2.7.26) and (2.7.16) indicate. The nonlinear saturation $\gamma_N W$ in this critical gradient problem (such as the Bohm scaling) is much below the level often discussed in the theoretical literature that discusses robust turbulent saturation (such as the gyro-Bohm scaling), in which the time scale of t_1 can not come in. In the critical gradient theory, the steady-state saturation of W is

$$W^{(\text{sat})} = \frac{\gamma_0}{\gamma_N} (\mu - \mu_c), \tag{2.7.28}$$

at the order of ε. Because the temporal evolution of W in Eq. (2.7.27) is in a slow scale (t_1), we can couple Eq. (2.7.27) with the slow scale transport equation (2.7.19). In many works where the level of $\mu - \mu_c$ is high, there exists no time scale separation between t_0 and t_1 and thus no transport separation between Eqs. (2.7.18) and (2.7.19).

The steady-state in Eq. (2.7.18) [over the time scale t_1 (beyond t_0), i.e. $\nabla_0 \cdot Q_0 = 0$] leads to

$$-r\chi_i^{(0)}T_0\mu = \text{const} = \psi(r_1). \tag{2.7.29}$$

The constancy of Eq. (2.7.29) means a constant over the spatial scale εL, but beyond that scale it may be a slowly varying function of r_1, and then written as $\psi(r_1)$. The amount of ψ determines a magnitude of heat flux, which needs an additional equation, as is considered by Eq. (2.7.19). These are the consequences of the self-organized critical state due to the presence of the extended modes over εL, as the spatial smallness expansion parameter is ε. With these considerations, the transport equation coupled to Eq. (2.7.27) is derived from Eq. (2.7.19) as

$$\frac{3}{2}\frac{\partial T_0}{\partial t_1} = \frac{1}{r}\frac{\partial}{\partial r_0}\left[r\chi_i^{(0)}\left(\frac{\partial T_1}{\partial r_0} + \frac{\partial T_0}{\partial r_1}\right) + r\chi_i^{(1)}\frac{\partial T_0}{\partial r_0}\right] + \frac{1}{r}\frac{\partial}{\partial r_1}\left(r\chi_i^{(0)}\frac{\partial T_0}{\partial r_0}\right) + \frac{P_{in}(r_1)}{n_i}, \tag{2.7.30}$$

where the first bracketed terms arose from $\nabla_0 \cdot Q_1$ and the second bracketed term from $\nabla_1 \cdot Q_0$ and the source term S is given by the input (or fusion burning) power per unit volume P_{in}. In Eq. (2.7.30) the term $\partial T_1/\partial r_0$ may be put to be zero, as the temperature T_1 quickly relaxes to zero in time t_0 according to Eq. (2.7.18).

One important distinction of the self-organized critical gradient theory (Kishimoto et al., 1996) is that the slow time scale of the Landau-Ginzburg equation describing the wave growth and damping. As discussed earlier, unless the instability is near marginal, such an assignment of the timescale is not possible. A second distinct property of the present theory is that a slow time scale is now tied to a nonlocal spatial scale so that the equations describe the nonlocal spatial behavior: We have to now recognize three distinct spatial scales for the problem: $\Delta r_s, \Delta r_g, L$, (where Δr_s is the small eddy size, Δr_g is the global mode scale, and L the global scale).

When a stationary solution of Eq. (2.7.30) is sought, we integrate Eq. (2.7.30) with LHS $= 0$ radially from (r_1) from the surface of the plasma $(r = a)$. A typical solution (Kishimoto et al., 1996) is given as a formal solution

$$-r\chi_i^{(0)}T_0(r_1)\mu^2(r_1) = C\exp\left(-\int_a^r \frac{r_1 P_{in}(r_1)}{n_i\psi(r_1)}dr_1\right), \tag{2.7.31}$$

where $C = \mu(a)\psi(a)$ is the boundary value. Equation (2.7.31) is an implicit integral equation in which ψ is related to μ and $\chi_i^{(0)}$ through Eq. (2.7.29) and $\chi_i^{(0)}$ is related to W through Eq. (2.7.14). The quantity W is now governed by Eq. (2.7.27). Thus Eqs. (2.7.27) (with $\partial/\partial t_1 = 0$) and (2.7.31) constitute a complete set of coupled equations that self-consistently describe a steady state realized by the interplay between plasma fluctuations and heat transport in a self-organized critical plasma. The temperature gradient profile is shown in Fig. 2.15 along with the case for the stellar case of Sec. 2.7.1.

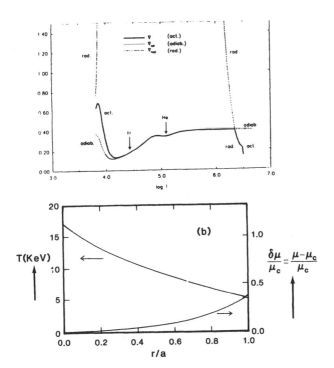

FIGURE 2.15 The temperature gradient in the solar atmosphere and that in a tokamak plasma. (a) The solar temperature gradient (after Spruit, 1977), showing it very close to but slightly above the marginal stability. The deviation from marginality gets greater toward the surface. (b) The tokamak plasma temperature (T) gradient (μ) as a function of the torus minor radius (r). Again the gradient μ is close to the marginality (μ_c) and $\delta\mu$ is a fraction of unity and gets greater toward the surface $(r = a)$ (Kishimoto et. al., 1996)

Extended modes that are developed due to the gradient of some equilibrium plasma quantities such as the density, temperature, pressure, magnetic fields in the direction parallel to the gradient tend to saturate at such a low level just to sustain the overall energy balance. Thus it is stuck close to the marginal stability. This is because the extended modes far above marginality would destroy the equilibrium profile of plasma quantities so fast that the equilibrium wold not be maintained. Localized modes can be allowed, on the other hand, to grow into highly nonlinear, turbulence states. The above example of toroidal metric caused extended drift waves has been analyzed in this situation. Most MHD modes, including convective instabilities, are also extended, thus they behave similarly. These modes relax the equilibrium profile near (or at) the marginal stability. The temporal dependence

105

in Eq. (2.7.30) can describe phenomena such as the heat pulse propagation, the transport related pulsation, and perhaps the edge-localized modes (ELM). The relaxation oscillations in such a phenomenon have the transport time scale (t_1). It is tempting to think that such analysis might be applicable to the analysis of stellar pulsations, often observed in older stars such as Cepheids, but also observed in our Sun (helioseismic oscillations) (Cox, 1980). We see another application of marginal stability in the *magnetic rotational instability* (see Sec. 4.2.6).

2.8　General Relativistic Plasma Physics

Many astrophysical phenomena of recent interest are believed to be tied to a black hole (ranging from stellar scales to active galactic nuclei scales). Strong gravitational effects on the electromagnetic waves, for example, have been extensively studied and observational link to such phenomena as gravitational lensing has been established. However, there have been little systematic studies of plasma physics in the influence of strong gravity. This is in part EM waves in vacuum in strong gravity is already sufficiently challenging and in part many people assume that little mass may be around a black hole. Here to rectify the paucity of this understanding and opening a virgin field, we summarize a recent formulation of general relativistic plasma physics in (i) high frequency electromagnetic radiation from a plasma around a black hole in Sec. 2.8.1, and (ii) low frequency magnetic effects (MHD equilibria etc.) around a black hole in Secs. 2.8.2–2.8.3. Through these it is already clear a host of unexpected new wave phenomena are possible (Sec. 2.8.1) and, further, we find that (M)HD equilibria around a black hole allow steady-state existence of matter within three Schwarzschild radii under appropriate conditions, whose input on gamma ray burst is possible.

2.8.1　Electromagnetic Waves in a Black Hole Atmosphere

We formulate general relativistic plasma physics in the $3 + 1$ *formalism* of Thorne and MacDonald (1982). The following summarizes how to conformalize Maxwell's equations from a $3+1$ *Schwarzschild metric*. Maxwell's equations in flat space (non-relativistic coordinates) are

$$\nabla \cdot \mathbf{E} = 4\pi \rho, \tag{2.8.1}$$

$$\nabla \cdot \mathbf{B} = 0, \tag{2.8.2}$$

$$\frac{1}{c}\frac{\partial}{\partial t}\mathbf{E} = \nabla \times \mathbf{B} - 4\pi \mathbf{J}, \tag{2.8.3}$$

$$\frac{1}{c}\frac{\partial}{\partial t}\mathbf{B} = -\nabla \times \mathbf{E}. \tag{2.8.4}$$

This is the form in which most physicists are used to seeing. They can also be expressed in four-dimensional geometric language, which is how they are most readily adapted to *general relativity*. Once the relativistic field equations have been solved, however, it is useful to project the results into $3 + 1$ coordinate language (Thorne, *et al.*, 1982, 1986) since they strongly resemble their conventional form. The $3 + 1$ Maxwell's equations for the expanding cosmos metric have been formulated in Holcomb and Tajima (1989). Maxwell's equations in the Schwartzschild metric are, on the other hand,

$$\nabla \cdot \mathbf{E} = 4\pi\rho, \tag{2.8.5}$$

$$\nabla \cdot \mathbf{B} = 0, \tag{2.8.6}$$

$$\frac{1}{c}\frac{\partial}{\partial t}\mathbf{E} = \nabla \times (\alpha\mathbf{B}) - 4\pi\alpha\mathbf{J}, \tag{2.8.7}$$

$$\frac{1}{c}\frac{\partial}{\partial t}\mathbf{B} = -\nabla \times (\alpha\mathbf{E}), \tag{2.8.8}$$

where the general relativistic metric effect manifests through the *lapse function*, α

$$\alpha = \sqrt{1 - \frac{2M}{r}}, \tag{2.8.9}$$

where M is the mass of the *black hole* (or normalized into half the Schwarzschild radius r_s), and all quantities are measured by a local *fiducial observer* (FIDO). Thus the vector quantities given are neither covariant nor contravariant. So the curl of a vector $\mathbf{V}$ in the three-metric $ds^2 = h_1^2 dq_1^2 + h_2^2 dq_2^2 + h_3^2 dq_3^2$ can be written as

$$\mathbf{V} = \frac{1}{h_1 h_2 h_3} \begin{vmatrix} \hat{q}_1 h_1 & \hat{q}_2 h_2 & \hat{q}_3 h_3 \\ \frac{\partial}{\partial q_1} & \frac{\partial}{\partial q_2} & \frac{\partial}{\partial q_3} \\ h_1 V_1 & h_2 V_2 & h_3 V_3 \end{vmatrix}, \tag{2.8.10}$$

and the divergence is just

$$\nabla \cdot \mathbf{V} = \frac{1}{h_1 h_2 h_3} \left[\frac{\partial}{\partial q_1}(V_1 h_2 h_3) + \frac{\partial}{\partial q_2}(V_2 h_3 h_1) + \frac{\partial}{\partial q_3}(V_3 h_1 h_2) \right]. \tag{2.8.11}$$

In an orthonormal coordinate system, $h_i = 1$ for all i, and these expressions reduce to their familiar form. In the case of a slowly (or not) rotating black hole, the three-metric is

$$ds^2 = -\alpha^2 dt^2 + \alpha^{-2} dr^2 + r^2 d\theta^2 + r^2 \sin^2\theta \, d\phi^2. \tag{2.8.12}$$

This is the metric to use when applying the $3 + 1$ paradigm and operators in the new, modified Maxwell's equations.

Assuming propagation exclusively in the $\pm\hat{r}$-direction, we get

$$\boldsymbol{\nabla} \times (\alpha E^{\hat{\theta}}\hat{\theta}) = \frac{\alpha}{r}\frac{\partial}{\partial r}(r\alpha E^{\hat{\theta}})\hat{\phi}, \tag{2.8.13}$$

for the θ polarization. We can combine this with the wave equation

$$\frac{1}{c^2}\frac{\partial^2}{\partial t^2}\mathbf{E} = -\boldsymbol{\nabla} \times [\alpha\boldsymbol{\nabla} \times (\alpha\mathbf{E})] - 4\pi\alpha\frac{\partial}{\partial t}\mathbf{J}, \tag{2.8.14}$$

and, for the moment, allow

$$\mathbf{J} = 0. \tag{2.8.15}$$

The resulting wave equation is (Daniel and Tajima, 1997)

$$\frac{1}{c^2}\frac{\partial^2}{\partial t^2}E^{\hat{\theta}} = -\frac{\alpha}{r}\frac{\partial}{\partial r}\left[\alpha^2\frac{\partial}{\partial r}(\alpha r E^{\hat{\theta}})\right]. \tag{2.8.16}$$

Now we can make the substitution

$$E_\xi^{\hat{\theta}} = \alpha r E^{\hat{\theta}}, \tag{2.8.17}$$

and the change in the coordinate variable to the conformalized coordinate ξ as

$$\frac{\partial}{\partial\xi} = \alpha^2\frac{\partial}{\partial r}. \tag{2.8.18}$$

This allows us to rewrite the wave equation to be

$$\frac{1}{c^2}\frac{\partial^2}{\partial t^2}E_\xi^{\hat{\theta}} = -\frac{\partial^2}{\partial\xi^2}E_\xi^{\hat{\theta}}. \tag{2.8.19}$$

If we make the following substitutions,

$$\rho_\xi = r\alpha^2\rho \tag{2.8.20}$$

$$\mathbf{E}_\xi = r\alpha\mathbf{E} \tag{2.8.21}$$

$$\mathbf{B}_\xi = r\alpha\mathbf{B} \tag{2.8.22}$$

and

$$\mathbf{J}_\xi = r\alpha^2\mathbf{J}, \tag{2.8.23}$$

then we can rewrite Maxwell's equations to be in their "flat" form, without the relativistic α terms and having $\mathbf{E}_\xi$ instead of $\mathbf{E}$, etc.

In the Schwarzschild metric Eq. (2.8.12), let us zoom in to the vicinity of the horizon $r - r_s = \rho$:

$$\left(1 - \frac{r_s}{r_s + \varrho}\right)^{-1}d\varrho = d\xi. \tag{2.8.24}$$

108

where $r_s = 2\,\mathrm{M}$. This can be transformed to

$$\frac{r_s + \varrho}{\varrho} d\varrho = d\xi \qquad (2.8.25)$$

which yields the result the so-called conformalized coordinate ξ

$$\xi = \varrho + r_s ln(\varrho) + \text{ const.} \qquad (2.8.26)$$

This changes the wave equation (2.8.16) to

$$\frac{1}{c^2}\frac{\partial^2}{\partial t^2}E^{\widehat{\theta}} = -\frac{\alpha}{r_s}\frac{\partial}{\partial\varrho}\left[\alpha^2\frac{\partial}{\partial\varrho}(\alpha r_s E^{\widehat{\theta}})\right], \qquad (2.8.27)$$

and instead of the substitution equation (2.8.17) we use the substitution

$$E^{\widehat{\theta}}_\xi = \alpha E^{\widehat{\theta}} \qquad (2.8.28)$$

to arrive at Eq. (2.8.27), since the r_s cancels out. Now we can investigate the properties of this sort of conformalism (Daniel and Tajima, 1997). Maxwell's equations become 'flat':

$$\nabla \cdot \mathbf{E}_\xi = 4\pi\rho_\xi \qquad (2.8.29)$$

$$\nabla \cdot \mathbf{B}_\xi = 0 \qquad (2.8.30)$$

$$\frac{1}{c}\frac{\partial}{\partial t}\mathbf{E}_\xi = \nabla \times \mathbf{B}_\xi - 4\pi\mathbf{J}_\xi \qquad (2.8.31)$$

$$\frac{1}{c}\frac{\partial}{\partial t}\mathbf{B}_\xi = -\nabla \times \mathbf{E}_\xi. \qquad (2.8.32)$$

This is exactly the kind of transformation that is desired. These equations with this choice of variables appear to be flat and the results can be transformed back into the original variables, if one wishes. Note, however, that the charge and current densities are modified according to Eqs. (2.8.20) and (2.8.23). These expressions will reflect in the expressions of plasma and cyclotron frequencies.

When we use the conformalized set of equations, we need to keep track of how they relate to the real physics that might be observed far from the hole. Note that as $r \to \infty$, $\alpha \to 1$, so the local physics far from the hole is "flat," and the time far from the hole is universal time. The conformalized ξ-space is the same far from the hole as the local physics far from the hole. If we model, for example, some phenomena occurring near the hole from which is emitted some observable light wave that propagates far from the hole, the frequency in the model is the same as that which would be observed. When a pair of ee^+ annihilate and emit γ whose energy 1 MeV is locally measured in τ, for example, the main concern is dealing with the proper time, τ, of the phenomenon near the hole, where the effects of time dilation are significant. Since $d\tau = \alpha dt$, then $\omega_\tau = \alpha^{-1}\omega_t$, where ω_τ is the locally measured frequency, and ω_t is the frequency measured in space as well as the frequency measured far from the hole. In the following ω_t is simply written as ω.

The dispersion relation provides a tool with which to analyze wave propagation. We start out by linearizing the wave equation (2.8.14).

$$\mathbf{B} = \mathbf{B}_0 + B_1 \hat{z}$$

$$\mathbf{E} = 0 + E_1 \hat{y}$$

$$\mathbf{J} = 0 + J_{1y} \hat{y} + J_{1z} \hat{z}$$

$$\mathbf{v}_i = v_{i1y} \hat{y} + v_{i1z} \hat{z}$$

$$\mathbf{v}_e = v_{e1y} \hat{y} + v_{e1z} \hat{z}$$

$$\mathbf{k} = k_x \hat{x}$$

$$c^2 \mathbf{k} \times \mathbf{k} \times \mathbf{E}_1 = 4\pi i \omega \mathbf{J}_1 + \omega^2 \mathbf{E}_1. \tag{2.8.33}$$

This yields the dispersion relation for a positron-electron plasma

$$\omega^2 - c^2 k^2 = 2\omega_p^2 \left(1 - \frac{\Omega_e^2}{\omega^2}\right)^{-1}, \tag{2.8.34}$$

where the plasma frequency and the cyclotron frequency have been modified from the conformalization transformation as

$$\omega_p^2 = \frac{4\pi \alpha^2 n_e e^2}{m}, \quad \Omega_e = \frac{\alpha e B}{mc}. \tag{2.8.35}$$

The preceding analysis is true even for the conformalized equations, though in this case, the conformalized constant ω_{p_ξ} and constant Ω_{e_ξ} translate to varying quantities in the fiducial coordinates. More specifically, the constant ω_{p_ξ} implies the fiducial density $\rho_r = \alpha^{-2} \rho_\xi$ varies spatially as α^{-2}. The converse is true, as well. Also, if we assume a particular fiducial density contour, say an exponentially decreasing density in $\hat{r}$, the contour in $\hat{\xi}$ will be the product of the exponential with the α^2 factor. In this manner, we can perform a simple WKB analysis of wave propagation, since the conformalization allows the linearization of Maxwell's equations in the Schwarzschild metric. Now, α increases from zero to one monotonically as ξ (or r) increases. For a light pulse travelling outward from the black hole, this means the 'effective' plasma frequency is increasing. Thus a pulse of frequency $\omega < \omega_{p\hat{r}}$ will hit a 'cutoff' point as it travels outward, and bounce back in. This is obviously a nonphysical case, since the constant density cannot be justified near a black hole, but it does serve to illustrate a piece of the physics which occurs: that the gravitational redshift can serve to lower the frequency of a photon to below the plasma frequency—a not unexpected effect. These effects include resonance, mode conversion due to the radially changing spatial metric (through $\alpha(r)$) and plasma equilibria (through $n(r), B(r), T(r)$ etc.). (See Fig. 2.16.)

110

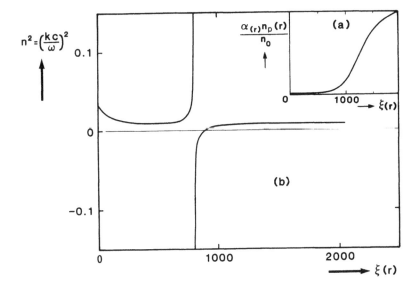

FIGURE 2.16 A schematic profile of ω_{pf}^2 through $\alpha(r)$ and $n(r)$ (a) and $n^2 = (kc/\omega)^2$ (b) (n is the refractive index of the plasma). The internal modes generated from the turbulent near horizon plasma may try to propagate outwards, but encounter resonance and then perhaps mode-converts out into the outside, which may be observable from the Earth (Daniel & Tajima, 1997).

The resultant spatial variation of the density and the time-constant magnetic field in the electromagnetic wave spectroscopy arising from or near a black hole can play an important role in the γ-ray burst phenomenon and this has been analyzed by using this technique (Daniel and Tajima, 1997). In particular coupled with the intrinsic volatile plasma (see next Sec. 2.8.2 and 2.8.3) near the black hole horizon, these electromagnetic wave properties may be a key to understand some class of γ-ray bursts.

2.8.2   General Relativistic MHD

In this subsection we consider low-frequency (and thus fluid) plasma dynamics near a black hole. Thus we wish to formulate moment equations of plasma dynamics in the 3+1 formalism. We begin with a spacetime metric $g_{\mu\nu}$ and choose a time-like vector field n^μ whose flow forms a congruence in the spacetime 4-manifold. The flow of n^μ defines the world lines of particles at "rest." To achieve the $3+1$ decomposition, we follow the method developed by Holcomb and Tajima (1989) but more closely by that of Tarkenton (1966), where we project all tensors onto n^μ and into the space-like hypersurface orthogonal to n^μ. This leads to a decomposition of the metric into two parts:

$$g_{\mu\nu} = \gamma_{\mu\nu} - n_\mu n_\nu,$$

where $\gamma_{\mu\nu}$ is the positive-definite metric on the space-like hypersurfaces. The mixed components of γ form the space-like projection operator. This completes our specification of kinematics; we now turn to dynamics.

All the dynamics that we will consider come from energy-momentum conservation, $T^{\mu\nu}_{;\nu} = 0$, where $T^{\mu\nu}$ is the energy-momentum tensor and the subscripted semicolon denotes covariant differentiation with respect to $g_{\mu\nu}$. The projection of the conservation law onto n^μ yields an equation that we interpret as energy conservation, while the projection with γ^τ_μ produces momentum conservation.

Energy Conservation:

$$D_t \epsilon + \theta \epsilon + \frac{1}{\alpha^2} \nabla \cdot (\alpha^2 \mathbf{S}) + \left(\sigma_{ij} + \frac{1}{3} \theta \gamma_{ij} \right) W^{ij} = 0; \tag{2.8.36}$$

Momentum Conservation:

$$D_t \mathbf{S} + \frac{4}{3} \theta \mathbf{S} + \underline{\sigma} \cdot \mathbf{S} + \epsilon \mathbf{g} + \frac{1}{\alpha} \nabla \cdot (\alpha \underline{W}) = 0, \tag{2.8.37}$$

where $\epsilon = n_\mu T^{\mu\nu} n_\nu$ is the local energy density; $S^\tau = \gamma^\tau_\mu T^{\mu\nu} n_\nu$ is the momentum vector; and $W^{\sigma\tau} = \gamma^\sigma_\mu T^{\mu\nu} \gamma^\tau_\nu$ is the stress tensor. $D_t A^\mu = \gamma^\mu_\tau A^\tau_{;\nu} n^\nu$ is the Fermi-Walker time derivative; $\theta = n^\mu_{;\mu}$ is the expansion of the time-like *congruence* and $\sigma_{\mu\nu} = \gamma^\beta_\mu \gamma^\delta_\nu (n_{\beta;\delta} + n_{\delta;\beta}) - \frac{1}{3} \gamma_{\mu\nu} \theta$ is its shear. α is the lapse function and $\mathbf{g} = -\nabla \ln \alpha$ is the local gravitational acceleration. ∇ is the spatial covariant derivative derived from $\gamma_{\mu\nu}$. These equations govern all the dynamics occurring on the fixed background. Once the energy-momentum tensor is specified, all the parameters are fixed.

The last bit of general formalism we need to record is the behavior of the specific force electrodynamics. The electromagnetic field, $F^{\mu\nu}$, obeys Maxwell's equations, whose covariant form is

$$F^{\mu\nu}_{;\nu} = 4\pi J^\mu,$$

$$F^{*\mu\nu}_{;\nu} = 0,$$

where $*^{\mu\nu}$ is the dual to $F^{\mu\nu}$ and J^μ is the 4-vector current. Given the space-time splitting based on n^μ, we can define local electric and magnetic fields by projections of $F^{\mu\nu}$: $E^\mu = F^{\mu\nu} n_\nu$ and $B^\mu = \frac{1}{2} \epsilon^{\mu\rho\sigma\tau} n_\rho F_{\sigma\tau}$. We can make a similar decomposition of the current: $\rho_e = -n_\mu J^\mu$ is the electric charge density and $j^\mu = \gamma^\mu_\nu J^\nu$ is the electric current density. We can write Maxwell's equations in their 3 + 1 forms as we did in Sec. 2.8.1.

All the dynamics of an ideal fluid show up in the energy-momentum tensor for the fluid. In a general frame, this takes the form

$$T^{\mu\nu} = (e + p) u^\mu u^\nu + p g^{\mu\nu}$$

112

where u^μ is the 4-velocity of the fluid measured in our frame, e is the internal energy function of the fluid, including the rest-mass energy, and p is the pressure in the fluid and we've put $c = 1$.

There are three projections which are

$$\epsilon = n_\mu T^{\mu\nu} n_\nu = \Gamma^2(e + p) - p = \omega - p,$$

$$\mathbf{S} = \omega\mathbf{v}, \underline{\underline{W}} = p\underline{\gamma} + \omega\mathbf{v} \otimes \mathbf{v},$$

where $\omega = \Gamma^2 nw = \Gamma^2(e + p)$ is a local measure of the fluid's enthalpy, w being the enthalpy per particle, n is the particle number density and the special relativistic parameters $u^\mu n_\mu = \Gamma = (1 = v^2)^{-1/2}$ and $\gamma_\tau^\mu u^\tau = \Gamma\mathbf{v}$, $\mathbf{v}$ being the fluid's 3-velocity. Inserting these expressions into the energy and momentum conservation laws Eqs. (2.8.49) and (2.8.50), we arrive at the general relativistic versions of the Euler equations:

$$\frac{1}{\alpha}\frac{\partial\omega}{\partial t} + \frac{1}{\alpha}\nabla\cdot(\alpha\omega\mathbf{v}) = \omega\mathbf{v}\cdot\mathbf{g} + \frac{1}{\alpha}\frac{\partial p}{\partial t}, \tag{2.8.38}$$

and

$$\omega\left(\frac{1}{\alpha}\frac{\partial\mathbf{v}}{\partial t} + \mathbf{v}\cdot\nabla\mathbf{v}\right) = -\nabla p + \omega\mathbf{g} - \mathbf{v}\left(\omega\mathbf{v}\cdot\mathbf{g} + \frac{1}{\alpha}\frac{\partial p}{\partial t}\right). \tag{2.8.39}$$

The unusual source terms in these equations arise from the fact that we view the fluid's motion not from its local comoving frame. We can interpret the new terms in the energy conservation (2.8.38); the first represents the work done by the fluid against gravity, and the second involves the rate at which pressure changes, another work term. The new terms in the momentum equation (2.8.39) arise in the same way as the energy equation, since we used the energy equation to pull ω out of the derivatives on the left-hand side.

Equations (2.8.38) and (2.8.39) are not closed. To close the system, we must specify a thermodynamic relation between ω and p. Typically, the relation is between p and ω, necessitating an equation governing n. This is the usual continuity equation that states that particles can neither be created nor destroyed. This requirement takes the covariant form $(nu^\mu)_{;\mu} = 0$, which we must decompose according to our space-time splitting. The result for our Schwarzschild coordinates is

$$\frac{1}{\alpha}\frac{\partial(\Gamma n)}{\partial t} + \frac{1}{\alpha}\nabla\cdot(\alpha\Gamma n\mathbf{v}) = 0. \tag{2.8.40}$$

This equation, plus the equation of state and the first law of thermodynamics, $dw = \frac{1}{n}dp + Tds$, where T and s are the fluids temperature and entropy, respectively, close the system.

We now want to supplement the ideal fluid energy-momentum tensor with electromagnetic fields. The standard form for the electromagnetic energy-momentum tensor is [c.f. Jackson (1975), Landau and Liftchitz (1975)]

$$T^{\mu\nu} = \frac{1}{4\pi}\left(F^{\mu\sigma}F^\nu_\sigma + \frac{1}{4}\gamma^{\mu\nu}F^{\sigma\tau}F_{\sigma\tau}\right)$$

113

which we can write in terms of the fields as

$$T^{\mu\nu} = \frac{1}{4\pi}\left[\frac{1}{2}n^{\mu}n^{\nu}(E^2 + B^2) - (E^{\mu}E^{\nu} + B^{\mu}B^{\nu}) + \frac{1}{2}\gamma^{\mu\nu}(E^2 + B^2)\right.$$

$$\left.(n^{\nu}\epsilon^{\mu\alpha\beta\gamma} - n^{\mu}\epsilon^{\nu\alpha\beta\gamma})E_{\alpha}B_{\beta}n_{\gamma}\right].$$

Using Maxwell's equations or directly in terms of the fields, we can show that the divergence of the electromagnetic energy-momentum tensor takes the form

$$T^{\mu\nu}_{;\nu} = -F^{\mu\nu}J_{\nu}. \qquad (2.8.41)$$

This is the simplest starting point for deriving the 3 + 1 contributions to the conservation laws.

Projecting Eq. (2.8.41) onto n^{μ}, we find the electromagnetic contribution to energy conservation to be $\mathbf{j} \cdot \mathbf{E}$. Projecting the same relation into the spatial hypersurface, we obtain the momentum conservation part: $\rho_e\mathbf{E} + \mathbf{j} \times \mathbf{B}$. Using these, the modified Euler equations and the equation of motion are obtained. These equations are not closed. Supplementing them with thermodynamic relations as in the fluid case almost closes the system. We need an additional relation that specifies the electromagnetic current density. We will take this constitutive relation to represent the assumption of perfect conductivity. This leads to the condition (see Anile, 1989, Dixon, 1978)

$$F^{\mu\nu}u_{\nu} = 0,$$

which has the 3 + 1 form $\mathbf{E} + \mathbf{v} \times \mathbf{B} = 0$. We now have 18 equations: 3 from momentum conservation, 6 from Ampére's and Faraday's laws, 3 from the perfect, conductivity relation, 1 each from energy and particle number conservation, both electromagnetic Gauss' laws, the equation of state and the first law of thermodynamics; we also have 18 unknowns: $\mathbf{v}, \mathbf{E}, \mathbf{B}, \mathbf{j}, n, w, p, T, s, \rho_e$.

We summarize all the relevant equations below (Tarkenton, 1996).

Energy Conservation:

$$\frac{1}{\alpha}\frac{\partial w}{\partial t} + \frac{1}{\alpha}\nabla \cdot (\alpha w\mathbf{v}) = w\mathbf{v} \cdot \mathbf{A} + \mathbf{j} \cdot \mathbf{E} + \frac{1}{\alpha}\frac{\partial p}{\partial t}, \qquad (2.8.42)$$

where $\mathbf{A}$ is the acceleration.

Momentum Conservation:

$$w\left(\frac{1}{\alpha}\frac{\partial \mathbf{v}}{\partial t} + \mathbf{v} \cdot \nabla\mathbf{v}\right) = -\nabla p + w\mathbf{A} + \rho_e\mathbf{E} + \mathbf{j} \times \mathbf{B} - \mathbf{v}\left(w\mathbf{v} \cdot \mathbf{A} + \mathbf{j} \cdot \mathbf{E} + \frac{1}{\alpha}\frac{\partial p}{\partial t}\right) \qquad (2.8.43)$$

Maxwell's Equations [2.8.5)–(2.8.8)]:

114

$$\nabla \cdot \mathbf{E} = 4\pi \rho_e,$$

$$\nabla \cdot \mathbf{B} = 0,$$

$$\frac{1}{\alpha}\frac{\partial \mathbf{E}}{\partial t} = \frac{1}{\alpha}\nabla \times (\alpha\mathbf{B}) - 4\pi\mathbf{j},$$

$$\frac{1}{\alpha}\frac{\partial \mathbf{B}}{\partial t} = \frac{1}{\alpha}\nabla \times (\alpha\mathbf{E}).$$

Particle Number Conservation:

$$\frac{1}{\alpha}\frac{\partial(\Gamma n)}{\partial t} + \frac{1}{\alpha}\nabla \cdot (\alpha\Gamma n\mathbf{v}) = 0 \qquad (2.8.44)$$

Thermodynamics:

$$f(n, s) = 0 \qquad (2.8.45)$$

Perfect Conductivity Relation:

$$\mathbf{E} + \mathbf{v} \times \mathbf{B} = 0 \qquad (2.8.46)$$

Consider a fluid configuration under the assumptions of static ($\mathbf{v} = 0$) equilibrium. The governing equations reduce to essentially 2: momentum conservation and the perfect gas equation of state. These are

$$\nabla p = w\mathbf{A}, \qquad (2.8.47)$$

and

$$w = \frac{\gamma}{\gamma - 1}T + m, \qquad (2.8.48)$$

where γ is the polytropic index of the fluid, and m is the mass of a hydrogen atom. (We have set k_B and c both equal to 1 here.) The solution to these equations is

$$p = p_0 e^{-\int \frac{dr}{H(r)}}, \qquad (2.8.49)$$

where the scale height $H(r)$ is

$$H(r) = \frac{1}{(\frac{\gamma}{\gamma-1} + \frac{m}{T})\nabla \ln \alpha} = \frac{2\alpha r^2 T}{mR_s(1 + \frac{T}{m}\frac{\gamma}{\gamma-1})}.$$

For low temperatures, $T \ll m$, the scale height simplifies to

$$H(r) = \frac{2\alpha r^2 T}{R_s m}.$$

Notice that as $r \rightarrow \infty$, the low temperature H returns to its Newtonian form $T/mg = Tr^2/GMm$.

We seek time-independent, axisymmetric solutions to the equations of general relativistic hydrodynamics. Using stationarity and axisymmetry we may put $\partial_t = 0 = \partial_\phi$ which turns the particle conservation law into

$$\frac{\alpha}{r^2}\frac{\partial}{\partial r}\left(r^2\alpha^2\rho v^{\widehat{r}}\right) + \frac{1}{r\sin\theta}\frac{\partial}{\partial\theta}\left(\sin\theta\alpha^2\rho v^{\widehat{\theta}}\right) = 0,$$

where the hats on the components indicate physical components and we have defined $\rho = \Gamma n/\alpha$, the proper number density. We can satisfy this equation identically if we let

$$v^{\widehat{r}} = \frac{1}{\alpha^2\rho r\sin\theta}\left(\frac{1}{r}\frac{\partial\psi}{\partial\theta}\right), \tag{2.8.50}$$

$$v^{\widehat{\theta}} = \frac{1}{\alpha^2\rho r\sin\theta}\left(\alpha\frac{\partial\psi}{\partial r}\right), \tag{2.8.51}$$

for a *poloidal stream function* ψ. Notice that these equations define $v^{\widehat{r}}$ and $v^{\widehat{\theta}}$ implicitly, since ρ contains a factor of Γ. We now want to recast the energy and momentum equations in terms of this stream function.

Using our assumptions of stationarity and axisymmetry, we can write

$$0 = \nabla\cdot\left(\alpha^2\Gamma^2 nw\mathbf{v}\right) = \alpha\Gamma w\nabla\cdot(\alpha\Gamma n\mathbf{v}) + \alpha\Gamma n\mathbf{v}\cdot\nabla(\alpha\Gamma w),$$

$$= \alpha\Gamma n\mathbf{v}\cdot\left[\alpha\Gamma T\nabla s + \alpha^2\Gamma^3 w\mathbf{v}\times\nabla(\mathbf{v}/\alpha)\right],$$

$$= \alpha^2\Gamma^2 nT(\mathbf{v}\cdot\nabla s).$$

This implies that $\mathbf{v}_p\cdot\nabla s = 0$, where $\mathbf{v}_p = (v^{\widehat{r}}, v^{\widehat{\theta}})$ is the poloidal velocity. This means that $s = s(\psi)$; so the entropy is constant on the surfaces of constant ψ. This is our first *flux function*, and it represents the conservation of thermal energy.

The next application of the momentum equation is to extract another surface function. Axisymmetry reduces this to $\mathbf{v}_p\cdot\nabla(\alpha\Gamma w) = 0$, which implies that $\alpha\Gamma w = J(\psi)$, is the *Bernoulli flux function*. It represents the conservation of mechanical energy.

We see that surface functions seem to be associated with conserved quantities. This lead us to search for another surface function related to the conservation of angular momentum, which follows from our axisymmetric assumption. There are two components of the momentum equation left from which to extract this surface function. The result is

$$\mathbf{v}\times\nabla\times\left(\frac{\mathbf{v}}{\alpha}\right) = \frac{1}{\alpha^5\rho^2(r\sin\theta)^2}\nabla\psi\left(\tilde{\Delta}_v\psi - \frac{1}{\rho}\nabla\rho\cdot\nabla\psi\right)$$

$$- \frac{1}{r\sin\theta}\mathbf{v}_p\cdot\nabla\left(\frac{r\sin\theta v^{\widehat{\phi}}}{\alpha}\right)\mathbf{e}^{\widehat{\phi}} + \frac{v^{\widehat{\phi}}}{r\sin\theta}\nabla\left(\frac{r\sin\theta v^{\widehat{\phi}}}{\alpha}\right) \tag{2.8.52}$$

where

$$\tilde{\Delta}_v \psi \cong \alpha^4 \frac{\partial}{\partial r} \left(\frac{1}{\alpha^2} \frac{\partial \psi}{\partial r} \right) + \frac{\sin\theta}{r^2} \frac{\partial}{\partial \theta} \left(\frac{1}{\sin\theta} \frac{\partial \psi}{\partial \theta} \right) \tag{2.8.53}$$

is the general relativistic *Grad-Shafranov operator*. (See Sec. 2.5 and Sec. 4.3.3). If we contract the momentum equation with $\mathbf{v}^{\widehat{\phi}}$, we obtain

$$\mathbf{v}_p \cdot \nabla \left(\frac{r \sin\theta}{\alpha} v^{\widehat{\theta}} \right) = 0 \tag{2.8.54}$$

which implies that $\frac{r \sin\theta}{\alpha} v^{\widehat{\theta}} = L(\psi)$; this flux function represents the conservation of angular momentum.

Finally, we contract the momentum equation with $\nabla \psi$ and we obtain an equation for ψ itself. The result is

$$\tilde{\Delta}\psi - \frac{1}{\rho} \nabla\rho \cdot \nabla\psi = \alpha^4 \rho^2 \left[(r \sin\theta)^2 \frac{J'}{\Gamma^2 J} + \alpha^2 LL' \right] - \alpha^4 (r \sin\theta)^2 \frac{\rho}{J} nTs' \tag{2.8.55}$$

where the $()'$ denotes a derivative with respect to ψ. This equation is a type of equation that arises in magnetized plasma physics known as the Grad-Shafranov equation. See Chap. 4 for more details. Equations like it have shown up in hydrodynamics before [c.f. Lamb (1984)]. It is important to note that plasma and its flows are not precluded even in the near-horizon space ($r < 3R_s$), where corpuscular movements are unstable. These new equilibria emerge because the plasma behaves as a (collisional entity called) gas (fluid). We will have even more equilibria, including magnetic fields in Sec. 2.8.3.

2.8.3   Schwarzschild Equilibria*

We now present some numerical solutions to the full isentropic Grad-Shafranov equation (2.8.55) and analyze the resulting equilibrium. Then we move on to discuss two equilibrium configurations in detail. These are essentially thick-disk or dust-tori type equilibria and they may play a role in the dynamics of γ-ray bursters and other high-energy phenomena (Sec. 4.3.5).

Since the gas comprising the disk is likely to be ionized, its electromagnetic properties will play an important dynamical role. We concentrate on these properties in this section. We develop a formalism analogous to the fluid treatment given above and examine some analytical and numerical solutions to the equilibrium equations.

Starting with the equations in Tarkenton's (1996) "Governing Equations" in Sec. 2.8.2, we can reduce everything down to a single equation for a scalar function, following a standard plasma physics technique. This function is now the *poloidal magnetic flux*, ψ, now defined by the following equations:

$$B^{\widehat{r}} = \frac{1}{r \sin\theta} \frac{1}{r} \frac{\partial \psi}{\partial \theta}, \tag{2.8.56}$$

and

$$B^{\hat{\theta}} = \frac{1}{r \sin\theta} \alpha \frac{\partial \psi}{\partial r}. \tag{2.8.57}$$

From the ϕ-component of the infinite conductivity relationship, Eq. (2.8.46), we see that the poloidal velocity, $\mathbf{v}_p$ must be proportional to the poloidal magnetic field, $\mathbf{B}_p$. We can express this relationship as $\mathbf{v}_p = \kappa \mathbf{B}_p$. Combining this with the particle number conservation equation (2.8.44), we find that

$$\mathbf{B}_p \cdot \nabla(\alpha \Gamma n \kappa) = 0,$$

which implies that $\alpha \Gamma n \kappa$ is a constant on surfaces of constant ψ. This implies that the function

$$F = 4\pi \alpha \Gamma n \kappa = 4\pi \alpha^2 \rho \kappa$$

is a flux function, i.e. $F = F(\psi)$, related to the conservation of particles, and $\rho = \Gamma n/\alpha$ is the proper number density.

The remaining two components of the infinite conductivity equation give the electric field:

$$-\mathbf{E} = \mathbf{v} \times \mathbf{B} = -\left(v^{\hat{\phi}} - \kappa B^{\hat{\theta}}\right) \frac{1}{r \sin\theta} \nabla\psi.$$

Using this in Faraday's Law leads us to

$$\nabla\psi \times \nabla \left(\frac{\alpha}{r \sin\theta}\left(v^{\hat{\phi}} - \kappa B^{\hat{\phi}}\right)\right) = 0$$

or that

$$\frac{\alpha}{r \sin\theta}\left(v^{\hat{\phi}} - \kappa B^{\hat{\phi}}\right) = G(\psi). \tag{2.8.58}$$

This is another flux function related to the electric field.

We write the electromagnetic work term as

$$\mathbf{E} \cdot \mathbf{j} \cong \mathbf{E} \cdot \nabla \times (\alpha \mathbf{B}) = \frac{G}{\alpha} \nabla\psi \times (\alpha \mathbf{B}) = \frac{G}{\alpha} \mathbf{B}_p \cdot \nabla(\alpha r \sin\theta B^{\hat{\phi}})$$

The *relativistic magnetized Grad-Shafranov* operator is now

$$\Delta^* \psi = \frac{\partial}{\partial r}\left(\alpha^2 \frac{\partial \psi}{\partial r}\right) + \frac{\sin\theta}{r^2} \frac{\partial}{\partial \theta}\left(\frac{1}{\sin\theta} \frac{\partial \psi}{\partial \theta}\right). \tag{2.8.59}$$

The combination $\alpha \Gamma w$ was the Bernoulli flux function. It will not be a flux function here due to the electromagnetic terms on the right. Mobarry and Lovelace (1986) introduce the notation $J_F = \alpha \Gamma w$ which we will follow here. We can use the momentum equation to extract several surface functions and the magnetized Grad-Shafranov equation. We use G for the relativistic magnetized Grad-Shafranov equation:

$$0 = \left(1 - \frac{J_F F^2}{4\pi \rho \alpha^4} - \frac{(r \sin\theta)^2 G^2}{\alpha^2}\right) \Delta^* \psi + \nabla \left(1 - \frac{J_F F^2}{4\pi \rho \alpha^4} - \frac{(r \sin\theta)^2 G^2}{\alpha^2}\right) \cdot \nabla\psi$$

$$+\left(\frac{J_F F F'}{4\pi\rho\alpha^4}+\frac{(r\sin\theta)^2}{\alpha^2}GG'\right)|\nabla\psi|^2+\frac{1}{\alpha^2}\left(H+\frac{r\sin\theta}{\alpha}v^{\hat\phi}FJ_F\right)\left(H'+\frac{r\sin\theta}{\alpha}v^{\hat\phi}FJ'_F\right)$$

$$+4\pi(r\sin\theta)^2\rho\left[J'+\frac{r\sin\theta}{\alpha}v^{\hat\phi}J_FG'\right]-4\pi(r\sin\theta)^2nTs',\tag{2.8.60}$$

where we have defined J through other flux functions $R_E, G, H,$ and F as

$$J=\left(1-\frac{r\sin\theta}{\alpha}v^{\hat\phi}G\right)J_F$$

$$=\frac{F_E-GH}{F},\tag{2.8.61}$$

which is another flux function. Mobarry and Lovelace (1986) arrived at the same equation using a slightly different approach; they have analyzed the mathematical character of the operator and have shown that the Grad-Shafranov equation is hyperbolic if the potential flow speed is super-Alfvénic.

We now look at two analytic solutions of the Grad-Shafranov equation (2.8.60). We consider the homogeneous or vacuum case which corresponds to a purely poloidal magnetic field; it will be important for boundary conditions on the inhomogeneous equation, just as in the fluid case. The second type of analytic solution we will compute is a simple inhomogeneous problem.

The vacuum Grad-Shafranov equation takes the form

$$\Delta^*\psi=\frac{\partial}{\partial r}\left(\alpha^2\frac{\partial\psi}{\partial r}\right)+\frac{1-x^2}{r^2}\frac{\partial^2\psi}{\partial r^2}=0,\tag{2.8.62}$$

where we have introduced the new independent variable $x=\cos\theta$. This equation separates, just like the fluid case. The separated ordinary differential equations are

$$\frac{d}{dr}\left(\alpha^2\frac{dR}{dr}\right)-\frac{k}{r^2}R=0;\tag{2.8.63}$$

$$(1-x^2)\frac{d^2X}{dx^2}+kX=0.\tag{2.8.64}$$

The angular equation is precisely the same as the fluid case and so the solutions are the associated Legendre functions:

$$X(x)=\sqrt{1-x^2}\,P_\ell^1(x);\tag{2.8.65}$$

$$X(x)=\sqrt{1-x^2}\,Q_\ell^1(x),\tag{2.8.66}$$

with $k=\ell(\ell+1)-2$. The radial equation is almost the same as the fluid case; the only difference is the sign in front of the first derivative term:

$$r(1-r)\frac{d^2R}{dr^2}-\frac{dR}{dr}+kR=0.\tag{2.8.67}$$

119

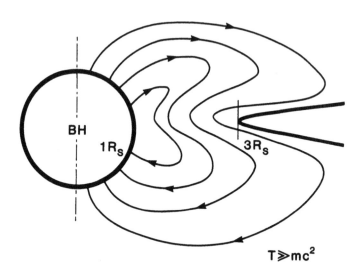

BH

1R$_s$

3R$_s$

T$\geqslant$mc^2

FIGURE 2.17 Schematic flux lines of a equilibrium near a black hole *a la* from the solution of Eq. (2.8.69).

Thus, the singularity structure is identical to the fluid case, permitting the series solution

$$R = \frac{1}{r^s} \sum_{n=0}^{\infty} \frac{a_n}{r^n}, \tag{2.8.68}$$

where

$$s = \frac{1}{2} \pm \sqrt{\ell(\ell+1) - \frac{7}{4}},$$

$$a_{n+1} = a_n \left(\frac{(n+s)}{(n+s+2) - \ell(\ell+1) + 2} \right).$$

This series converges for $1 < r < \infty$ and diverges for $r = 1$ according to Gauss' test, where we have set $R_s = 1$. The nonrelativistic solution is simply the indicial part, being identical to the nonrelativistic fluid solution, since as $\alpha \to 1$, both the fluid and magnetic homogeneous Grad-Shafranov equations become identical. With an appropriate boundary condition we can construct the equilibrium as in Fig. 2.17.

The next type of equation we consider is one that has $F = 0, G = 0, H = $ constant, $s = $ constant, $J = k\psi$, and $\rho = $ constant. This leads to equation

$$\Delta^* \psi = \kappa r^2 (1 - x^2), \tag{2.8.69}$$

120

with κ being the collection of all the constants. This equation is not separable, but we can apply Hill's ansatz to extract the solution. Starting with $\psi = r^2(1 - x^2)/f(r)$ and inserting this into (2.8.69), we find

$$(\alpha^2 r^2 f') + \frac{4}{r}(\alpha^2 r^2 f') = \kappa r^2, \tag{2.8.70}$$

which has the solution

$$f = \int \left(\frac{\kappa}{7} \frac{r}{\alpha^2} + \frac{c_1}{r^6 \alpha^2} \right) dr + c_0,$$

where c_1 and c_0 are the constants of integration. The two fundamental solutions we will use are

$$I_1 = \int \frac{r}{\alpha^2} dr = \int \frac{r^2}{r-1} dr = \frac{1}{2} r^2 + r + \ln(r\alpha^2)$$

and

$$I_2 = \int \frac{1}{r^6 \alpha^2} dr = \int \frac{1}{r^5(r-1)} dr = \frac{1}{4r^4} + \frac{1}{3r^3} + \frac{1}{2r^2} + \frac{1}{r} + \ln \alpha^2,$$

where we have set $R_s = 1$. Both of these functions are singular on the horizon; however, the linear combination $I_- = I_1 - I_2$ is finite on the horizon. I_1 is obviously divergent at infinity. $I_2 \sim \ln(1 - 1/r) + 1/r \sim 1/2r^2 + \mathcal{O}(1/r^3)$ as $r \to \infty$ and so $r^2 I_2 \to 1/2$ is finite at infinity. Thus, $I_2^- = I_2 - 1/2$ is zero at infinity. Thus, the most convenient solutions are I_- and I_2^-. Therefore, we can write the solution to Eq. (2.8.69) as

$$\psi = r^2(1 - x^2) \left(\frac{\kappa}{7} I_- + c_1 I_2^- + c_0 \right). \tag{2.8.71}$$

This is a nontrivial solution to the general relativistic Grad-Shafranov equation for magnetic flux of Hill's vortex type valid arbitrarily close to the horizon. It is of interest that magnetic fields in equilibria near the horizon do exist. These fields along with the existence of plasma and its flows (Sec. 2.8.2) may play an important role in a variety of astrophysical phenomena, including γ-ray bursts.

Problem 2-1: List spatial hierarchy in plasma physics and characterize each spatial scale that occurs. Contrast this with temporal hierarchy in plasma physics. Suggest what characterizes the sizes in Fig. 2.9.

Problem 2-2: For a collisionless plasma, where $g \ll 1$ or the wavelength $\ell \gg \lambda$ (the mean free path) (these two are independent conditions, unless $k \sim k_D$), let us consider the phase velocity of plasma waves in comparison with electron and ion thermal velocities v_e and v_i. Note that if the phase velocity ω/k is of the order of v_e or v_i, the coupling of the wave and particles is strong and the fluid description breaks down. Show waves with high frequencies $\omega/k > v_e$ and the collective motion may be described by the fluid. On the other hand waves with low frequencies, such as the Alfvén wave (v_A), typically $v_e > v_A > v_i$, and show that this is the case when $1 > \beta > m/M$. These waves, once again, are describable by the fluid.

Problem 2–3: For a magnetized plasma, the MHD waves in the transverse direction have wavenumber up to ρ_i^{-1}, where ρ_i is the ion Larmor radius. Count the number of available MHD modes in the system and compare with the total degrees of freedom $3N$, where N is the number of particles.

Problem 2–4: Show that the following equations hold for relaxed e^+e^-P plasmas:

$$\frac{\partial}{\partial t} \mathbf{H} = \nabla \times [\mathbf{H} \times (\nabla \times \mathbf{H})],$$

$$\frac{\partial}{\partial t} \nabla^2 \mathbf{H} = \nabla \times [(\nabla^2 \mathbf{H}) \times (\nabla \times \mathbf{H})].$$

Validate the assumption $\partial/c\partial t\, \mathbf{E} \ll \nabla \times \mathbf{B}$ from the above equation(s). Obtain the range of validity for the neutrality condition:

$$\frac{1}{4\pi e \delta n^\bullet} \nabla \cdot \mathbf{E} = -\frac{B_0^2}{4\pi \delta n^\bullet m c^2} (\mathbf{H} \cdot \nabla^2 \mathbf{H} + (\nabla \times \mathbf{H})^2)$$

and relate it to the *Beltrami condition* $\nabla \times \mathbf{H} = \alpha \mathbf{H}$.

Problem 2–5: Show that $\nabla^2 h = P[p]$ in Sec. 2.5.3.

Problem 2–6: Show that a two-dimensional viscous fluid has the velocity spectrum in wavenumber (α, β are inverse temperatures)

$$\langle v_k^2 \rangle = \frac{1}{\beta + 2\alpha k^2} \frac{1}{\rho},$$

and the frequency spectrum

$$\langle v_\omega^2 \rangle = \frac{1}{4\rho\alpha} \frac{1}{\omega},$$

using the fluctuation-dissipation theorem. The latter is the so-called $1/f$ *spectrum*. Show also that this result is equivalent to the two-dimensional magnetized plasma ($\mathbf{E} \times \mathbf{B}$) velocity spectra (Taylor, 1974).

Problem 2–7: Show $\nabla\psi \times (\alpha \mathbf{B}) = \mathbf{B}_p \times \nabla(\alpha r \sin\theta B^\phi)$ in Sec. 2.8.3.

Problem 2–8: Revisit the Kolmogorov turbulent spectrum using the renormalization group theory, where the group theoretical technique is applied to the point that the invariances of physics in scales in the inertial domain (Longscope & Sudan, 199?).

Problem 2–9: In many astrophysical phenomena the spectra show the power law dependence. For example, from the Sun the solar granule population with respect to the size, the magnetic field strength and the AR size distribution of the Sun, and the solar X-ray spectra as well as the flare energy and frequency distribution. These all show the power law dependence. From accretion disks the X-ray and optical luminosities show (often) the power law dependence. The cosmic ray energy shows the power law dependence [Fig. 2.13(a)]. The population of corpuscular objects with respect to their mass shows such [Fig. 2.13(c)]. Are these power-law dependences coming from individual and distinct reasons or from some common underlying physical reason, such as the fractal nature, self-similarity due to self-organized criticality, etc.?

References

Alfvén, H., Nature **150**, 405 (1942).

Anile, A.M. *Relativistic Fluids and Magnetofluids*, (Cambridge Univ. Press, Cambridge, 1989).

Babcock, H.D., Astrophys. J. **130**, 304 (1959).

Batchelor, G.K., *An Introduction to Fluid Dynamics*, (Cambridge Univ. Press, Cambridge, 1967).

Bak, P., Tang, C., and Wiesenfeld, K., Phys. Rev. A **38**, 364 (1988).

Benzi, R., Colella, M., Briscolini, M. and Santangelo, P., Phys. Fluids A **4**, 1036 (1992).

Berk, H.L. *et al.*, Phys. Fluids **24**, 1758 (1986).

Bernstein, I.B., Frieman, E.A., Kruskal, M.D., and Kulsrud, R.M., *Proceedings of the Royal Society*, vol. A244, p. 17 (1958).

Bernstein, I.B., Phys. Rev. **109**, 10 (1958).

Biskamp, D., and Welter, H., Phys. Fluids B **2**, 1787 (1990).

Biskamp, D., Welter, H., and Walter, M., Phys. Fluids B **2**, 3024 (1990).

Bohm, D., Burhop, E.H.S., and Massey, H.S.W. in *The Characteristics of Electrical Disharges in Magnetic Fields*, eds. Guthrie, A., and Wakerling, R.K. (McGraw Hill, New York, 1949) p. 13.

Bogoliubov, B., *Studies in Statistical Mechanics*, eds. deBaer, J., and Uhlenback, G.E., (North-Holland, Amsterdam, 1962), vol. 1, p. 1.

Brachet, M., Meneguzzi, M., Politano, H., and Sulem, P.L., J. Fluid Mech. **194**, 333 (1988).

Carnevale, G.F., McWilliams, J.C., Pomeau, Y. Weiss, J.B., and Young, W.R., Phys. Rev. Lett. **66**, 2735 (1991).

Chandrasekhar, S., Rev. Mod. Phys. **15**, 1 (1943).

Chandrasekhar, S., *Hydrodynamic and Hydromagnetic Stability* (Oxford: Clarendon Press, 1961), p. 384.

Chen, F.F. *Introduction to Plasma Physics*, (Plenum, New York, 1984).

Cox, J.P., *Theory of Stellar Pulsation* (Princeton Univ. Press, Princeton, 1980).

Daniel, J., and T. Tajima, Phys. Rev. D. (1997).

Davidson, B., Proc. Roy. Soc. Lond. Ser. A **225**, 252 (1954).

De Vaucouleurs, G., Science **167**, 1203 (1970).

Dixon, W.G. *Special Relativity*, (Cambridge Univ. Press, Cambridge, 1978).

Drake, J.F., and Lee, Y.C., Phys. Fluids **20**, 1341 (1977).

Drake, J.F., Phys. Fluids **30**, 2429 (1987).

Edwards, S.F., and Taylor, J.B., Proc. R. Soc. Lond. A **336**, 257 (1974).

Elsasser, W.M., Phys. Rev. **79**, 183 (1950).

Feigenbaum, M., J Stat. Mech. **19**, 25 (1978). (198).

Field, G.B., Astrophys. J. **142**, 531 (1965).

Fischer, D., and Duerbeck, H., *Hubble*, (Springer-Verlag, 1996).

Furth, H.P., Killeen, J., and Rosenbluth, M.N., Phys. Fluids **6**, 459 (1963).

Fyfe, D., Montgomery, D., and Joyce, G., J. Plasma Phys. **17**, 369 (1977).

Ginet, G.P., and Sudan, R.N., Phys. Fluids **30**, 1667 (1987).

Gribbin, J., and Rees, M., *Cosmic Coincidences*, (Bantam, New York, 1989).

Hillas, A.M., Ann. Rev. Astron. Astrophys.**22**, 425 (1984).

Holcomb, K., and Tajima, T., Phys. Rev. D **40**, 3809 (1989).

Horiuchi, R., and Sato, T., Phys. Fluids **31**, 1142 (1988).

Horton, W., Plasma Phys. Contr. Fusion **27**, 937 (1985).

Horton, W., and Ichikawa, Y.H., *Chaos and Structures in Nonlinear Plasmas*, (World Scientific, Singapore, 1996).

Ichimaru, S., *Basic Principles of Plasma Physics, A Statistical Approach* (Benjamin, Reading, 1973).

Ichimaru, S., *Statistical Plasma Physics*, (Addison-Wesley, Reading, 1993).

Ikezi, H., Phys. Fluids **29**, 1764 (1986).

Isichenko, M., and Marnachev, Sov. Phys. JETP **66**, (1987).

Itoh, K., and Itoh, S.-I., to be published in Plasma Phys. Contr. Fusion (1996).

Jackson, J.D. *Classical Electrodynamics* (John Wiley and Sons, Inc., New York, 1975).

Jha, R., Kaw, P.K., Mattoo, S.K., Rao, C.V.S., Saxena, Y.C., and ADITYA Team, Phys. Rev. Lett. **69**, 1375 (1992).

Joyce, G., and Montgomery, D., J. Plasma Phys. **10**, 107 (1973).

Kadomtsev, B.B., *Plasma Turbulence*, (Academic Press, New York, 1965).

Kinney, R.M., Tajima, T., McWilliams, J.C. and Petviashvili, N., Phys. Plasmas **1**, 260 (1994).

Kinney, R.M., McWilliams, J.C., and Tajima, T., Phys. Plasmas **2**, 3623 (1995).

Kishimoto, Y., Tajima, T., M. LeBrun, G. Furnish, and W. Horton, Phys. Plasmas **3**, 1289 (1996).

Kolmogorov, A.N., Dok. Akad. Nauk SSSR [Sov. Phys. Dokl.] **30**, 299 (1941).

Kraichnan, R.H., and Montgomery, D., Rep. Prog. Phys. **45**, 547 (1980).

Krall, N., and Trivelpiece, A., *Principle of Plasma Physics*, (McGraw-Hill, New York, 1973), p. 78.

Kucera, *et al.*, (1997).

Kulsrud, R.M., and Anderson, S., Astrophys. J. **396**, 606 (1992).

Lamb, F.K. *High Energy Transients in Astrophysics (AIP Conference Proceedings 115)* Ed. S.E. Woosley (American Institute of Physics, New York, 1984), p. 179.

Landau, L.D., and Lifshitz, E.M., *Fluid Mechanics*, (Pergamon, New York, 1978).

Longcope, D.W., and Sudan, R.N., Phys. Fluids B $3(8)$, 1945 (1991).

Mathewson, D.S., and Ford, V.L., Mon. Not. R. Astr. Soc. **74**, 139 (1970).

Matthaeus, W.H., Stribling, W.T., Martinez, D., Oughton, S., and Montgomery, D., Physica D **51**, 531 (1991).

Matthaeus, W.H., Stribling, W.T., Martinez, D., Oughton, S., and Montgomery, D., Phys. Rev. Lett. **66**, 2731 (1991).

McWilliams, J.C., *et al.*, *Eddies and Marine Science*, ed. Robinson, A. (Springer-Verlag, Berlin, 1983).

McWilliams, J.C., J. Fluid Mech. **146**, 21 (1984).

Meerson, B., Rev. Mod. Phys. **68**, 215 (1996).

Mikhailovskii, A.B., *Theory of Plasma Instabilities*, (Consultants Bureau, New York, 1974), vol. 1, p. 160.

Mima, K., Horton, W., Tajima, T., and Hasegawa, A., in *Nonlinear Dynamics and Particle Acceleration*, eds. Y.H Ichikawa and T. Tajima (American Institute of Physics, NY, 1990) p. 27.

Moffatt, H.K., J. Fluid Mech. **159**, 359 (1985).

Moffatt, H.K., Zaslavsky, G.M., Conte, P., and Tabor, M., *Topological Aspects of the Dynamics of Fluids and Plasmas* (Kluwer, Dordrecht, 1992).

Moffatt, H.K., in *Turbulence and Nonlinear Dynamics in MHD Flows* (1992).

Mobarry, C.M., and Lovelace, R.V.E., Astrophys. J. **309**, 455 (1986).

Montgomery, D., and Joyce, G., Phys. Fluids **17**, 1139 (1974).

Montgomery, D., Turner, L., and Vahala, G., Phys. Fluids **21**, 757 (1978).

Montgomery, D., Turner, L., Vahala, G., J. Plasma Phys. **21**, 239 (1979).

Montgomery, D., Shan, X., and Matthaeus, W.H., Phys. Fluids A **5**, 2207 (1993).

Oakes, M.E., Michie, R.B., Tsui, K.H., and Copeland, J.E. J. Plasma Phys. **21**, 205 (1979).

Oliveira, S. and T. Tajima, Prev E **51**, 4287 (1995).

Onsager, L., Nuovo Cime. Suppl. **6**(9), 279 (1949).

Orszag, S., and Tang, C., J. Fluid Mech. **90**, 129 (1979).

Parker, E.N., Astron. Astrophys. **191**, 245 (1974).

Parker, E.N., *Spontaneous Current Sheets in Magnetic Fields with applications to Stellar X-rays*, (Oxford Univ. Press, Oxford, 1994).

Parma *et al.*, (1987).

Petviashvili, V., and Yankov, V.V., *Review of Plasma Physics* (Consultants Bureau, New York, 1989), Vol. 14.

Pouquet, A., J. Fluid Mech. **88**, 1 (1978).

Prandtl, L., *Essentials of Fluid Dynamics*, (Blakie, London, 1952).

Pumir, A., and Siggia, E., Phys. Fluids **0**(6), 1606 (1987).

Royer, A.J., Math. Phys. **25**, 2873 (1984).

Sagdeev, R.Z., and Galeev, A.A., *Nonlinear Plasma Theory*, (Benjamin, New York, 1969).

Scalo, J.M., *Protostars and Planets II*, eds. Black, D.C., and Matthews, M.S., (Univ. of Arizona Press, 1985) p. 201.

Seyler, Jr., S.E., Phys. Fluids **19**, 1336 (1976).

Sharp (1985).

Shibata, K. *et al.*, Astrophys. J. **388**, 471 (1989).

Silk, J., *The Big Bang*, (Freeman, New York, 1984).

Spiegel, E.A., in *Stellar Evolution* eds. R.F. Stein and A. Cameron (Plenum, NY, 1966) p. 143.

Spruit, H.C., Ph.D. Dissertation; also Solar Phys. **55**, 3 (1977). (1977).

Steinolfson, R.S., and Tajima, T., Astrophys. J. **322**, 503 (1987).

Stenflo, J.O., Solar Phys. **32**, 41 (1973).

Stribling, W.T., and Mathaeus, W.H., Phys. Fluids B **2**, 1979 (1990).

Swinney, H., and Golub, J.P, Phys. Today **31**, (August) 41 (1978).

Swinney, H., and Golub, J.P., ed. Joseph, D.D., *Hydrodynamic Instabilities and Transition to Turbulence*, "Stability of Fluid Motions," L-L *Fluid Mechanics*, Chap. 3 (198?).

Tajima, T., and Leboeuf, J.-N., Phys. Fluids **23**, 884 (1980).

Tajima, T., *Computational Plasma Physics* (Addison-Wesley, Redwood City, CA, 1989) p. 486.

Tajima, T. and Taniuti, T., Prev A **42**, 3587 (1990).

Tarkenton, G., Ph.D. Dissertation, UT-Austin (1996).

Tasso, H., Trans. Theory Stat. Phys. **16**, 231 (1987).

Taylor, J.B., Phys. Lett. A **40**, 1 (1972).

Taylor, J.B., Phys. Rev. Lett. **32**, 199 (1974).

Taylor, J.B., Rev. Mod. Phys. **58**, 741 (1986).

Thorne, K.S., and MacDonald, D., Mon. Not. R. Astr. Soc. **198**, 339 (1982).

Thorne, K.S., Price, and MacDonald, D. *Black Holes: The Membrane Paradigm*, 1986).

Weinberg, S., *The First Three Minutes* (Basic Books, New York, 1977).

Wilson and Penzias, Nature (1965).

Yusef-Zadesh, F., Morris, M., and Chance, O., Nature **310**, 557 (1984).

Yusef-Zadeh, F., in *The Center of the Galaxy* (Kluwer, Boston, 1988), p. 243.

Zaidman, E., and Tajima, T., Astrophys. J. **338**, 1139 (1989).

Zakharov, Sov. Phys. J. **35**, 968 (1972).

Chapter 3

Fundamental Processes in Hydrostatic and Magnetostatic Objects

3.1 Dynamo

Why are magnetic fields present in planets, stars, accretion disks, galaxies, and universe? In the case of very young objects, such as collapsing protostellar clouds, magnetic fields are introduced from parents objects and we do not need to consider the generation mechanism of magnetic fields in such young objects, as we will discuss it in detail in Chapter 4. However, magnetic fields must be generated somewhere in parents objects. Furthermore, in the case of well-aged objects such as the Earth and the Sun, we know that the magnetic fields are not steady; the memory of the primordial magnetic field has been lost, and the polarity reversal occurs periodically or intermittently. This is the reason why we have to consider the generation mechanism of magnetic fields in *hydrostatic objects*, i.e. *well-aged objects* such as the Sun and stars. The most promising mechanism to generate magnetic fields in these objects is *dynamo* mechanism, which enables magnetic fields to be amplified from very weak seed magnetic fields to the observed strong magnetic fields by the interaction of fluid motion (differential rotation and convective turbulence) with magnetic fields. The dynamo mechanism, in principle, can work also in accretion disks and galactic disks, since there are strong differential rotation and turbulent motions in these objects. The dynamo is the ultimate origin of enormous magnetic activity observed in the Sun, stars, and even of some vigorous activity found in accretion disks and galaxies.

In spite of the recent progress of the dynamo theory and numerical modeling, the dynamo problem still remains unsolved. Therefore, in this chapter, we will discuss some basic points of the dynamo theory, the present status of the theory and numerical modeling of solar and galactic dynamos, and the basic processes in magnetoconvection to understand how magnetic fields interact with convective fluid motion (as a key process of the dynamo theory).

3.1.1 Introduction to Dynamo Theory

3.1.1.1 Kinematic and Dynamic Dynamo

The equation which describes the time evolution of magnetic fields is the induction equation,

$$\frac{\partial \mathbf{B}}{\partial t} = \text{rot}(\mathbf{v} \times \mathbf{B}) + \eta \nabla^2 \mathbf{B}. \tag{3.1.1}$$

If the plasma velocity $\mathbf{v}$ is a given function of $\mathbf{x}$ and t, Eq. (3.1.1) becomes a linear equation. The dynamo in such a case is called *kinematic dynamo* or *linear dynamo*. The kinematic dynamo is applied to the situation where $\mathbf{J} \times \mathbf{B}$ force is much smaller than other forces. However, in many astrophysical situations, this is not necessarily the case, and we must solve the momentum equation including $\mathbf{J} \times \mathbf{B}$ force term

$$\rho \left[\frac{\partial \mathbf{v}}{\partial t} + (\mathbf{v} \cdot \nabla)\mathbf{v} \right] = -\nabla p + \frac{1}{4\pi} \text{rot} \, \mathbf{B} \times \mathbf{B} - \rho \nabla \Psi. \tag{3.1.2}$$

Since this equation includes the term with $\mathbf{B}$, the velocity $\mathbf{v}$ is a function of $\mathbf{B}$. Hence in this case Eqs. (3.1.1) becomes a nonlinear equation. In this sense, the dynamo process determined by (3.1.1) and (3.1.2) is called *nonlinear dynamo* or *dynamic dynamo*.

3.1.1.2 Cowling's Theorem

Let us now prove the fundamental theorem in the dynamo theory, called *Cowling's theorem* (Cowling, 1934); "*A steady axisymmetric magnetic field cannot be maintained.*"

Here, we assume that the axisymmetric magnetic fields have both poloidal (B_r and B_z) and toroidal (B_φ) components in cylindrical geometry.

Consider Ohm's law,

$$\frac{\mathbf{j}}{\sigma} = \mathbf{E} + \frac{\mathbf{v}}{c} \times \mathbf{B}. \tag{3.1.3}$$

From the assumption of axisymmetry, we have a circle path C where $B_r = B_z = 0$ (see Fig. 3.1). Integrating Eq. (3.1.3) on C, we have

$$\oint_c \frac{\mathbf{j}}{\sigma} \, ds = \oint_c \mathbf{E} \cdot ds + \frac{1}{c} \oint_c (\mathbf{v} \times \mathbf{B}) \cdot ds. \tag{3.1.4}$$

Using Stoke's theorem, Eq. (3.1.4) is rewritten as

$$\oint_c \frac{j_\varphi}{\sigma} \, ds = \int_S \text{rot} \, \mathbf{E} \cdot d\mathbf{S} + \frac{1}{c} \oint_c (\mathbf{v} \times \mathbf{B})_\varphi \cdot ds. \tag{3.1.5}$$

The first term in the right-hand side in Eq. (3.1.4) becomes 0, because $\text{rot} \, \mathbf{E} = \partial \mathbf{B}/\partial t = 0$ from steady assumption. The second term becomes

$$\oint_c (\mathbf{v} \times \mathbf{B})_\varphi \cdot ds = \oint_c (v_z B_r - v_r B_z) ds. \tag{3.1.6}$$

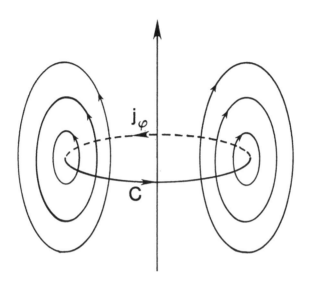

FIGURE 3.1 Axisymmetric magnetic field cannot be maintained (Cowling's anti-dynamo theorem).

Since $B_r = B_z = 0$ on the path C, this term also vanishes. Thus, we have

$$\oint_C \frac{j_\varphi}{\sigma}\, ds = 0, \tag{3.1.7}$$

so that $j_\varphi = 0$ on C. Since j_φ should not vanish on C to maintain the poloidal field, this means that the steady axisymmetric field cannot be maintained.

We can understand the Cowling's theorem also from Eq. (3.1.1) as follows. Assuming incompressibility for fluid, and noting that magnetic field $\mathbf{B}$ is written as

$$\mathbf{B} = \nabla \times (A_\varphi \mathbf{e}_\varphi) + B_\varphi \mathbf{e}_\varphi = \left(\frac{\partial A_\varphi}{\partial z},\ B_\varphi,\ -\frac{1}{r}\frac{\partial}{\partial r}(r A_\varphi) \right), \tag{3.1.8}$$

Eq. (3.1.1) is rewritten as

$$\frac{\partial A_\varphi}{\partial t} = -\frac{\mathbf{v}_p}{r} \cdot \nabla(r A_\varphi) + \eta \left(\Delta - \frac{1}{r^2} \right) A_\varphi, \tag{3.1.9}$$

133

$$\frac{\partial B_\varphi}{\partial t} = r(\mathbf{B}_p \cdot \nabla)\left(\frac{v_\varphi}{r}\right) - r(\mathbf{v}_p \cdot \nabla)\left(\frac{B_\varphi}{r}\right) + \eta\left(\Delta - \frac{1}{r^2}\right)B_\varphi, \qquad (3.1.10)$$

where $\mathbf{v}_p$ and $\mathbf{B}_p$ are poloidal components of velocity and magnetic field vectors. These equations show that B_φ can be generated from B_p [Eq. (3.1.10)], but $A_\varphi(B_p)$ cannot be generated from B_φ if axisymmetry $(\partial/\partial\varphi = 0)$ is assumed [Eq. (3.1.9)].

3.1.1.3 Parker's (1955) Suggestion

These considerations suggest that *nonaxisymmetric velocity field must be present in order to maintain a steady axisymmetric magnetic field*, which led Parker (1955) to suggest the following idea. That is, in the solar convection zone, there are a lot of rising convective elements or blobs. When the blob rises, it expands due to the decrease of ambient gas pressure, and the expansion leads to the rotation of the blob due to Coriolis force. This rotation of the blob produce the poloidal field from the toroidal field as illustrated in Fig. 3.2. Each blob produces a small loop of magnetic field with nonvanishing projection on the meridional plane. This projected poloidal loop is represented by a localized vector potential. The sum of many such local azimuthal vector potentials contributes an overall mean azimuthal vector potential, representing the mean poloidal field.

The scenario was justified by considering the cyclonic displacement of the fluid and field to take place in brief bursts, followed by an extended period of time in which the small scale magnetic fields decayed away tearing on the manifold.

Parker modeled the net effect of many convective cells expressing the result as

$$\frac{\partial A_\varphi}{\partial t} \cong \alpha B_\varphi. \qquad (3.1.11)$$

The resulting MHD equation for A_φ is, then, including the contribution of the cyclonic convection

$$\frac{\partial A_\varphi}{\partial t} + \frac{\mathbf{v}_p}{r}\cdot\nabla(rA_\varphi) = \alpha B_\varphi + \eta\left(\Delta - \frac{1}{r^2}\right)A_\varphi. \qquad (3.1.12)$$

The effect represented by the term αB_φ is called α *effect*, which enables to generate poloidal fields B_p from toroidal fields B_φ. On the other hand, the effect to produce B_φ from B_p by the differential rotation (the first term in Eq. (3.1.10)) is called ω *effect*. A formal treatment can be found in Parker (1979). The dynamo including both α and ω effects is called $\alpha\omega$ *dynamo*, while the dynamo including only α effect is called α^2 *effect*.

3.1.1.4 Turbulent Dynamo

Steenbeck, *et al.* (1966) developed the theory of the turbulent dynamo, which is often referred to as *mean-field electrodynamic approach*. Consider $\mathbf{B} = \mathbf{B}_0 + \mathbf{b}, \mathbf{V} = \mathbf{V}_0 + \mathbf{v}$, where the quantities with subscript 0 denote the mean field variables, and $\mathbf{b}$ and $\mathbf{v}$ are small deviations from the mean field values. We assume $\bar{\mathbf{b}} = \bar{\mathbf{v}} = 0$, where a bar means a spatial (temporal)

134

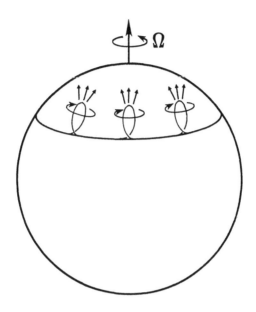

FIGURE 3.2   Parker (1955)'s cyclonic dynamo.

average. Inserting $\mathbf{B} = \mathbf{B}_0 + \mathbf{b}$ and $\mathbf{V} = \mathbf{V}_0 + \mathbf{v}$ into Eq. (3.1.1) and taking spatial average, we have

$$\frac{\partial \mathbf{B}_0}{\partial t} = \mathrm{rot}(\mathbf{V}_0 \times \mathbf{B}_0) + \mathrm{rot}(\overline{\mathbf{v} \times \mathbf{b}}) + \eta \nabla^2 \mathbf{B}_0, \qquad (3.1.13)$$

$$\frac{\partial \mathbf{b}}{\partial t} = \mathrm{rot}(\mathbf{v} \times \mathbf{B}_0 + \mathbf{V}_0 \times \mathbf{b} + \mathbf{v} \times \mathbf{b} - \overline{\mathbf{v} \times \mathbf{b}}) + \eta \nabla^2 \mathbf{B}_0. \qquad (3.1.14)$$

Now, we neglect the terms $\mathbf{v} \times \mathbf{b} - \overline{\mathbf{v} \times \mathbf{b}}$ in Eq. (3.1.14). This approximation is called *first order smoothing* or *quasilinear approximation*. Furthermore, if $\mathbf{V}_0 = 0$, Eq. (3.1.14) reduces to

$$0 = \mathrm{rot}(\mathbf{v} \times \mathbf{B}_0) + \eta \nabla^2 \mathbf{b} \qquad (3.1.15)$$

in steady state. This equation is rewritten as

$$0 = (\mathbf{B}_0 \cdot \nabla)\mathbf{v} + \eta \nabla^2 \mathbf{b}. \qquad (3.1.16)$$

135

Hence,

$$\mathbf{b} \sim \frac{l^2}{\eta} (\mathbf{B_0} \cdot \boldsymbol{\nabla}) \mathbf{v}.$$

Then,

$$\overline{\mathbf{v} \times \mathbf{b}} \sim B_0 v^2 l / \eta \sim \alpha B_0,$$

where

$$\alpha \sim \frac{l}{\eta} v^2. \tag{3.1.17}$$

We write

$$\overline{\mathbf{v} \times \mathbf{b}} = \alpha \mathbf{B_0}, \tag{3.1.18}$$

so that Eq. (3.1.13) becomes

$$\frac{\partial \mathbf{B_0}}{\partial t} = \mathrm{rot}(\mathbf{V_0} \times \mathbf{B_0}) + \mathrm{rot}(\alpha \mathbf{B_0}) + \eta \nabla^2 \mathbf{B_0}. \tag{3.1.19}$$

This is essentially the same as the Eq. (3.1.12).[1]

3.1.1.5 Dynamo Waves

We consider the local Cartesian coordinate on the surface of the star as shown in Fig. 3.3. Let us assume that $\partial/\partial y = 0$, $v_x = v_z = 0$. Eqs. (3.1.10) and (3.1.12) are written as

$$\left[\frac{\partial}{\partial t} - \eta \nabla^2 \right] B_y = \frac{dv_y}{dz} \frac{\partial A_y}{\partial x}, \tag{3.1.20}$$

$$\left[\frac{\partial}{\partial t} - \eta \nabla^2 \right] A_y = \alpha B_y, \tag{3.1.21}$$

where we used $\varphi \to y$. Assuming that

$$B_y \propto \exp\left[\omega t + i(k_x x + k_z z) \right], \tag{3.1.22}$$

we obtain the following dispersion relation

$$\omega = -\eta k^2 \pm (1 + i) \left[\frac{\alpha k_x}{2} \frac{dv_y}{dz} \right]^{1/2} \tag{3.1.23}$$

$$= \eta k^2 \left[-1 \pm (1 + i) N_D^{1/2} \right], \tag{3.1.24}$$

where $k^2 = k_x^2 + k_z^2$ and

$$N_D \equiv \frac{\alpha k_x}{2\eta^2 k^4} \frac{dv_y}{dz} \tag{3.1.25}$$

---

[1]The $\alpha\omega$ dynamo is not the only possible turbulent dynamo. There is a different type of turbulent dynamo, e.g., cross-helicity dynamo (see Yoshizawa and Yokoi, 1993; Yokoi, 1996).

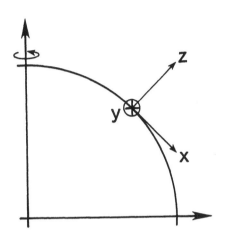

FIGURE 3.3 Local Cartesian coordinate in the Sun (for understanding propagation of dynamo waves).

is called *dynamo number*. The physical meaning of N_D is

$$N_D \simeq \left(\frac{\tau_{\text{decay}}}{\tau_{\text{dynamo}}}\right)^2, \tag{3.1.26}$$

where τ_{decay} and τ_{dynamo} are time scale of field decay and dynamo action;

$$\tau_{\text{decay}} = (\eta k^2)^{-1}, \tag{3.1.27}$$

$$\tau_{\text{dynamo}} = \left(\alpha k_x \frac{dv_y}{dz}\right)^{-1/2}. \tag{3.1.28}$$

The dispersion relation (3.1.23) indicates that we have a wave with the phase velocity

$$v_{ph} = -\frac{\text{Im}(\omega)}{k} = -\left[\frac{\alpha k_x}{2k^2}\frac{dv_y}{dz}\right]^{1/2} \tag{3.1.29}$$

137

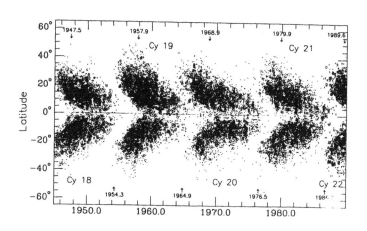

FIGURE 3.4 Butterfly diagram for latitudinal distribution of sunspots as a function of time (Harvey, 1993).

if $dv_y/dz > 0$. This wave is called *dynamo wave* (Parker, 1955). If $|N_D| > 1$, there are growing solutions. When $N_D > 0$ $(dv_y/dz > 0)$, the wave propagates northward ($v_{ph} < 0 \rightarrow$ negative x). On the other hand, when $N_D < 0$ $(dv_y/dz < 0)$, the wave propagate southward. Since observation of solar cycle shows southward propagation (migration of sunspot, see Fig. 3.4), this means $dv_y/dz < 0$, or the rotational velocity decreases outward. [Yoshimura (1975) found that the dynamo waves propagate along the iso-rotation surface.]

In order to have a stable amplitude, we assume $\text{Re}(\omega) = 0, \text{Im}(\omega) \neq 0$. Then, we have

$$\eta k^2 = \left[\frac{\alpha k_x}{2} \frac{dv_y}{dz} \right]^{1/2}.$$

Hence, the period $\tau (\simeq$ time scale of magnetic field amplification) is

$$\tau \simeq \left[\frac{\alpha k_x}{2} \frac{dv_y}{dz} \right]^{-1/2} = \frac{1}{\eta} \left(\frac{\lambda}{2\pi} \right)^2. \tag{3.1.30}$$

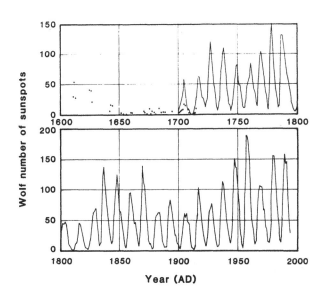

FIGURE 3.5 The Wolfe number of sunspots. The sold curves are the data from the Zurich Observatory and the Sunspot Number Data Center (at Bruxelles). The dots are after J.A. Eddy, Science **192**, 1189 (1976).

If we take $\eta = \eta_t \sim l v_0$, $l \sim 3 \times 10^8$cm, $v_0 \sim 10^4$cm/s, $\lambda \sim 5 \times 10^{10}$cm, we obtain

$$\tau \sim 20\,\text{yr}.$$

This is comparable to the observed period 22 yr of solar cycle (Fig. 3.5).

3.1.1.6 Results of Kinetic Dynamo*

In the previous subsection, we have studied local properties of *kinematic $\alpha\omega$ dynamo*, by deriving the local dispersion relation Eq. (3.1.24). In order to examine whether the kinematic $\alpha\omega$ dynamo is actually working in the Sun and galaxies, we need to solve a global eigenvalue problem of the kinematic $\alpha\omega$ dynamo equations (3.1.10) and (3.1.12), assuming appropriate boundary conditions, and need to compare solutions with observed global patterns of magnetic fields in the Sun and galaxies. Here, we show some examples of such solutions for

139

the Sun and galaxies (See Parker, 1979; Krause, Radler, and Rudiger, 1993; Proctor and Gilbert, 1994; for review of this kind of work).

Figure 3.6 shows one example of global solution of the kinematic $\alpha\omega$ dynamo equation in a spherical geometry, assuming appropriate functional form of α and ω (Stix, 1976). We see that global patterns of magnetic fields, their oscillatory behavior, and the latitudinal migration of generated fields are similar to those of the actually observed solar magnetic fields.

Figure 3.7 shows the theoretical butterfly diagram based on the global solution of the kinematic $\alpha\omega$ dynamo equation (Yoshimura, 1975). Compare this figure with the observed *butterfly diagram* in Fig. 3.4. The observed pattern of 22-yr butterfly diagram is remarkably well reproduced. Yoshimura (1975) assumed two zones for α-distribution based on the hypothetical global convection pattern, and succeeded to reproduce poleward migration of magnetic flux near the pole region. This is also consistent with observed poleward migration of dark filament. Yoshimura (1978) then extended the model to the somewhat ad hoc nonlinear dynamo, assuming the artificial nonlinear term representing the time delay of the back reaction of generated magnetic field to the convection motion. This nonlinear model showed the intermittent magnetic field generation such as long term weak field phase followed by normal periodic (strong field) phase, which reminds us the *Maunder minimum* (Eddy, 1976).

Extending the Parker (1971)'s *Galactic dynamo model*, Sawa and Fujimoto (1986) showed that not only the circular field but also the *bi-symmetric-spiral* (BSS) field can be excited by dynamo action. Figure 3.8 shows one of their results indicating 3D distribution of magnetic fields in a galaxy, which explains the observed BSS magnetic field configuration such as shown in Fig. 3.9. Chiba and Tosa (1989) obtained the similar results, but from different approach. That is, they solved the dynamo equation coupled with the effect of spiral arm where the gas and magnetic field are compressed due to the galactic shock, and found that the dynamo mechanism is actually enhanced in the spiral arm. The reader should be referred to Kronberg (1994) and Beck *et al.* (1996) for a review of more recent works in this field.

3.1.1.7　Present Status of Dynamo Theory*

The kinematic solar dynamo models have reproduced many observed properties of global patterns of solar magnetic fields as described in Sec. 3.1.1.6. Hence, before the mid '70s people had thought that the solar dynamo theory was successful with only minor disagreements between theoretical modeling and observations, which would be resolved with better knowledge of the solar differential rotation and with a realistic turbulence model (Schmitt, 1993). However, it has become clear that there are some fundamental difficulties in the solar dynamo theory.

First, Parker (1975) pointed out that an isolated magnetic flux tube in the convection zone rises rapidly due to its strong magnetic buoyancy (see Sec. 3.2) with a time scale of the order of a month, much too short for the dynamo to support the field. Evidence that changes in X-ray emission properties of stars at the low-mass end of the main sequence is coincident with the onset of fully-convective interiors (Rosner and Vaiana, 1980) led to suggestions that the magnetic activity of solar-type stars was not tied to field amplification

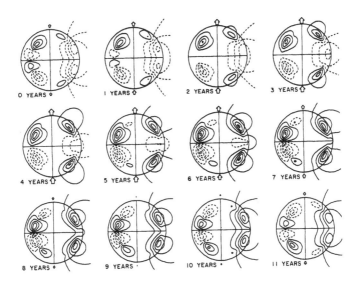

FIGURE 3.6 Oscillatory kinematic $\alpha\omega$ dynamo. The meridional cross-sections show contours of constant toroidal field strength on the *left*, and poloidal lines of force on the *right*. *Arrows* indicate strength and sign of the polar field, the time scale is adjusted to 11 years for each half-cycle. (from Stix, 1976).

zone itself, but rather to dynamo activity in a "shell dynamo," possibly located at the base of the convection zone or below (Spiegel and Weiss, 1980; Rosner and Vaiana, 1980). More recently, *helioseismological* observations have revealed that the main convection zone rotates with the solar surface with no significant radial differential rotation (Libbrecht 1988, Gough *et al.*, 1993; Fig. 3.10), inconsistent with the predicted differential rotation from the kinematic dynamo theory (Sec. 3.1.1.6).

From these difficulties, many people abandoned the notion of the *convection zone dynamo* and instead have begun to consider the *overshoot layer dynamo*, i.e., the dynamo operating in an overshoot layer at the base of the convection zone, where magnetic buoyancy is small and there is a significant radial gradient of rotation (e.g., Spiegel and Weiss, 1980; Schmitt and Rosner, 1983; Schüssler, 1983; Schmitt, Rosner, and Bohn, 1984).

Furthermore, nonlinear studies (e.g., Cattaneo and Vainshtein, 1991; Vainshtein and Rosner, 1991; Vainshtein *et al.*, 1991; Rosner and Weiss, 1992; Cattaneo, Hughes, and Kim,

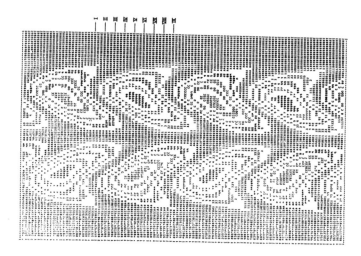

FIGURE 3.7   Yoshimura (1975)'s solar dynamo model.

1996; Cattaneo and Hughes, 1996) have shown that the back-reaction of generated magnetic fields is important at much lower field strengths than the equipartition strength. That is, the precise nature of the back-reaction is not simply suppression of fluid motions, but rather a much more subtle effect on the "memory" of the fluid motions, i.e., what is changed is the Lagrangian fluid correlation function, in such a way that the Eulerian velocity correlation function is hardly changed at all (Cattaneo 1994). This means that the basic assumption of the kinematic dynamo theory is no longer justified.

Schüssler (1980, 1983) questioned the validity of mean field dynamo, and proposed the *flux tube dynamo*, where the magnetic field is localized in a flux tube which is different from a passive, diffuse field assumed in the mean field dynamo theory. This view originally came from the observed properties of solar magnetic fields such that magnetic fields are concentrated in intense flux tubes and that the evolution of active regions is well understood by the flux tube picture. In fact, recent calculations on the rise of flux tubes in the convection zone (e.g., Moreno-Insertis, 1986; D'Silva and Choudhuri, 1991; Fan *et al.*, 1993) explain many observed properties of sunspots such as sunspot tilts, east-west asymmetry, and so on,

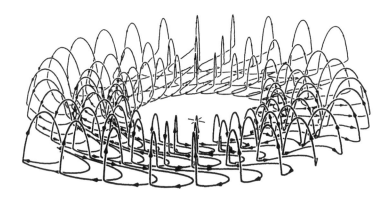

FIGURE 3.8 Fujimoto and Sawa (1990)'s galactic dynamo model.

supporting the flux tube picture. This picture is also consistent with the general tendency for a filamentary structure of high Reynolds number, high β MHD turbulence which is discussed in Secs. 2.3 and 2.4.

In the field of basic dynamo research, it has been shown that there are two types of dynamo. One is the *fast dynamo* (e.g., Finn and Ott, 1988) [2], and the other is the *slow dynamo*. This classification is in some sense similar to the *fast reconnection* (e.g., Petschek model) vs. the *slow reconnection* (e.g., Sweet-Parker model), i.e., the former takes place in a time scale independent of the magnetic Reyolds number, while the latter occurs in a time scale which depends on the resistivity. Both fast dynamo and fast reconnection are characteristic features of magneto-plasmas at high magnetic Reynolds number ($R_m \gg 1$), because the magnetic field tends to have intermittent (or fractal) structures at $R_m \gg 1$ so that small scale structures necessary for dissipation can be created in such high R_m plasmas.

---

[2]The original prototype of the fast dynamo is "Stretch-Two-Fold" Dynamo discussed in the paper by Vainshtein and Zeldovich (1972). See also Vainshtein, Sagdeev, and Rosner (1996) for more recent work on this subject.

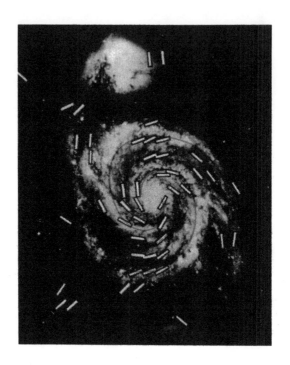

FIGURE 3.9   Magnetic fields in M51 (Tosa and Fujimoto, 1978).

The fast dynamo can occur when the flow has the chaotic structure; a simplest example of such flow is the 3D, incompressible, periodic *"ABC flow"* (named after Arnold, Beltrami, and Childress). (See also Sec. 5.2.) By performing numerical simulations of this ABC flow, Galloway and Frisch (1986) observed the exponential growth of initial magnetic field for certain ranges of R_m, and the field growth was found to be concentrated in the regions of chaotic streamlines which are concentrated on the separatices, where

$$\nabla \left(\frac{p + v^2}{2} \right) = \omega \times \mathbf{v} = 0.$$

Here, p is the gas pressure and $\omega = \nabla \times \mathbf{v}$ is the vorticity.

Although the galactic dynamo theory has much developed in recent years as discussed in previous section, it has difficulties similar to those of the solar dynamo theory, i.e., the assumption of kinematic and/or mean field dynamo might not be correct. Further, the galactic dynamo theory has its own difficulties; it has to take into account the effects of supernovae, multi-phase gases, cosmic rays, and so on. The nonlinear modeling including these processes inherent to the galactic dynamo is highly desired in the future. (Note that the time scale problem is much easier in the galactic dynamo than in the solar dynamo; i.e., in the former, the convection or the Alfvén time is comparable to the rotation time, while in the latter, the convection time is much longer than the rotation time. In the solar dynamo problem, even the modeling of the non-magnetic convection is a difficult task. Absence of this difficulty is an advantage of the galactic dynamo problem.) The dynamo action during or before galaxy formation will be discussed in Chap. 5.

As for the nonlinear modeling of the accretion disk dynamo, there are some developments recently. Brandenburg et al. (1996) found that the magnetic field is maintained in accretion disks at the level of local plasma β of 10–100 as a result of nonlinear evolution of the Balbus-Hawley instability (Sec. 4.2; Hawley et al., 1996). This saturation of the β value is consistent with MHD turbulence studies (Sec. 2.4; Kinney et al., 1995) that show, broadly speaking, two attractors, one in a *high β attractor* ($\sim 10 - 100$) and the other in a *low β attractor* (~ 1). Once the system stands with initial conditions in one basin, it will end up with an attractor of that basin. This kind of nonlinear study has just begun, and more work is needed to establish the accretion disk dynamo theory.

As for the *geo-dynamo*, on the other hand, Glatzmaier and Roberts (1995) recently succeeded to reproduce the reversal of geomagnetic fields, carrying out nonlinear 3D Bousinesque MHD numerical simulations (see also Kageyama et al., 1995 for a related work). The reason of this success is that the geo-dynamo occurs in a low R_m conducting liquid core; i.e., the geo-dynamo belongs to the slow dynamo.

In summary, although there were some outstanding developments especially on the nonlinear modeling of the geo-dynamo, the current situation of the solar (as well as galactic and accretion disk) dynamo theories is far from a final resolution.

3.1.2 Magneotconvection

As we discussed in the previous section, it is now believed that the dynamo in the Sun is operating near the base of the convection zone or, more likely, in a thin region of convective overshoot. The generated magnetic flux tube rises due to both magnetic and thermal buoyancies to emerge the surface of the Sun, making sunspots and associated active regions. The surface of the Sun (just below the photosphere) is the most strongly convective layer, so that the interaction between magnetic flux tubes and convection occurs most strongly there, generating various activity in the upper atmosphere. Altogether, we find that the convection in the presence of magnetic fields, i.e. *magnetoconvection*, is a very fundamental process for understanding the dynamo, the formation of flux tubes, and the origin of activity in the Sun (and possibly other astrophysical objects such as accretion disks). Hence, in this subsection, we will study the magnetoconvection, starting from the quick survey of non-magnetic convection.

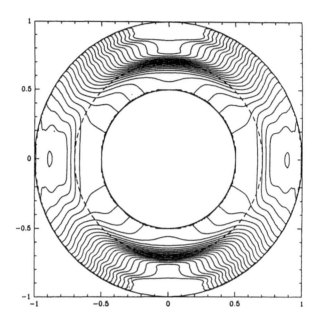

FIGURE 3.10 Solar internal rotation based on helioseismology (Gough *et al.*, 1993).

3.1.2.1 Convective Instability

Let's first consider the condition for the occurrence of thermal convection when there is no magnetic field. Suppose that a fluid element rises over an infinitesimally small distance dz through the gravitationally stratified gas layer in hydrostatic equilibrium. The blob expands because of the decrease in gas pressure in the external gas layer, so that the density of the blob decreases. If the density of the blob becomes smaller than that of the external gas, then the buoyancy force acts upward, and hence the blob continues to rise, i.e. the *convective instability* sets in.

Now calculate the density and gas pressure in the blob;

$$\rho_{in}(z + dz) = \rho(z) + \delta\rho, \tag{3.1.31}$$

$$p_{in}(z + dz) = p(z) + \delta p, \tag{3.1.32}$$

where $\delta\rho < 0, \delta p < 0$ for $dz > 0$, and the physical quantities with δ correspond to the variation of quantities in the blob due to expansion. The relation between $\delta\rho$ and δp is found from conservation of entropy as

$$\delta p/p = \gamma \delta \rho/\rho. \tag{3.1.33}$$

On the other hand, the density and gas pressure of the external gas layer at $z + dz$ are written as

$$\rho_{ex}(z + dz) = \rho(z) + d\rho, \tag{3.1.34}$$

$$p_{ex}(z + dz) = p(z) + dp, \tag{3.1.35}$$

where

$$dp = \frac{dp}{dz}dz, \qquad d\rho = \frac{d\rho}{dz}dz. \tag{3.1.36}$$

Remember that the external gas layer is in hydrostatic equilibrium,

$$\frac{dp}{dz} = -\rho g. \tag{3.1.37}$$

From pressure balance between inside and outside of the blob, we have

$$\delta p = dp, \tag{3.1.38}$$

i.e. from (3.1.33) and (3.1.37)

$$\frac{\gamma p}{\rho}\delta\rho = \frac{dp}{dz}dz, \tag{3.1.39}$$

so that

$$\delta\rho = \frac{\rho}{\gamma p}\frac{dp}{dz}dz. \tag{3.1.40}$$

The condition for the occurrence of the convection is

$$\delta\rho < d\rho. \tag{3.1.41}$$

Using (3.1.36) and (3.1.40), this becomes

$$\frac{dp}{dz} < \frac{\gamma p}{\rho}\frac{d\rho}{dz}. \tag{3.1.42}$$

If we use the equation of state $p = \rho R_g T/\mu$ and (3.1.37), the equation (3.1.42) can be further rewritten as

$$-\frac{dT}{dz} > \frac{(\gamma - 1)}{\gamma}\frac{\mu g}{R_g} = -\left(\frac{dT}{dz}\right)_{ad}, \tag{3.1.43}$$

where $(dT/dz)_{ad}$ is the adiabatic temperature gradient. This equation (3.1.43) is called the *Schwarzschild criterion* for *convective instability*.

If we remember that the entropy S is

$$S \propto \ln p\rho^{-\gamma}, \tag{3.1.44}$$

147

the equation (3.1.42) or (3.1.43) is rewritten also as

$$dS/dz < 0. \tag{3.1.45}$$

That is, as is well known, the necessary condition for occurrence of thermal convection is that *entropy should decrease with height*.

Next, let's derive the growth rate of the convective instability. The equation of motion of the blob raised by the infinitesimally small height dz is written as

$$\rho \frac{d^2 dz}{dt^2} = \Delta \rho g, \tag{3.1.46}$$

where

$$\Delta \rho = \delta \rho - d\rho. \tag{3.1.47}$$

From (3.1.40), we find

$$\Delta \rho = \left(\frac{\rho}{\gamma p} \frac{dp}{dz} - \frac{d\rho}{dz} \right) dz. \tag{3.1.48}$$

Then, (3.1.46) becomes

$$\frac{d^2 dz}{dt^2} = -N^2 dz, \tag{3.1.49}$$

where

$$N^2 = \frac{g}{T} \left(\frac{dT}{dz} - \left(\frac{dT}{dz} \right)_{ad} \right). \tag{3.1.50}$$

Consequently, if $N^2 < 0$, we have an instability, $dz \propto \exp(iNt)$. The growth rate is of order of the inverse of the gravitational free fall time over the temperature scale length if

$$|dT/dz| \gg |dT/dz|_{ad}. \tag{3.1.51}$$

On the other hand, if $N^2 > 0$, the instability does not occur. Instead, the blob undergoes the oscillation which produces the *internal gravity wave*. N is called the *Brunt Väisälä frequency*. Although the above derivation is heuristic, this is exactly valid for a short wavelength perturbation (i.e. $k_x H \gg 1$; Defouw, 1976).

3.1.2.2 Thermal (Rayleigh-Bernard) Convection

In actual systems, such as laboratory fluid convection experiments, *viscosity* and *heat conduction* significantly affect the onset of the convection. The viscosity decelerates a rising element by viscous force, while the heat conduction tends to eliminate the temperature difference between the blob and the external medium, thus tends to eliminate the buoyancy force. The condition for the onset of the thermal convection is determined by the *Rayleigh number* (see Sec. 2.7.1)

$$R = \frac{g \alpha \Delta T d^3}{\kappa \nu}, \tag{3.1.52}$$

where α is the volume coefficient

$$\alpha = -\frac{1}{\rho}\frac{d\rho}{dT},$$

(3.1.53)

is the coefficient of expansion, d is the depth of the gas layer, κ is the thermal conductivity, ν is the viscosity. The physical meaning of the Rayleigh number is

$$R \simeq \frac{\tau_{\text{cond}}\tau_{\text{vis}}}{\tau_{\text{buoy}}^2} = \frac{\text{heat} - \text{viscous time scale}}{\text{buoyancy time scale}},$$

(3.1.54)

where

$$\tau_{\text{cond}} = d^2/\kappa,$$

(3.1.55)

$$\tau_{\text{vis}} = d^2/\nu,$$

(3.1.56)

$$\tau_{\text{buoy}} = \left(\frac{d}{(\delta\rho/\rho)g}\right)^{1/2}.$$

(3.1.57)

Only when R is larger than the critical Rayleigh number R_c, the convection can occur. The actual value of R_c depends on the boundary condition, the shape of the cell pattern, etc., but the minimum R_c is about 650 (Chandrasekhar, 1961).

3.1.2.3 Solar Convection

At the surface of the Sun, we can see the convection cell pattern, called *granules*. The "typical" size of the cell is about 1000 km, and the convection velocity is of order of a few km/s. Since both the Rayleigh number and Reynolds number are much larger than 1, the granular convection is the turbulent thermal convection. Figure 3.11 shows the white light photograph of granules, and Fig. 3.12 denotes the power spectrum of the convection pattern, showing that the spectrum obeys the Kolmogorov law [Spruit *et al.* (1990) for a review of solar convection].

Since the size of the granule is larger than the pressure scale height of the photosphere, the density variation is large inside the convection cell. That is, the convection is the *compressible* convection. Recently, the numerical simulation of compressible convection showed that the supersonic convection flow (and associated shock) can occur near the top of the convection layer [Cattaneo *et al.* (1990) for 2D simulation and Malagoli *et al.* (1990) for 3D simulation]. The existence of such supersonic convection flow has been actually observed in granule convection (Nesis *et al.*, 1992).

3.1.2.4 Incompressible Magnetoconvection

Let us now consider the simplest situation of the *magnetoconvection*, to see how magnetic fields are stretched and deformed by the convection, when magnetic fields are weak (i.e. magnetic energy is much smaller than the convection kinetic energy). Taking a two-dimensional

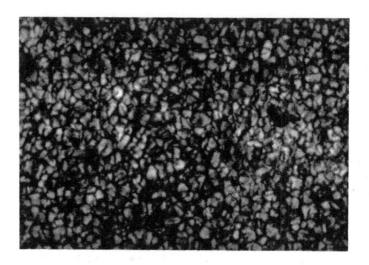

FIGURE 3.11   Solar convection, *granules* (Kawaguchi, 1980).

Cartesian box and *Buoussinesq fluid*,[3] Galloway and Weiss (1981) numerically studied this problem. Figure 3.13 shows one of their results, and show typical evolution of the magneto-convection. From this figure, we can see two important effects;

(1) *magnetic flux expulsion* (from the convection cell),

(2) *magnetic flux concentration* (near the boundary of the cell).

The flux concentration near the boundary can be examined using even simpler kinematic theory as in the following. Assume that the steady convection with the velocity field

$$v_x = -Ux/L, \tag{3.1.58}$$

$$v_z = Uz/L, \tag{3.1.59}$$

in the presence of uniform magnetic field $\mathbf{B} = (0, 0, B_z)$ (see Fig. 3.14). This velocity fields represents the flow towards the x-axis, and produces strong B_z along the x-axis. The

---

[3] *Buoussinesq fluid* is an approximate incompressible fluid though the effect of buoyancy (or temperature perturbation) is included.

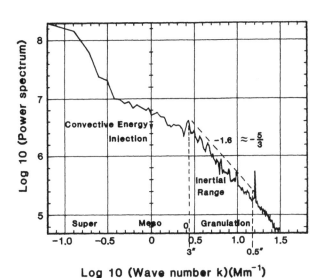

FIGURE 3.12 Power spectrum of granular convection turbulence (from Muller, 1989).

induction equation with diffusion term is

$$\frac{\partial B_z}{\partial t} = -\frac{\partial}{\partial x}(V_x B_z) + \eta \frac{\partial^2 B_z}{\partial x^2}. \tag{3.1.60}$$

Then the solution of this equation with prescribed velocity field (3.1.58) and (3.1.59) becomes

$$B_z = B_0 R_m^{1/2} g(t) \exp\left[-\frac{R_m}{2}\left(\frac{x}{L}\right)^2 g^2(t)\right], \tag{3.1.61}$$

where

$$g(t) = \left[1 + (R_m - 1)\exp(-\frac{2U}{L}t)\right]^{-1/2}, \tag{3.1.62}$$

and $R_m = UL/\eta$ is the magnetic Reynolds number. As time goes on, g tends to be close to 1. Hence the final field strength near $x = 0$ becomes

$$B_{z,\text{max}} \simeq B_0 R_m^{1/2}, \tag{3.1.63}$$

151

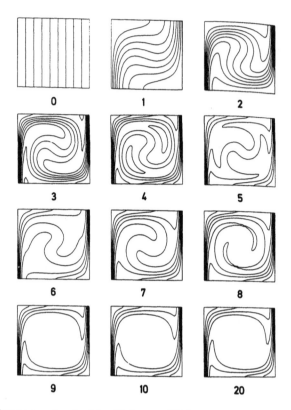

0	1	2
3	4	5
6	7	8
9	10	20

FIGURE 3.13 2D Buoussinesq convection simulation (Galloway and Weiss, 1981).

as long as the magnetic energy is smaller than the flow kinetic energy. The effective width of the concentrated flux is $\sim LR_m^{-1/2}$. In the case of 2D axisymmetry, the final field strength becomes

$$B_{z,\text{max}} \simeq B_0 R_m, \tag{3.1.64}$$

which is much larger than 2D Cartesian case because of geometrical converging effect.

In the framework of 2D axisymmetric Buoussinesq magnetoconvection, Galloway and Moore (1979) studied the dependence of the convection pattern as a function of magnetic field strength, or the *Chandrasekhar number*

$$Q = \frac{B_0 d^2}{4\pi \rho_0 \nu \eta}. \tag{3.1.65}$$

(This is the same as the square of the *Hartman number*.) In their simulation, the Reyleigh number is fixed to be 2–50 times of the critical Reyleigh number, and the parameter Q was changed from 0.001 to 7000. They found that the solutions are classified into the following four regime (see also Fig. 3.15);

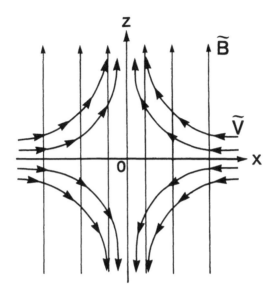

FIGURE 3.14 Magnetic field concentration by convection flow.

1) *kinematic regime* $(Q = 0.01)$
2) *dynamically limited regime* $(Q = 100)$
3) *mixed steady-oscillatory regime* $(Q = 1500)$
4) *oscillatory regime* $(Q = 7000)$

Oscillation in the regime 4) is, of course, a result of the restoring magnetic tension force. The reader should be referred to Proctor and Weiss (1982), Hughes and Proctor (1988) for a review of the related works.

3.1.2.5 Compressible Magnetoconvection*

Recent development of supercomputer has enabled to study the full 2D or 3D compressible magnetoconvection (Hulburt and Toomre, 1988; Cattaneo *et al.*, 1990; Brandenburg *et al.*, 1990) even including dynamo effect. There are a number of important effects coming from compressibility. One is the evacuation of magnetic flux tube and resulting magnetic flux concentration by gas pressure (Hanami and Tajima, 1991); (Fig. 3.16). This effect is very

153

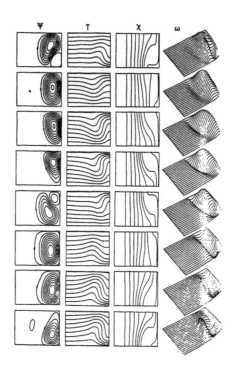

FIGURE 3.15   2D Boussinesq convection simulation (Galloway and Moore, 1979).

important in relation to the origin of intense magnetic flux tube in the solar photosphere (see next sub-sub-section). There is another effect of compressibility, i.e. magnetic buoyancy. The effect of magnetic buoyancy will be discussed in Sec. 3.2 in detail. In relation to the dynamo effect, Brandenburg *et al.* (1990) found from 3D MHD simulation of magnetoconvection under influence of rotation that the α effect is opposite to that expected from the first order smoothing approximation. This subject, including dynamo effect, is one of the most important subjects in solar and astrophysical MHD.

3.1.2.6   Formation of Intense Flux Tubes*

One of the most important discovery in solar physics in recent 20 years is that the general magnetic field in the solar photosphere is concentrated into kilo gauss small scale intense flux tubes with a size of only 100–200 km (Stenflo, 1973; Parker, 1979; Spruit and Roberts, 1983; Solanki, 1994).

Why and how are solar magnetic fields concentrated into such small scale intense flux

154

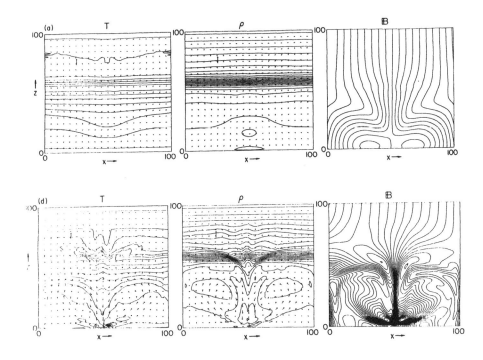

FIGURE 3.16 Compressible magnetoconvection (Hanami and Tajima, 1991).

tubes? Spruit and Zweibel (1979) attacked this problem and found that the *flux tube convective instability* occurs in vertical fields in the superadiabatic gas layer, if initial β is larger than $\beta_c \sim 1.5 - 1.8$, the precise value of which depends on the depth of the lower boundary. Only the tube with $\beta < \beta_c$ is stable for the instability. This means that any weak vertical field in the superadiabatic layer cannot be present and will be concentrated into isolated, intense flux tube, thus explaining the observations. Spruit (1979) then showed that the tube in the final equilibrium state has actually a kilo Gauss field, and called the process of the intense flux tube formation the *convective collapse* (Parker, 1978; Webb and Roberts, 1978; Unno and Ando, 1979).

Nonlinear development of the flux tube convective instability has been numerically studied by Hasan (1985) who claimed that the final state is an oscillatory state. However, Takeuchi (1993) recently showed that if the proper treatment is made for the lower boundary of numerical simulation, then the final state is not an oscillatory state, but a quasi-static equilibrium state.

Knölker *et al.* (1988) developed the 2D numerical model of the intense flux tube. Shibata

155

et al. (1990c, 1991) studied the convective collapse of emerging magnetic flux tubes, and found that the emerging equipartition field ($\sim 500\,\mathrm{G}$) can be intensified to a kG field by the convective collapse a few – 10 minutes after emergence of flux tubes into the photosphere. Further, they found strong supersonic downflow is created inside the flux tube in the low chromosphere as a result of the convective collapse.

3.2 Magnetobuoyancy

In the previous subsection, we have seen how magnetic fields are created in the astronomical objects, and how magnetic fields are stretched, deformed, and transported by convective motion driven by *thermal buoyancy*. In a compressible fluid, we have another important buoyancy, i.e. *magnetic buoyancy*. It is the magnetic buoyancy to ultimately expel the magnetic flux from the Sun, stars, accretion disks, and galaxies. Since ejected magnetic flux causes enormous activity at the surface of these objects, we can say that the magnetic buoyancy is the ultimate driver of the magnetic activity in the astronomical object. In this subsection, we shall discuss the properties of magnetic buoyancy, especially various characteristics of the *Parker instability* (i.e. undular mode of the *magnetic buoyancy instability*), such as the linear stability properties, the nonlinear evolution, and application to the formation of the interstellar clouds and solar emerging flux.

3.2.1 Magnetic Buoyancy

3.2.1.1 What is Magnetic Buoyancy?

It is now well established that sunspots are formed by the emergence of magnetic flux tubes from the convection zone (Zwaan, 1985; 1987). What is the force to raise the flux tube to the surface? (See Fig. 3.17.)

Suppose an isolated flux tube embedded in nonmagnetized plasma with density ρ_e and temperature T under uniform gravitational acceleration g. The temperature inside the tube is assumed to be the same as that of outside the tube. Let's further assume that the tube is thin, i.e. the radius of the flux tube is much smaller than the local pressure scale height

$$H = \frac{kT}{\mu m g} = \frac{R_g T}{\mu g}. \tag{3.2.1}$$

Here R_g is the gas constant, and μ is the mean molecular weight. In this case, the pressure and density around the tube is almost uniform, and the equilibrium configuration of magnetic flux tube is determined by the balance of the total pressure between inside and outside of the tube,

$$p_i + B^2/8\pi = p_e, \tag{3.2.2}$$

where p_i and p_e are internal and external gas pressures, respectively. Then, the density inside the tube, ρ_i, becomes smaller than the density ρ_e outside of the tube, i.e., from equation of

156

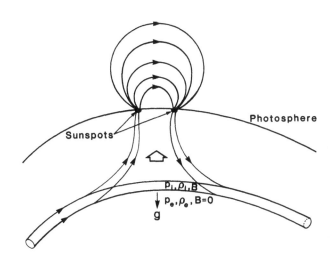

FIGURE 3.17 Isolated flux tube in the convection zone of the sun.

state $p = \rho R_g T / \mu$, we find

$$\rho_i = \rho_e - \frac{B^2}{8\pi} \frac{\mu}{R_g T} < \rho_e. \tag{3.2.3}$$

Hence the tube suffers the buoyancy force

$$\Delta \rho g = B^2 / (8\pi H). \tag{3.2.4}$$

Since this buoyancy force originates from magnetic field, this is called *magnetic buoyancy* (Parker, 1955). This is the fundamental force to raise the flux tube to the surface of the Sun and stars, and is the ultimate origin of magnetic activity of the Sun and stars. The magnetic buoyancy also plays an important role to transfer magnetic flux from galactic gas disks and accretion disks to their halo and corona.

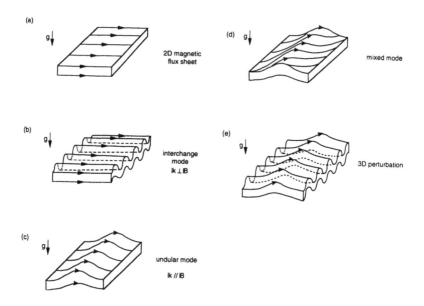

FIGURE 3.18 Interchange mode and undular (Parker) mode of magnetic buoyancy instability.

3.2.1.2 Magnetic Buoyancy Instability and Parker Instability

Note that a horizontal isothermal, isolated flux tube cannot be in equilibrium. On the other hand, a 2D isothermal flux *sheet* can be in equilibrium. Even in this case, however, the sheet often becomes unstable because of magnetic buoyancy. Such instability is called *magnetic buoyancy instability* (Hughes and Proctor, 1988).

There are two kinds of magnetic buoyancy instability. One is the case where the wavenumber vector **k** is perpendicular to the magnetic field vector **B**. This is called *interchange mode*, and is also sometimes called flute instability or magnetic Rayleigh-Taylor instability or Kruskal-Schwarzschild (1954) instability. The other is the mode occurring when the wavenumber vector is parallel to the magnetic field vector, which is called *undular mode* (see Fig. 3.18 and Table 3.2). This is called *ballooning instability* in fusion plasma physics (see Fig. 3.19), and *Parker instability* (1966) in astrophysics.

TABLE 3.1 Magnetic Buoyancy Instabilities (z is in the direction opposite to gravity)

mode	$\mathbf{k} - \mathbf{B}$ relation	necessary condition	wavelength
interchange mode (flute) (magnetic Rayleigh-Taylor) (Kruskal-Schwarzschild)	$\mathbf{k} \perp \mathbf{B}$	$\frac{d}{dz}\left(\frac{B}{\rho}\right) < 0$	arbitrary λ
undular mode (ballooning) (Parker)	$\mathbf{k}\|\mathbf{B}$	$\frac{d}{dz}B < 0$	$\lambda > \lambda_c \sim 10H$

3.2.1.3   Instability Condition for Interchange Mode

We follow the derivation by Acheson (1979). Consider the plasma penetrated by the horizontal magnetic field under uniform gravitational field. We assume that magnetic field lines are not bent, i.e. remain straight. Hence the problem is two-dimensional. Suppose that a 2D fluid element rises over a distance dz. Then the blob expands because of the decrease in total pressure. From conservation of mass and magnetic flux, we have

$$\rho A = (\rho + \delta\rho)(A + \delta A), \tag{3.2.5}$$

$$BA = (B + \delta B)(A + \delta A), \tag{3.2.6}$$

where A is the area of the fluid element, the physical quantities with δ correspond to the variation of these quantities due to expansion. From these equations, it follows that

$$\delta B/B = \delta\rho/\rho. \tag{3.2.7}$$

Pressure balance equation becomes

$$\delta p + B\delta B/4\pi = dp + BdB/4\pi, \tag{3.2.8}$$

where dp and dB are differences of these quantities between original height and new height in the external medium, i.e. $dp = p(z + dz) - p(z)$ etc. Conservation of entropy is written,

$$\delta p/p = \gamma\delta\rho/\rho, \tag{3.2.9}$$

From Eqs. (3.2.8) and (3.2.9) we have

$$\left(\frac{\gamma p}{\rho} + \frac{B^2}{4\pi\rho}\right)\delta\rho = dp + \frac{BdB}{4\pi}. \tag{3.2.10}$$

159

Instability can occur if the density in the blob is smaller than that of outside, i.e.

$$\delta\rho < d\rho. \tag{3.2.11}$$

Hence from equations (3.2.10) and (3.2.11), we find

$$\frac{dp}{dz} + \frac{B}{4\pi}\frac{dB}{dz} < (C_s^2 + V_A^2)\frac{d\rho}{dz}. \tag{3.2.12}$$

The equation (3.2.12) is rewritten as

$$V_A^2\frac{d}{dz}\ln\left(\frac{B}{\rho}\right) < -\frac{N^2}{g}C_s^2, \tag{3.2.13}$$

where N is the *Brunt-Väisälä frequency* defined as

$$N^2 = \frac{g}{\gamma}\frac{d}{dz}\ln(p\rho^{-\gamma}). \tag{3.2.14}$$

This is the general instability condition for interchange mode of magnetic buoyancy instability for arbitrary entropy distribution.

Problem 3–1: Show that when there is no magnetic field, the equation (3.2.13) reduces the *Schwarzschild criterion* for *convective instability*;

$$-\frac{dT}{dz} > -\left(\frac{dT}{dz}\right)_{ad} = \frac{(\gamma-1)}{\gamma}\frac{\mu g}{R_g}. \tag{3.2.15}$$

If the plasma is neutral for convective instability, we get

$$\frac{d}{dz}\ln\left(\frac{B}{\rho}\right) < 0. \tag{3.2.16}$$

This means that the magnetic field strength has to decrease with height faster than the density to induce the interchange mode magnetic buoyancy instability.

3.2.1.4 Instability Condition for Undular Mode (Parker Instability)

Using energy principle, a simple condition was derived by Newcomb (1969) for the undular mode;

$$-\frac{d\rho}{dz} < \frac{\rho^2 g}{\gamma p}. \tag{3.2.17}$$

Actually this condition is the most general instability condition applicable to any instability caused by buoyancy (gravitational acceleration). Since our problem assumes magnetostatic equilibrium

$$\frac{d}{dz}\left(p + \frac{B^2}{8\pi}\right) = -\rho g \tag{3.2.18}$$

for unperturbed state, we can rewrite above condition into the following one;

$$\frac{B}{4\pi}\frac{dB}{dz} < -\frac{\gamma p N^2}{g}. \tag{3.2.19}$$

If the plasma is convectively neutral ($N = 0$), then we find

$$\frac{dB}{dz} < 0. \tag{3.2.20}$$

From this relation, we find that *magnetic buoyancy drives an instability if the basic field strength decreases with height*. This is much broader condition than that for the interchange mode. Physical meaning of this is as follows. If the magnetic field decreases with height, plasmas can be lifted up by magnetic pressure gradient against the gravitational force. So there is a free (gravitational potential) energy in such a situation. The undular instability enables plasmas to fall down along field lines by bending magnetic field lines, so that the gravitational free energy can be released.

Parker (1966) emphasized an importance of this instability in the Galactic disk, including an effect of cosmic ray pressure, and succeeded to explain the formation of interstellar cloud complexes. Hence this mode is usually called *Parker instability* in astrophysics literatures, and hereafter the undular mode is called simply the Parker instability.

3.2.1.5 Ballooning instability

The *ballooning instability* is the ideal MHD instability driven by gas pressure, often occurring in magnetically confined fusion plasmas such as the Tokamak plasma. Since plasmas are confined in curved magnetic fields in these magnetic fusion plasmas, the gas pressure plays a role similar to that of the gravitational acceleration. That is, the centrifugal force acting on charged particles running at thermal speed along curved magnetic field lines is similar to the gravitational acceleration. Furthermore, the magnetic tension force acts as a stabilizing agent. From these reasons, the ballooning instability is physically very similar to the Parker instability. (Similarly, the pressure driven flute instability is physically analogous to the interchange mode of the magnetic buoyancy instability or the magnetic Rayleigh-Taylor instability.)

Let us derive the criterion of the instability (i.e., critical $\beta = \beta_c$). We consider a torus plasma with minor and major radii a and R with helically twisted magnetic fields (B_p, B_t), where B_p is the poloidal magnetic fields and B_t is the toroidal magnetic fields (Fig. 3.19). The instability occurs if

$$\frac{1}{\gamma} < \tau_c, \tag{3.2.21}$$

where γ is the growth rate of the ballooning instability and is of order of

$$\gamma \simeq (gk)^{1/2}. \tag{3.2.22}$$

Here g is the effective acceleration

$$g \simeq V_{th}^2/R, \tag{3.2.23}$$

V_{th} is the thermal speed of plasmas, k is the wavenumber of order of $1/a$. Thus the left hand side of Eq. (3.2.21) is the *effective free fall time*, while the right-hand side is the Alfvén time over one turn of a field line,

$$\tau_c = \frac{qR}{V_A},$$ (3.2.24)

where

$$V_A = B_t(4\pi\rho)^{-1/2},$$ (3.2.25)

and

$$q = \frac{aB_t}{RB_p}.$$ (3.2.26)

q is called the safty factor. The left-hand side in Eq. (3.2.21) is also understood as the destabilizing *gravitational (or buoyancy) force*, and the right-hand side is the stabilizing *magnetic tension force*.

From these relations, we find that the instability occurs if

$$\beta = (V_{th}/V_A)^2 > \beta_c = \frac{a}{q^2R}.$$ (3.2.27)

It is interesting to note that in the case of the Parker instability we have *critical wavelength* (see Sec. 3.2.2.1).

$$\lambda > \lambda_c \sim H = V_A^2/g,$$ (3.2.28)

(where we neglected a factor of 5–10), whereas in the case of the ballooning instability we have *critical β*. What is the reason of this difference?

The origin of this difference is in the formula of the effective gravitational acceleration in Eq. (3.2.23), $g \simeq V_{th}^2/R$. Inserting this to Eq. (3.2.22) and (3.2.21) yields

$$kV_{th}^2/R > V_A^2/(q^2R^2).$$ (3.2.29)

Since $q = aB_t/(RB_p)$, and $V_A \propto B_t$, we have

$$a^2k > V_{Ap}^2/g,$$ (3.2.30)

where $V_{Ap} = B_p/(4\pi\rho)^{1/2}$. If we write $\lambda \sim a \sim 1/k$ and $H \sim V_{Ap}^2/g$, then we find

$$\lambda > H.$$ (3.2.31)

This is the relation for the Parker instability [Eq. (3.2.28)]. In fusion plasmas, the wavelength is limited by the size of the device (a in this case), so that we have the critical β instead of the critical wavelength.

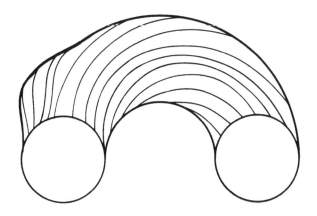

FIGURE 3.19 Ballooning instability in a magnetically confined torus plasma (e.g., Tokamak). Major radius (R) and minor radius (a) of a torus are shown.

3.2.2 Parker Instability

3.2.2.1 Critical Wavelength

The Parker instability occurs when magnetic fields are disturbed to undulate. Hence the magnetic tension force acts as a stabilizing force. Only when the buoyancy force becomes larger than the magnetic tension force, the Parker instability occurs;

$$\Delta\rho g > \frac{B^2}{4\pi r},\tag{3.2.32}$$

where r is the curvature radius of the field line.

Consider an isolated flux tube embedded in a field-free medium, undulating with a wavelength λ along the flux tube. In this case, magnetic buoyancy force is given by

$$\Delta\rho g = \frac{B^2}{8\pi H},\tag{3.2.33}$$

163

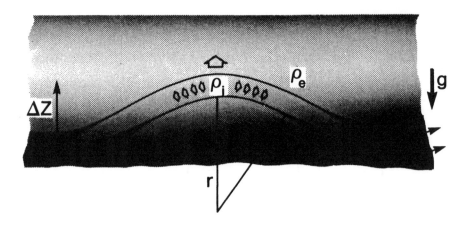

FIGURE 3.20 Undulating flux tube (or sheet).

as shown before (Eq. 3.2.4). Since the magnetic tension force is of order of $B^2/(4\pi\lambda)$, the instability condition becomes

$$\lambda > 2H. \tag{3.2.34}$$

Consequently, there is a critical wavelength below which the Parker mode is stable and the critical wavelength is of order of the local pressure scale height. As discussed earlier, above isothermal isolated flux tube is not in equilibrium, and hence these calculations are not exact (buoyancy force is too large).

Now consider a flux sheet in equilibrium and assume that both sound speed C_s and Alfvén speed V_A are constant. The unperturbed state of plasmas and magnetic field are given by the following equations,

$$p/p_0 = \rho/\rho_0 = B^2/B_0^2 = \exp(-z/\Lambda), \tag{3.2.35}$$

where

$$\Lambda = (C_s^2 + V_A^2/2)/g. \tag{3.2.36}$$

164

In this case, buoyancy force appears as a result of perturbation. Suppose that a flux tube (or sheet) is raised up over a small distance Δz so that plasmas fall down along an oblique magnetic flux tube (sheet) to a new equilibrium (Fig. 3.20). Since the new equilibrium is obtained from a hydrostatic balance along the flux tube (sheet) and there is no magnetic force along the flux tube, the density at the top of the raised portion of the tube becomes

$$\rho_i(\Delta z) \simeq \rho_0 \exp(-\Delta z/H) \simeq \rho_0(1 - \Delta z/H), \qquad (3.2.37)$$

where

$$H - C_s^2/g. \qquad (3.2.38)$$

On the other hand, the density outside the tube (sheet) at the same height (Δz) is

$$\rho_e(\Delta z) \simeq \rho_0 \exp(-\Delta z/\Lambda) \simeq \rho_0(1 - \Delta z/\Lambda). \qquad (3.2.39)$$

Hence, the net density depression at the top of the loop is

$$\Delta\rho \simeq \rho_e(\Delta z) - \rho_i(\Delta z) \simeq \rho_0\Delta z(\frac{1}{H} - \frac{1}{\Lambda}). \qquad (3.2.40)$$

The curvature radius r is rewritten using the wavelength λ as

$$r \simeq \left(\frac{\lambda}{4}\right)^2 \frac{2}{\Delta z}. \qquad (3.2.41)$$

Then after some manipulation, the condition for occurrence of the Parker instability $\Delta\rho g > B^2/4\pi r$ becomes

$$\lambda^2 > \lambda_c^2 = 16\Lambda^2/(1 + 1/\beta), \qquad (3.2.42)$$

where β is the ratio of gas pressure to magnetic pressure.

Finally the instability condition for wavelength becomes

$$\lambda > \lambda_c = 4\Lambda/(1 + 1/\beta)^{1/2}. \qquad (3.2.43)$$

An exact treatment (Parker, 1966, 1979) shows the dispersion relation for $k_y\Lambda \gg 1$ as follows

$$(2/\beta + \gamma)\Omega^4 + [(4/\beta)(1/\beta + \gamma)(k_x^2\Lambda^2 + 1/4) + \gamma - 1]\Omega^2$$

$$+(2/\beta)k_x^2\Lambda^2[(2/\beta)\gamma k_x^2\Lambda^2 - (1 + 1/\beta)(1 + 1/\beta - \gamma)] = 0, \qquad (3.2.44)$$

where

$$\Omega = \frac{\omega\Lambda}{C_s} \qquad (3.2.45)$$

and y is the direction perpendicular to both vertical and magnetic fields. From this, we find the exact critical wavelength for $k_y\Lambda \gg 1, \gamma = 1$ as

$$\lambda_c \simeq 8^{1/2}\pi\Lambda/(1 + 1/\beta)^{1/2}. \qquad (3.2.46)$$

Hence above very rough treatment gives fair answer in order-of-magnitude. The critical wavelength for the Parker instability tends to be constant for weaker magnetic field,

$$\lambda_c \sim 9\Lambda \sim 9H \qquad \text{for weak magnetic field } (\beta \gg 1), \tag{3.2.47}$$

while it increases with increasing field strength if the temperature of the gas layer is fixed, i.e.,

$$\lambda_c \sim 9H\beta^{-1/2} \qquad \text{for strong magnetic field } (\beta \ll 1). \tag{3.2.48}$$

3.2.2.2 Growth Rate and Critical γ

Parker (1966) solved the linear instability problem for magnetized gas layer with constant V_A, C_s under constant g, and found that the maximum growth rate becomes

$$\omega \sim 0.3V_A/\Lambda, \tag{3.2.49}$$

for $0.5 < \beta$ and $k_y \gg 1$ where y is the direction perpendicular to magnetic field lines (Parker, 1979). It is found that when $\beta \sim 1$,

$$\omega \sim 0.2V_A/H \sim 0.2(g/H)^{1/2}. \tag{3.2.50}$$

This rate is nearly comparable to the inverse of the free fall time $(H/g)^{1/2}$, so that we can understand that the Parker instability is driven by the release of gravitational energy. This character is similar to that of the Rayleigh-Taylor instability in which the growth rate is of order of

$$\omega_{RT} \sim (g/\lambda)^{1/2}, \tag{3.2.51}$$

where λ is the wavelength of perturbation parallel to the surface between dense and rarefied gas layers.

Figure 3.21 shows the growth rate for more realistic gravitational potential (nonuniform g) calculated by Horiuchi et al. (1988). They found that the growth rate and critical wavelength for non-uniform g are essentially similar to those calculated by Parker (1975) for uniform g. Figure 3.21 shows that (1) ω is largest for $k_y = \infty$, though ω for $k_y = 0$ is not so different from that of $k_y = \infty$. (2) Most unstable wavelength is of order of $2\lambda_c$. The most interesting finding by Horiuchi et al. (1988) is that the most unstable mode is not the mirror symmetric mode (even mode) with respective to the equatorial plane as assumed by Parker (1966), but is the *glide-reflection-symmetric mode* (odd mode), i.e. the mode in which perturbed magnetic field lines pass through the equatorial plane (Fig. 3.22) (Giz and Shu, 1993)

There is one more interesting character in the Parker instability, i.e. there is a *critical* γ. That is, in order to excite the Parker instability, the adiabatic index γ must satisfy the following condition (from equation (3.2.46));

$$\gamma < \gamma_c = 1 + 1/\beta, \tag{3.2.52}$$

when $k_y = \infty$. Physically, this means that the Parker instability cannot occur for incompressible plasma ($\gamma = \infty$) and require strong compressibility. This can be easily understood because the Parker instability is driven by the downflow along the field lines.

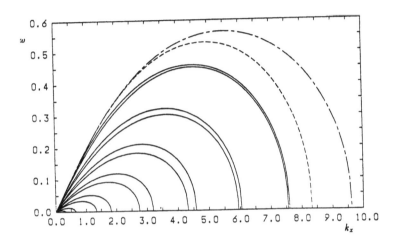

FIGURE 3.21 Linear growth rate of the Parker instability in the case of non-uniform gravitational field (from Horiuchi *et al.*, 1988). The parameters are $\beta = 1, \varepsilon = GM/r_0(c_s^2 + V_A^2/2) = 25$ and $k_y = 0.0$ (solid curves). For comparison, the growth rate for $g = 0.385 =$ constant and $k_y = k_z = 0.0$ is shown by the dashed curve. The dash-dotted curve denotes the growth rate of the fundamental mode when $k_y \rightarrow \infty$.

Let us briefly discuss the application of the Parker instability to the solar and galactic problems. In the case of the Sun, it is generally believed that the magnetic flux tubes rise from the base of the convection zone by the Parker instability (Spruit and van Ballooijen, 1982; Moreno-Insertis, 1986). At the base of the convection zone (depth of 10^5km), the temperature is about 1MK so that the pressure scale $H \sim 4 \times 10^4$ km. Since $\beta \gg 1$, we find $\lambda_c \sim 4 \times 10^5$ km. This seems to explain the apparent mean distance between active regions (Fig. 1.11). On the other hand, in galaxies, the scale height is about $\sim$ 100pc and $\beta \sim 1$. Hence the critical wavelength is about 1 kpc. This may explain the mean distance of giant molecular cloud complex on spiral arms (Fig. 3.23).

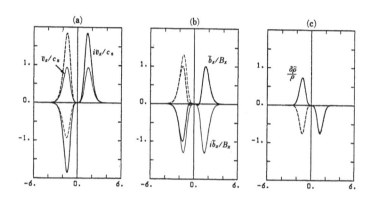

FIGURE 3.22 2D eigenfunction of the Parker instability in the non-uniform gravitational field (Horiuchi *et al.*, 1988).

3.2.2.3 Nonlinear Evolution of the Parker Instability: Most Unstable Mode

In the previous sections, we discussed various properties of the Parker instability. However those are all based on the linear instability theory. What happens in the nonlinear stage? Matsumoto *et al.* (1988) performed the 2D nonlinear simulation of the Parker instability for the first time, and confirmed the earlier conjecture by Parker (1966) and Mouschovias (1974) on the formation of large interstellar condensations at the bottom of magnetic pocket. They further discovered that the velocity of the downflow along rising loop exceeds the local sound speed to form shock waves when the downflow collides with the already accumulated cloud at the bottom of the magnetic pocket. This is similar to accretion shock appeared in protostar formation.

Matsumoto *et al.* (1988) adopted the simple gravitational potential produced by a point mass,

$$g(z) = GMz/(r_0{}^2 + z^2)^{3/2}, \tag{3.2.53}$$

which is the vertical component of the gravitational acceleration near the equator of the disk

168

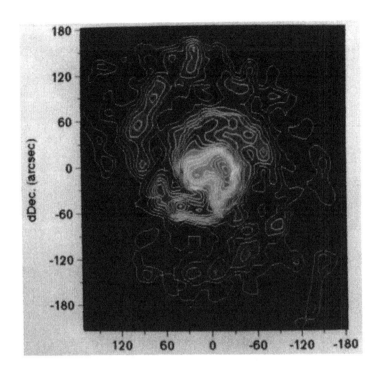

FIGURE 3.23 Molecular clouds in M51 observed with the Nobeyama 45 m radio telescope (from Nakai *et al.*, 1994).

rotating around the point mass, M, where r_0 is the distance from the point mass. This is exactly what is expected for accretion disk. In the case of galaxies, this is not an exact model, but at least the region near the equator can be applied to galaxies, because the gravitational acceleration near the disk in galaxies is approximately in proportion to the height from the disk. They assumed further (1) frozen in (2) isothermal perturbation $\gamma = 1$, (3) inviscid, (4) neglecting effects of cosmic rays, rotation (both Coriolis force and shear motion), curvature, and self-gravity. As the initial model, they adopted isothermal magnetostatic layer with $C_s = \text{constant}$ and $V_A = \text{constant}$. This model is exactly the same as that adopted by Horiuchi *et al.* (1988) for the linear instability analysis. The free parameters in this model are β and $\epsilon = GM/r_0/(C_s^2 + V_A^2/2)$.

Figure 3.24 shows a typical evolution of nonlinear evolution of the Parker instability. The initial plasma $\beta = 1$ and $\epsilon = 6$. As time goes on, magnetic field lines undulate so that the mass slides down along the bending magnetic field lines to form a dense cloud at the bottom of the magnetic pocket. On the other hand, at the top of the magnetic loop, the density decreases very much. The rise velocity of the magnetic loop is of order of $0.3 - 0.4V_A$ and

169

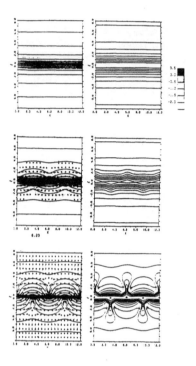

FIGURE 3.24 Nonlinear evolution of the Parker instability, the initial plasma $\beta = 1$ and $\epsilon = 6$. Note that shock waves are formed at the foot of the downflow along the magnetic loop (from Matsumoto *et al.*, 1988).

the maximum velocity of the downflow becomes comparable to $1 - 2V_A$ which exceeds the local sound speed. Hence when the downflow collides with the dense cloud at the pocket of field lines, a shock wave is formed. This is approximately a standing shock, but gradually propagate upward along the loop. A remarkable feature is a thin spur-like feature created just above the cloud. This may correspond to radio or dust spurs observed in our Galaxy (Sofue, 1973, 1976). Physically this is essentially the same as the *prominence* (especially the same as the *Kippenhahn-Schlüter model* of prominence). Fig. 3.25 shows time variation of various energies. Here these energies are defined as follows:

$$E_k = \int \int \frac{1}{2}\rho(V_x^2 + V_z^2)dxdz, \qquad \text{(kinetic energy)} \qquad (3.2.54)$$

$$E_m = \int \int \frac{1}{8\pi}\rho(B_x^2 + B_z^2)dxdz, \qquad \text{(magnetic energy)} \qquad (3.2.55)$$

170

$$E_h = \int \int p \ln p \, dx dz, \qquad \text{(heat energy)} \qquad (3.2.56)$$

$$E_g = \int \int \rho \psi \, dx dz, \qquad \text{(gravitational energy)} \qquad (3.2.57)$$

$$T = E_k + E_m + E_h + E_g, \qquad \text{(total energy)} \qquad (3.2.58)$$

where ψ is the gravitational potential, and *heat energy* is used here in place of internal energy because isothermal equation of state is assumed here (Mouschovias, 1974). From Fig. 3.25 we find that the gravitational energy considerably decreases while the magnetic energy does not change much. This means that the Parker instability develops owing to the release of the gravitational energy of the mass supported by the magnetic field. We find also that the kinetic energy attains only a small fraction of the released gravitational energy. This is because a large part of the released gravitational energy goes into the work to compress the gas. We further notice that the total energy integral T begins to decrease after $t = 7.67$. This corresponds to the formation of shock waves, because in isothermal gas the heat energy is not conserved if shock waves are produced (Matsumoto *et al.*, 1990).

Problems 3–2: Prove that (1) the total energy T is conserved if shock wave is not produced, and that (2) T decreases if shock waves are produced.

Figure 3.26 shows the growth of velocity perturbation (horizontal velocity) of the Parker instability simulated by Matsumoto *et al.* (1988) which is the same model as shown in Fig. 3.24. In the linear regime, the simulation agrees fairly well with linear theory. It is interesting to note that the saturation occurs when the velocity becomes comparable to the initial Alfvén speed.

Interestingly, even when the initial plasma β is larger than 1, i.e. magnetic field strength is smaller than gas pressure, we find the similar saturation at the initial Alfvén speed. Figure 3.27 shows an example of such case ($\beta = 10$). In this case the Alfvén speed is less than the sound speed so that there is no shock formation, and instead nonlinear oscillations take place. In fact the total energy is nearly conserved in this case (see Fig. 3.28), while other energies change similarly to the case of shock wave formation (low β case). It is also interesting to note that similar nonlinear oscillation is found in nonlinear stage of the resistive kink instability (Nakajima, 1987).

Why saturation occurs at Alfvén velocity? The reason is as follows. Initially we assume the gas layer is in magnetostatic equilibrium with the scale height $\Lambda = (C_s^2 + V_A^2/2)/g$. After the Parker instability occurs, the gas slides down along the field lines. Plasmas find a new hydrostatic equilibrium with $H = C_s^2/g$ along the field lines. Hence the released gravitational energy, which becomes kinetic energy of downflows, is determined by the height difference between two scale heights, $\Lambda - H$;

$$\rho V^2/2 \simeq \rho g(\Lambda - H) = \rho V_A^2/2. \qquad (3.2.59)$$

Hence the downflow velocity becomes comparable to the Alfvén speed.

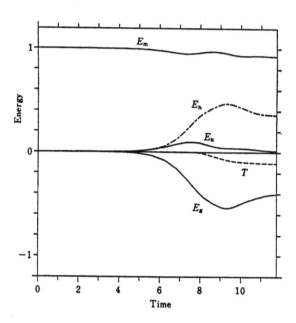

FIGURE 3.25 Time variation of various energies in the nonlinear evolution of the Parker instability for the case shown in Fig. 3.24 (Matsumoto *et al.*, 1988).

3.2.2.4 Condition for Shock Waves Formation and Nonlinear Oscillation*

What is the condition for the formation of shock wave or nonlinear oscillation? To examine this, Matsumoto *et al.* (1990) studied the nonlinear evolution of the Parker instability changing the wavelength of the initial perturbation. They found that shock wave formation occurs when

$$\lambda > \lambda_{cs} \simeq (3.5\beta + 6)\Lambda_m, \tag{3.2.60}$$

where $\Lambda = C_s^2(1+1/\beta)/g_{\max}$, $g_{\max} = 0.385 GM/r_0^2$, while nonlinear oscillations arise if $\lambda < \lambda_{cs}$ (see Fig. 3.29).

Note that this critical wavelength λ_{cs} [4] is larger than the most unstable wavelength $\lambda_{\max}$ for $\beta > 3$. Consider then what happens if we input random perturbation to the gas layer. In this case, the most unstable mode first develops. In low $\beta(< 3)$ disks, the most unstable

---

[4]Note that this critical wavelength λ_{cs} is not the same as the critical wavelength λ_c for instability discussed before.

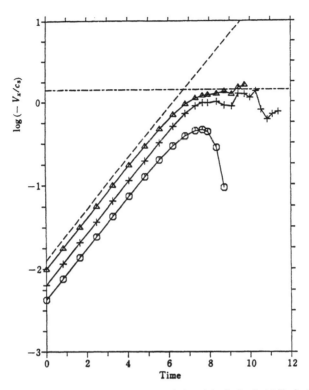

FIGURE 3.26 Growth of velocity perturbation in the nonlinear evolution of the Parker instability for the case shown in Fig. 3.24 (Matsumoto *et al.*, 1988).

wavelength is longer than above critical wavelength, so that shock wave formation occurs and hence the free energy dissipates at the shock front, leading the system to a new quasi static equilibrium state. In high β disks, on the other hand, the most unstable wavelength is smaller than the critical wavelength so that the nonlinear oscillation appears. During the nonlinear oscillation, the nonlinear coupling occurs between different wave modes, generating longer wavelength modes. As time goes on, the energy is gradually transferred to longer wavelength modes. When the wavelength becomes comparable to the critical wavelength λ_{cs}, shock waves are produced, i.e. the nonlinear oscillation stops, leading to quasi static layer with magnetic loops whose length determined by equation (3.2.60). Summarizing these results, the length of magnetic loops in final equilibrium state is roughly given by $\max(\lambda_{cs}, \lambda_{max})$, i.e.,

$$\lambda_{\text{loop}} \simeq \begin{cases} 6\pi[\beta/(\beta+2)]^{1/2}\Lambda_m = 6\pi(\beta+1)\beta^{-1/2}(\beta+2)^{-1/2}H_m, & (\text{for } \beta < 3) \\ (3.5\beta+6)\Lambda_m = (3.5\beta+6)(1+1/\beta)H_m, & (\text{for } \beta > 3) \end{cases} \qquad (3.2.61)$$

173

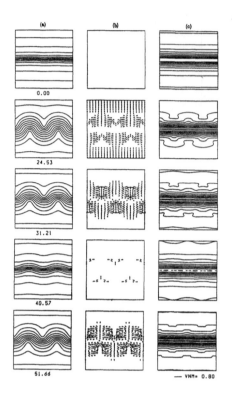

FIGURE 3.27 Nonlinear evolution of the Parker instability in the case of weak magnetic field ($\beta = 10$) (Matsumoto *et al.*, 1988). Note that in this case shock waves are not formed, but the nonlinear oscillation occurs.

where $H_m = C_s^2/g_{\max}$ and $g_{\max} = 0.385GM/r_0^2$. When applying this result to the actual accretion disks and galaxies, we can assume H_m approximately corresponds to the thickness of the disk when $\beta > 1$. This result would be important to estimate the actual length of magnetic loops produced by the Parker instability in accretion disks and in galactic disks.

3.2.2.5 Self-Similar Evolution*

Shibata *et al.* (1989a, 1989b) found a self-similar expansion of a magnetic loop in the nonlinear stage of the Parker instability. They performed 2D nonlinear simulation of the Parker instability of an isolated magnetic flux sheet embedded in a field free gas layer, with the motivation to explain the emergence of magnetic flux tubes in the solar atmosphere. The key point of their model was that they assumed a computational region including many scale heights in vertical direction to simulate flux emergence from the photosphere to the corona. Actually solar observations show that the density decreases from 10^{-7} g cm^{-3} in

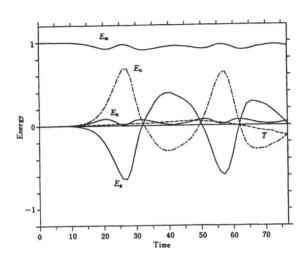

FIGURE 3.28 Time evolution of various energies in the nonlinear evolution of the Parker instability in the case of weak magnetic field ($\beta = 10$) (Matsumoto et al., 1988).

the photosphere to 10^{-14} g cm^{-3} or less in the corona. Such a highly stratified gas layer is a characteristic feature of celestial objects confined by gravitational force, which is the origin of many interesting physical processes. MHD numerical simulation in such a highly stratified gas layer is, however, one of the most difficult numerical simulations in astrophysics. Shibata et al. (1989a, 1989b) and subsequent works (Shibata et al., 1989a, 1989b, 1992; Kaisig et al., 1990; Nozawa et al., 1992; Matsumoto et al., 1992, 1993; Yokoyama and Shibata, 1994a, 1994b) succeeded to perform the nonlinear simulation of the Parker instability (and related problems) in such a highly stratified gas layer.

Let us now discuss numerical simulations of Shibata et al. (1989a, 1989b) in more detail. Initial condition of Shibata et al. (1989a, 1989b) is shown in Fig. 3.30. The gas layer is initially in magnetostatic equilibrium and consists of a cold isothermal plasma layer (i.e. photosphere/chromosphere), which is partly permeated by horizontal isolated magnetic flux sheet in $z_0 < z < z_0 + D$, and a hot isothermal, non-magnetized plasma layer (i.e., corona) above the cold layer. Hereafter, the units of length, velocity, and time are H, C_s, and H/C_s, where

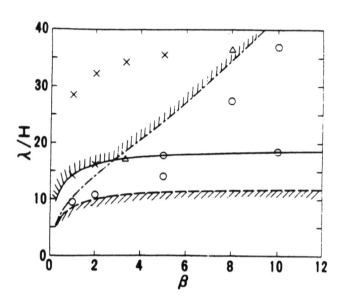

FIGURE 3.29 Diagram showing the condition of shock wave formation and the nonlinear oscillation in the nonlinear evolution of the Parker instability. The ordinate is the horizontal wavelength of the initial perturbation in unit of the scale height of the disk, and the abscissa is $\beta = p_g/p_{\mathrm{mag}}$. Circles denote the models showing nonlinear oscillations, and crosses represent those where the shock waves are formed. The models denoted by the triangles show both the shock waves and the nonlinear oscillations. Dashed curve denotes the lower limit of the wavelength of an unstable perturbation. Solid curve denotes the wavelength of the most unstable mode. Dash-dotted curve denotes $(v_x/C_s)_{\mathrm{max}}/(b_z/B_x)_{\mathrm{max}} = 1$. The upper shaded region shows the expected size of magnetic loops (from Matsumoto et al., 1990).

$H = C_s^2/(\gamma g)$ is the scale height and C_s is the sound speed in the cold layer. The temperature distribution is taken to be

$$T(z) = T_1 + (T_2 - T_1)\left[\tanh\left((z - z_{tr})/w_{tr}\right) + 1\right]/2, \qquad (3.2.62)$$

where T_2/T_1 (which is assumed to be $= 25$ here) is the ratio of the temperature in the hot layer to that in the cold layer, $z_{tr}(= 18H)$ is the height of the transition layer between the cold and hot layers, w_{tr} is the temperature scale height in the transition layer ($= 0.6H$ for all our calculations, where H is the pressure scale height of the cold layer).

176

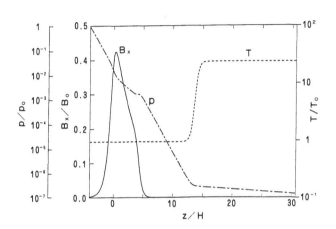

FIGURE 3.30 Initial condition of the nonlinear simulation of the Parker instability of an isolated magnetic flux sheet embedded in the solar photosphere. The magnetic field strength B_x, gas pressure p, and temperature T as a function of height z/H, where H is the pressure scale height in the photosphere/chromosphere. Initial β in the flux sheet is 1 (from Shibata *et al.*, 1989b).

Shibata *et al.* (1989a, 1989b) assume that the magnetic field is initially parallel to the horizontal plane; $\mathbf{B} = (B_x(z), 0, 0)$, and is localized in the cold layer. The distribution of magnetic field strength is given by

$$B_x(z) = [8\pi p(z)/\beta(z)]^{1/2}, \tag{3.2.63}$$

where

$$\beta(z) = \beta_*/f(z), \tag{3.2.64}$$

$$f(z) = [\tanh((z - z_0)/w_0) + 1][-\tanh((z - z_1)/w_1) + 1]/4. \tag{3.2.65}$$

Here, β_* is the ratio of gas pressure to magnetic pressure at the center of the magnetic flux sheet, z_0 and $z_1 = z_0 + D$ are the heights of the lower and upper boundary of the magnetic flux sheet, D is the vertical thickness of the magnetic flux sheet. It is assumed that $z_0 = 4H, D = 4H, w_0 = w_1 = 0.5H, w_2 = 0.6H, \beta_* = 1$.

On the basis of the above initial plasma β distribution, the initial density and pressure distributions are numerically calculated by the equation of static pressure balance

$$\frac{d}{dz}\left[p + \frac{B_x^2(z)}{8\pi}\right] + \rho g = 0. \tag{3.2.66}$$

Periodic boundary is assumed for $x = 0$ and $X_{max}(= 80)$, conducting wall boundary for $z = 0$, and free boundary for $z = Z_{max}(= 35)$.

The MHD equations with $\gamma = 1.05$ are solved numerically by using modified Lax-Wendroff scheme with artificial viscosity. To initiate the instability, small-amplitude perturbations with $\lambda = 20H$ are added to the system in the finite horizontal domain ($X_{max}/2 - \lambda/2 < x < X_{max}/2 + \lambda/2$).

Figure 3.31 shows the time evolution of magnetic lines of force, the velocity field, and the density distribution. As the magnetic loop rises, the gas slides down along the loop. Spikes of dense regions are created on the valleys of the undulating field lines, whereas the rarefied regions are produced around the top of magnetic loops. The most salient characteristics in the nonlinear stage ($t > 40$) is the approximate *self-similar pattern* of magnetic loop expansion; the rise velocity of the magnetic loop and the velocity of downflow along the loop increase with height as the loop expands and ascends. Figure 3.32 shows some physical quantities at $x = X_{max}/2$ (midpoint of the magnetic loops), indicating approximate self-similar behavior as a function of height. We also find

$$V_z = az, \tag{3.2.67}$$

$$V_A = bz, \tag{3.2.68}$$

where $a \simeq 0.06$ (for $t < 60$), $b \simeq 0.3$, and z is the height measured from $z_0(= 4)$. [5] On the other hand, we find the density and magnetic field strength have the power-law distribution;

$$\rho \propto z^{-4}, B_x \propto z^{-1}. \tag{3.2.69}$$

Figure 3.33 shows the time evolution of the Lagrangian displacement of a test particle at the midpoint of the loop in the simulation results. In the initial stage, the growth rate of the perturbation amplitude agrees well with linear theoretical values ($\omega_l = 0.12$). The amplitude increases exponentially with time even in the nonlinear stage ($t > 40$); $z \propto \exp(\omega_n t)$ and

$$\omega_n \simeq a \simeq 0.06 \simeq \omega_l/2. \tag{3.2.70}$$

This characteristic that the nonlinear growth rate $\omega_n = a$ becomes comparable to half the linear growth rate holds in a wide parameter range (Shibata *et al.*, 1989a) including the case

---

[5]Note that hereafter we often use z as the height from $z_0 = 4$, the base of the magnetic flux sheet, instead of the height from the bottom of the computing box when discussing the self-similar behavior of the magnetic flux expansion.

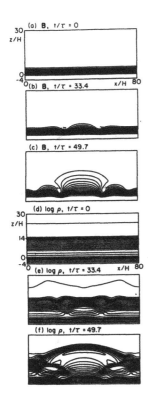 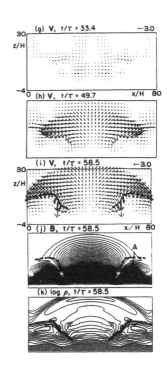

FIGURE 3.31 The nonlinear simulation results of the Parker instability triggered by the localized perturbation in the isolated flux sheet in Fig. 3.30. (a) The magnetic field lines $B = (B_x, B_z)$, (b) the velocity vector $V = (V_x, V_z)$, (c) density contours ($\log \rho$). Note that the magnetic flux expands approximately self-similarly (from Shibata *et al.*, 1989b).

of the convective-Parker instability (Nozawa *et al.*, 1992) and in the 3D Parker instability coupled with interchange mode (Matsumoto *et al.*, 1993; see Sec. 3.2.2).

We also find that the Alfvén speed in the rising loop increases with height. This means that the plasma β decreases with height and explains why we have a low β corona inside magnetic loop. As magnetic loop expands, it tends to be current free since both thermal and gravitational forces become smaller than the magnetic force as the loop rises (see equations (3.2.80) and (3.2.81).

Although Shibata *et al.* (1989a) found these results assuming isolated magnetic flux initially, almost the same results are found also by assuming continuous magnetic field distribution as an initial condition (Kamaya *et al.*, 1996).

179

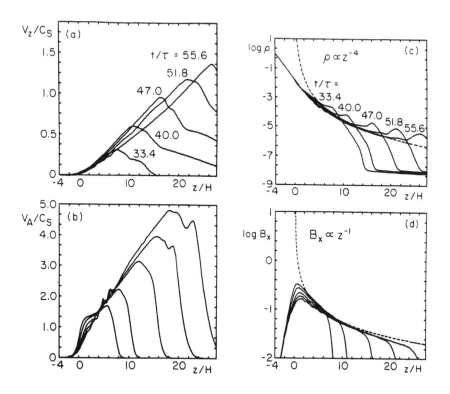

FIGURE 3.32 1D distribution in z of some physical quantities at the midpoint of the expanding magnetic loop in the case of the nonlinear simulation model of the Parker instability shown in Fig. 3.31. (a) The vertical velocity V_z, (b) the local Alfvén speed V_A, (c) the horizontal magnetic field ($\log B_x$), (d) the density ($\log \rho$) at $x = X_{max}/2 = 40$. The numbers attached to the curves correspond to the following time (in unit of H/C_s); (1) $t = 42.1$, (2) 49.6, (3) 57.6, (4) 64.6 (from Shibata *et al.*, 1989b).

3.2.2.6   Theory of Nonlinear Parker Instability*

A. Nonlinear instability

   Shibata *et al.* (1989a, 1990a) found an approximate 1D self-similar solution which explains numerical simulation results very well. Following Shibata *et al.* (1990b), we shall derive a self-similar solution of the problem by analytical method. We have the following relation from Eq. (3.2.67);

$$\partial V_z/\partial \tau = \partial V_z/\partial t + V_z \partial V_z/\partial z = aV_z, \qquad (3.2.71)$$

where τ is the time in Lagrangian coordinates, while t and z are the Eulerian coordinates. This leads to

$$V_z(\xi, \tau) = a\xi \exp(a\tau), \qquad (3.2.72)$$

180

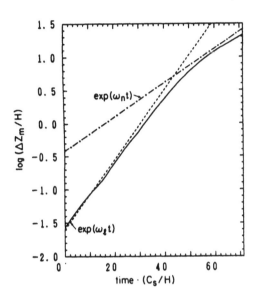

FIGURE 3.33 The time evolution of Lagrangian displacement ($\Delta z_m = z_m(t) - z_m(0)$; solid curve) of a test particle. The dashed line shows the linear growth with $\omega_l = 0.121 C_s/H$, and the dash-dotted line shows the nonlinear growth with $\omega_n = 0.06 C_s/H$ (from Shibata *et al.*, 1990a).

where

$$\xi = z \exp(-a\tau) \tag{3.2.73}$$

is the Lagrangian coordinate.

We assume the quasi one-dimension (1D) for the problem, i.e. we consider only vertical (z) variation of the physical quantities at the midpoint of the loop. The basic equations of our quasi-1D problem are

$$\frac{\partial \rho}{\partial t} = -\frac{\partial}{\partial z}(\rho V_z) - \frac{\partial}{\partial z}(\rho V_x), \tag{3.2.74}$$

$$\rho\left(\frac{\partial V_z}{\partial t} + V_z\frac{\partial V_z}{\partial z}\right) = -\left[\frac{\partial}{\partial z}\left(\frac{B_x^2}{8\pi}\right) + \frac{B_x^2}{4\pi R}\right], \tag{3.2.75}$$

181

$$\frac{\partial B_x}{\partial t} = -\frac{\partial}{\partial z}(B_x V_z), \tag{3.2.76}$$

where R is the radius of curvature of field lines at the midpoint of the magnetic loop. The last term on the right hand side of Eq. (3.2.75) is in a simplified phenomenological form in order to keep the variation in one (z) direction, and simply shows the tension effect due to the curvature of the field line. Here we neglect the gas pressure and the gravitational forces in Eq. (3.2.75), and the reason will become clear after the self-similar solutions are found. The neglect of B_z-related term in Eq. (3.2.76) may be justified because $B_z \ll B_x$ near the midpoint of the magnetic loop. A central Ansatz of the present quasi-1D model is

$$\partial(\rho V_x)/\partial x = (N-1)\partial(\rho V_z)/\partial z, \tag{3.2.77}$$

where N is assumed to be constant. The left hand term in Eq. (3.2.77) corresponds to fluid leaking along the field line away from the midsection of the loop because the bent loop allows the fluid to escape under the gravitational influence. Here, we measure the amount of matter leakage in the horizontal direction in terms of the vertical flow motion. If $N = 1$, no leakage arises, which corresponds to a pure 1D motion. In order to have matter leakage, $N < 1$. $N - 1$ is a parameter that measures severity of matter leakage in the x-direction. Since the matter leakage is due to the gravity along the magnetic loop, it can also be said that the parameter N is a function of the gravitational acceleration. (Note that V_x is determined from the equation of motion in x-direction (*along* the loop).) Therefore even if the gravity term is out of the basic equations (3.2.74) $-$ (3.2.76), it is implicitly included if we use the equation (3.2.77). We further assume $R = cz$ and $c = $ constant, which is a manifestation of the self-similar evolution of the spatial pattern of the loop, as observed in Fig. 3.31(c).

Under these assumptions, a particular self-similar solution, that satisfies the empirical velocity functions (3.2.67) and (3.2.72) and quasi-1D MHD equations (3.2.74)–(3.2.76), is found;

$$\rho = \sigma_1 \xi^{-4-2q} \exp(-4a\tau) = \sigma_1 z^{-4-2q} \exp(2qat), \tag{3.2.78}$$

$$B_x = b_1 \xi^{-1-q} \exp(-a\tau) = b_1 z^{-1-q} \exp(qat), \tag{3.2.79}$$

where $q = 3N/[2(1-N)]$, $\sigma_1 = \sigma(\xi = 1)$, $b_1 = a[4\pi r_1/(q+1-1/c)]^{1/2}$. The simulation results (3.2.69) [see Figs. 3.32 (c) and (d)] indicate the analytical solution with $N = q = 0$, which leads to a steady solution in Eulerian coordinates in Eqs. (3.2.78)–(3.2.79). The self-similar solution leads to

$$\frac{\rho g}{\partial/\partial z(B_x^2/8\pi)} \propto z^{-1} \tag{3.2.80}$$

and

$$\frac{\partial p/\partial z}{\partial/\partial z(B_x^2/8\pi)} \simeq (C_s/V_A)^2 \propto z^{-2}. \tag{3.2.81}$$

Hence, as the magnetic loop rises, both forces decrease more rapidly than the magnetic force; as long as above force ratios are less than unity at $t = 0$, the neglect of the gravitational

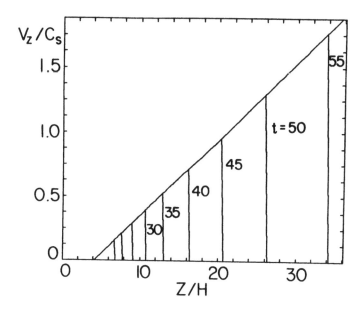

FIGURE 3.34 Approximate quasi-1D self-similar solution; velocity (Shibata *et al.*, 1989a).

and the gas pressure forces in the nonlinear evolution is valid, while the nonlinear growth rate a is found to be related to g. The semi-empirical self-similar solutions are illustrated in Fig. 3.34.

B. General self-similar analysis and solutions

In this subsection, we shall show that there is another class of self-similar solutions with power-law time dependence, in addition to the empirical solution with exponential time dependence discussed in the previous subsection (Shibata, *et al.*, 1990a) We further consider why V_z is proportional to z. Table 3.2 summarizes solutions found in this subsection. One of the power-law solutions has the characteristics that $V_z \propto z/t, \rho \propto z^{-4}, B_x \propto z^{-1}$, and explains the behavior of simulation results after $t > 47$ in Fig. 3.32; after the magnetic loop enters the hot layer.

In contrast to the previous subsection and Shibata *et al.* (1989a), we here use the cylindrical coordinate (r, θ, y), where r is the radial distance from the point $(X_{max}/2, z_0)$, z_0 is the base

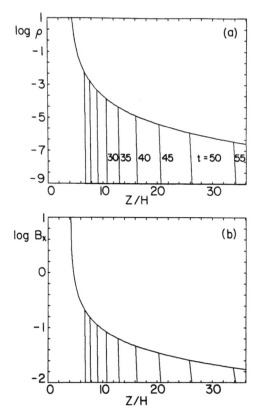

FIGURE 3.35 Approximate quasi-1D self-similar solution; density and magnetic field for $N = 0$ (Shibata *et al.*, 1989a).

height of the initial magnetic flux sheet, and θ is the angle measured from the vertical line towards the positive x-direction. (Hence we have $r \simeq z$ for $|\theta| \ll 1$.) This is because the magnetic field configuration tends to have the current free configuration of $(B_r, B_\theta, B_y) \propto (1/r, 0, 0)$ as time proceeds, so that the equation becomes simpler in the cylindrical coordinate than in the Cartesian coordinate. We also find a new solution, a force-free solution, which are not found in the previous analysis.

We now write the quasi-1D equations (3.2.74)–(3.2.76) in the cylindrical coordinate.

$$\frac{\partial \rho}{\partial t} = -\frac{\partial}{r\partial r}(r\rho V_r) - \frac{\partial}{r\partial \theta}(\rho V_\theta), \tag{3.2.82}$$

$$\rho\left(\frac{\partial V_r}{\partial t} + V_r \frac{\partial V_r}{\partial r}\right) = -\left[\frac{\partial}{\partial r}\left(\frac{B_\theta^2}{8\pi}\right) + \frac{B_\theta^2}{4\pi R}\right], \tag{3.2.83}$$

184

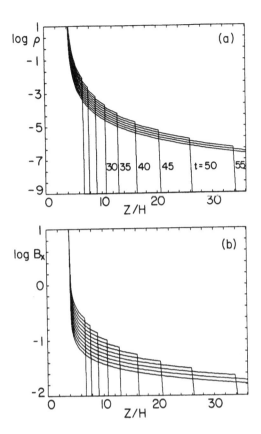

FIGURE 3.36 Approximate quasi-1D self-similar solution; density and magnetic field for $N = -0.5$ (Shibata *et al.*, 1989a).

$$\frac{\partial B_\theta}{\partial t} = -\frac{\partial}{\partial r}(B_\theta V_r), \tag{3.2.84}$$

and we assume again $R = cr$, and

$$\frac{\partial}{r\partial \theta}(\rho V_\theta) = (N-1)\frac{\partial}{r\partial r}(r\rho V_r). \tag{3.2.85}$$

As in the standard self-similar analysis (Zeldovich and Raizer, 1967; Sedov, 1959), we seek a general self-similar solution by adopting the new dimensionless independent variable

$$\zeta = \frac{r}{Z(t)}, \tag{3.2.86}$$

$$\tau = \frac{t}{t_0}, \tag{3.2.87}$$

185

where $Z(t)$ is the scale function, with dimension of length, and t_0 is a normalization constant for the time. We further assume

$$V_r = \dot{Z}(\tau)v(\zeta)/t_0, \tag{3.2.88}$$

$$\rho = \Sigma(\tau)\sigma(\zeta), \tag{3.2.89}$$

$$B_\theta = (4\pi\Sigma\dot{Z}^2)^{1/2}b(\zeta)/t_0, \tag{3.2.90}$$

where v, σ, and b are nondimensional quantities with only ζ-dependence, and the dot represents $d/d\tau$. Substituting equations (3.2.86)–(3.2.90) into equations (3.2.82)–(3.2.84), we have

$$\frac{\dot{\Sigma}}{\Sigma}\frac{Z}{\dot{Z}} + Nv' + N\frac{v}{\zeta} + (Nv - \zeta)\frac{\sigma'}{\sigma} = 0, \tag{3.2.91}$$

$$\frac{Z\ddot{Z}}{\dot{Z}^2}v + (v - \zeta)v' + \frac{b^2}{\sigma}\left(\frac{b'}{b} + \frac{1}{\zeta}\right) = 0, \tag{3.2.92}$$

$$\frac{Z\ddot{Z}}{\dot{Z}^2} + \frac{1}{2}\frac{\dot{\Sigma}}{\Sigma}\frac{Z}{\dot{Z}} + (v - \zeta)\frac{b'}{b} + v' = 0, \tag{3.2.93}$$

where $' \equiv d/d\zeta$.

We adopt the boundary condition for $v(\zeta)$

$$v(\zeta = 0) = 0 \tag{3.2.94}$$

because there is no magnetic flux below $\zeta = 0$, and thus the fluid is stationary there. Thus, $v(\zeta)$ may be written as a self-similar function that satisfies equation (3.2.94),

$$v(\zeta) = \zeta^\delta, \tag{3.2.95}$$

where δ is a positive real number. We now consider the solution of equations (3.2.91)–(3.2.93) for boundary condition (3.2.94) for both $Z\ddot{Z}/\dot{Z}^2 = 1$ and $\neq 1$.

1. Case $Z\ddot{Z}/\dot{Z}^2 \neq 1$: power-law solution

In this case, Z may be written as

$$Z = Z_0\tau^\alpha, \tag{3.2.96}$$

where Z_0 is a normalization constant. We also assume that Σ has a similar power-law dependence on τ,

$$\Sigma = \Sigma_0\tau^\beta, \tag{3.2.97}$$

where Σ_0 is a density normalization constant. This solution in this case (Table 3.2) is called *power-law solution* with a feature $V_z \propto t/z$, which explains the later evolution of the

186

magnetic loop expansion (e.g. see Fig. 10 of Shibata *et al.* (1989a), and Fig. 13 of Nozawa *et al.* (1992)).

One particular case $Z\ddot{Z}/\dot{Z}^2 = 0$ leads to *force-free solution* (Table 3.2), i.e. the solution which shows no acceleration in Lagrangian frame (zero inertial term). The force-free solution for $N = 1$ in Table 3.2 corresponds to a one-dimensional version of Low (1982)'s two-dimensional self-similar solution, although he did not discuss the solution with $N \neq 1$, i.e. the solution with downflows.

2. Case $Z\ddot{Z}/\dot{Z}^2 = 1$: exponential solution

In this case, we have the solution varying as

$$Z = Z_0 \exp(a\tau). \qquad (3.2.98)$$

We assume that the density Σ also has exponential time dependence,

$$\Sigma = \Sigma_0 \exp(n\tau). \qquad (3.2.99)$$

The solution in this case (Table 3.2) is called the *exponential solution*, which is essentially the same as the empirical solution in previous section, and explains numerical simulation results very well. (Note that $p = N/(1-N)$ is different from $q = 3N/2(1-N)$ in Eqs. (3.2.78) and (3.2.79) because the geometry is different.)

From now on, we will derive the exponential solution with $\delta = 1$. In this case, using (3.2.95),(3.2.98), (3.2.99), and $\delta = 1$, equations (3.2.91)–(3.2.93) becomes

$$\frac{n}{a} + 2N + (N-1)\zeta\frac{\sigma'}{\sigma} = 0, \qquad (3.2.100)$$

$$\zeta + \frac{b^2}{\sigma}\left(\frac{b'}{b} + \frac{1}{c\zeta}\right) = 0, \qquad (3.2.101)$$

$$2 + \frac{1}{2}\frac{n}{a} = 0. \qquad (3.2.102)$$

Thus for $N \neq 1$ we have

$$\frac{n}{a} = -4, \qquad (3.2.103)$$

$$\sigma = \sigma_1\zeta^{-4-2p}, \qquad (3.2.104)$$

and

$$b = b_1\zeta^{-1-p}, \qquad (3.2.105)$$

where $p = N/(1-N)$, and $b_1 = \{\sigma_1/[1/(1-N) - (1/c)]\}^{1/2}$. The solution is summarized as follows:

$$V_r = (aZ_0/t_0)\zeta \exp(a\tau) = (a/t_0)r, \qquad (3.2.106)$$

TABLE 3.2 Summary of self-similar solutions

$Z\ddot{Z}/\dot{Z}^2 \neq 1$ power-law	$\delta \neq 1$	$N \neq 0$	no solution
		$N = 0$	$V_r \propto \tau^{\delta/(1-\delta)}\zeta^\delta \propto r^\delta$ $B_\theta \propto \tau^{-\delta/(1-\delta)}\zeta^{-\delta} \propto r^{-\delta}$ $\rho \propto \tau^{-4\delta/(1-\delta)}\zeta^{-4\delta} \propto r^{-4\delta}$ $\alpha = 1/(1-\delta)$
$[Z \propto \tau^\alpha]$	$\delta = 1$	$N \neq 1$	$V_r \propto \tau^{\alpha-1}\zeta^\alpha \propto r/t$ $B_\theta \propto \tau^{-\alpha}\zeta^{-1-h} \propto t^{-h\alpha}r^{-1-h}$ $\rho \propto \tau^{2-4\alpha}\zeta^{-4-2h} \propto t^{2(1+h\alpha)}r^{-4-2h}$ where $h = (N - 1/\alpha)/(1-N)$
		$N = 1$	same as the force-free solution
force-free $[Z \propto \tau]$	$\delta = 1$	$N = 1$	$V_r \propto \zeta \propto r/t$ $B_\theta \propto \tau^{-1}\zeta^{-1} \propto r^{-1}$ $\rho \propto \tau^{-2}\zeta^{-\mu} \propto t^{\mu-2}r^{-\mu}$ $\alpha = 1, c = 1, \mu = $ arbitrary
		$N \neq 1$	$V_r \propto \zeta \propto r/t$ $B_\theta \propto \tau^{-1}\zeta^{-1} \propto r^{-1}$ $\rho \propto \tau^\beta\zeta^{\beta+w} \propto t^{-w}r^{\beta+w}$ $w = (\beta+2)N/(1-N)$ $\alpha = 1, c = 1, \beta = $ arbitrary
$Z\ddot{Z}/\dot{Z}^2 = 1$ exponential	$\delta \neq 1$	any N	no solution
	$\delta = 1$	$N \neq 1$	$V_r \propto \zeta \exp(a\tau) \propto r$ $B_\theta \propto \zeta^{-1-p}\exp(-a\tau) \propto r^{-1-p}\exp(pat)$ $\rho \propto \zeta^{-4-2p}\exp(-a\tau) \propto r^{-4-2p}\exp(2pat)$ where $p = N/(1-N)$
$[Z \propto \exp(a\tau)]$		$N = 1$	no solution

$$\rho = \Sigma_0\sigma_1\zeta^{-4-2p}\exp(-4a\tau) = \Sigma_0\sigma_1\left(\frac{r}{Z_0}\right)^{-4-2p}\exp\left[2pa\frac{t}{t_0}\right], \qquad (3.2.107)$$

$$B_\theta = B_0 b_1 \zeta^{-1-p}\exp(-a\tau) = B_0 b_1\left(\frac{r}{Z_0}\right)^{-1-p}\exp\left[pa\frac{t}{t_0}\right]. \qquad (3.2.108)$$

As we already mentioned, these solutions are essentially the same as the empirical self-similar solution in (3.2.78) and (3.2.79).

Problem 3–3: Derive the power-law solution and the force free solution in Table 3.2.

Note: $\zeta = r/Z(\tau)$ is Lagrangian coordinate, r is Eulerian coordinate (height measured from the base of the initial magnetic flux tube), $r \simeq z$ for $|\theta| \ll 1$, and δ is the exponent of

I.	linear stage	**linear instability** $$\omega_l \simeq 0.2(1+2\beta_*)^{-1/2}C_s/H \quad \text{(for } \beta_* \le 2)$$ $$V_z \propto \exp(\omega_l t)$$
II.	nonlinear stage (A)	**exponential expansion** (nonlinear instability) $$V_z \simeq az \sim (\omega_l/2)z$$ (i.e. $V_z \propto \exp(at) \sim \exp(\omega_l t/2)$ in Lagrangian frame) $$V_A \simeq bz \simeq 5az$$ $$\rho \propto z^{-4}, \; B_x \propto z^{-1}$$
III.	nonlinear stage (B)	**power-law expansion** $$V_z \simeq az/t \quad (\alpha > 1)$$ (i.e. $V_z \propto t^{(\alpha-1)}$ in Lagrangian frame) $$\rho \propto z^{-4}, \; B_x \propto z^{-1}$$
III.′	nonlinear stage (B′)	**force-free expansion** $$V_z \simeq z/t$$ (i.e. $V_z = $ constant in Lagrangian frame) $$\rho \propto z^{-4}, \; B_x \propto z^{-1}$$ $$V_{max} \simeq (1+2\beta_*)^{1/2}C_s$$ at $z > z \simeq 10(1+2\beta_*)H$
IV	final stage	a) **Quasi-static state:** If there is a hot (high pressure) corona or strong magnetic fields above the loop, the expansion stops at $$z_d/H = (1+1/\beta_{cor})^{-1/2}(p_{g,cor}/p_{g,0})^{-1/2}e^{-1/2}$$ $$= (1+1/\beta_{cor})^{-1/2}\exp[z_{cor}/(2H)]e^{-1/2}$$ b) **Unlimited expansion:** If there is no hot (high pressure) corona or if $p_{mag,loop} > p_{ambient}$, the loop expansion continues to infinity

velocity function ($V_r \propto r^\delta$).

3.2.2.7 Physical Interpretation

A. Classification of evolutional stage by curvature radius of magnetic field lines

On the basis of numerical and analytical results obtained in previous sections, we can construct a rough model of linear-nonlinear evolution of the undular mode of the magnetic buoyancy instability. A key physical parameter is R, a curvature radius of magnetic field lines at the midpoint of the loop. In the linear regime, R decreases with time (or height), $R_l = (\lambda/2\pi)^2/z$, while R increases with time (height), $R_n = cz$, in the nonlinear regime (Fig. 3.37). The transition from the linear regime to the nonlinear regime occurs when the two radii become equal, $R_l = R_n$. This occurs at the height $z = z_c = \lambda/(2\pi c^{1/2})$, where the curvature radius is $R = R_c = c^{1/2}\lambda/2\pi$. In the simulation model, (Figs. 3.31-3.33) $z_c \simeq 3$, because $\lambda = 20$ and c is of order of unity. The actual curvature radius in the simulation results at $z \sim z_c$ is not equal to R_c, but is somewhat larger than R_c and is nearly constant ($\simeq R_0$) near z_c. This stage (called stage II) corresponds to the nonlinear stage characterized by the exponential expansion of magnetic loops (Fig. 3.37). When R begins to increase with height, the nature of the magnetic loop expansion gradually changes from the exponential to the power-law type expansion. In section E, we will calculate the second critical height z_t beyond which the solution has the power-law time dependence.

B. Non-equilibrium in magnetic flux in nonlinear stage of the undular magnetic buoyancy instability in isothermal layer

We here discuss why the magnetic flux expansion initiated by the undular instability does not decelerate in an isothermal layer in the nonlinear stage. First, we should note that the vertical thickness (D) of our flux sheet is not thin; $D = 4H > H$, where $H = C_s^2/(\gamma g)$ is the pressure scale height. Therefore, the thin tube approximation ($D \ll H$) cannot be used; i.e. $B \neq$ constant inside the flux sheet. Secondly, we note $\beta = 1$ initially, so that the magnetic pressure cannot be assumed to be a perturbation. For these reasons, we have to consider internal structure of the flux sheet as shown in Fig. 3.38. The expanding flux sheet may be approximated by a current free (potential) magnetic field because plasma pressure (or, equivalently plasma β) is decreased inside the flux sheet owing to the downflow along the field line. If the potential field is approximated by the field illustrated in Fig. 3.39, the field strength at the midpoint of the loops is $B_x \propto \exp(-z/H_m)$, where $H_m = \lambda/\pi = 6.4H$. Thus the magnetic pressure decreases with height; $p_m \propto \exp(-2z/H_m) = \exp[-z/(3.2H)]$. On the other hand, the gas pressure outside the magnetic flux decreases with height as $p_g \propto \exp(-z/H)$. Consequently, if $\beta = 1$ at the base of the flux sheet, the magnetic pressure at the top of the loop cannot be balanced with the outside gas pressure. This is the reason why there is no saturation (or non-equilibrium) in the nonlinear stage as long as the gas layer is isothermal. There is a critical horizontal wavelength, $\lambda_c = 2\pi H$, beyond which there is no equilibrium state. In other words, the necessary condition for the non-equilibrium is

$$\lambda > \lambda_c = 2\pi H. \tag{3.2.109}$$

In our case, $\lambda = 20H$, and hence this condition is satisfied. [For more exact treatment, see Low (1981). It is interesting to note that the above condition is the same as that for the thin tube (Parker, 1979).]

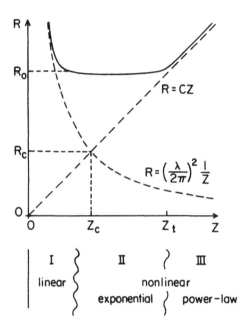

FIGURE 3.37 The schematic picture of the curvature radius R of a magnetic field line as a function of the height z, and the classification of the evolutionary stages: I (linear), II (nonlinear; exponential time dependence), and III (nonlinear; power-law time dependence). The solid curve shows the actual (but not exact) curvature radius of a rising magnetic loop, the dashed curve with $R = (\lambda/2\pi)^2/z$ corresponds to the curvature radius of a loop in the linear regime, and another dashed line ($R = cz$) shows that of the nonlinear stage. From Shibata et $al.$ (1990a)

C. Nonlinear evolution in stage II: exponential expansion

In this subsection, we consider the nonlinear evolution of magnetic loop expansion in stage II around $z \sim z_c$ (Fig. 3.37), and derive basic formulae found in numerical and analytical studies in previous sections, using an idealized model illustrated in Fig. 3.40.

We first consider the motion of a fluid element at s along the magnetic loop, where s is the distance from the loop top along the loop. If the gas pressure is neglected, the fluid element falls freely along the loop due to gravity. The equation of motion along the loop is written as

$$dV_s/dt = d^2s/dt^2 = g\sin\theta \simeq g\theta \simeq gs/R, \tag{3.2.110}$$

191

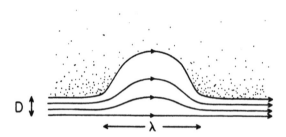

FIGURE 3.38 Schematic illustration of the expanding magnetic flux as a result of the undular mode of the magnetic buoyancy instability with wavelength λ. Note that the thickness of the sheet D is much larger than the pressure scale height H outside the flux sheet. From Shibata *et al.* (1990a).

for $\theta \ll 1$. If $R \simeq R_0$ is nearly constant in time as discussed in section **A.**, we find

$$s \propto \exp(\Omega t), \tag{3.2.111}$$

$$V_s = \Omega s \propto \exp(\Omega t), \tag{3.2.112}$$

where

$$\Omega = (g/R_0)^{1/2}. \tag{3.2.113}$$

We next consider the vertical motion of a local part of the loop (shaded area in Fig. 3.40) near the midpoint of the loop, and assume $V_z = az$ in Eulerian coordinate and thus $V_z \propto \exp(at)$ in Lagrangian coordinate on the basis of numerical results in previous section. The cross-sectional area of the loop is A, and by assumption $V_z = dA/dt$ and $A \propto \exp(at)$.

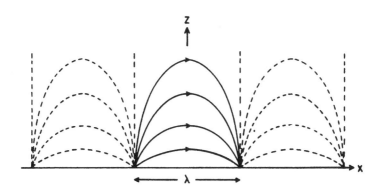

FIGURE 3.39 An example of the potential magnetic field when the horizontal wavelength is fixed to λ even in the nonlinear stage. From Shibata *et al.* (1990a).

From conservations of magnetic flux and mass, we have $B_s = 1/A \propto \exp(-at)$ and $\rho As =$ constant, respectively. Then the density inside the shaded area decreases with time as

$$\rho \propto \exp[-(\Omega + a)t]. \tag{3.2.114}$$

The relation between a and Ω is found from the consideration of the equation of motion in z-direction,

$$\frac{\partial V_z}{\partial t} = -g - \frac{1}{\rho}\frac{\partial p}{\partial z} - \frac{1}{\rho}\left[\frac{\partial}{\partial z}\left(\frac{B_s^2}{8\pi}\right) + \frac{B_s^2}{4\pi R}\right]. \tag{3.2.115}$$

The term in the left hand side scales as $\exp(at)$ by the assumption $V_z \propto \exp(at)$. The first term in the right-hand side, i.e. the gravitational force term, is constant in time. The second term (the gas pressure force) scales as $\propto \exp(-at)$, and the third term (the magnetic pressure gradient force) $\propto \exp[(\Omega - 2a)t]$. Since initially the first and second terms are comparable to the third term because $\beta = 1$, the first and second terms can be neglected

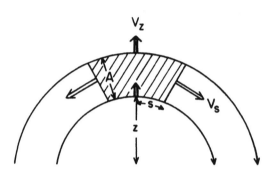

FIGURE 3.40 An idealized model for the rising magnetic loop with downflow along it due to gravity. From Shibata *et al.* (1990a).

in the later stage if $\Omega > 2a$. The fourth term (the magnetic tension force) is in proportion to $\exp[(\Omega - a)t]$ if $R = R_0 =$ constant. Although the magnetic tension term increases with time more rapidly than the magnetic pressure term, the former is much smaller than the latter in the early stage, and hence may be neglected before it becomes comparable to the magnetic pressure term. In later phase of Stage II, the tension term has the time dependence of $\propto \exp[(\Omega - 2a)t]$, which is the same as that of the magnetic pressure. Altogether, we find that the left hand side ($\propto \exp(at)$) should balance with the third term in the right-hand side ($\propto \exp[(\Omega - 2a)t]$), and thus $a = \Omega - 2a$, or

$$a = \frac{\Omega}{3} = \frac{1}{3} \left(\frac{g}{R_0} \right)^{1/2}. \tag{3.2.116}$$

From this relation, we find that the parameter N introduced in Eq. (3.2.77) should be equal to 0;

$$N = 1 + \frac{\partial(\rho V_x)/\partial x}{\partial(\rho V_z)/\partial z} = 1 + (3az^{-4})/(-3az^{-4}) = 0, \qquad (3.2.117)$$

because $\rho \propto \exp[-(\Omega + a)t] = \exp(-4at) \propto z^{-4}, V_x = \Omega s \simeq \Omega x = 3ax, V_z = az$. Since $\Omega = (g/R_0)^{1/2}$ is of order of the linear growth rate (ω_l) of the undular instability, the nonlinear growth rate $\omega_n = a$ is also of order of the linear growth rate; i.e. $\omega_n \simeq \Omega/3 \sim \omega_l/3$. This explains the numerical results found in Fig. 3.33. Consequently, one can understand that the exponential time dependence of the physical quantities in the Lagrangian coordinate found in numerical and self-similar solutions originates from the exponential time dependence of the downflow speed along the loop due to gravity as shown in equation (3.2.112).

D. Nonlinear evolution in stage III: power-law expansion

As the magnetic tension force becomes comparable to the magnetic pressure force, the nature of the expansion changes. That is, the curvature radius increases with height, and the whole magnetic configuration becomes close to that of the potential magnetic field produced by a line current at $(x, z) = (X_{max}/2, 0)$;

$$B_x \propto z/(\Delta x^2 + z^2), \qquad (3.2.118)$$

$$B_z \propto -\Delta x/(\Delta x^2 + z^2), \qquad (3.2.119)$$

where $\Delta x = x - X_{max}/2$. The distribution of B_x at the midpoint of the loop ($\Delta x = 0$) in Stage II already has the same distribution $B_x \propto 1/z$ as that given by Eq. (3.2.118).

We now show the reason why the expansion law changes from the exponential to the power-law one. Suppose that the curvature radius R is in proportion to s. In this case, we have

$$\frac{d^2 s}{dt^2} \simeq g\frac{s}{R} \simeq g_0, \qquad (3.2.120)$$

where $g_0 = gs/R = $ constant. Then we find $s = g_0 t^2/2$ and

$$V_s = g_0 t = 2s/t. \qquad (3.2.121)$$

This has the power-law time dependence, and this may be the origin of the power-law expansion of the magnetic flux. On the other hand, the equation of motion vertical to the loop is

$$dV_z/dt = -\frac{1}{\rho}\left[\frac{\partial}{\partial z}\left(\frac{B^2}{8\pi}\right) - \frac{B^2}{4\pi R}\right]. \qquad (3.2.122)$$

If the right-hand side is exactly equal to zero, i.e. if the magnetic field is exactly the potential field, we have also the power-law solution for V_z;

$$V_z = V_0 = z/t. \qquad (3.2.123)$$

195

These solution satisfies the mass continuity equation

$$\frac{\partial \rho}{\partial t} = -\frac{\partial}{\partial s}(\rho V_s) - \frac{\partial}{\partial z}(\rho V_z) - \frac{\rho V_z}{R},$$ (3.2.124)

if

$$\rho = \rho_0 z^{-4},$$ (3.2.125)

where ρ_0 is a constant. Note that we assume $dz = dr$, $ds = rd\theta$, and $R = r$ in Eq. (3.2.82) where r is the radial distance from the origin $(x, z) = (X_{max}/2, 0)$ in the cylindrical coordinate (r, θ). The distribution of B_z in the potential field [Eqs. (3.2.118) and (3.2.119)]) also satisfies the induction equation [see Eq. (3.2.76)]. Consequently, we find that the power-law solution discussed here corresponds to one of the power-law solutions discussed in Table 3.2 (the case of $\beta = -4$ and $N = 0$ in force free solution).

E. The second critical height z_t and the maximum velocity of the magnetic loop

We calculate the second critical height z_t between Stages II and III (Fig. 3.37). This height corresponds to the height where the curvature radius ($R_n = cz$) of magnetic field lines in nonlinearly expanding magnetic flux becomes equal to R_0. Thus we have

$$z_t = R_0/c.$$ (3.2.126)

To calculate z_t as a function of β_* and H, we use a relation between the nonlinear and linear growth rates,

$$\Omega/3 \sim \omega_l/2.$$ (3.2.127)

By using Eq. (3.2.113), $\Omega = (g/R_0)^{1/2}$, and the approximate analytical expression for the linear growth rate (for $\beta_* \leq 2$), $\omega_l \simeq 0.2(1 + 2\beta_*)^{-1/2} C_s/H$, we find

$$R_0 \simeq 10(1 + 2\beta_*)H.$$ (3.2.128)

When $\beta_* \simeq 1$, we find $R_0 \simeq 30H$ and thus $z_t \simeq 30H/c \sim 30H$ if $c \sim 1$. This value seems to be consistent with the approximate critical height found in numerical simulations.

Since the Stage III has not yet been studied in detail by nonlinear simulations because of some numerical difficulties, it may be premature to discuss the properties of the power-law solution shown in the previous subsection. Therefore we here discuss only that there is a maximum velocity (V_{max}) of the magnetic loop expansion, if the analysis in section **D.** is correct and hence the velocity of the magnetic loop in a Lagrangian frame is nearly constant in Stage III (see Eq. [3.2.123]). If the gas layer is isothermal in an infinite space, this velocity corresponds to the terminal velocity, and is simply evaluated as

$$V_{max} = az_t \simeq \frac{1}{2}\omega_l R_0 \simeq (1 + 2\beta_*)^{1/2} C_s,$$ (3.2.129)

for $\beta_* \leq 2$. Thus the maximum velocity is $\sim 1.7 C_s$ when $\beta_* = 1$, consistent with simulation results.

3.2.2.8 Comparison with Rayleigh-Taylor Instability

It is known that the rise velocity V_b of the bubble observed in the laboratory (Sharp, 1984) and in the ionosphere (Ossakow and Chatuvedi, 1978) tend to be steady in the Lagrangian frame and is in proportion to the radius R_b of the bubble;

$$V_b = a_3 R_b, \tag{3.2.130}$$

where

$$a_3 \simeq (1/3 - 1/2) \times (g/R_b)^{1/2} \tag{3.2.131}$$

is of the order of the linear growth rate of the Rayleigh-Taylor instability. This is similar to the rise velocity of the Parker-unstable magnetic loop (Shibata et al., 1989a, 1990a), $V_z = az \simeq aR$, where R is the curvature radius of the magnetic loop, $a \simeq \omega_l/2$ and ω_l is the linear growth rate. However, the rise velocity of the Parker-unstable magnetic loop is not steady in the Lagrangian frame, but increases exponentially with time in the stage II (i.e. the nonlinear-exponential stage; see Fig. 3.37).

This *nonlinear instability* in the Lagrangian frame is also observed in the exact solution found by Ott (1972) for the Rayleigh-Taylor instability of a thin, cold gas layer, which is supported against gravity by a hot gas, with a second hot gas above the thin layer (see Fig. 3.41). In this case, the growth rate in the nonlinear stage is exactly the same as that in the linear stage. Physically, this is because cold gas in the thin layer freely falls along the curved interface between two hot gases, and mathematically, because the nonlinear basic equations become linear ones in the Lagrangian frame (Dawson, 1959; Davidson, 1972; see also Sec. 4.1).

Let us consider this problem briefly below. Assume that the thin, cold gas layer has surface density σ_0. From the pressure balance, we have

$$p_b = p_a - \sigma_0, \tag{3.2.132}$$

where p_a and p_b are gas pressures below and above the thin layer, and are assumed to be constant. It is also assumed that the hot gas layer are so tenuous that the inertia of hot gas layer does not affect dynamics of the thin layer. Suppose then the layer is disturbed to undulate. Taking the Lagrangian coordinate ξ along the layer, and Cartesian coordinate (x, y) such that the initial gas layer is parallel to the x direction, and y is in the vertical direction opposite to the gravitational acceleration, the gravitational force acting on a fluid element between ξ_0 and $\xi_0 + d\xi_0$ in y direction is written as $g\sigma_0 d\xi_0$. The pressure force acting on the element in y direction becomes $(p_a - p_b)(\partial x/\partial \xi_0)d\xi_0$. Then the equation of motion in y direction becomes

$$\sigma_0 d\xi_0 \frac{\partial^2 y}{\partial t^2} = (p_a - p_b)\frac{\partial x}{\partial \xi_0}d\xi_0 - g\sigma_0 d\xi_0, \tag{3.2.133}$$

and hence using (3.2.132) we get

$$\frac{\partial^2 y}{\partial t^2} = g\frac{\partial x}{\partial \xi_0} - g. \tag{3.2.134}$$

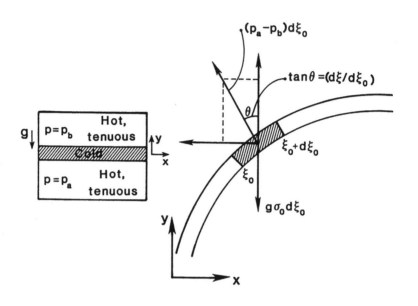

FIGURE 3.41 The Rayleigh-Taylor instability of a thin cold layer.

Similarly, the pressure force acting on the element in x direction is $-(p_a - p_b)(\partial y/\partial \xi_0)d\xi_0$, so that the equation of motion in x direction becomes

$$\sigma_0 d\xi_0 \frac{\partial^2 x}{\partial t^2} = -(p_a - p_b)\frac{\partial y}{\partial \xi_0}d\xi_0. \tag{3.2.135}$$

Using (3.2.132), this is written as

$$\frac{\partial^2 x}{\partial t^2} = -g\frac{\partial y}{\partial \xi_0}. \tag{3.2.136}$$

Combining equations (3.2.134) and (3.2.136), we finally find

$$\frac{\partial^4 x}{\partial t^4} = -g^2\frac{\partial^2 x}{\partial \xi_0^2}. \tag{3.2.137}$$

This is a *linear* equation, and hence easily solvable, yielding a special exact solution;

$$x = \xi_0 - A\exp(\omega t)\cos k\xi_0, \tag{3.2.138}$$

198

$$y = A \exp(\omega t) \sin k\xi_0, \tag{3.2.139}$$

where

$$\omega = (kg)^{1/2}, \tag{3.2.140}$$

and k is the wavenumber of the disturbance. In this case, the nonlinear growth rate ω is exactly the same as the growth rate of the linear (Rayleigh-Taylor) instability.

Although the nonlinear growth rate in the Parker unstable magnetic loop is not exactly equal to the linear growth rate, the involved physics is common between the Parker unstable magnetic loop and Ott's problem; the exponential growth in the nonlinear stage is due to the gravitational free fall *along* the magnetic loop. That is, the equation of motion along magnetic loop in the former problem is written as $d^2s/dt^2 \simeq (g/R)s$ as shown in Eq. (3.2.110), where s is the horizontal distance from the midpoint of the loop. This equation has the exponential solution with the growth rate of $(g/R)^{1/2}$. In addition to this character of *nonlinear instability*, this solution has the *self-similar* property, which is not in Ott's solution.

Finally, we note that the magnetic flux expansion considered in the previous section is similar to the free expansion of a magnetized plasma into vacuum in some sense, because the magnetic pressure at the top of the magnetic flux is much larger than the ambient gas pressure. There is, however, one important difference between the Parker unstable case and the pure one-dimensional expansion of a magnetized plasma where there is no gravity and hence the plasma flow vector is exactly perpendicular to field lines; in the latter case, the expansion velocity is constant, while in the Parker case the velocity increases with time (or height) in the exponential phase (Stage II in Fig. 3.37). This is because the magnetic loop is evacuated by the downflow due to gravity along the loop, which increases the local Alfvén speed inside the loop. If there is a high pressure plasma above the expanding magnetic flux, the acceleration of the magnetic loop ceases soon, and the entire flux is decelerated to find a final equilibrium state (Shibata *et al.*, 1989a, 1989b, see also Table 3.3).

Problem 3–4: Calculate the rise speed U of a steadily rising non-magnetic bubble with a radius R under uniform gravitational acceleration g. Assume incompressible and inviscid fluid (Davies and Taylor, 1950).
(Ans. $U = (gR)^{1/2}/2$.)

3.2.2.9 Application to Solar Emerging Flux*

There are a lot of observational evidence that sunspots are formed by the emergence of magnetic flux tubes from below the photosphere (Zwaan, 1987). Newly emerged magnetic flux is called *emerging flux*, and the region of emerging flux is called *emerging flux region* or *EFR*. The observed characteristics of emerging flux are summarized as follows (see also Fig. 3.42).

(1) The first appearance of emerging flux in the solar photosphere is a tiny bipolar magnetic element observed in magnetograms, or a *granular dark lane* observed in the white

Schematic Picture of Emerging Flux Region (EFR)

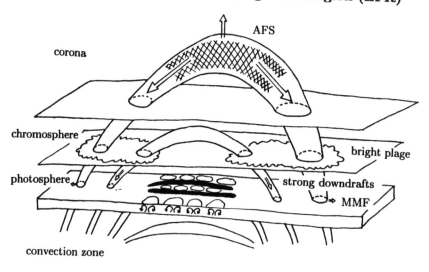

FIGURE 3.42 Schematic illustration of emerging flux in the solar atmosphere (from Shibata *et al.*, 1989b).

light granules in the photosphere (Brants, 1985). The width of each dark lane is about 1000 km, and magnetic field strength is about 500±300 G (Brants, 1985). Faculae or pores are observed near the edges of dark lanes.

(2) The next observational evidence for emerging flux in the photosphere is *strong downdrafts* near pores (Kawaguchi and Kitai, 1976; Brants and Steenbeck, 1985b). The downflow speed parallel to the line-of-sight direction is 1–2 km/s at the photospheric level. Since the tube is nearly horizontal between two polarities, the actual downflow speed along the loop may be much larger (Shibata, 1980). At this stage, the rise velocity of emerging flux is very small in the photosphere, and observations give a velocity of less than $1 - 2\,\text{km/s}$ (Kawaguchi and Kitai, 1976; Brants and Steenbeck, 1985; Chou and Wang, 1987; Tarbel *et al.*, 1988). However, this may be due to low spatial resolution, and the actual rise velocity may be even larger.

(3) The first chromospheric response to emerging flux is Hα (or Ca II) *brightening* or *bright plages* (Bumba and Howard, 1965; Born, 1974; Glackin, 1975). On the other hand, Kurokawa *et al.* (1988) found that the first manifestation of emerging flux in Hα is a *surge*

200

activity (Kurokawa and Kawai, 1993).

(4) The most well-known observational signature of emerging flux is *arch filament*. The emerging flux region showing arch filaments is often called *arch filament system* or simply *AFS*. The first arch of an AFS appears in Hα from 1 hour to about 1.5 hour following the initial Hα brightening (Glackin, 1975; Kawaguchi and Kitai, 1976). The rise velocity of arch filaments is $10 - 15$ km/s (Bruzek, 1967; 1969; Chou and Zirin, 1988), and the downflow along the filament amounts to $30 - 50$ km/s. The length of the filament (as well as the size of an AFS) is (1–3) $\times 10^4$ km. The lifetime of individual filaments is about 20 minutes (Bruzek, 1967). At this stage, *moving magnetic features* (MMFs) with a horizontal velocity of 0.5–2 km/s are frequently observed at the foot of the arch filaments in magnetograms.

These observational characteristics of emerging flux region are quite nicely explained by the numerical simulation results of the nonlinear evolution of the Parker instability (Shibata *et al.*, 1989a, 1989b, 1990a).

For example, the quite different observed rise velocities of magnetic flux (loop) in the photosphere and chromosphere such as $V_{rise} \simeq 1 \sim 2$ km/s in the photosphere ($z < 400$ km), $V_{rise} \simeq 10 \sim 15$ km/s in the chromosphere ($z \sim 3000 \sim 10000$ km) are nicely explained by the following formula found from the numerical simulation (and self-similar theory),

$$V_z = \omega_n z \sim 0.06 C_s (z/H) \sim 12 \text{ km/s } \left(\frac{z}{4000\text{km}}\right) \left(\frac{H}{200\text{km}}\right)^{-1} \left(\frac{C_s}{10\text{km/s}}\right). \qquad (3.2.141)$$

Observed downflow speed along the rising magnetic loop in the chromosphere ($30 - 50$ km/s) and in the photosphere ($1 - 2$ km/s) are also well explained by the simulation results.

Furthermore, the bright plages near the footpoint of the arch filament may be explained by the shock heating at the footpoint of the rising magnetic loop. The total energy dissipated at the shock is estimated to be $\sim 10^{25}$ erg/s (assuming the area of 10^{18} cm^2), which is comparable to the observed value (Svestka, 1976).

Finally, the surge activity often observed in EFRs is explained as a result of reconnection between emerging flux and pre-existing magnetic field as shown by the numerical simulation of reconnection associated with emerging flux (Shibata *et al.*, 1992; Yokoyama and Shibata, 1995), which will be discussed in Sec. 3.3.2.

3.2.2.10   Three Dimensional Effect: Coupling with Interchange Mode

What will happen for the nonlinear evolution of the Parker instability in real 3D space? In 3D space, the interchange mode can couple with the undular (Parker) mode (Fig. 3.43). Matsumoto and Shibata (1992) studied this effect with full 3D simulation, and found that the resulting magnetic structure is a *interleaved structure* for both galactic case and solar case. It is also found that the *vortex motion* occur around the rising magnetic loops, which then produce *magnetic twist* or *torsional Alfvén waves* (Fig. 3.44). Interestingly, even with such interleaved structure, the overall dynamics such as the self-similar expansion, is similar to that found from 2D simulations, as long as the magnetic flux *sheet* is assumed initially.

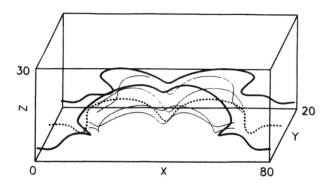

FIGURE 3.43 Cartoon showing the Parker instability coupled with interchange mode (Matsumoto *et al.*, 1993).

Matsumoto *et al.* (1993) also studied 3D evolution of the emergence of the isolated magnetic flux tube, and found that the tube expands not only to the upward direction but also to the horizontal direction perpendicular to the tube. Such horizontal expansion decreases internal magnetic pressure significantly, and so the rise motion of the tube tends to be suppressed when the tube size (or the total magnetic flux of the tube) is smaller. It is found that there is a threshold magnetic flux for the tube to rise into the corona, which is about 0.3×10^{20}Mx. Interestingly, this agrees with observationally known threshold flux for the formation of arch filament system (Zwaan, 1987).

It should be noted that the linear growth rate of the interchange mode is much larger than that of the undular (Parker) mode. Thus, in the 3D evolution of the Parker instability of a magnetic sheet, the interchange mode first grows to create filamentary structures near the surface of the sheet. In the fully nonlinear stage, however, the interchange mode saturates soon, whereas the undular (Parker) mode does not saturate (as we have seen in 2D case). The reason of the early saturation of the pure interchange mode is as follows. The rise velocity of the magnetic tube is determined by the local Alfven speed. In the case of the

202

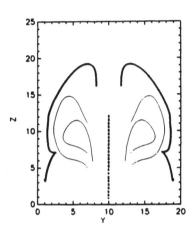

FIGURE 3.44 3D simulation of the Parker instability coupled with interchange mode (Matsumoto *et al.*, 1993).

pure interchange mode, the tube remains straight, so that $B \propto \rho$. Hence,

$$V_A \propto B/\rho^{1/2} \propto \rho^{1/2}.$$

Since the density ρ decreases with height as the tube rises and expands, the Alfven speed V_A *decreases* with height. On the other hand, in the case of the undular (Parker) mode, the Alfven speed *increases* with height, because the density decreases much faster than the magnetic field strength owing to the gravitational downflow along the rising magnetic loop. This is why the undular (Parker) mode is more important than the interchange mode in the fully nonlinear stage, even if the linear growth rate of the undular mode is much smaller than that of the interchange mode.

In summary, the interchange mode plays an important role to create filamentary structures of magnetic fields, though the overall nonlinear dynamics is governed by the physics of the undular (Parker) mode (Matsumoto and Shibata 1992, Matsumoto et al. 1993).

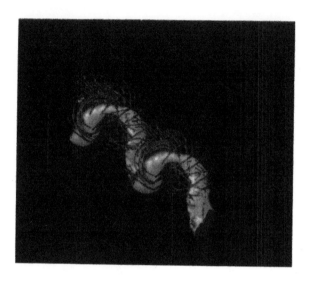

FIGURE 3.45   3D simulation of emergence of twisted magnetic flux tube (from Matsumoto *et al.*, 1996).

3.2.2.11   Emergence of Twisted Flux Tube*

Observations (Kurokawa, 1989; Leka, 1994a) suggest that in some active regions the emerging flux is already twisted before emergence. Matsumoto *et al.* (1996), studied the nonlinear evolution of the coupling of the Parker instability with the kink instability as a model of emergence of a twisted flux tube. They found that the resulting magnetic flux tube show the double helix pattern (Fig. 3.45) and that their results explain various observations such as the peculiar motion of sunspots and the apparent sheared S structure of coronal loops found by *Yohkoh* (Fig. 3.46).

3.2.2.12   Effect of Rotation and Shearing Motion*

It is known (e.g., Shu, 1974; Hanawa *et al.*, 1992b) that the Coriolis force has a stabilizing effect on the Parker instability and tends to reduce the linear growth rate. It is interesting, however, to note that the critical wavelength remains the same even when there is a Coriolis force, and that the Coriolis force cannot stabilize the Parker instability completely.

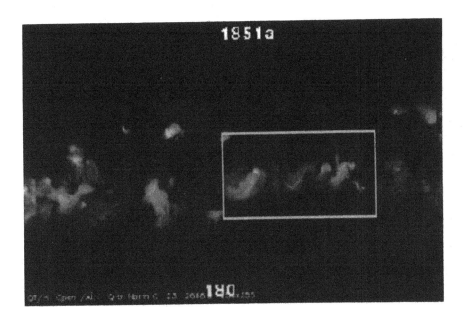

FIGURE 3.46   Yohkoh observations of solar corona showing sheared S structure (from Matsumoto *et al.*, 1997).

Chou *et al.* (1997) performed 3D MHD simulation to study the effect of Coriolis force on the nonlinear evolution of the Parker instability of an isolated magnetic flux tube in the isothermal gas layer which is rotating rigidly at angular speed Ω. They found that in the nonlinear stage, the rising magnetic flux tube is twisted by the Coriolis force to form a globally twisted flux tube (Fig. 3.47). This mechanism explains the observed tilt of bipolar sunspot groups.

Shibata and Matsumoto (1991), on the other hand, noted that the local twist may be generated in the downflow along the rising magnetic loop in addition to the global twist, since the downflowing blob or cloud contracts during the course of the downflow and suffer from the Coriolis force, similar to usual contracting cloud in our Galaxy (Fig. 3.48). The sense of the rotation of the contracting cloud is the same as that of the rotation of the Galaxy. The twist accumulates in the valley of the undulating magnetic field lines, where the giant molecular cloud is formed. They applied this mechanism to explain the helical magnetic twist observed in a giant molecular cloud in the Galaxy (Uchida *et al.*, 1990). Turbulent magnetic fields often observed in the valley of undulating magnetic fields [e.g. M31 (Beck,

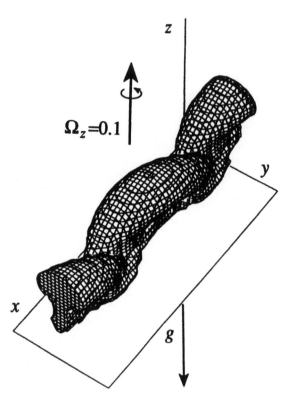

FIGURE 3.47 Twisted flux tube formed by the nonlinear evolution of the Parker instability in the presence of Coriolis force (Chou *et al.*, 1995).

1989) may also be explained by this mechanism.] The twisted magnetic field thus created is, in turn, favorable for the formation of large scale cloud complex since the twist (magnetic shear) stabilize the small scale interchange mode (Hanawa *et al.*, 1992a). It should be noted here that *the direction of this local twist is opposite to that of the global twist*. Hence, this mechanism may explain the origin of magnetic twist observed in the solar active regions (Pevtsov *et al.*, 1994) and the filaments (Rust, 1994).

In actual rotating systems, there is a shear flow, i.e., differential rotation. What is the effect of shearing flow on the Parker instability? Shibata, *et al.* (1990a) studied this effect in a situation such that the horizontal magnetic flux (which is unstable for the Parker instability if there is no shear flow) is suffering from the amplification by the shear flow in an accretion disk. Usually the field amplification by the shear flow leads to instability or loss of equilibrium (as is well known in theoretical research of solar flares). However, against the expectation, this case leads to effective *stabilization* of the Parker instability, since the shearing time scale is shorter than the Parker time scale. As a result of this dynamic stabilization, the

206

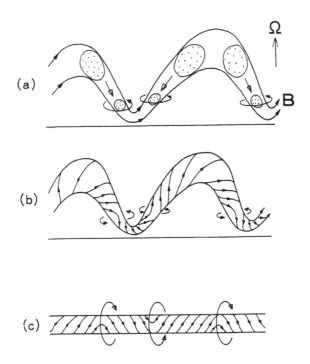

FIGURE 3.48 The formation of the local twist by the effect of the Coriolis force on the contracting downflowing blob (or cloud). (Shibata and Matsumoto, 1991).

magnetic field cannot escape from the disk, and hence the disk would evolve into the *low* β disk in which various violent phenomena (such as flares) can occur *inside* the disk. On the other hand, if the shearing effect is weak, magnetic flux can rapidly escape from the disk to form high β disk, similar to conventional picture of an accretion disk. In this case, the disk-corona structure may be similar to the solar interior-corona structure. Shibata *et al.* (1990b) discussed the possibility of these two types of accretion disks, *low* β disk and *high* β disk. Recently, Mineshige *et al.* (1995) applied this idea to the observed two states of accretion disks, low state and high state (Miyamoto *et al.*, 1995), and proposed that low states correspond to the *low* β disk. (See Sec. 4.2.3.8) (As for a linear stability analysis of the Parker instability in a shearing flow parallel to the magnetic field (Fogglizo and Tagger, 1994).)

The shearing motion, on the other hand, can give destabilizing effect[6] on magnetized gas

---

[6]Even the Parker instability can be nonlinearly triggered by the shearing motion if the gas layer is linearly

layer, i.e. magneto-rotational instability (or Velikhov-Chandrasekhar instability or Balbus-Hawley instability). This instability gives one answer on the origin of viscosity in the accretion disk (Hawley et al., 1995; Matsumoto and Tajima, 1995; Brandenburg et al., 1995; see Sec. 4.2.3). Since this instability occurs in the wavelength shorter than the Parker unstable wavelength, many interesting questions arise. Once this instability develops, does the disk magnetic field become turbulent and have only short scale? In that case, is the Parker instability completely suppressed? If the initial magnetic field is not so weak as in the case of star forming regions, what will happen? The nonlinear coupling between the magneto-rotational instability and the Parker instability in realistic 3D situation should be studied further.

3.3 Magnetic Reconnection

The magnetic field in a plasma and the plasma displacement perpendicular to it obey the same convective equation (Landau and Lifshitz, 1960) in a perfectly conducting plasma. The equation of motion for an electron component of the fluid in the magnetic field under the massless electron approximation is

$$\mathbf{E} + \mathbf{v}_e \times \mathbf{B}/c = 0, \tag{3.3.1}$$

where $\mathbf{v}_e$ is the electron fluid velocity and the inertia term in $d\mathbf{v}_e/dt$ is neglected since it is proportional to the electron mass. The pressure effect is dropped here for simplicity. If $\mathbf{v}_e$ represents electron velocity, this equation reduces to the zeroth order electron guiding-center equation. By the non-relativistic Lorentz transformation of electric field on the frame moving with the velocity $\mathbf{v}_e$ the above equation is put in the form

$$\mathbf{E} + \mathbf{v}_e \times \mathbf{B}/c = \mathbf{E}' = 0. \tag{3.3.2}$$

This indicates that the electric field in the frame of reference moving with the fluid is zero and that the fluid experiences no electric acceleration. This is a manifestation of vanishing electric resistivity in the conducting fluid (plasma)

$$E' = \eta J' = 0 \tag{3.3.3}$$

which constitutes the ideal magnetohydrodynamic condition in which the plasma and the field move together. For example, in the shear Alfvén wave the fluid and field vibrate together perpendicular to the magnetic field (and the propagation) direction (Alfvén, 1963). The time scale τ_A of the ideal MHD motion such as in the Alfvén oscillations is determined by the length a over which such perturbation travels divided by the Alfvén velocity v_A: $\tau_A = a/v_A$.

A mathematical possibility in which a partial detachment of the plasma from the magnetic field lines takes place, may be envisioned by having a non-vanishing resistivity $\eta \neq 0$ which is often called Ohm's law

$$\mathbf{E} + \mathbf{v}_e \times \mathbf{B}/c = \eta \mathbf{J}. \tag{3.3.4}$$

---

stable (Kaisig et al., 1990).

This yields a second (and usually much longer) time scale of the resistive time τ_r and is characterized by the second term of Faraday's equation

$$\frac{\partial \mathbf{B}}{\partial t} = \nabla \times (\mathbf{v}_e \times \mathbf{B}) + \eta \nabla^2 \mathbf{B}. \tag{3.3.5}$$

The temporal change of flux Φ through any closed contour that moves with the material motion $\mathbf{v}_e$ is expressed as

$$\frac{d\Phi}{dt} = \int \frac{\partial \mathbf{B}}{\partial t} \cdot d\mathbf{S} + \oint \mathbf{v}_e \times d\boldsymbol{\ell} \cdot \mathbf{B} = \int \left[\frac{\partial \mathbf{B}}{\partial t} - \nabla \times (\mathbf{v}_e \times \mathbf{B}) \right] \cdot d\mathbf{S}$$

$$= \eta \int \nabla^2 \mathbf{B} \cdot d\mathbf{S}. \tag{3.3.6}$$

Or we have

$$\frac{d\Phi}{dt} = \eta \nabla^2 \Phi. \tag{3.3.7}$$

If $\eta = 0$ (perfectly conducting fluid or the ideal MHD fluid) and the fluid motion is characterized by Eq. (3.3.1), the right-hand side of Eq. (3.3.7) vanishes and the flux change through the moving fluid is zero i.e. the magnetic flux is tied to the fluid. On the other hand, when $\eta \neq 0$ (resistive MHD fluid), the magnetic flux diffuses out of the contour according to Eq. (3.3.6). Thus the resistive (diffusion) time is given by $\tau_r = a^2/\eta$ with a being the characteristic length of diffusion.

In the model resistive MHD equations the parameter that governs the system besides the Alfvén time is the ratio of the resistive time to the Alfvén time, $S = \tau_r/\tau_A$, the so-called magnetic Reynolds number or the Lundquist number. The ideal MHD system is dissipationless, while the resistive MHD system is dissipative in that the magnetic energy is converted into kinetic energy via the resistivity. In many applications the magnetic Reynolds number is quite large, perhaps 10^7 in a present day tokamak plasma and $10^{11} \sim 10^{13}$ in a solar coronal loop plasma. In these problems it has been either observed or postulated that the magnetic energy configuration changes its topology accompanied by the magnetic field line reconnection which involves the resistive process, and the magnetic energy is converted into kinetic energy in a time much shorter than the resistive time. It is particularly challenging to plasma physics theory that in some instances in the solar flares' impulsive phase the energy conversion from magnetic to kinetic forms seems to take place in a matter of Alfvén time. This is to say that such a process is some $10^{11} - 10^{13}$ times faster than the resistive time. A similarly abrupt energy release is observed in the laboratory, e.g. in the tokamak plasma disruption and sawteeth oscillations, which seem to take place in only a matter of the Alfvén time scale. The theoretical community has continued to quest for a resolution of this spectacularly fast magnetic reconnection process for more than three decades.

In this subsection, we first overview classical observations and models of solar flares as a good example of explosive magnetic energy release caused by magnetic reconnection, and then dicuss the theory of reconnection in some detail. The observations and understanding of solar flares have been greatly advanced by X-ray solar physics satellite Yohkoh; a number of evidences of magnetic reconnection have been found in flares as well as in various coronal phenomena. We will see how these new observations are going to be understood by th theory of reconnection. Finally, the most difficult issue in the reconnection theory; i.e., the collisionless conductivity, will be discussed in Sec. 3.3.4.

3.3.1 Flares as Violent Magnetic Energy Release

3.3.1.1 What is a Flare?

Solar flares are among the most energetic and enigmatic phenomena in the solar atmosphere. Large amounts of energy ($10^{29} - 10^{32}$ erg) are suddenly released in the corona, accelerating great quantities of nonthermal particles and heating coronal and chromospheric plasmas, resulting in transient brightenings in throughout the electromagnetic spectrum, including vigorous Hα brightening. The maximum temperature of superhot plasmas in largest flares is estimated to be $3 - 4 \times 10^7$ K, and the maximum energy of nonthermal electrons and protons reach 1 MeV and 1 GeV, respectively, whereas the maximum velocity of bulk plasma flows amounts to a few 10^3 km/s.

Figure 3.49 shows time variation of intensity of various electromagnetic waves emitted from flares, ranging from gamma rays to radio waves. The simplest definition of a flare may be the following; *a transient phenomenon showing a rapid increase (followed by either rapid or gradual decay) in some electromagnetic waves.* In this definition, the absolute value of the peak intensity itself does not matter. Hence we call micro-flares for transient phenomena with energy of order of $10^{25} - 10^{27}$ erg, and nano-flares for those with energy of $10^{22} - 10^{24}$ erg (Parker, 1988). Normally the *rapid* time scale corresponds to 10 sec–10 min, but we will see later that the time scale is also relative one.

Historically, flares were found as sudden brightenings in Hα monochromatic images and (rarely) in white light images. Later flares have been observed by radio waves and X-rays. From detailed imaging observations of flares, we find that most flares occur in active regions, i.e. in the region of strong magnetic field near sunspots. It is now established that the source of energy of flares is magnetic energy stored in active regions.[7] As for more detailed description of classical observations of flares, the reader should be referred to Svestka (1976), Kundu and Woodgate (1986), Tanaka (1987), and Zirin (1988).

Problem 3–5: Calculate the total magnetic energy stored near sunspot ($B = 2000$ G) with size of 3×10^4 km. Is it possible to explain the energy of largest flares?

3.3.1.2 Morphology of Flares

In Hα (Fig. 3.50), large flares often show bright two ribbons; such flares are called two-ribbon flares. The two ribbons have length of 10^5 km and width of some 10^4 km, and are parallel to the polarity-inversion line which suggests that they are footpoints of arcades overlying the

---

[7]Rarely, flares occur in quiet regions without sunspots, which are called "spot-less flares." In such case, the source of energy of flares is magnetic energy stored in "magnetic structures around spot-less flares."

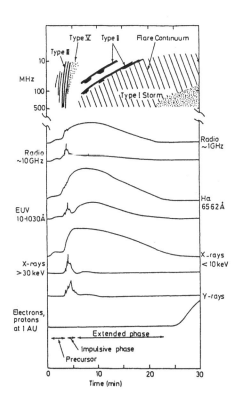

FIGURE 3.49 Time variation of various electromagnetic radiations from a typical flare (from Dulk *et al.*, 1985).

inversion line. Large two ribbon flares are often associated with filament eruptions and/or coronal mass ejections (CMEs) (see Fig. 3.51).

On the other hand, there are many flares which are not associated with CMEs; such flares are relatively compact, and have magnetic field configurations which do not seem to change much. Although some of literatures refer these flares as simple loop flares, we simply adopt the term *compact flares*, because they are not necessarily simple loops. The linear size of Hα bright patches of these flares is 1000–a few 10^4 km.

In this section, we shall use the terms *CME related flares* and *compact flares* in place of "large two-ribbon flares" and "simple loop flares." Then we shall discuss a few models for *CME related flares* and *compact flares* separately, and discuss their merits and demerits (or remaining problems). Note that it is not possible to discuss all theoretical models of flares, because, as is well known (Priest, 1982), *there are as many flare theories as there are flare theorists.* (When this was told in one international conference, someone commented "there are *more* flare theories because even one theoretician proposes many theories"!)

It should be, however, remembered here that new observations of flares by *Yohkoh* has

211

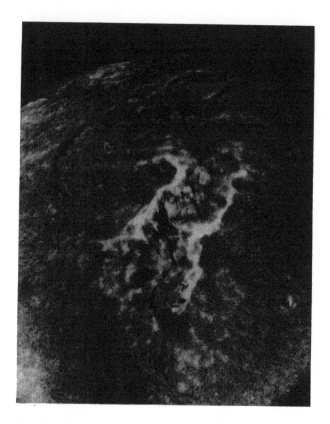

FIGURE 3.50 Hα image of two ribbon flares.

revealed that these two types of flares show many common properties, and hence opened the possibility of unification of these two distinct types of flares, as will be discussed in detail in Sec. 3.3.3.

3.3.1.3 Models of CME Related Flares*

It was once considered that CMEs are a kind of blast waves or shocks produced by sudden pressure enhancements in flares (e.g., Wu *et al.*, 1978; Dryer, 1979). It has become clear, however, that some of large flares are preceded by the start of CME (Harrison, 1986; Fig. 3.53). That is, flares are not the origin of CMEs, but both flares and CMEs seem to have the same origin, which may be a kind of MHD instability (or loss of equilibrium) occurring in the global magnetic configuration of the corona.

CME related flares are often associated with the filament eruptions. Such filament eruptions have been believed to occur as a result of the MHD instability (e.g. the kink instability) in a twisted flux rope in the chromosphere or in the corona (Hirayama, 1974, 1991; Moore

212

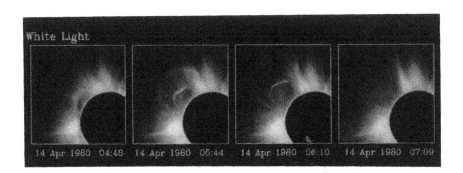

FIGURE 3.51 Coronal mass ejection (CME).

and Roumeliotis, 1992). Even if the filament eruption is not observed in Hα, it is possible that a twisted flux rope embedded in a hot coronal plasma becomes unstable to the MHD instability, initiating a flare.

Sakurai (1976) first studied the three dimensional (3D) nonlinear evolution of the kink instability, and applied the results to the dynamical motion of eruptive prominences (see also a related 3D simulation performed by Zaidman and Tajima, 1989).

There are three possibilities to trigger the MHD instability in a twisted flux rope:

(1) Time variation of the global magnetic field configuration around the twisted rope, such as evolution induced by slow shearing motions at the footpoints of the global magnetic fields (Low, 1981; Mikic *et al.*, 1988; Sakurai, 1989). (This corresponds to a change in the condition at the outer boundary of the twisted tube.)

(2) Time variation of the twisted flux tube (sheared field) itself, such as the twisting of the tube at the footpoint (Steinolfson and Tajima, 1987), or reconnection leading to an unstable twisted tube (Sturrock, 1989).

(3) Change in the twisted tube's lower boundary condition, such as the interaction of

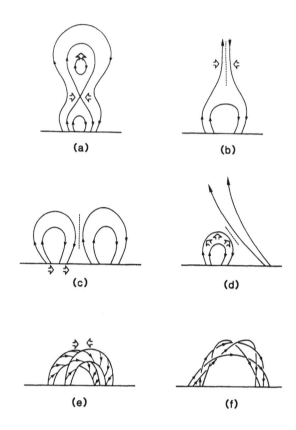

FIGURE 3.52 Various magnetic configurations considered in various flare models. (a) A model including filament eruption or CMEs (Hirayama, 1974, Sturrock, 1989, Moore and Roumeliotis, 1992), (b) a model including helmet streamer-like current sheet (Kopp and Pneuman, 1976), (c) a model including moving bipoles (Sweet, 1958), (d) en emerging flux model (Heyvaerts *et al.*, 1977), (e) a loop-loop interaction model (Gold and Hoyle, 1960), and (f) Parker's nanoflare model.

emerging flux with the filament from below (Rust, 1972; Heyvaerts *et al.*, 1977).

It is often argued that photospheric shear motions at the footpoints of the global magnetic field configuration generates the twisted flux tube or the sheared magnetic field configuration, which eventually lead to instability or the loss of equilibrium as in (1) and (2) above. It is, however, possible to interpret the observed development of sheared magnetic field configurations as the emergence of the twisted magnetic flux tube (Tanaka, 1987; Kurokawa, 1989). Since the energy density of the turbulent convective motion in the deep interior of the convection zone is much larger than that in the photosphere, the flux tube is much more easily twisted and sheared in the convection zone than in the photosphere. Hence it is very possible that *twisted flux tubes formed deep inside the convection zone are the ultimate source of flares,* and that *the occurrence of flares is controlled by the emergence of such twisted flux*

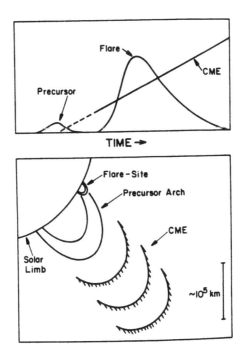

FIGURE 3.53 The temporal relation between the start of a CME and an associated flare (Harrison, 1986). Harrison writes "A coronal arch of scale-length several times 10^5 km brightens in soft X-rays (precursor). At this time a CME is launched and it appears to propagate directly from the arch. Some tens of minutes later a flare occurs in one foot of the arch."

tubes.

Future observations should clarify this point, i.e. whether or not shear motions of footpoints of magnetic loops (arcades) actually occur. If so, there should be a pronounced global velocity field around the footpoints of the magnetic structures. On the other hand, if the sheared configuration in the chromosphere or in the corona is a result of the emergence of already twisted flux tube from the convection zone, there should be an upward component in the velocity field between the two footpoints of the loop. The results of Hanaoka and Kurokawa (1989) and Kaisig *et al.* (1990) imply that the latter can be indirectly inferred from downflow velocities along the filament. Furthermore, the detailed horizontal velocity distribution around the footpoints of the magnetic loops may help to distinguish the two hypotheses, i.e. the *shear motion hypothesis* and the *emerging flux hypothesis*. That is, if the velocity of the footpoint of a loop is very different from that of the ambient non-magnetic

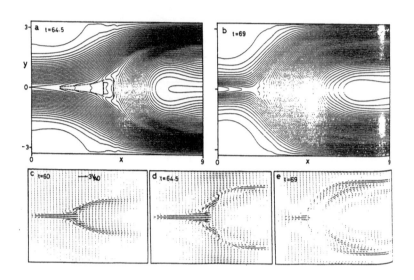

FIGURE 3.54 Magnetic fields and plasma-flow configurations in Ugai's (1987) 2D MHD numerical simulation of loop heating by the magnetic reconnection. Both the slow shocks (dotted lines) and fast shocks (dashed) are clearly visible. The dot-dash line shows the loop front.

plasmas, the emerging flux hypothesis would be favorable.

One classical scenario for CME related flares is that following a filament eruption, previously closed field lines are opened up, and create a current sheet; i.e. a helmet-streamer type field configuration (Fig. 3.52) is created. If the magnetic reconnection occurs successively in such a current sheet, the outward expansion of the two Hα ribbons is naturally explained (Carmichael, 1964; Sturrock, 1968; Hirayama, 1974; Kopp and Pneuman, 1976). This kind of model is now called *CSHKP* model after the names of these five pioneers (Sturrock, 1992; Svestka, 1992; Shibata, 1996).

Forbes and Priest (1983) performed a 2D MHD numerical simulation of magnetic reconnection occurring in a vertical current sheet line-tied at the photosphere, and found that a fast shock is created just below the downwardly directed reconnection jet. Figure 3.54 shows the simulation results by Ugai (1987) of loop heating by magnetic reconnection, where not only a slow shock but also a fast shock at the loop top are clearly shown. Cargil and Priest

(1983) extended the model by Kopp and Pneuman (1976) to include the effect of joule heating at the slow shock front.

Recently, it has been noticed that the infinitely long current sheet in the helmet-streamer-type configuration has a maximum energy state (in the force-free approximation) (Aly, 1991), and cannot be reached by the MHD instability of the force-free field consisting of sheared closed loops against the expectation of the previous work (Barnes and Sturrock, 1972). In order to exceed such maximum energy state, we must have either (a) the closed magnetic field lines (magnetic islands) detached from the base boundary, or (b) the effect of high gas pressure. On the other hand, it has been argued that such field configuration including magnetic islands (or plasmoids) can be created in a closed field system without passing through an infinitely long current sheet (Mikic *et al.*, 1988; Biskamp and Welter, 1989; Forbes, 1990; Kusano *et al.*, 1994, 1995; Linker and Mikic, 1995). Note that Hα filament and/or CME are interpreted as such magnetic islands (or plasmoids). The transition to such system involving reconnection and ejection of plasmoids may be a key process in flares.

Finally, we shall summarize the remaining fundamental questions for CME related flares.

- How are sheared field configurations formed? Are these due to shearing by convective motion at the footpoint of the magnetic loop? Or are these simply due to emergence of twisted flux tubes? Or are these results of magnetic reconnection?

- What is the physics of the triggering of flares? Is this the MHD instability or loss of equilibrium? If it is the instability, is it ideal MHD instability or resistive MHD instability?

- What is the role of magnetic islands or plasmoids? Are these the necessary ingredients to induce flares? Or are these simply biproducts of flares?

- Is magnetic reconnection really occurring in flares? If so, is it important in energy release in flares? And if so, how and why does magnetic reconnection occur?

3.3.1.4 Models of Compact Flares

A. Emerging flux model

To explain compact flares, Heyvarets *et al.* (1977) developed the emerging flux model, in which the flare is produced by magnetic reconnection between emerging loop and the overlying pre-existing coronal field.

Forbes and Priest (1984) first performed two-dimensional MHD numerical simulations of magnetic reconnection between emerging flux and a coronal field, by taking R_m to be much smaller than the actual solar value. Although their simulation results show many interesting nonlinear processes, they did not consider the effect of the gravitational acceleration. Since the main force raising the emerging flux is magnetic buoyancy (Parker, 1979), the gravitational force is fundamentally important to the emerging flux model. Thus Shibata

217

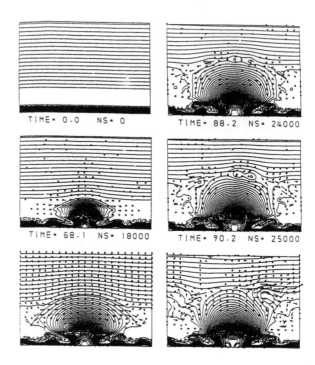

FIGURE 3.55 Magnetic field lines and velocity vectors in a typical example of magnetic reconnection between emerging flux and an overlying coronal field (Shibata *et al.*, 1992a). The times are in units of $\tau \simeq 20$ sec, and the horizontal and vertical sizes of the computing box are 16000 km and 10800 km. The scale of the velocity vector is shown below the frame of time $= 96.1$ in units of C_s; VNM $= 6.0$ indicates that the arrow with the length of this line has the velocity of $6.0 C_s \simeq 60$ km s^{-1}. Note that a dense filament in the neutral sheet rises with velocity ~ 10 km s^{-1} over $t/\tau \simeq 75.1 - 81.7$. Magnetic reconnection starts after the filament gas in the sheet drains down at $t/\tau \simeq 88$. Note the formation of three magnetic islands at $t/\tau = 88.2$, their rapid coalescence at $t/\tau = 90.2$, and their subsequent jetting along the neutral sheet with supermagnetosonic speed. Hence fast shocks are created along both edges of the neutral sheet.

et al. (1989a, 1989b, 1990a, 1990b), and Nozawa *et al.* (1992) have constructed more realistic models of the emerging flux incorporating the gravitational acceleration (see Fig. 3.55). Their models explain many observed features of emerging flux, such as the rise velocity ($10 \sim 15$ km s^{-1}) of the arch filament and the downflow speed ($30 \sim 50$ km s^{-1}) along the filament (Bruzek, 1969; Chou and Zirin, 1988) and the small rise velocity (~ 1 km s^{-1}) of the photospheric emerging flux (Kawaguchi and Kitai, 1976; Zwaan, 1987). (See Shibata *et al.*, 1991 for a review of their studies on emerging flux.)

Figure 3.55 (Shibata et al., 1992b) shows a typical example of numerical simulation results of reconnection between a realistic emerging flux and an overlying coronal field, assuming small $R_m (\sim 1000)$. In this model, the resistivity is assumed to be a function of the current density and mass density, simulating an anomalous resistivity. The results show that: (1) the reconnection starts after the most of mass in the filament fallen, (2) several magnetic islands are created in the current sheet (i.e. impulsive bursty reconnection), (3) these islands dynamically coalesce with each other via the coalescence instability (Tajima and Sakai, 1989a, 1989b), (4) the islands and neutral sheet plasmas are accelerated along the sheet up to $\sim$ Alfvén speed just outside the sheet, which exceeds the local magnetosonic speed in the sheet and hence the fast shocks are created at both edges of the current sheet. Heating by the slow and fast shocks, as well as in the current sheet, may account for the X-ray bright points associated with emerging flux (Golub et al., 1976). (See Yokoyama and Shibata, 1994, 1995, 1996, and Sec. 3.3.2.7 and 3.3.3.5 for new models.)

Uchida and Sakurai (1977) suggested, in their emerging flux model, that the very short time scale ($\sim$ dynamical time scale) of the impulsive phase of flares may be explained by the dynamical transition to a lower energy interleaved state induced by the 3D MHD interchange instability of the current sheet between the emerging flux and overlying coronal field (see also Sakurai and Uchida (1977). Since the collapse is the ideal MHD process, there is no difficulty arising from long current dissipation time (or large R_m, see section 3.3.2). They also suggest that continued current dissipation in the interleaved book-page structure can provide an explanation of the later decay phase. In order to see whether these processes work well or not, 3D MHD simulations are necessary.

B. Sheared Loop and Loop Coalescence Models*

Spicer (1977) has presented the sheared loop model to explain simple loop (compact) flares. He suggested that the nonlinear mode coupling and the multiple tearing modes significantly enhance the reconnection rate compared with the single tearing mode instability. Although these processes might play a fundamental role in flares, there are still some problems associated with the reconnection process, which will be discussed in Sec. 3.3.2.

Gold and Hoyle (1960) first considered the interacting (coalescence) loop model of flares (Fig. 3.52). More recently, Tajima et al. (1982, 1987), Sakai and Tajima (1986) have presented a more refined loop coalescence model (Fig. 3.56) based on the concept of explosive reconnection as discussed later. They succeeded in explaining the very fast rise time of the impulsive phase of flares, and the rapid amplitude oscillations (Figs. 3.63) found in hard X-rays, gamma rays, and microwaves (Nakajima et al., 1983; Sakai and Ohsawa, 1987; Tajima et al., 1989).

C. Unwinding Magnetic Twist Jet and Non-Reconnection Loop Flare Model*

Applying the mechanism of the acceleration of cosmic jets (Uchida and Shibata, 1985; Shibata and Uchica, 1985, see Sec. 4.3) to solar jets, Shibata and Uchida (1986a) proposed a magnetodynamic mechanism for the acceleration of solar jets, such as surges and EUV jets

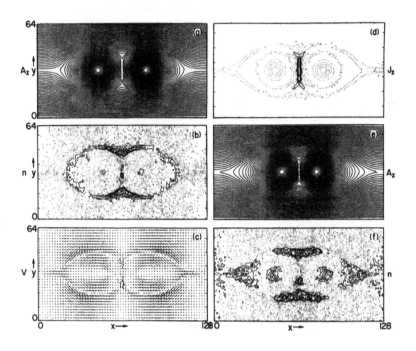

FIGURE 3.56 Spatial pattern of various physical quantities before and during explosive coalescence (from Tajima *et al.*, 1987).

(Brueckner, and Bartoe, 1983; see also Dere, 1994). They call this mechanism the *sweeping-magnetic-twist mechanism*, or the *sweeping-pinch mechanism*, because the acceleration is due to the $\mathbf{J} \times \mathbf{B}$ force in an unwinding (propagating or sweeping) magnetic twist (i.e. non-linear torsional Alfvén wave) and the pinching occurs in association with the propagation of the nonlinear magnetic twist (Fig. 3.57b). This model explains very well the rotating eruption of an untwisting filament observed by Kurokawa *et al.* (1988). The origin of the magnetic twist is attributed to processes occurring deep in the convection zone, where the plasma beta is very high so that the flux tube is easily twisted by turbulent convective motion. They proposed that such sudden relaxation of magnetic twist would be realized by the magnetic reconnection between twisted flux tube and untwisted flux tube (Fig. 3.58).[8]

---

[8]Recently, this process has been numerically studied by Okubo *et al.* (1997) and Karpen *et al.* (1997). See also Canfield *et al.* (1996) and Schneider *et al.* (1995) for related observations. (It is interesting to note that in the model of cosmic jets ejected from the accretion disks (Uchida and Shibata, 1985; Shibata and Uchida, 1986b), the magnetic twist is created by the rotation of the accretion disks.)

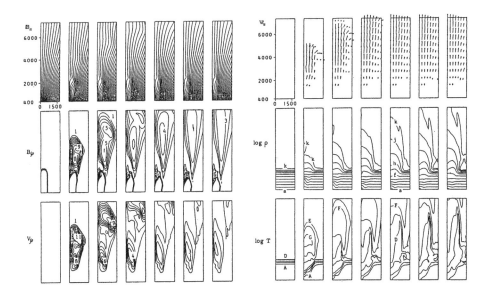

FIGURE 3.57 The unwinding magnetic-twist-jet model by Shibata and Uchida (1986a). These figures show the results of a 2.5D axisymmetric MHD numerical simulations of the dynamical relaxation of a nonlinear magnetic twist in the upper chromosphere; from the top to the bottom in the left column, the poloidal field lines ($\mathbf{B}_\parallel$), the toroidal field (B_φ) and the azimuthal (rotational) velocity (V_φ) contours; in the right column, the velocity vectors ($\mathbf{V}_\parallel$), the density ($\log \rho$) and the temperature ($\log T$) contours in a logarithmic scale. The horizontal and vertical sizes of the computing box are 1600 km and 7000 km, respectively. Times are in units of seconds, and the maximum velocity of the jet is about 400 km s^{-1}. The scale of the velocity vector is shown above the frame of $t = 0$ of $\mathbf{V}_\parallel$. The contour level step width for $\log \rho$ and $\log T$ is 0.5. Note that the jet spins about the z-axis with a rotation velocity of $\sim 60 \sim 200$ km s^{-1}. Note also that a hot region ($\sim 5 \sim 10 \times 10^6$ k, denoted by the large letter F and G) appears and propagates just ahead of the dense jet.

Uchida and Shibata (1988) then extended this mechanism to loop flares; if two magnetic-twist-jets are launched separately from the footpoints of the loop, and if the sense of the magnetic twists is opposite each other a very hot region appears as a result of strong shock formation when the two twists collide at the loop top. Gradual heating continues because of the dynamical relaxation of the magnetic twist which successively propagates into the region near the top of the loop (see also Uchida and Shibata (1990)). An interesting point in this

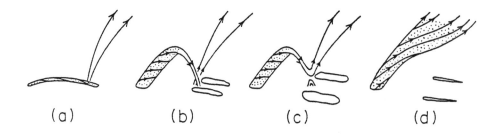

(a)　　　　　(b)　　　　　(c)　　　　　(d)

FIGURE 3.58 The sudden relaxation of magnetic twist by the magnetic reconnection between a twisted flux tube and an untwisted flux tube.

scenario is that it does not require any reconnection process, and thus it is free from the difficulty coming from long current dissipation time (or large R_m). However, there remains a fundamental question on how such two magnetic-twist-jets are launched simultaneously from both footpoints of the loop.

3.3.2　Theory of Magnetic Reconnection

As we have discussed at the beginning of this section, it is very difficult to dissipate magnetic (current) energy in a short time scale, since the electrical conductivity is very high in solar coronal plasmas. This is the basic difficulty in flare (and coronal) physics. Actually this difficulty applies not only to the Sun but also to fusion plasmas, magnetospheric plasmas, and general astrophysical plasmas.

3.3.2.1 Magnetic Diffusion

Let us consider the simplest case in which magnetic field (or current) dissipates only by resistivity. In this case, the diffusion time becomes

$$\tau_D = L^2/\eta \simeq 10^{14} \quad (L/10^9 \text{cm})^2 (T/10^6 \text{K})^{3/2} \text{ sec}, \tag{3.3.8}$$

where L is the typical size of a flare ($= 10^9$ cm), η is the magnetic diffusivity ($= c^2/(4\pi\sigma)$) and becomes

$$\eta \simeq 10^{13} T^{-3/2} \simeq 10^4 \quad (T/10^6 \text{K})^{-3/2} \text{cm}^2 \text{s}^{-1}, \tag{3.3.9}$$

for coulomb collision. This is often called classical resistivity or Spitzer (1962) resistivity. The diffusion time becomes 10^{14} sec ~ 3 million years (!) for typical coronal parameters. Even if we take into account anomalous resistivity of 10^6 times larger than the classical resistivity, it is difficult to explain flare time scale ($10 - 100$ sec).

Here the dynamical time scale is defined as the Alfvén transit time τ_A, which is given by

$$\tau_A = L/V_A \simeq 1(L/10^9 \text{cm})(B/100\text{G})^{-1}(n/10^9 \text{cm}^{-3})^{1/2} \text{ sec}, \tag{3.3.10}$$

$$V_A = B/(4\pi\rho)^{1/2} \simeq 10000 \quad (B/100\text{G})(n/10^9 \text{cm}^{-3})^{-1/2} \text{ km/s}. \tag{3.3.11}$$

Here V_A is the Alfvén speed. The magnetic Reynolds number is defined by

$$R_m = \frac{\text{magnetic diffusion time}}{\text{Alfvén time}} = \tau_D/\tau_A = V_A L/\eta \tag{3.3.12}$$

$$\simeq 10^{14} \quad (L/10^9 \text{cm})(T/10^6 \text{K})^{3/2}(B/100\text{G})(n/10^9 \text{cm}^{-3})^{-1/2}. \tag{3.3.13}$$

As already mentioned above, explosive magnetic energy release occurs even though $R_m \gg 1$ in every kind of hot plasmas. This puzzle is considered to be one of the most difficult and fundamental problem in natural science.

3.3.2.2 Sweet–Parker Model

Sweet (1958) and Parker (1957) have developed a simple theory of steadily driven reconnection, taking into account the effect of the plasma flow (Fig. 3.59).

Consider a current sheet (or diffusion region) with a thickness of w and length L, with an inflow into the current sheet at the velocity v_i. Outflow velocity is taken to be v_o. We assume steady state and incompressibility. From mass conservation, we have

$$Lv_i = wv_o. \tag{3.3.14}$$

The induction equation is written in integral form as

$$v_i B = \eta B/w. \tag{3.3.15}$$

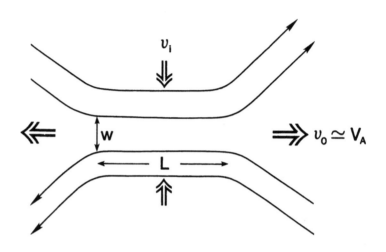

FIGURE 3.59 Sweet–Parker reconnection model.

The equation of motion perpendicular to the magnetic field becomes that of total pressure balance between inside of current sheet ($B^2/8\pi \ll p$) and outside ($B^2/8\pi \gg p$), because the flow velocity is smaller than v_A.

$$p = B^2/8\pi. \tag{3.3.16}$$

The equation of motion parallel to magnetic field becomes

$$\rho v_o dv_o/dy = -dp/dy. \tag{3.3.17}$$

Integrating this equation, we get

$$\rho v_o^2/2 \simeq p \simeq B^2/8\pi. \tag{3.3.18}$$

It follows from this equation that

$$v_o \simeq V_A. \tag{3.3.19}$$

That is, the velocity of outflow (*reconnection jet*) is of order of the Alfvén speed. Furthermore, we find

$$v_i = R_m^{-1/2} V_A, \tag{3.3.20}$$

$$\tau_{SP} = (\tau_D \tau_A)^{1/2} = R_m^{1/2} \tau_A = L^{3/2}/(\eta V_A)^{1/2} \tag{3.3.21}$$

$$\simeq 10^7 (L/10^9 \text{cm})^{3/2} (T/10^6 \text{K})^{3/4} (B/100\text{G})^{-1/2} (n/10^9 \text{cm}^{-3})^{1/4} \text{ sec}. \tag{3.3.22}$$

Hence the time scale of Sweet–Parker reconnection still strongly depends on the magnetic Reynolds number (as $R_m^{1/2}$) and slow. Numerically, the time scale of Sweet–Parker reconnection is three months for classical resistivity and 3 hours for anomalous resistivity, and still larger than the time scale of flares. The thickness of diffusion region is then

$$w \simeq R_m^{-1/2} L, \tag{3.3.23}$$

and becomes 100 cm for classical resistivity, and about 1 km for anomalous resistivity.

3.3.2.3   Petschek Model

Petschek (1964) noted that considering the effect of slow mode MHD shock (or wave) on the region outside the diffusion region greatly increases reconnection rate up to

$$v_i \sim \left(\frac{\pi}{8}\right) \frac{V_A}{\ln(R_m)} \sim 0.01 - 0.1 V_A \tag{3.3.24}$$

(see also Sonnerup, 1970), nearly independent of R_m.

Here let us study the fundamental properties of the Petschek model (Fig. 3.60). For simplicity, we assume the velocity of inflow to the X-point is uniform, and the angle θ of the slow shock measured from the current sheet is small, $\theta \ll 1$. Then, mass conservation becomes

$$v_i = \theta V_A, \tag{3.3.25}$$

where incompressibility is again assumed. Similar to Sweet–Parker model, the velocity of outflow (reconnection jet) is of order of the Alfvén velocity in this model. Since the Sweet–Parker model is valid also for the diffusion region around the central X-point, we have

$$v_i = R_{m_*}^{-1/2} V_A. \tag{3.3.26}$$

Here $R_{m_*} = \ell V_A/\eta$ is the magnetic Reynolds number defined by the length of the diffusion region ℓ. In the problem with characteristic size L, we have to define the magnetic Reynolds number by $R_m = L V_A/\eta$. Then we get the following relation for the inflow speed

$$v_i = \left(\frac{L}{\ell}\right)^{1/2} R_m^{-1/2} V_A. \tag{3.3.27}$$

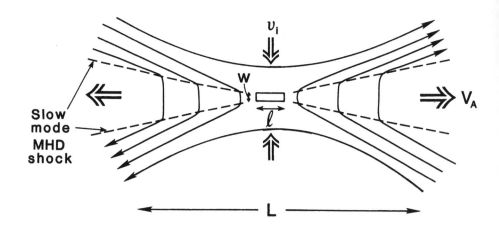

FIGURE 3.60 Petschek model for fast reconnection.

If ℓ becomes sufficiently small, or if (L/ℓ) becomes sufficiently large, we can find finite value for v_i even if R_m is large. This is an essential point of the Petschek model. Here, of course when $\ell \simeq L$, the reconnection becomes Sweet–Parker type (See Table 3.1). How small is the length of diffusion region in Petschek model to get fast reconnection ($v_i \sim V_A$)? In order for that, the length ℓ must be as small as

$$\ell = \theta^{-2} R_m^{-1}. \tag{3.3.28}$$

Therefore, we find $\ell \sim 0.001$ cm (!!), for $\theta \sim 0.1$, $L \simeq 10^9$ cm, $R_m \sim 10^{14}$ (classical resistivity), and $\ell \sim 10^3$ cm $= 10^4$ m for anomalous resistivity model.

Here we should note that the gyro–radius of ions in coronal plasmas is

$$r_g = cm_i v/(eB) \simeq 10 \left(\frac{T}{10^6 \mathrm{K}}\right)^{1/2} \left(\frac{B}{100\mathrm{G}}\right)^{-1} \mathrm{cm}. \tag{3.3.29}$$

Hence the length scale of the Petcheck model becomes smaller than the ion gyroradius in the case of classical resistivity, though this is very unlikely situation. Even in the case of

TABLE 3.4 Comparison of Three Magnetic Reconnection Mechanisms

model	τ/τ_A	v_i/V_A	w/L	ℓ/L
diffusion	R_m^{+1}	R_m^{-1}	1	1
Sweet–Parker	$R_m^{+1/2}$	$R_m^{-1/2}$	$R_m^{-1/2}$	1
Petschek	R_m^{0}	$R_m^{0}\theta$	$R_m^{-1}\theta^{-1}$	$R_m^{-1}\theta^{-2}$

Note: $R_m =$ magnetic Reynolds number, $\tau = L/v_i =$ time scale of the reconnection, $\tau_A = L/V_A =$ Alfvén time, $\theta = v_i/V_A$, $w =$ width of diffusion region, $\ell =$ length of diffusion region, $L =$ size of the whole system.

anomalous resistivity, the lenth scale (10 m) is much smaller than the characteristic size of flares (10^4 km).

Although the Petschek model is a very attractive idea, the size of the diffusion region is very small, so that the question arises whether the single Petcheck type reconnection controls the entire flare process or not (Kahler *et al.*, 1980).

3.3.2.4   Numerical Simulations of Reconnection

Original Petschek model was only an approximate theory. The general characteristics of Petschek model, such as slow shock formation and fast reconnection, has been confirmed by self-consistent numerical simulations by Ugai and Tsuda (1977) and later by Sato and Hayashi (1979). The latter stressed the importance of externally *driven reconnection* in causing the sudden energy release, and suggested that any driven reconnection occurs independently of the initial R_m, even for very large R_m.

However, using numerical simulations of incompressible driven reconnection, Biskamp (1986) showed that Petschek's model for fast reconnection does not arise under uniform resistivity in the limit of large R_m. On the other hand, Priest and Forbes (1987) developed a unified theory of steadily driven reconnection including both Petschek regime and the flux-pile-up regime. The latter case arises when the inflow speed exceeds that of the Petschek regime, so that magnetic flux piles up just outside the current sheet, creating a long sheet. Forbes and Priest (1987) suggested that the long current sheet appearing in the flux-pile-up regime is unstable to the tearing mode, resulting in nonsteady "impulsive bursty reconnection." Scholer (1989) presented a somewhat different view, concurring with Biskamp (1986), that fast steady reconnection may occur if the resistivity is spatially limited so that the length of the diffusion region is sufficiently small, supporting some related argument by Ugai (1987).

More recently, Priest and Forbes (1987) succeeded to construct a unified theory which explains why Biskamp's simulation leads to slow reconnection, showing that the boundary condition adopted by Biskamp's simulation is the cause of slow reconnection, and demonstrated that the boundary condition is essential to determine whether the reconnection is fast or slow. But this seems to be too strong argument, because Ugai (1992) and Yokoyama and Shibata (1994) showed some examples of driven reconnection in which the reconnection

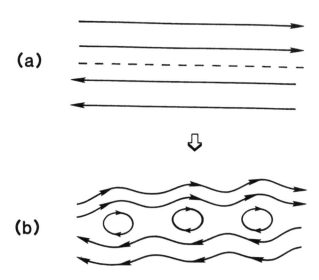

(a)

(b)

FIGURE 3.61 Tearing instability (from Steinolfson and van Hoven, 1984).

rate is not uniquely determined by the driving process. As we will see later, the nonuniformity of current sheet arising from various causes, such as nonuniform resistivity (Yokoyama and Shibata, 1994), nonuniform current distribution (Tajima and Sakai, 1986), etc. has an essential key to solve the question what is the condition for the fast reconnection.

3.3.2.5 Tearing Instability

Until now, we have mainly considered the steady reconnection process. However, the current sheet itself is unstable for the formation of magnetic islands as shown in Fig. 3.61, and the magnetic islands grow unsteadily. This is called the *tearing instability*.

Furth, *et al.* (1963) (famous as FKR) have developed tearing instability theory in an attempt to explain the abrupt disruption of magnetically confined fusion plasmas. According to their analysis, the growth rate of the tearing instability is of order of

$$\omega \sim \tau_{D,*}^{-3/5} \tau_{A,*}^{-2/5} \sim \tau_{A,*}^{-1} \alpha^{-2/5} (R_{m,*})^{-3/5} \tag{3.3.30}$$

228

for large $R_{m,*}$ and long wavelength $\lambda > 2\pi a$, where

$$\tau_{D,*} = \frac{a^2}{\eta},$$ (3.3.31)

$$\tau_{A,*} = \frac{a}{V_A},$$ (3.3.32)

$$R_{m,*} = \frac{\tau_{D,*}}{\tau_{A,*}} = \frac{aV_A}{\eta},$$ (3.3.33)

$$\alpha = ka = 2\pi a/\lambda,$$ (3.3.34)

and a is the thickness of the current sheet. Steinolfson and van Hoven (1984) numerically analyzed the linear tearing instability, without using the so called *constant ψ approximation* which was used in the FKR theory, and found that (Fig. 3.62), for long wavelengths

$$\omega_{max} \sim \tau_{A,*}^{-1}\alpha^{2/3}(R_{m,*})^{-1/3},$$ (3.3.35)

$$\alpha_{max} \sim R_{m,*}^{-1/4}.$$ (3.3.36)

Note that this growth rate is faster than that found by FKR. Thus

$$\omega_{max} \sim \tau_{A,*}^{-1}R_{m,*}^{-1/2},$$ (3.3.37)

for most unstable wavelength.

Horton and Tajima (1988) analytically derived the similar results by examining the linear *driven tearing instability* adopting the boundary condition such that the flow velocity eigenfunction does not disappear at the boundary.

Let us now apply above results to the actual solar corona. If we assume $a = 10^4$ km, the magnetic Reynolds number becomes $R_{m,*} \sim 10^{14}$. Then we find

$$\alpha_{max} \sim 10^{-3.5}, \quad \text{i.e.,} \quad \lambda_{max} \sim 2 \times 10^4 a \sim 2 \times 10^8 \text{km} > R_\odot.$$ (3.3.38)

Hence the (most unstable) tearing instability cannot be applied to the solar corona. On the other hand, if we assume $a = 1$ km, we have $R_{m,*} \sim 10^{10}$, and

$$\alpha_{max} \sim 10^{-2.5}, \quad \text{i.e.,} \quad \lambda_{max} \sim 2 \times 10^3 a \sim 2 \times 10^3 \text{km},$$ (3.3.39)

and the growth time becomes

$$\tau_{tearing} \sim 10 \text{ sec}.$$ (3.3.40)

So in this case the tearing instability can occur in the solar corona, and will form multiple magnetic islands with a size of $\sim 2 \times 10^3$ km in the coronal current sheet.

Problem 3–6: Consider the tearing instability in the Sweet–Parker current sheet. Prove that the growth rate of the tearing instability scales

$$\omega_{SP-tearing,max} \sim \tau_A^{-1}R_m^{1/4}.$$

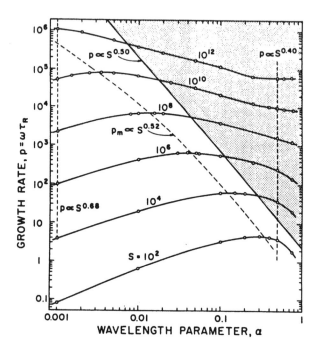

FIGURE 3.62 The growth rate of the tearing instability as a function of wavenumber (Steinolfson and van Hoven, 1984).

Here $\tau_A = L/V_A$ and $R_m = LV_A/\eta$. L is the length of the diffusion region (current sheet). (Hint: Note the relation $R_{m,*} = (a/L)R_m$ and $a/L = R_m^{-1/2}$ from Sweet–Parker theory.) Note that the growth rate *increases* with increasing the magnetic Reynolds number. Prove also that

$$\frac{\lambda}{L} = 2\pi R_m^{-3/8}.$$

Thus when $R_m \sim 10^{14}$, we find $\omega_{\text{SP-tearing,max}}\tau_A \sim 10^{-3.5}$. If we apply this result to the solar corona $L \sim 10^4\,\text{km}$ and $V_A \sim 10000\,\text{km/s}$, then we find $\tau_{\text{SP-tearing,max}} \sim 3 \times 10^{-4}\,\text{sec}$, and $\lambda \sim 0.5\,\text{km}$. That is, we have many small magnetic islands in the long current sheet.

3.3.2.6 Coalescence Instability and Explosive Reconnection Model

In order to convert magnetic energy into kinetic energy rapidly and by a substantial amount, it seems necessary that the bulk of the available magnetic energy must participate in the conversion process. The resistive heating in the *X-point* alone is too meager because the

230

available magnetic energy at the X-point is small by itself. On the other hand, the ideal MHD instabilities such as the kink instability and the coalescence instability are the processes that involve the bulk current redistribution in a matter of the Alfvén time scale. In the previous section, we considered basic processes of fast reconnection and in the present section we consider nonlinear *driven reconnection* triggered as a secondary process by a primary instability. We consider the *coalescence instability* as the primary instability because (i) although it is an ideal MHD instability in the linear sense, it would not nonlinearly evolve if there were no resistive (non-ideal MHD) effect; (ii) it involves a large amount of conversion of magnetic to kinetic energies in a short time. The coalescence instability starts from the *Fadeev equilibrium* which is characterized by the current localization parameter ϵ_c: The equilibrium toroidal current (in the z-direction) is given as

$$J_z = B_{ox}k(1 - \epsilon_c^2)(\cosh ky + \epsilon_c \cos kx)^{-2}. \tag{3.3.41}$$

The parameter ϵ_c varies from 0 to 1 with small ϵ_c corresponding to a weak localization and ϵ_c close to unity corresponding to a peaked localization; in the limit where $\epsilon_c \rightarrow 1$ the current distribution becomes the delta function.

When ϵ_c is as small as 0.3, the rate of reconnection was that of Sweet-Parker (see earlier part of this section). When ϵ_c is larger than 0.3 but smaller than 0.8, the reconnection rate experiences two phases (Bhattacharjee *et al.*, 1983). This emergence of two phases is similar to the case of the driven reconnection (Brunel *et al.*, 1982). The intensity of coalescence and the rate of subsequent reconnection are controlled by just one parameter, the current localization (ϵ_c). For the case where ϵ_c=0.7, the second phase showed the reconnected flux ψ increasing as t^α with $1 < \alpha$. This indicates that the more the current localizes, the faster the reconnection becomes. This leads to a question: Can the reconnected flux ψ increase explosively as $(t_0 - t)^{-\alpha}(\alpha > 0)$ triggered by the coalescence instability? The answer from computer simulation is: yes it can, when ϵ_c is further increased. The computational results obtained by Tajima *et al.* (1982, 1987) have shown the explosive process and some of its properties.

Figure 3.63 displays the time history of various field and particle quantities observed in the simulation in which, after the initial transient, the phase of coalescence of two magnetic islands commences. It is seen in Figs. 3.63(a)–(c) that around $t = 27$ the magnetic and electrostatic field energies shoot up explosively as well as the ion temperature in the direction of coalescence (the x-direction). It is also seen in Fig. 3.63(a)–(c) that (i) after the explosive increase of the field energies and temperature, this overshooting results in synchronous amplitude oscillations of all these quantities with the period being approximately the compressional Alfvén period; and (ii) superimposed on these overall amplitude oscillations is a distinct double-peak structure in the electrostatic field energy and the ion temperature. It was as in Figs. 3.63(d)–(f) that (i) the magnetic energy explodes as $(t_0 - t)^{-8/3}$; (ii) the electrostatic energy explodes as $(t_0 - t)^{-4}$; and (iii) the ion temperature in the coalescing direction explodes as $(t_0 - t)^{-8/3}$ until saturation due to overshooting sets in, where t_0 is the explosion time measured here to be $t_0 \sim 27$ in this run.

It is very interesting to observe the existence of an explosive process (or instability) (see

231

also Sec. 4.3.5.) and its indices of explosion (the exponent to the time) that governs the explosive magnetic process (the *magnetic collapse*) and this can be modeled as follows.

During explosive coalescence, there is no specific scale length. The scale length characterizing the current sheet varies continuously in time without deformation of global structure of current sheet. That is, the relations (laws) that govern the explosive coalescence themselves are invariant under a changing time scale. This was the manifestation of the presence of self-similarity in the system during explosive coalescence. This may be called universality in time, as opposed to the conventional universality in space such as in Kolmogorov's turbulence spectrum (Kolmogorov, 1941; see also Sec. 2.3). A similar situation also arises in the general theory of relativity in which the scale factor a plays a role in the Hubble expansion of the universe. Such a physical situation may be best described by *self-similar solutions* in which scale factors vary continuously.

These self-similarity scale factors $a(t)$ and $b(t)$ are introduced as follows,

$$v_{ex} = \frac{\dot{a}}{a} x, \tag{3.3.42}$$

$$v_{ix} = \frac{\dot{b}}{b} x, \tag{3.3.43}$$

where a dot represents the time derivative. Following the observed simulation results, an ansatz imposed here is that the velocities are linear in x, which implies that particles flow in the opposite direction around the center of current sheet, $x = 0$. The scale factors a and b will be determined from the above basic equations. From the continuity equations of electrons and ions, we obtain

$$n_e = n_0/a, \tag{3.3.44}$$

$$n_i = n_0/b, \tag{3.3.45}$$

where n_0 is a constant. Equations (3.3.44) and (3.3.45) show that the densities of ions and electrons are nearly homogeneous in space and vary only in time during coalescence. The self-similar solutions obtained here are local solutions in space whose properties are dominated by the physical process near the current sheet. We therefore neglect the higher order terms in space proportional to x^3 and higher hereafter. The current J_z in the sheet is nearly constant. This means that as n is nearly constant, v_z is also approximately constant in space. Neglecting the term with x^3 in the magnetic induction equation, we obtain

$$\frac{B_0(t)}{\lambda} = \frac{4\pi e n_0}{c} \left(\frac{v_{iz}^{(0)}}{b} - \frac{v_{ez}^{(0)}}{a} \right), \tag{3.3.46}$$

where it is assumed the magnetic-field B_y varies as $B_y = B_0(t)\frac{x}{\lambda}$, where λ is the magnetic field scale length.

From the y-component of Ampére's law and the z-component of equation of motion for electrons we obtain

$$\dot{B}_0 = 2c\frac{E_{z1}}{\lambda}, \tag{3.3.47}$$

$$E_{z1}\frac{x^2}{\lambda^2} + \frac{\dot{a}}{a}\frac{B_0(t)}{\lambda c}x^2 = 0, \tag{3.3.48}$$

$$\frac{\partial v_{ez}^{(0)}}{\partial t} = -\frac{e}{m_e}E_{z0}, \tag{3.3.49}$$

$$E_z = E_{z0}(t) + E_{z1}(t)\frac{x^2}{\lambda^2}. \tag{3.3.50}$$

From these equations (3.3.42)–(3.3.50) following Tajima *et al.* (1989b), we derive the equation that governs the dynamics of the scale factor a (and b, where $b = a$ if the quasineutrality between ions and electrons holds) and various fields in terms of $a(t)$. The basic equation for $a(t)$ is now (γ the adiabatic gas constant)

$$\ddot{a} = -\frac{v_A^2}{\lambda^2 a^2} + \frac{c_s^2}{\lambda^2 a^\gamma}, \tag{3.3.51}$$

where

$$v_A^2 = \frac{B_{00}^2}{4\pi n_0(m_i + m_e)} \quad \text{and} \quad c_s^2 = \frac{(P_{0e} + P_{0i})}{(m_e + m_i)n_0}. \tag{3.3.52}$$

In Eq. (3.3.51) the first term of the right-hand side corresponds to the $J \times B$ term: this is the term that drives magnetic compression (collapse). The second term corresponds to the pressure gradient term: this term may eventually be able to balance the magnetic collapse when $\gamma = 3$. The condition $\gamma = 3$ means that the plasma compression takes place in a nearly one-dimensional fashion so that the degree of freedom of the system f becomes unity. The (MHD) fields are expressed as (Tajima *et al.*, 1989b):

$$B_y = \frac{B_{00}}{a^2}\frac{x}{\lambda} \tag{3.3.53}$$

$$E_x = \left(-\frac{m_i}{e_\lambda}\frac{v_A^2}{a^3} + \frac{P_{0e}}{e_\lambda a^4 n_0}\right)\frac{x}{\lambda} \tag{3.3.54}$$

$$E_z = -\frac{B_{00}\dot{a}x^2}{ca^3\lambda} - \frac{B_{00}m_e c\dot{a}}{4\pi n_0 e^2 \lambda a^2} \tag{3.3.55}$$

$$v_{ez} = -\frac{cB_{00}}{4\pi e n_0 \lambda a} \tag{3.3.56}$$

$$v_{ix} = v_{ex} = \frac{\dot{a}}{a}x \tag{3.3.57}$$

$$n_i = n_e = \frac{n_0}{a} \tag{3.3.58}$$

233

In the explosive phase we neglect the second term on the right-hand side of Eq. (3.3.56), which only acts to saturate the explosive collapse;

$$\ddot{a} = -\frac{v_A^2}{\lambda^2 a^2}. \tag{3.3.59}$$

Once the solution $a(t)$ is obtained from Eq. (3.3.59), these physical quantities behave as follows, which is valid in the explosive phase of the coalescence;

$$v_x = v_{ix} = v_{ex} = -\frac{2}{3}\frac{x}{(t_0 - t)}, \tag{3.3.60}$$

$$n = n_i = n_e = \left(\frac{2}{9}\right)^{1/3}\frac{\lambda^{2/3}n_0}{v_A^{2/3}(t_0 - t)^{2/3}}, \tag{3.3.61}$$

$$E_x = -\frac{2}{9}\frac{m_i}{e}\frac{x}{(t_0 - t)^2}, \tag{3.3.62}$$

$$B_y = \left(\frac{2}{9}\right)^{2/3}\frac{B_{00}\lambda^{1/3}x}{v_A^{4/3}(t_0 - t)^{4/3}}, \tag{3.3.63}$$

$$E_z = \frac{2}{3}\left(\frac{2}{9}\right)^{2/3}\frac{B_{00}\lambda^{1/3}x^2}{v_A^{4/3}c(t_0 - t)^{7/3}}$$

$$+\frac{2}{3}\left(\frac{2}{9}\right)^{1/3}\frac{B_{00}c}{\omega_{pe}^2\lambda^{1/3}v_A^{2/3}(t_0 - t)^{5/3}}. \tag{3.3.64}$$

These explosive behaviors upon coalescence, including the exponents, are in reasonable agreement with the observed simulation results in Fig. 3.63(d)–(f). Moreover, for the ion temperature we find its explosiveness as

$$T \sim \frac{1}{a^4} \sim \frac{1}{(t_0 - t)^{8/3}} \tag{3.3.65}$$

when $\gamma = 3$.

The nonlinear coalescence instability of current carrying loops heretofore discussed seems to be taking place in the magnetotail, and possibly related to the magnetospheric substorm process. It, moreover, explains many of the characteristics of solar flares such as their impulsive nature, simultaneous heating and high-energy particle acceleration, and amplitude oscillations of electromagnetic emission as well as the characteristic development of microwave images obtained during a flare. The main characteristics of the explosive coalescence are: (i) a large amount of impulsive increases in the kinetic energies of electrons and ions, (ii) simultaneous heating and acceleration of electrons and ions in high and low energy ranges, (iii) ensuing quasi-periodic amplitude oscillations in fields and particle quantities, (iv) the double peak (and triple peak) structures in these oscillations, and (v) the

234

characteristic break in energy spectra of electrons and ions. A single pair of currents as well as multiple currents (see the fractal nature of flux tubes in Sec. 3.3.7) may participate in the coalescence process, yielding varieties of phenomena. These physical properties seem to underlie some impulsive solar flares. In particular, double sub-peak structures in the quasi-periodic oscillations found in the time profiles of two solar flares on 1980 June 7 and 1982 November 26 are well explained in terms of the coalescence instability of two current loops. This interpretation is supported by the observations of two microwave sources and their interaction in the November 26 flare. In the following these observations of solar flares are briefly summarized in light of the theory of nonlinear coalescence.

In order to explain the rapid quasi-periodic particle acceleration of both electrons and ions observed in the 1980 June 7 flare, a likely mechanism for the impulsive energy release in solar flares is the current loop coalescence instability discussed above. During the coalescence of two current loops, magnetic energy stored by the plasma current is explosively transformed to plasma heating as well as to production of high energy particles within an Alfvén transit time across the current loop (which is about $1 \sim 10$ seconds for appropriate radius of the loop) through the magnetic reconnection process. Furthermore, the energy release is achieved in a quasi-periodic fashion whose periodicity depends on plasma parameters such as the plasma beta ratio (β), the ratio B_P/B_t between the poloidal (B_P: produced by the loop current) and the toroidal (B_t: potential field) components of the magnetic field, as well as the colliding velocity of two current loops that is determined mostly from its initial total plasma loop current profile. The plasma is heated up to ~ 60 times its initial temperature. At the same time, electrons and ions are accelerated simultaneously by the transverse electrostatic field which is produced during the explosive coalescence process.

In the following a comparison of the above simulation/theory with observations of two solar flares are presented: the 1980 June 7 event and the 1982 November 26 event. This should serve to provide an example to indicate the insight obtained from the cooperative study between simulation and observation. Both flares showed quasi-periodic amplitude oscillations with double sub-peak structure in hard X-ray and microwave time profiles. Since the two events vary widely in duration, source size, source height, etc., they provide a stringent test for examining the validity of the model of particle acceleration in solar flares in terms of the coalescence instability.

Explosive Coalescence—the 1980 June 7 Flare

We summarize below some essential points of the 1980 June 7 solar flare at 0312 UT [Fig. 3.63(g)–(i)] observation.

1. The flare was composed of seven successive pulses with a quasi-periodicity of about 8 seconds. Each of the pulses in hard X-rays, prompt γ-ray lines, and microwaves was almost synchronous and similar in shape.

2. Several microwave pulses consisted of double sub-peaks as vividly seen especially in the second and fourth pulses in Fig. 3.63(g). The first sub-peak coincided with the peak of the corresponding hard X-ray pulse [Fig. 3.63(h) and (i)], while the second sub-peak

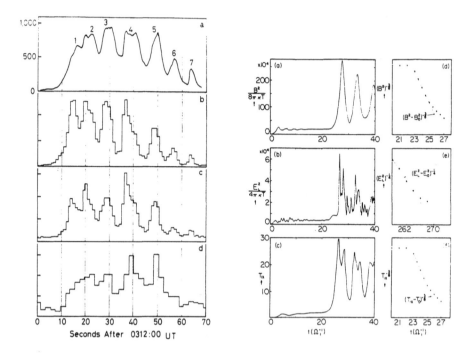

FIGURE 3.63 Current loop coalescence model of flares by Tajima *et al.* (1987). These figures show the explosive increase of field energies and temperature during the coalescence of two magnetic islands, based on the electromagnetic particle simulations. Note that the magnetic energy, $\sim B^2$, the electrostatic energy, $\sim E^2$, and the temperature, T, diverge as $(t_0 - t)^{-8/3}$, $(t_0 - t)^{-4}$, $(t_0 - t)^{8/3}$, respectively. Note also the vigorous, large amplitude oscillations of these quantities just after the explosive phase (Tajima *et al.*, 1987). Electromagnetic signals observed from the 1980 June 7 flare Tajima *et al.*, 1987).

coincides with the peak of the corresponding γ-ray pulse [Fig. 3.63(j)] and with the small hump in hard X-ray time profiles.

3. The starting times of hard X-rays, prompt γ-ray lines, and microwaves coincide within ± 2.2 seconds. Therefore, the acceleration of electrons (up to several MeV) and ions (up to several tens of MeV/nucleon) must have begun almost simultaneously. The time scales of the accelerations are less than 4 seconds.

4. The height of the microwave source was estimated to within 10 arcsec above the photosphere (Hα flare: N12°, W64°). The source had a small size of less than 5 arcsec in the east-west direction and showed no motion.

5. According to the Hα photographs taken at the Peking Observatory, the flaring region had two loops or two arcades of loops that appear to be in contact with each other,

one stretching in the east-west direction and the other in the north-south.

All the above-mentioned characteristics of this flare appear to imply that the coalescence of two current loops is the essential release mechanism of this flare. Indeed, the observed time history shown in Fig. 3.63(g)–(j) resembles those obtained from computer simulation in Figs. 3.63(a)–(f). Since the detail identification and comparison may be found in Tajima *et al.* (1989), here we stop short of going quantitative comparison and analysis of the observation.

3.3.2.7   Reconnection Driven by Magnetic Buoyancy Instability

Yokoyama and Shibata (1994) studied the role of the resistivity model for *magnetic reconnection driven by the magnetic buoyancy instability*, which is another example of the *self-consistent driven-reconnection* in a sense that the driving process is self-consistently determined by the internal physics.

Their study is based on the numerical simulation of 2D nonlinear MHD model of the emerging magnetic loop in the solar atmosphere (Shibata *et al.*, 1989a, 1990a; Nozawa *et al.*, 1992). In this model, the rise motion of a magnetic loop is initiated by the Parker instability (see Sec. 3.2), and does not saturate in the nonlinear stage if the the magnetic pressure of the loop is higher than the ambient total pressure; the rise velocity of the loop increases linearly with height until the loop's magnetic pressure balances with the ambient total pressure (Shibata *et al.*, 1989b, 1990a, Nozawa *et al.*, 1992). Since the nonlinear evolution of the undular instability is already well studied (Shibata *et al.*, 1989a, 1990a; Nozawa *et al.*, 1992), and it explains observations of solar emerging flux very well, it is convenient to extend this model to the reconnection problem. As for the details of the initial condition, see Sec. 3.2.2.5.

They assume two types of the resistivity model;

$$\text{(a)} \quad \eta = \eta_0 = \text{constant},$$

$$\text{(b)} \quad \eta = \begin{cases} \alpha(v_d/v_c - 1)^2, & (\text{for } v_d > v_c) \\ 0, & (\text{for } v_d < v_c) \end{cases}$$

where $v_d = j/\rho/(j_0/\rho_0)$ is the normalized drift velocity of the diamagnetic current in the neutral sheet, j is the current density, ρ is the mass density, v_c is the critical velocity above which anomalous resistivity sets in, and variables with suffix $_0$ denote normalizing constants. Here we assumed v_c and α are free parameters. They assume also that there is an upper limit for the resistivity, η_{max}. Equations are solved numerically by using a modified Lax-Wendroff scheme with artificial viscosity. The size of the computation box is $(X_{max}, Z_{max}) = (150, 60)$ and the total grid numbers in the box are 400×400 or 200×200. The boundary conditions are as follows: $z = 0$ is a rigid conducting boundary, $z = Z_{max}$ is a free boundary, and $x = 0, X_{max}$ are periodic boundaries. Note that the free boundary at $z = Z_{max}$ is located far from the reconnection region and hence does not affect the physics in the computation box.

Figure 3.64 shows the typical evolution of both (a) uniform resistivity model ($\eta_0 = 0.3$) and (b) anomalous resistivity model ($\alpha = 0.01, v_c = 1000, \eta_{max} = 1$). In the anomalous resistivity model (b), several magnetic islands are created in the current sheet by the tearing instability. These islands coalesce with each other by the coalescence instability, and are ejected from the current sheet at the Alfvén speed. After the ejection of islands, the current sheet suddenly collapses, leading to fast reconnection. There are several evidences of slow mode MHD shock extending from the neutral point (X-point) after the ejection of magnetic islands. The length of the current sheet is short and the plasma inflow is converging into the X point. Although the dynamics is nonsteady, the situation near the X-point is similar to that predicted by Petschek.

In the uniform resistivity model (a), on the other hand, there is no island formation nor slow mode MHD shock formation. The current sheet is long, inflow into the sheet is slow, and hence the reconnection is slow, though the velocity of the outflow along the sheet is about the Alfvén speed. The reconnection in this case is nearly the steady Sweet–Parker type (Fig. 3.64). In fact, the Sweet-Parker scaling holds very well; the reconnection rate $\eta j \propto \eta^{1/2}$, the current sheet thickness $\delta \propto \eta^{1/2}$, the current sheet length $\Delta \propto \eta^0$, and the field strength $B \propto \eta^0$.

Note that in both models, the initial velocity of the rising loop just before the collision with the overlying horizontal magnetic field is not small, of order of $0.1V_A$, where V_A is the local Alfvén speed in the loop just below the current sheet. Nevertheless, the fast reconnection does not immediately occur even in the anomalous resistivity model.

Figure 3.65 shows time evolution of several physical quantities for both anomalous resistivity model and uniform resistivity models ($\eta_0 = 0.1, 0.3, 1.0$). It is seen that maximum thermal energy, maximum kinetic energies for both hot and cool plasmas, maximum reconnection rate in the anomalous resistivity model are all larger than those in the uniform resistivity model. It is also seen that the timing of the peak of the reconnection rate for the anomalous resistivity model is nearly simultaneous with the peak of the kinetic energies, which corresponds to the ejection of magnetic island.

It should be noted that the maximum resistivity in the anomalous resistivity model is set to be 1.0, i.e. the same as the (constant) resistivity of the uniform resistivity model with $\eta_0 = 1.0$. Nevertheless, the maximum reconnection rate for the anomalous resistivity model is much larger (nearly double) than that for the uniform resistivity model.

The reason is as follows: (1) In anomalous resistivity model, the threshold of anomalous resistivity is high so that the reconnection is inhibited initially even if the current sheet is formed. During this stage, magnetic energy is stored around the current sheet, which is suddenly released once the reconnection starts. (2) In the uniform resistivity model, even if the islands are formed, they soon disappear because of high diffusion in the current sheet, whereas in the anomalous resistivity case, the islands do not disappear and can grow since the enhanced resistivity region is localized. Once the islands grow, the current sheet length is limited between each island, which then makes the high resistivity region small, and vice versa. The islands coalesces each other to make a bigger island, and the size of the final biggest island can be comparable to the size of the initial current sheet. When the biggest island is ejected, the current sheet collapses, generating large converging flow into the X-point, and hence producing a slow shock. Both the converging inflow and the slow mode MHD shock are characteristic features of Petschek type reconnection model.

Figure 3.66(a) and (b) show the time evolution of thermal energy of hot plasma and the

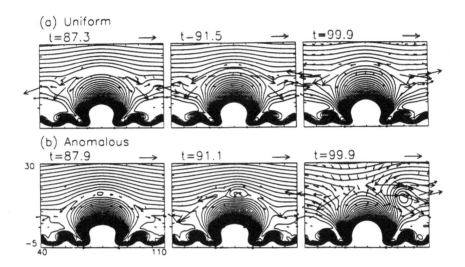

FIGURE 3.64 Time evolution of 2D distributions of magnetic field lines for (a) uniform resistivity model ($\eta_0 = 0.3$) and (b) anomalous resistivity model. The scale of the velocity vector is shown above each frame and the arrow with this size has the velocity of 5.0 (from Yokoyama and Shibata, 1994).

reconnection rate at the X-point for several parameter values of v_c. For low threshold values (smaller v_c), the reconnection is essentially the Sweet–Parker type, but for high threshold values, it is more similar to the Petschek type. It is remarkable that the maximum reconnection rate increases with increasing v_c and well exceeds the maximum value for uniform resistivity model. That is, as the reconnection is inhibited more and more (for longer duration), it becomes more violent and faster once it occurs. Figure 3.66(c) and (d) show similar time evolution for several α values. In this case, the reconnection rate is almost independent of α as expected for fast reconnection coupled with anomalous resistivity (Sato and Hayashi, 1979; Ugai, 1987).

In summary, their simulation shows that their self-consistent driven reconnection is not a simple driven reconnection because the reconnection rate is not uniquely determined by the driving process but strongly dependent on the resistivity model, i.e. on the local plasma condition. These results confirm the previous results (Ugai, 1992; Scholer, 1989) in a more

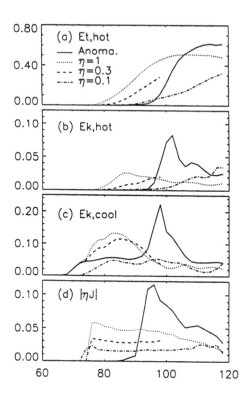

FIGURE 3.65 Time evolution of (a) the total thermal energy content involved in hot gas component and (b) the reconnection rate at the X-point for several parameter values of v_c. It is remarkable that the maximum reconnection rate increases with increasing v_c and well exceeds the maximum value for uniform resistivity model. (c),(d) The same as in (a) and (b), respectively, but for various values of α. In this case, the reconnection rate is almost independent of α (from Yokoyama and Shibata, 1994).

realistic situation, and verified the Biskamp's scenario (Biskamp, 1992) on phenomenological modeling of magnetic reconnection based on anomalous resistivity. They find that the coupling between the locally enhanced resistivity (due to *anomalous resistivity*) (see also Sec. 4.2.3.4) and the tearing instability leads to the rapid formation and ejection of magnetic islands (plasmoids), which is found to be a key physical process leading to fast reconnection. However, they also found the evidence of the slow shock and converging flow into the X-point after plasmoid ejection, both of which are characteristic of Petschek model, as discussed by Priest and Forbes (1987).

Furthermore, the increasingly explosive character of the reconnection with increasing threshold for anomalous resistivity is suggestive of the impulsive solar flares, which show a weak precursor heating that suddenly bursts into action to provide the flare. One can

240

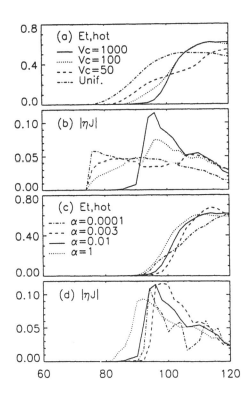

FIGURE 3.66 Time evolution curve for several physical values; (a) the total thermal energy content involved in hot gas component ($T \geq 1.2T_{hot}$), (b) the total kinetic energy of hot gas component, (c) the total kinetic energy of cool gas component ($T < 1.2T_{hot}$), (d) the reconnection rate ($E = \eta j$) measured at the X-point, for typical anomalous resistivity model and three uniform resistivity models; $\eta_0 = 1.0, 0.3, 0.1$ (from Yokoyama and Shibata, 1994).

imagine that the requirement that the drift velocity in the current sheet exceeds the electron thermal velocity for the onset of strong anomalous resistivity provides a high threshold and therefore a truly explosive outburst when that threshold is crossed.

Finally, it is interesting to note that the recent observations of solar flares with the soft and hard X-ray telescopes aboard *Yohkoh* have revealed that many energetic large flares show cusp-shaped configurations suggesting magnetic reconnection (Tsuneta, 1993a; Masuda, 1994a, 1994b, 1995), and that such energetic flares are often associated with eruption of an X-ray filament or plasmoids (Tsuneta, 1993; Shibata *et al.*, 1995; Ohyama and Shibata, 1996, 1997; Nitta, 1997). These observations seem to be consistent with their idea that the formation and ejection of plasmoids leads to fast reconnection.

3.3.2.8 Fractal Nature of Current Sheet

If the magnetic reconnection is really occurring in the solar flares, we need high anomalous resistivity as we have seen in the previous sections. Anomalous resistivity is a result of either microscopic plasma turbulence triggered by some plasma instability (see Sec. 2.7) or macroscopic MHD turbulence. The plasma instability theory shows that the anomalous resistivity occur if the drift velocity of electric current exceed ion thermal velocity (e.g. for the Lower Hybrid drift instability (Tanaka and Sato, 1981)) or electron thermal velocity. In the former case, it is found that the thickness of the current sheet must be smaller than the ion gyro-radius. In the solar corona, this length becomes of order of 1 m (see equation 3.3.29). (If we use the electron thermal speed as the threshold for occurrence of anomalous resistivity, we will find bigger difficulty. See Problem 3–7.) Hence we face the very fundamental difficulty if we want to apply the present anomalous resistivity theory to reconnection in solar flares. The actual size of flares is of order of 10^4 km, whereas the size of current sheet which can induce the reconnection is only of order of 1 m! How can we reconcile enormous gap between the small current sheet thickness and flare size? If we use the Petschek reconnection model, the size of the diffusion region is only $1 \text{ m} \times 10 - 100 \text{ m}$. Does such small diffusion region control the 10^4 km size flare?

One possible answer to these questions is to consider *fractal current sheet* as shown in Fig. 3.68. We suggest that there is a *global current sheet* in the neutral point of the flare region. The size of the global current sheet is less than the flare size 10^4 km, but may be larger than say 1 km. The global current sheet contains many small magnetic islands in 2D view, between which there are many small thin current sheets. We suggest that the size spectrum of the size of each magnetic island is *power law*, i.e. the current sheet has a *fractal* nature. In the smallest scale, the thickness of the current sheet is comparable to the ion gyro-radius, so that the reconnection can start from such smallest scale. The reconnection proceeds by the coalescence instability (Tajima *et al.*, 1987). As the reconnection goes on, the size of magnetic islands grows and grows, and finally the reconnection between the largest magnetic islands becomes possible. Since the time scale of each small scale reconnection is of order of the Alfvén time, the final time scale for the reconnection of the largest scale islands is also the Alfvén time.

In the idealized situation, this kind of fractal structure can occur in the coalescence instability as illustrated in Fig. 3.67. We suggest that this is the basic process of the coalescence of magnetic islands, and hence of the MHD turbulence.

Problem 3–7: Calculate the thickness of the current sheet to induce plasma turbulence whose threshold is the electron thermal speed.

Problem 3–8: Two current key questions in reconnection theory may be summarized as follows:

(1) What is the exact condition for fast reconnection that occurs (nearly) independently of R_m?

(2) Is the anomalous resistivity really necessary for fast reconnection? What is an actual anomalous resistivity?

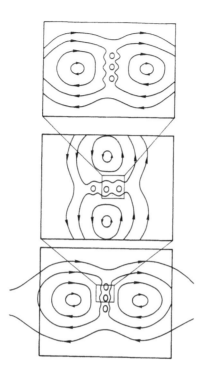

FIGURE 3.67 Cartoon showing the fractal nature of current sheet in the coalescence instability.

3.3.3 Yohkoh Observations of Flares*

The solar X-ray satellite, *Yohkoh*, was launched on Aug. 30, 1991, by the Institute of Space and Astronautical Science (ISAS) in Japan, as international collaboration project between Japan, US, and UK (Ogawara *et al.*, 1991, 1992). *Yohkoh* carries 4 instruments, Hard X-ray Telescope (HXT) (Kosugi *et al.*, 1991), Soft X-ray Telescope (SXT) (Tsuneta *et al.*, 1991), Bragg Crystal Spectrometer (BCS) (Culhane *et al.*, 1991), and Wide Band Spectrometer (WBS) (Yoshimori *et al.*, 1991), and has observed already more than 1000 flares, and more than a few million soft X-ray images. *Yohkoh*/SXT has revealed that *the solar corona is much more dynamic than had been thought*, i.e. the corona is full of transient loop brightenings, jets, global restructuring of magnetic fields, magnetic loop expansion, etc. (see reviews by Acton, 1992, Uchida, 1993; Tsuneta, 1993; Shibata, 1994; Hudson, 1994), suggesting that the solar corona is full of *magnetic reconnection.*

Here, we would like to summarize recent discoveries by *Yohkoh*/SXT (with some HXT results also), which show various evidences of magnetic reconnection associated with mass

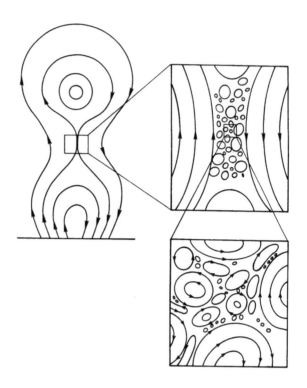

FIGURE 3.68 Hypothetical fractal current sheet in a solar flare.

ejection in flares.

Note that previously it has been believed that there are basically two types of flares as discussed in previous Sec. 3.3.1. Table 3.5 summarizes various "two types of flares" which strongly depends on the observing methods. (Note that the classification in Table 3.1 is not unique, and also that there are many intermediate types of flares. So the purpose of this table is to show reader that many observers have *believed* that there are two distinct class of flares, and that some of them considered that the difference of the type is an evidence of different physical mechanism of flare occurrence.) However, *Yohkoh* has revealed various common properties in these "two types of flares," which seem to be related to the common physical properties of magnetic reconnection.

3.3.3.1 LDE (Long Duration Events) Flares*

One of the biggest discoveries by *Yohkoh*/SXT is the discovery of cusp-shaped loop structures in LDE (Long Duration Events) flares. Figure 3.69(a) shows one beautiful example of this

244

TABLE 3.5 Two types of flares

observing method	larger flares	smaller flares
coronagraph and Hα	CME related flares	compact flares
Hα and skylab soft X-ray	two-ribbon flares	simple loop flares
Hα	eruptive flares	confined flares
Yohkoh soft X-rays	arcade flares	loop flares
soft/hard X-rays	LDE flares	impulsive flares

kind of flare, which occurred on Feb. 21, 1992, at the east limb (Tsuneta *et al.*, 1992a; Tsuneta, 1996; Forbes and Acton, 1996). This flare occurred a few hours after a large scale coronal eruption (possibly, CME), which created a helmet-streamer-like configuration, suggesting that a current sheet is formed as a result of global MHD instability. The apparent height of the loop and the distance between two footpoints of the loop increase gradually with time at a few km/s. This is nicely explained as the result of the successive reconnection in the current sheet above the loop, as described by the classical flare model for two-ribbon flares, which was developed by Carmichael (1964), Sturrock (1968), Hirayama (1974), and Kopp and Pneuman (1976). This kind of model is hereafter called *CSHKP model* (Sturrock, 1992). Modern version of this model has been developed by Forbes and Priest (1983) and others.

Tsuneta *et al.* (1992a) further found the following characters:

(1) The temperature distribution is somewhat chaotic in early phase (during and just after rise phase; < 30 min), while it is systematically higher at the outer edge of the loop (or it is lower at the inner part of the loop) in a later phase (> 1 hour).

(2) The gas pressure is highest at the top of the loop, where the temperature is rather low.

Both seem to be consistent with modern version of this class of flare model, because the low temperature at the inner region can be explained by the conductive and/or radiative cooling (Forbes and Malherbe, 1991; Yokoyama and Shibata, 1997), and the high pressure at the same region may be explained by the slow shock (Ugai, 1987).

More recently, from close examination of the SXT movie of this flare (Hudson, 1994), it is found that a small magnetic island (or plasmoid) with a size of a few 10^4 km is ejected at a few 100 km/s along the current sheet during the rise phase of the flare. It is likely that the ejection of the plasmoid triggered the flare. In fact, SXR images in the preflare phase show that there seem to be filamentary structures perpendicular to the current sheet, suggesting that the perpendicular magnetic field lines penetrate the current sheet, preventing the magnetic reconnection in the current sheet. The flare (possibly, reconnection) suddenly occurred after the plasmoid ejection.

3.3.3.2 Large Scale Arcade Formation*

Cusp-shaped loops or arcades which show similar evolutional feature to that of LDE flares have also been found in much larger spatial scale (Tsuneta *et al.*, 1992b; McAllister *et al.*,

245

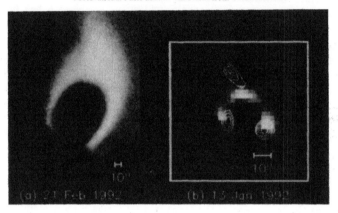

FIGURE 3.69 (a) LDE flare on Feb. 21, 1992 (from Tsuneta *et al.*, 1992a); (b) Impulsive flare on Jan. 13, 1992 (from Masuda *et al.*, 1994).

usually occur in association with disappearance of a dark filament. Tsuneta *et al.* (1992b) described an event associated with a disappearance of a polar crown filament on Nov. 12, 1991. This event has gradually increased its size for more than 20 hours to a size of 1.5 solar radius times 0.5 solar radius at maximum. Similar events (Fig. 3.70) occurred on Apr. 14, 1994, which was luckily reported by KSC *toban* using Email to the world, and the NOAA/SEL people then predicted the large geomagnetic storm successfully.

A large helmet streamer appearing after a filament eruption and CME is possibly a side view of this kind of large scale arcade formation. A beautiful example of such large helmet streamer formation occurred on Jan. 24, 1992, which was reported by Hiei *et al.* (1993). It is interesting to note that temperature is higher at outer edge of the cusp-shaped loops, similar to LDE flares. Note also that the X-ray intensity of these events is usually very low so that often these cannot be noticed from GOES X-ray light curve. For this reason, previously these events were not considered to be flares. However, *Yohkoh*/SXT has revealed that these large scale arcade formation are very similar to LDE flares from various points of

246

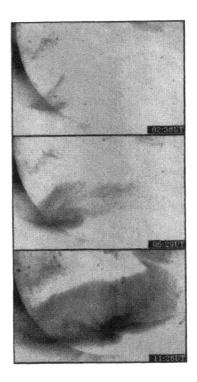

FIGURE 3.70 Giant arcade formation event on Apr. 14, 1994 (from McAllister *et al.*, 1995).

view (morphology, evolution such as apparent rise motion of arcade-loops, emission measure and temperature distribution pattern, etc.). Only difference may be the size and magnetic field strength, which can explain other differences, such as time scale, total released energy, emission measure, etc., using scaling law based on magnetic reconnection theory as discussed later. Consequently, we can now say that these events are one class of flares.

3.3.3.3 Impulsive Flares*

Though LDE flares and large scale arcade formation events show clear cusp-shaped loop structure suggesting magnetic reconnection, there is no such cusp-shaped structure in *impulsive flares* whose occurrence frequency is much more than LDE flares. The impulsive flares are bright in hard X-rays and show impulsive phase whose duration is short (< a few minutes), whereas the LDE flares are usually week in hard X-rays and do not necessarily show impulsive phase. The apparent shape of the impulsive flares in SXT images is *a simple loop*, as already found by Skylab. Are such impulsive, loop flares fundamentally different

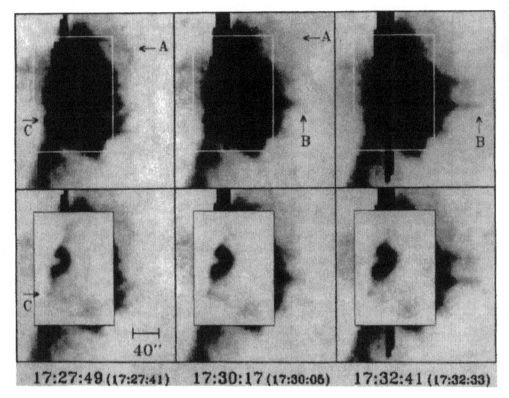

FIGURE 3.71   X-ray Plasmoid/Filament Eruption in Impulsive Flare on January 13, 1992 (from Shibata *et al.*, 1995).

from LDE flares ? (The term, *compact flares* or *confined flares* are often used to describe this class of loop flares, though these do not necessary correspond to exactly the same phenomena.) This led some theoreticians to consider the loop flare models which assume energy release occurring inside the loop (Spicer, 1977; Uchida and Shibata, 1988). The apparent lack of cusp-shaped structure of these impulsive flares in SXT images has been thought to be a negative evidence of reconnection model such as the CSHKP model.

Recently, Masuda *et al.* (1994, 1995) discovered with HXT that *in some of impulsive limb flares, a loop top hard X-ray (HXR) source appeared well above a soft X-ray (SXR) bright loop during the impulsive phase.* Figure 3.71 shows one typical example of such impulsive limb flare showing HXR loop top source. We can see in this figure that the HXR source is well above ($5'' - 10''$) the SXR loop. Although this loop top source is somewhat less bright than the two bright footpoint HXR sources, the time history of HXR intensity of loop top source is similar to those of footpoint sources (Masuda, 1994a). This indicates that an impulsive energy release did not occur within the soft X-ray loop but occurred above the loop. (Aschwanden *et al.* (1996) later confirmed this conclusion, using independent method,

248

loop. (Aschwanden *et al.* (1996) later confirmed this conclusion, using independent method, i.e., measurement of electron time-of-flight distance based on BATSE/CGRO HXR data.) This is a quite exciting discovery because bright soft X-ray loops were often considered to be an evidence of "loop flares" in which energy release occurs within the loop, as discussed above. (If the loop top source is thermal, its temperature is estimated to be as high as a few 100 million K.) One possible physical mechanism to produce such loop top hard X-ray source is *magnetic reconnection* occurring above the loop; i.e., a high speed jet is created through the reconnection and collides with the loop top, producing fast shock, superhot plasma and/or high energy electrons emitting hard X-rays. In this sense, the discovery of the loop top HXR source may open a possibility to unify two distinct classes of flares, *LDE flare* and *impulsive flare*, by the single mechanism, the magnetic reconnection.

If the reconnection hypothesis is correct, a plasmoid or a filament (loop) ejection is expected to occur associated with these impulsive flares, as suggested by the CSHKP model. Shibata *et al.* (1995) searched for such plasmoid or filament (loop) ejections in 8 impulsive limb flares which are selected in an unbiased manner by Masuda (1994), and found that *all these flares were associated with X-ray filament/plasmoid ejections* (see Figure 3.71). The following characteristics are found (Shibata *et al.*, 1995): (1) The velocity of the ejections is 50–400 km/s. (2) The size of the ejections is $4 - 10 \times 10^4$ km. (3) The SXR intensity of the ejections is $10^{-4} - 10^{-2}$ of the peak flare SXR intensity in the main bright SXR loop. A very weak SXR intensity of these ejections is the reason why these ejections have not *always* been seen on the *disk impulsive flares* around which the background SXR intensity is usually high. (4) The onset of the ejections is nearly simultaneous with the impulsive phase (see also Ohyama and Shibata, 1996, 1997). This holds also for multiple ejections. In the case of the 4-Oct-92 flare, the first and second eruptions are nearly simultaneous with the first and second impulsive peaks. (5) A small SXR bright point appeared during the impulsive phase about a few 10^4 km distant from the SXR loop. The bright point seems to be the footpoint of the large scale erupting loop.

3.3.3.4 A Unified Model of Flares*

These recent findings give further support for the magnetic reconnection hypothesis as illustrated in Figure 3.72. In this view (Shibata *et al.*, 1995; Shibata, 1996; Magara *et al.*, 1996; Ohyama and Shibata, 1997), the erupting features correspond to the plasmoid (i.e. a large scale helically twisted loop, in three dimensional view), similar to the LDE flares associated with the Hα filament eruption. A very faint SXR intensity of the erupting features implies that the electron density is not high in these features, of order of $10^9 - 10^{10}$cm^{-3}. If the volume of these features is $\sim 10^{29}$cm^3 (the length is 10^{10}cm and the cross-sectional area is $(3 \times 10^9$cm$)^2$), then the total kinetic energy of the eruptions is of order of $10^{28} - 10^{29}$ erg. This is an order of magnitude smaller than the total released energy during the impulsive phase, estimated from the HXR data by Masuda (1994). Hence *the eruptions are not the energy source of the flares, but simply triggered the flares*. Where does the flare energy come from? Shibata (1996) suggested that the energy is stored in the magnetic field around the current sheet and the plasmoid. On the basis of these considerations, he presented a unified

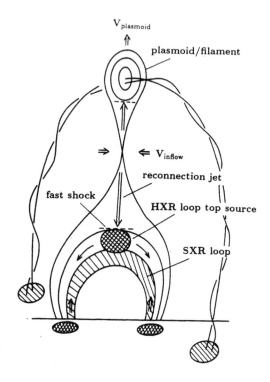

FIGURE 3.72 A Unified Model of Flares: Plasmoid-Driven Reconnection Model (from Shibata *et al.*, 1995; Shibata, 1996).

The model begins with the hypothesis that *the impulsive phase corresponds to the initial phase of plasmoid ejection*. From observations, one finds $V_{\text{plasmoid}} \sim 50 \sim 400$ km/s. Ejection of plasmoid induces a strong inflow into the X-point, which drives the fast reconnection. The velocity of *inflow* into the X-point is estimated to be

$$V_{\text{inflow}} \sim V_{\text{plasmoid}} \sim 50 - 400 \text{ km/s}, \qquad (3.3.66)$$

from the mass conservation law assuming that plasma density does not change much during the process. Since the Alfvén speed around the plasmoid is

$$V_A \simeq 3000 \left(\frac{B}{100\text{G}}\right) \left(\frac{n_e}{10^{10}\text{cm}^{-3}}\right)^{-1/2} \text{km/s}, \qquad (3.3.67)$$

where B is the magnetic flux density and n_e is the electron density. The Alfvén Mach number of the inflow becomes

$$M_A = V_{\text{inflow}}/V_A \sim 0.02 - 0.1 V_A. \qquad (3.3.68)$$

250

This is comparable to the inflow speed expected from the Petschek theory.

The magnetic reconnection theory predicts two oppositely directed high speed jets from the reconnection point at Alfvén speed,

$$V_{\text{jet}} \sim V_A \simeq 3000 \left(\frac{B}{100\text{G}}\right) \left(\frac{n_e}{10^{10}\text{cm}^{-3}}\right)^{-1/2} \text{ km/s.} \tag{3.3.69}$$

The downward jet collides with the top of the SXR loop, producing MHD fast shock, superhot plasmas and/or high energy electrons at the loop top, as observed in the HXR images. The temperature just behind the fast shock becomes

$$T_{\text{loop-top}} \sim m_i V_{\text{jet}}^2/(6k) \sim 2 \times 10^8 \text{ K} \left(\frac{B}{100\text{G}}\right)^2 \left(\frac{n_e}{10^{10}\text{cm}^{-3}}\right)^{-1}, \tag{3.3.70}$$

where m_i is the hydrogen ion mass and k is the Boltzmann constant. This explains the observationally estimated temperature of the loop top HXR source (Masuda *et al.*, 1994).

Similar physical processes are also expected for the upward directed jet (see Fig. 3.72). Indeed a SXR bright point is found during the impulsive phase somewhat far from the SXR loop. This bright point seems to be located at the footpoint of the erupting loop.

The time scale of the impulsive phase is determined from the duration of the strong inflow, which may be comparable to the traveling time of the plasmoid across its size, i.e.,

$$t_{\text{impulsive}} \sim \frac{L_{\text{plasmoid}}}{V_{\text{plasmoid}}} \sim 10 t_A \left(\frac{M_A}{0.1}\right)^{-1}$$

$$\sim 1 \text{ min} \left(\frac{M_A}{0.1}\right)^{-1} \left(\frac{B}{100\text{G}}\right)^{-1} \left(\frac{n_e}{10^{10}\text{cm}^{-3}}\right)^{1/2} \left(\frac{L_{\text{plasmoid}}}{2 \times 10^9 \text{cm}}\right), \tag{3.3.71}$$

where $t_A = L_{\text{plasmoid}}/V_A$. This is roughly consistent with the observed duration $(1 - 2\,\text{min})$ of one impulsive peak. (Note that the total duration of the impulsive phase ranges from 1 min to 10 min. Longer impulsive phase usually include multiple impulsive peaks.)

The magnetic energy stored around the current sheet and the plasmoid is suddenly released through reconnection into kinetic and thermal/nonthermal energies after the plasmoid is ejected. The magnetic energy release rate at the current sheet (with the length of $L_{cs} \sim L_{\text{plasmoid}} \simeq 2 \times 10^4$ km) is estimated to be

$$\frac{dE}{dt} = 2 \times L_{\text{plasmoid}}^2 B^2 V_{\text{inflow}}/4\pi$$

$$\sim 4 \times 10^{28} \text{ erg/s} \left(\frac{V_{\text{inflow}}}{100 \text{ km/s}}\right) \left(\frac{B}{100 \text{ G}}\right)^2 \left(\frac{L_{\text{plasmoid}}}{2 \times 10^9 \text{cm}}\right)^2. \tag{3.3.72}$$

This is comparable with the energy release rate during the impulsive phase, $4 - 100 \times 10^{27}$ erg/s, estimated from the HXR data (Masuda, 1994), assuming the lower cutoff energy as 20 keV.

251

The reason why the HXR loop top source is not bright in SXR is that the evaporation flow has not yet reached the colliding point and hence the electron density (and so the emission measure) is low. The key physical parameter discriminating impulsive flares and LDE flares (or impulsive phase and gradual phase) is the velocity of the inflow, V_{inflow}. If V_{inflow} is large, the reconnection is fast, so that the reconnected field lines accumulate very fast and hence the MHD fast shock (i.e. HXR loop top source) is created well above SXR loop which is filled with evaporated plasmas. On the other hand, if V_{inflow} is small, the reconnection is slow and hence the fast shock is produced at the SXR loop which has been already filled with evaporated plasmas. In that case, the density at the shocked region is high because of evaporation, and so the temperature behind the fast shock becomes

$$T_{\text{gradual-loop-top}} \sim \left(\frac{n_{e,jet}}{n_{e,loop}}\right) \frac{m_i V_{jet}^2}{6k} \sim 2 \times 10^7 \text{ K} \left(\frac{B}{100G}\right)^2 \left(\frac{n_{e,loop}}{10^{11}\text{cm}^{-3}}\right)^{-1}, \qquad (3.3.73)$$

which roughly agrees with temperatures found at the loop top in gradual phase of impulsive flares (Masuda, 1994) and in LDE flares (Tsuneta et al., 1992a).

In the case of impulsive flares, the inflow speed may decrease by an order of magnitude after the plasmoid ejection, because the current sheet length between the reconnected loop and the plasmoid increases much. This corresponds to gradual phase. The inductive electric field associated with the inflow and the jet is comparable to

$$E \sim V_{inflow}B \sim V_{jet}B_{jet} \sim 3 \times 10^3 \text{volt/m} \left(\frac{M_A}{0.1}\right)\left(\frac{B}{100G}\right)^2 \left(\frac{n_{e,jet}}{10^{10}\text{cm}^{-3}}\right)^{-1/2}, \qquad (3.3.74)$$

which may be more or less related to the particle acceleration process (though exact mechanism is not known yet) and decreases much in gradual phase since V_{inflow} decreases much. This explains why the particle acceleration is less efficient in the gradual phase than in the impulsive phase. On the other hand, in LDE flares and large scale arcade formation, the size scale is large and the field strength is small, so that the Alfvén speed, the inflow speed, the inductive electric fields are smaller and the Alfvén time is longer than in impulsive flares. This is why these flares are slow and there is no bright HXR emission.

3.3.3.5 Transient Brightenings (Microflares) and X-ray Jets*

As mentioned in the beginning of Sec. 3.3.3, Yohkoh/SXT found that the corona is full of transient brightenings (Shimizu et al., 1992), and X-ray jets (Shibata et al., 1992b; Strong et al., 1992), both of which are new discoveries by Yohkoh. Shimizu et al. (1992, 1994) studied the transient brightenings in active regions in detail, and found that active region transient brightenings (ARTBs) usually show a single or multiple loops, the total thermal energy content in one transient brightening is $10^{25} \sim 10^{29}$ erg, time scale is $1 \sim 10$ min, and the loop length is $0.5 \sim 4\times10^4$ km. They further found that ARTBs correlate well with GOES C-class or sub-C class flares so that ARTBs are considered to be a spatially resolved soft X-ray counterpart of hard X-ray microflares (Lin, 1984). Using BCS data, Watanabe (1994) found that these sub-C class flares show maximum temperatures of order of 10^7 K, which are not so

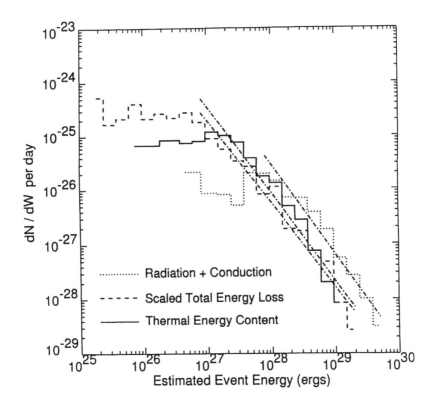

FIGURE 3.73 Histogram of occurrence frequency of microflares (Shimizu, 1995)

different from those of larger flares. Morphology of ARTBs, such as multiple loop structures, is suggestive of magnetic reconnection due to *loop-loop interaction* (Gold and Hoyle, 1960; Tajima *et al.*, 1987; Sakai and Koide, 1992), though clear evidence of interaction between two loops has not yet been found until now. It is possible that the magnetic reconnection similar to that occurring in larger flares produces two neighboring loops as a result of reconnection, which can be seen as "multiple loop brightenings". Although the observational evidence of reconnection in ARTBs are not enough at present, Shimizu *et al.* (1992) found an interesting statistical property of ARTBs. That is, the number of ARTBs, N, as a function of their total thermal energy content, W, scales as a single power law

$$\frac{dN}{dW} \propto W^{-1.5\sim-1.6},\tag{3.3.75}$$

where W ranges from 10^{27} to 10^{29} erg. Since this relation is essentially the same as that of larger flares and HXR microflares (Hudson, 1991), it is likely that the same physical mechanism causes ARTBs as in larger flares. This is also consistent with the finding by

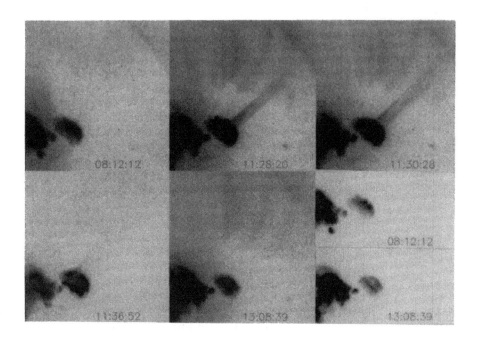

FIGURE 3.74 X-ray jet of 11 Nov. 1992 and a Reconnection Model (from Shibata *et al.*, 1992b).

Watanabe (1994) that temperature of microflares is not so different from those of larger flares.

In contrast to ARTBs, there are many observational evidences of reconnection for *X-ray jets*. X-ray jets are defined as transitory X-ray enhancements with an apparent collimated motion (Shibata *et al.*, 1992b) (Fig. 3.74), and occur in association with small flares (microflares to subflares) which occur in active regions (ARs), emerging flux regions (EFRs), or X-ray bright points (XBPs). The occurrence frequency is more than 20 per month between November, 1991 and May 1992. Shimojo *et al.* (1996) compiled 100 jets during this period, and studied statistical property of jets. They found that average length and (apparent) velocity of jets are $\simeq 1.7 \times 10^5$ km and $\simeq 200$ km/s. Shibata *et al.* (1992b, 1993, 1994a, 1994b, 1996) found several cases in which the footpoint AR changed their shape or morphology during a jet, which can be an indirect evidence of reconnection in the AR. Shibata *et al.* (1993) noted that jets are often ejected from EFRs as a result of interaction between emerging flux and coronal magnetic field (some of which are clearly seen in SXT full Sun movie), and that there are basically two types of interaction of emerging flux with coronal fields (see Fig. 3.75);

(1) *Anemone-Jet* type: When an emerging flux appears in coronal holes, a vertical jet is ejected from an EFR. During the jet, a small loop flare occurs in the EFR. An EFR (or an AR) looks like a sea-anemone and hence is called an anemone-AR.

(2) *Two-Sided-Loop (or Jet)* type: When an emerging flux appears in quiet region, two horizontal jets (or loops) are produced both sides of an EFR.

These features are explained very well by magnetic reconnection model developed by Yokoyama and Shibata (1995, 1996). They performed two dimensional MHD numerical simulation of reconnection between emerging flux and coronal field (Shibata *et al.*, 1992a), extending the pioneering work by Heyvaerts *et al.* (1977). Shimojo *et al.* (1996) found also that many smaller jets ejected from XBPs show a converging shape or an upside down Y shape, and that the brightest parts of the small flare associated with jets are not just at the footpoint of jets. These features are very similar to those found for larger jets, *anemone-jet*, and hence could be an indirect observational evidence of magnetic reconnection even if the footpoints of these jets are not spatially resolved well. Canfield *et al.* (1996) found some examples showing both X-ray jets and H_α surges in the same direction, and found several new signatures of magnetic reconnection in these surges; i.e., converging footpoints and moving blue shifts.

3.3.3.6 Toward A Further Unified Model of "Flares"*

2D MHD numerical simulations by Shibata *et al.* (1992a) and Yokoyama and Shibata (1995, 1996) have shown that multiple magnetic islands are created in the current sheet between emerging flux and overlying coronal field. Dense and cool plasmas are confined in these islands, while ambient rarefied plasmas are heated much outside the islands (see Figs. 3.55 and 3.76). These islands coalesce each other to make bigger islands because of coalescence instability (Tajima *et al.*, 1987). Finally, a biggest island (plasmoid), whose size is comparable to the length of the original current sheet, is rapidly ejected from the current sheet, creating an instantaneous vacuum and hence rapid inflow into the X-point (collapse of current sheet). Thus, after the ejection of the island (plasmoid), sudden collapse of the current sheet occurs to drive fast reconnection with a pair of slow shocks that are characteristic of Petschek model (1964). If there is a perpendicular field component, a magnetic island (a plasmoid) would be seen as a twisted loop. Ejection of twisted loop is similar to those often observed in larger eruptive flares (discussed before).

In this sense, physical processes occurring in small scale reconnection seen in small flares (microflares and subflares) may be similar to those in large scale reconnection seen in larger flares (LDE flares and impulsive flares). In both cases, if the current sheet is long enough, the coupling between anomalous resistivity and nonlinear tearing instability leads to the formation of magnetic islands (plasmoids with helically twisted field lines), and the ejection of plasmoids triggers the rapid collapse of the current sheet, leading to very fast reconnection. New observations by *Yohkoh* have shown that the mass ejection (plasmoids or jets) in association with flares is much more universal than had been thought, which seems to support this hypothesis (Shibata, 1996).

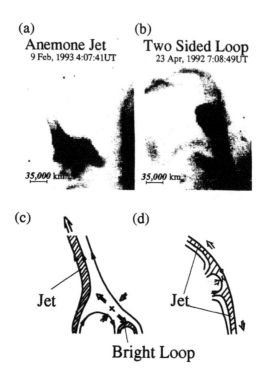

(a)
Anemone Jet
9 Feb, 1993 4:07:41UT

(b)
Two Sided Loop
23 Apr, 1992 7:08:49UT

35,000 km

35,000 km

(c)

(d)

Jet

Jet

Bright Loop

FIGURE 3.75 Two types of interaction between emerging flux and overlying coronal magnetic fields (from Yokoyama and Shibata, 1995, 1996).

explaining both larger and smaller flares in fundamentally the same physics, as summarized in Table 3.8. In this model, the start of a story is the global MHD instability (or loss of equilibrium) which creates a current sheet. In largest flares, this corresponds to CME, while it could be emerging flux driven by magnetic buoyancy instability in smaller flares. Any other instability can be a candidate if it creates a current sheet. The point is that the fast reconnection does not necessarily begin immediately after the instability. As shown by Yokoyama and Shibata *et al.* (1994), the fast reconnection can delay depending on the local plasma condition (such as the presence of perpendicular field penetrating the current sheet and the condition for occurrence of anomalous resistivity) even if the current sheet is compressed by the global instability. *Yohkoh* observations also have shown such examples [e.g. LDE flare on Feb. 21, 1992 (Tsuneta *et al.*, 1992a)]). Observations show that the impulsive phase or the rise phase is nearly simultaneous with rapid ejection of plasmoids (X-ray/Hα filament eruption). Hα surge (and/or X-ray jets) often observed in subflares may correspond to such plasmoid ejection. However, loop-like ejection is rare for smaller surges

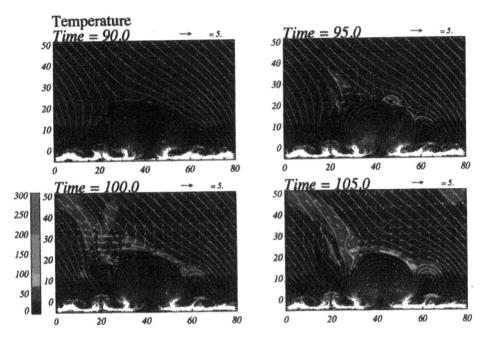

FIGURE 3.76 Numerical Simulation of Reconnection between Emerging Flux and Coronal Field (Yokoyama and Shibata 1995, 1996)

and jets. Why? The answer may be as follows (Shibata, 1997): the length of current sheet in smaller flares is very limited as shown in Fig. 3.77, and hence an ejected island (plasmoid loop) collides and reconnect with the neighboring flux tube immediately after the ejection. Thus the ejected loop cannot emerge directly from the current sheet, and in turn the cool dense plasmas originally contained in the loop can now flow along the neighboring flux tube. This may correspond to Hα surges.

3.3.3.7 Summary and Remaining Questions

In this subsection, we have summarized various new observational findings by *Yohkoh*, with emphasis upon observational evidences of magnetic reconnection. Some of key observational findings are summarized in Tables 3.6 and 3.7. The point is that various flare-like event ranging from very small microflares to very large arcade formation events can be understood by the same physical process, *magnetic reconnection*. The wide range of total flare energy,

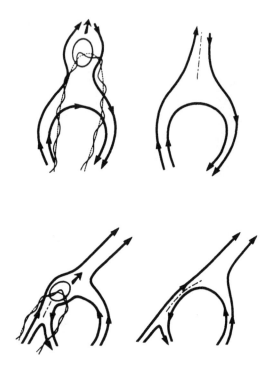

FIGURE 3.77 Further Unified model of flares, microflares, and X-ray jets (Shibata, 1996)

from 10^{26} erg to 10^{32}, is simply explained by the available magnetic energy contained in the relevant volume as in the following way,

$$E_{\text{flare}} \sim L^3 B^2/(8\pi) \sim 4 \times 10^{32} \, \text{erg} \left(\frac{B}{100 \, \text{G}}\right)^2 \left(\frac{L}{10^{10} \text{cm}}\right)^3. \qquad (3.3.76)$$

On the other hand, the time scale of the flare ranges from 1 min to 1 day. If we normalized it by the Alfvén time, it becomes

$$t_{\text{flare}} \sim 10 \sim 100 \, t_A. \qquad (3.3.77)$$

This is similar to time scale observed in magnetospheric substorm and in explosive phenomena in laboratory fusion plasma, and is also similar to that expected from fast reconnection theory.

On the basis of these new observations, a *unified model* of LDE flares and impulsive flares, and even a *grand unified model* explaining both larger flares (LDE and impulsive

258

TABLE 3.6 Comparison of Various "Flares"

"flare"	size (L) (10⁴ km)	time scale (t) (sec)	energy (erg)	mass ejection
microflares (ARTBs)	$0.5 - 4$	$60 - 600$	$10^{26} - 10^{29}$	jet/surge
impulsive flares	$1 - 10$	$60 - 3 \times 10^3$	$10^{29} - 10^{32}$	X-ray/Hα filament eruption
LDE flares	$10 - 40$	$3 \times 10^3 - 10^5$	$10^{30} - 10^{32}$	X-ray/Hα filament eruption
large scale arcade formation	$30 - 100$	$10^4 - 2 \times 10^5$	$10^{29} - 10^{32}$	X-ray/Hα filament eruption

TABLE 3.7 Comparison of Various "Flares" (continued)

"flare"	B (G)	n_e (cm⁻³)	V_A (km/s)	$t_A = L/V_A$ (sec)	t/t_A
microflares	100	10^{10}	3000	5	$12 - 120$
impulsive flares	100	10^{10}	3000	10	$6 - 300$
LDE flares	30	2×10^9	2000	90	$30 - 10^3$
large scale arcade formation	10	3×10^8	1500	400	$25 - 500$

flares) and smaller flares (microflares, subflares, and X-ray jets), have been put forward (Shibata, 1996) which include fast reconnection driven by plasmoid ejection as a key process (Table 3.8). But of course, this is only a first trial to understand the complex "flares" as simple as possible, and we need more detailed observations such as high spatial resolution observations ranging from X-ray to optical regime, especially on smaller flares (microflares and X-ray jets). In fact, there are still not enough observational evidence of reconnection in microflares because of lack of high spatial resolution in *Yohkoh*. No one knows "true" velocity of X-ray jets at present due to lack of Doppler-shift measurement. Even in large flares, high speed reconnection jet (> 2000 km/s) have not yet been found by *Yohkoh*. All these remaining puzzles and problems would be an important subject in the future solar mission.

TABLE 3.8 Further Unified Scheme of Solar "Flares"

physical process	large flares (LDE, impulsive)	small flares (sub-, microflares)
global MHD instability (driving force) $\Downarrow$	ex. CME	ex. emerging flux (magnetic buoyancy)
current sheet formation $\Downarrow$	X- or Y- type configuration	interaction with overlying or ambient field
(anomalous resistivity + nonlinear tearing) $\Downarrow$		
plasmoid ejection $\Downarrow$	Hα/X-ray filament eruption	Hα surge and/or X-ray jets
very fast reconnection $(t \sim 10 t_A)$ $\Downarrow$	impulsive or rise phase	(impulsive or) rise phase
(particle acceleration) $\Downarrow$	(HXR/SXR double footpoints, HXR loop top)	(SXR double footpoints)
fast reconnection $(t \sim 100 t_A)$	gradual or decay phase (SXR loop/arcade, Hα loop)	gradual or decay phase (SXR loop)

3.3.4 Collisionless Conductivity*

3.3.4.1 Collisionless Plasma and Memory Decay*

The electrical conductivity depends on the coherence time for acceleration of the particle in the electric field. In kinetic theory this coherence time is determined precisely by the decay of the two-time velocity correlation function for the particle. In the presence of collisions this decay time is simply the mean time between collisions. In the presence of a spectrum of waves it is the scattering time of the particle's momentum by the waves. In the absence of waves and collisions the velocity correlation function generally oscillates without decay showing the reversibility and long-time memory of the particles. In the presence of a sufficiently nonuniform magnetic field, however, the Lorentz force itself $\Delta p = q v \times B(x) \Delta t$ could produce stochastic motion giving rise to a decay of the correlation function showing that the particles have a finite memory or correlation time just as in the case of collisions or wave scattering. Indeed one may argue that the strongly nonuniform $B(x)$ is the limit of a zero frequency wave spectrum that is producing the energy-conserving momentum scattering. For the very simple magnetic field structures that are considered here we adopt the Hamiltonian and study its stochasticity acting as the momentum scattering process. We find that the strongly nonuniformity of the field gives rise to the stochastic orbits of the particles and the decay of the two-time velocity correlation function. This is a bit surprising because we learn that Hamiltonian system not only conserves energy but also entropy. Thus the chaos resulting in the Hamiltonian system is a subtle concept and can defy our intuition. Then we introduce ensembles of particles to obtain the proper weighting of the various types of orbits and calculate the effective conductivity of the (collisionless) plasma. We compare the formulas derived here with the Lyons and Speiser (1985) formula and the simple idealized limit of the unmagnetized Landau damping conductivity introduced by Galeev and Zelenyĭ (1976) for the tearing mode calculation.

We consider the mechanism for the conductivity that exists even in the absence of collisions (that are known to increase entropy).

The particles in a magnetized plasma continually exchange their momentum mv with the magnetic field B through the Lorentz force acceleration. The momentum change $\Delta(mv)$ is clearly seen from the conservation of the canonical momentum $P = mv + qA$ which is locally conserved in the gyromotion. In the local gyromotion the momentum exchange is completely reversible with $\delta p = 0$ over each gyrorevolution. In the case of bouncing between magnetic mirrors (i.e. the local nonuniform magnetic field) the particle momentum changes reversibility after each complete cycle. When the particle motion becomes stochastic, however, the particle has a finite memory time. In this case the particle "forgets" exactly where and when to return the borrowed momentum. The result is an irreversible scattering of the momentum of the particle due to its interaction with the inhomogeneous magnetic field. The "forgetting" or loss of memory process is described by the decay of the two-time velocity correlation function along the single particle orbit. The well-known pitch angle jumps which accumulate with repeated mirroring to become pitch angle diffusion are a particular example of the momentum scattering due to stochasticity. The pitch angle diffusion of electrons from collisionless electron orbits is introduced by Büchner and Zelenyĭ (1987) for a model of electron dissipation producing a collisionless electron tearing mode. Electron orbits are generally in the adiabatic regime compared with ion orbits in the case of typical geomagnetic tail plasma or other plasmas with greater radius of curvature (such as the solar helmet, loop structure etc.). The importance of quiet time ion precipitation from pitch angle jumps in crossing the plasma neutral sheet is shown by Sergeev and Tsyganenko (1987).

The coupling of the two degrees of freedom in x and z are described by the effective potential

$$V_{\text{eff}}(x, z; p_y) = \frac{q^2 B_{z0}^2}{2m} \left(\frac{z^2}{2a} - \frac{B_z}{B_{z0}} x + \frac{p_y}{q B_{z0}} \right)^2.$$

Now oscillations in the z-direction in the effective potential occur at the characteristic frequency

$$\omega_{bz} = \left(\frac{v \omega_{cx0}}{a} \right)^{1/2} \tag{3.3.78}$$

with the excursion distance given by

$$\Delta z = \frac{v}{\omega_{bz}} = \left(\frac{va}{\omega_{cx0}} \right)^{1/2}. \tag{3.3.79}$$

This is a typical situation for chaos of two coupled oscillators. The details of this stochasticity, the temporal decay of correlations and their derived dissipation (or conductivity) one found in Horton and Tajima (1988, 1990).

The charged particle dynamics in our system is governed by the Hamiltonian

$$H = \frac{p_z^2}{2m} + \frac{1}{2m} \left(p_x - \frac{q B_y z}{c} \right)^2 + \frac{1}{2m} \left(p_y + \frac{q B_{z0}}{c} \frac{z^2}{2L} - \frac{q B_z x}{c} \right)^2. \tag{3.3.80}$$

In this case the constants of motion are the energy $H = \frac{1}{2}mv^2$, and the canonical momentum p_y. Energy is conserved because we are assuming no electric field; however, it is important to note that, unlike the case with $B_y = 0$, if a constant electric field E_y is assumed to exist, there is no reference frame transformation where E_y can be shut down. In the case where $B_y = 0$, the motion can be decomposed into an oscillatory motion in the z-direction with characteristic frequency ω_{bz} and the cyclotron motion about B_z with frequency $\omega_{cz} = q B_z / mc$. Equation (3.3.80) may be decomposed into a Hamiltonian $H_z(p_z, z) = \frac{p_z^2}{2m} + \frac{1}{2m} \left(p_y + \frac{q B_{z0} z}{2ac} \right)^2$ and another $H_x(p_x, x) = p_x^2 + \frac{q^2 B_z^2}{2mc^2} x^2$ and a coupling term $B_{z0} B_z yz$. Both oscillators are nonlinearly coupled, and stochasticity on the system arises due to the resonance overlapping of the two motions. The north-south frequency is given by

$$\omega_{bz} = \left(\frac{v_{th} \omega_{cx0}}{L} \right)^{1/2} = \left(\frac{\rho}{L} \right)^{1/2} \omega_{cx0}, \tag{3.3.81}$$

where $v_{th} = (2H/m)^{1/2}$ is the thermal speed, $\omega_{cx0} = q B_{z0}/mc$, and $\rho = v_{th}/\omega_{cx0}$ is the Larmor radius about B_{z0}. Here we take $v = v_{th}$ for the reference values.

The three important parameters for the system are: (1) the ratio of field component normal to the current sheet and the asymptotic limit of the lobe field,

$$b_z \equiv \frac{B_z}{B_{z0}}, \tag{3.3.82}$$

(2) the ratio of the cross-tail magnetic field and the normal field,

$$b_y \equiv \frac{B_y}{B_z}, \tag{3.3.83}$$

(3) the ratio of the Larmor radius and the current sheet thickness,

$$\epsilon \equiv \frac{\rho}{L}. \tag{3.3.84}$$

It is also necessary to introduce the parameter

$$\kappa \equiv \frac{\omega_{cz}}{\omega_{bz}} = \frac{b_z}{\epsilon^{1/2}}. \tag{3.3.85}$$

Taking $1/\omega_{bz}$ as the unit time and $(\rho L)^{1/2}$ as the unit length, the Hamiltonian (3.3.80) takes the dimensionless form

$$h = \frac{p_z^2}{2} + \tfrac{1}{2}(p_x - \kappa b_y z)^2 + \tfrac{1}{2}\left(p_y + \tfrac{1}{2}z^2 - \kappa x\right)^2$$

$$= \frac{1}{2}, \tag{3.3.86}$$

where $h \equiv H/(m\rho L \omega_{bz}{}^2)$.

3.3.4.2 Collisionless Tearing Instability*

We are interested in the mobilities and conductivities in response to tearing-like perturbations,

$$\delta A_y(x, z, t) = \delta A_y(z)e^{\gamma t}\cos(k_x x - \omega t)$$

$$= \mathrm{Re}\left[\delta A_y(z)e^{i(k_x x - \omega t - i\gamma t)}\right]. \tag{3.3.87}$$

In particular, we are concerned with the response functions for the central plasma sheet for $\omega \lesssim \omega_{bz}$ and $k_x v_{th} \ll \omega_{cz}$. In this low-frequency/long-wavelength domain the global orbits determine what is called the bounce-averaged response function. With the fluid response component of current δj^{MHD} being given (Horton and Tajima, 1991) by

$$\delta j_y{}^{\mathrm{MHD}}(x, z, t) = \frac{-1}{\mu_0}\frac{B_x''(z)}{B_x(z)}\delta A_y(x, z, t) \tag{3.3.88}$$

where the prime means derivation with respect to z, Ampére's law is expressed as

$$\nabla^2 \delta A_y = -\mu_0 \delta j_y = -\mu_0\left(\delta j_y{}^{\mathrm{irr}} + \delta j_y{}^{\mathrm{MHD}}\right), \tag{3.3.89}$$

leading to a self-consistent field equation

$$\left[\frac{d^2}{dz^2} - k_x{}^2 - \frac{B_x''(z)}{B_x(z)}\right]\delta A_y(x, z, t) = -\mu_0 \delta j_y{}^{\mathrm{irr}}(x, z, t); \tag{3.3.90}$$

where

$$\delta j_y{}^{\mathrm{irr}} = \frac{q^2}{m v_{th}{}^2}\int d\mathbf{v} f_0 \int_{-\infty}^t dt' v_y(t) v_y(t') e^{i(k_x x(t') - \omega t')} E_y. \tag{3.3.91}$$

The right-hand side of (3.3.90) is negligible everywhere, except in a region of width $\Delta z \sim (\rho L)^{1/2}$ about the plane $z = 0$, in which δj_y is such as to give $\langle \delta j_y E_y \rangle$ irreversible heating of the current sheet plasma (Doxas et al., 1990). In this layer the conductivity may be given as

$$\sigma_{\alpha\beta} = \frac{n_0 q^1}{m v_{th}^2} \int \alpha \mathbf{v}\, f(\mathbf{v}) v_\alpha v_\beta \pi \delta(\omega - \mathbf{k} \cdot \mathbf{v})$$

$$= \frac{n_0 q^2}{m v_{th}^2} \left\langle \int^\infty d\tau C_{\alpha\beta}(\tau \cdot \mathbf{v}) e^{i\omega\tau} \right\rangle. \tag{3.3.92}$$

where $C_{\alpha\beta}(\tau, \mathbf{v})$ is the two-time velocity correlation function defined by

$$C_{\alpha\beta}(\tau, \mathbf{v}) = \lim_{T \to \infty} \frac{1}{T} \int_{-T/2}^{T/2} dt\, v_\alpha(t) v_\beta(t - \tau). \tag{3.3.93}$$

This is another manifestation of the fluctuation-dissipation theorem (See also Sec. 5.3).

Chaotic scattering and stochastic diffusion of single particle trajectories occurs well within the current sheet, which is the region where there is a strong gradient of the magnetic field. Outside this region, the motion is essentially the (integrable) cyclotron motion along the field lines. Because of the chaotic motion, the correlation function decays inside the current sheet and oscillates without decay outside the region. Similarly, because of the continuum mixing of frequencies produced by the chaotic motion, the main contribution to the power spectrum comes from the motion within the current sheet. Outside the current sheet, the only essential contribution to the power spectrum is at the resonance.

Charged particles are coherently accelerated only in the region of weak $B_x(z)$ field. Outside the acceleration region, the cyclotron motion about $B_x(z)$, is faster than the north-south bounce frequency, ω_{bz}. Inside the acceleration region, $\omega_{bz} > \omega_{cx}(z)$, the bounce motion is faster than the cyclotron motion. We determine the width of the acceleration region Δ by finding the value of the z coordinate for which the cyclotron frequency,

$$\omega_{cx}(z) = \omega_{cx0} \frac{|z|}{L}, \tag{3.3.94}$$

is equal to the frequency of the north-south oscillatory motion,

$$\omega_{cx}(z = \Delta) = \omega_{bz}. \tag{3.3.95}$$

Thus the thickness of the acceleration region or resistive layer is given by

$$\Delta = (\rho L)^{1/2}. \tag{3.3.96}$$

It is important to note that for $|z| < \Delta$, the local Larmor radius about $B_x(z)$,

$$\rho(z) = \frac{v_\perp}{\omega_{cx0} \frac{|z|}{L}}, \tag{3.3.97}$$

264

would be larger than $|z|$, and thus no cyclotron motion occurs. Conversely, for $|z| > \Delta$, we have that $\rho(z) < |z|$ and the distorted cyclotron orbit is formed with a substantial ∇B drift.

The correlation function can be evaluated by its temporal decay time τ_c as

$$\sigma_{dc} = \frac{n_0 q^2}{2 m v_{th}{}^2} \left(\langle \overline{v_{\perp y}^2} \rangle \tau_{c\perp} + \langle \overline{v_{\parallel y}^2} \rangle \tau_{c\parallel} \right), \tag{3.3.98}$$

where

$$\langle \overline{v_{\perp y}^2} \rangle \equiv \langle C_{\perp y}(\tau=0, \mathbf{v}) \rangle, \langle \overline{v_{\parallel y}^2} \rangle \equiv \langle C_{\parallel y}(\tau=0, \mathbf{v}) \rangle, \tag{3.3.99}$$

and

$$\tau_{c \frac{\perp}{\parallel}} \equiv \frac{1}{\langle \overline{v_{\frac{\perp}{\parallel} y}^2} \rangle} \left\langle \int_0^\infty d\tau \, C_{\frac{\perp}{\parallel} y}(\tau, \mathbf{v}) \right\rangle. \tag{3.3.100}$$

The contribution from the perpendicular motion is a modified form of the Lyons-Speiser (1985) conductivity

$$\sigma_{LS} = \frac{n_0 q^2}{m} \tau_{LS}, \tag{3.3.101}$$

where the Lyons-Speiser correlation time is

$$\tau_{LS} = \frac{2}{\pi |\omega_{cz}|}. \tag{3.3.102}$$

On the other hand, as discussed in Horton and Tajima (1990), the Lyons-Speiser conductivity obtains its value only in the acceleration region Δ, not in the whole current sheet characterized by the half thickness L. Thus, it is necessary to reduce the Lyons-Speiser conductivity (3.3.101) by a factor $\Delta/L = \epsilon^{1/2}$, that is,

$$\sigma_\perp = \frac{n_0 q^2}{m} \frac{2 \epsilon^{1/2}}{\pi |\omega_{cz}|} \tag{3.3.103}$$

when referring to the conductivity of the current sheet. Recall that the fundamental definition of the conductivity in (3.3.103) is nonlocal in space and time.

A rough estimate of the effect of B_y on the tearing mode growth rate γ can be obtained in the following way. The outer solution is given by White et al. (1977)

$$\delta A_y(z) = \delta A_y(0) e^{-k_x |z|} \left[1 + \frac{1}{k_x L} \tanh \left(\frac{|z|}{L} \right) \right], \tag{3.3.104}$$

which satisfies the boundary condition $\delta A_y(z) \to 0$ for $|z| \to \infty$. In order to couple the inner and outer regions it is necessary to introduce the matching parameter

$$\Delta' \equiv \left. \frac{d \ln(\delta A_y(z))}{dz} \right|_{-\Delta_i}^{\Delta_i} \approx \frac{2}{k_x}(1 - k_x^2 L^2). \tag{3.3.105}$$

265

The contribution from the inner layer is obtained integrating Eq. (3.3.105) from $-\Delta_i$ to Δ_i, assuming that the solution is approximately constant ($\delta A_y(z) \approx \delta A_y(0)$) in the resistive layer and using the matching parameter (3.3.105). Thus we have

$$\Delta'_{k_x}\delta A_y(0) = -\mu_0 \int_{-\Delta_i}^{\Delta_i} \delta j_y^{irr} dz \approx 2\mu_0\gamma\Delta_i \left[\sigma_i + \left(\frac{\rho_e}{\rho_i}\right)^{1/2}\sigma_e\right]\delta A_y(0), \qquad (3.3.106)$$

where we have taken into account that the electron conductivity σ_e acquires its value only in the small region $|z| < \Delta_e$.

Considering the contribution from the inner layer, we have

$$\sigma_i + \left(\frac{\rho_e}{\rho_i}\right)^{1/2}\sigma_e \approx \frac{n_0 q^2}{m|\omega_{cz}|}(a_1 + a_2\kappa_i\kappa_e b_y^2), \qquad (3.3.107)$$

and substituting (3.3.107) into (3.3.106) we get the growth rate

$$\gamma = \frac{m|\omega_{cz}|}{2\mu_0 n_0 q^2}\frac{\Delta'_{k_x}}{\Delta_i\left[a_1 + a_2\kappa_i\kappa_e b_y^2\right]}. \qquad (3.3.108)$$

Furthermore, taking

$$n_0 = \frac{B_{x0}^2}{2\mu_0(T_i + T_e)}, \qquad (3.3.109)$$

and substituting into (3.3.108), we obtain for the growth rate

$$\frac{\gamma}{\omega_{cx0}} = b_z \left(\frac{\rho_i}{L}\right)^{3/2}\left(1 + \frac{T_e}{T_i}\right)\frac{L\Delta'_{k_x}}{a_1 + a_2\kappa_i\kappa_e b_y^2}, \qquad (3.3.110)$$

where the stabilizing effect of the electrons when $B_y \neq 0$ is evident.

It is interesting to note that in the case of a sheared slab geometry with $B_z = 0$, the presence of a finite B_y reduces the tearing growth rate (Drake and Lee, 1977). Assuming unmagnetized ion motion in the singular layer, have found the growth rate

$$\gamma = \frac{k_x v_{the}\Delta'_{k_x}}{2\pi^{1/2}k_0^2 l_s}, \qquad (3.3.111)$$

where $k_0^{-1} = c/\omega_{pe}$ is the collisionless skin depth, and l_s is the magnetic shear length

$$l_s = \frac{B_y}{(dB_x/dz)_{z=0}} = L\frac{B_y}{B_{x0}}. \qquad (3.3.112)$$

From (3.3.110) and (3.3.111) we show that γ is reduced with increasing values of B_y for fixed B_{x0} and L.

In conclusion, we demonstrated that the thermal (velocity) fluctuations in a plasma emersed in a nonuniform plasma give rise to (even in the absence of collisions) dissipation due to the chaotic nature of particle orbits. This was shown through the fluctuation-dissipation theorem (Kubo, 1957; Cable and Tajima, 1992) in a nonuniform plasma. We see that this enhanced dissipation (or reduced conductivity) leads to an enhanced growth rate of reconnection rate (or the tearing instability). The rate is greatest in the absence of the magnetic field component B_y. For example, in the Earth's magnetosphere plasma $B_y = 0$. In this case the Hamiltonian chaos of orbits leads to a maximum dissipation. A recent experiment by Ono et al. (1996) shows that coalescence of two current-carrying magnetic flux tubes and thus reconnection are much more rapid when the two flux tubes have opposite (antiparallel) helicity than the case of two coparallel helicity tubes. This experimental observation may be interpreted as follows: the former case corresponds to $B_y \sim 0$ as the field lines cross nearly 180°, while the latter has a narrower angle. The very rapid (Alfvén time scale) reconnection and a large ion heating were observed in the former case in these experiments. These findings are significant in understanding the magnetic energy conversion and release in a rapid manner, appearing in a variety of astrophysical settings.

Problem 3–9: Discuss the differences between the geo-dynamo and the solar dynamo. What are the values of magnetic Reynolds number, time scales, ratio of kinetic to magnetic energies, and so on, in these dynamos? (See, e.g., Parker, 1979.)

Problem 3–10: Explain the mechanism of reversals of the magnetic field direction in the context of the $\alpha - \omega$ type solar dynamo theory. Is this model of the oscillating dynamo applicable to the galactic dynamo? (see, e.g., Sawa and Fujimoto, 1986) What is the mechanism of reversals of magnetic fields in the geo-dynamo theory? (See, e.g., Glatzmaier and Roberts, 1995.)

Problem 3–11: Prove that the phase speed of the MHD wave propagating along an isolated magnetic tube embedded in a field-free plasma is

$$C_T = \left(\frac{C_s^2 V_A^2}{C_s^2 + V_A^2} \right)^{1/2},$$

where C_s is the sound speed and V_A is the Alfvén speed. (Use the *thin tube approximation*, i.e., a tube radius $\ll$ local pressure scale height. See, e.g., Spruit and Zweibel, 1979.)

Problem 3–12: The Parker instability is driven by the gravitational acceleration and thus can take place in a usual accelerating field. Discuss an application of the Parker instability to supernova remnants, and estimate the size and mass of a cloud (or a clump) produced by this instability. (see, e.g., Baierlein, 1983).

Problem 3–13: In galaxies, the *cosmic ray pressure* is comparable to the magnetic pressure and the gas (or turbulent) pressure, so that it influences the dynamics of interstellar gas through the coupling with magnetic fields. Discuss the effects of the cosmic ray pressure on the linear and nonlinear evolution of the Parker instability in galaxies. (See, e.g., Parker, 1968).

Problem 3–14: In the solar corona, hot plasmas are confined in magnetic loops. Write the energy balance equation along the loop including the radiative cooling, thermal conduction, and (unknown) heating term, and then solve the equation with other equations (hydrostatic equilibrium and perfect gas law). Using these solutions, prove the following relation;

$$T_{max} \propto (pL)^{1/3},$$

where T_{max} is the temperature at the loop top and p is the gas pressure which is nearly uniform when the length L of the coronal loop is shorter than the pressure scale height. (See, e.g., Rosner, Tucker, Vaiana, 1978).

Problem 3–15: In many solar flares, the temperature reaches a few 10 MK or more. In such high temperatures, the role of thermal conduction becomes important, and we cannot neglect it even in the magnetic reconnection process. It has been shown (Forbes and Malherbe 1988, Yokoyama and Shibata, 1997) that under the condition of solar flares, a slow mode MHD shock in the Petscheck type reconnection is dissociated into a conduction front and an isothemal slow shock. Explain why this happens, and estimate the temperature behind the conduction front, assuming the Spitzer conductivity.

Problem 3–16: When the energy released in a flare is transported to the upper chromosphere, cool chromospheric plasmas are suddenly heated up and expand into the corona owing to enhanced gas pressure. This process, called the *chromospheric evaporation*, takes place along magnetic fields, and can be studied by one dimensional hydrodynamics along the loop (e.g., Mariska, 1992). Estimate the gas pressure of the evaporation flow and the flare loop temperature before the radiative cooling becomes effective. (Hint: Use the approximate relations that enthalpy flux $\sim$ thermal conduction flux for deriving the gas pressure of the evaporation flow, and that heating $\sim$ thermal conduction flux for estimating the flare loop temperature (See, e.g., Fisher and Hawley, 1990). When the heating is due to magnetic reconnection, derive the flare temperature as a function of magnetic field strength, electron density, and flare loop length.

Problem 3–17: Discuss possible mechanisms of particle acceleration in solar flares. Is it possible to apply the particle acceleration mechanism known for shock waves to solar flares? (See, e.g., Blandford and Ostriker, 1980; Terasawa and Scholer, 1989.)

Acheson, D.J., Solar Phys. **62**, 23 (1979).

Acton, L., *et al.*, Science **258**, 618 (1992).

Alfvén, H., *Cosmical Electrodynamics* (Clarendon Press, Oxford, 1963).

Aly, J.J., Astrophys. J. **375**, L61 (1991).

Aschwanden, M.J., *et al.*, Astrophys. J. **464**, 985 (1996).

Barnes, C.W., and Sturrock, P.A., Astrophys. J. **174**, 659 (1972).

Beck, R., Loiseau, N., Hummel, E., Berkhuijsen, E.M., and Wielebinski, R., Astron. Astrophys. **222**, 58 (1989).

Beck, R. *et al.*, Ann. Rev. Astron. Astrophys. **34**, 155 (1996).

Beirlein, R., Mon. Not. R. Astr. Soc. **205**, 669 (1983).

Bhattacharjee, A., Brunel, F., and Tajima, T., Phys. Fluids **26**, 3332 (1983).

Biskamp, D., and Welter, H., Phys. Rev. Lett. **44**, 1069 (1980).

Biskamp, D., Phys. Fluids **29**, 1420 (1986).

Biskamp, D., and Welter, H., Solar Phys. **120**, 49 (1989).

Biskamp, D., *Nonlinear Magnetohydrodynamics*, (Cambridge Univ. Press, 1992).

Blandford, R.P., and Ostriker, J.P., Astrophys. J. **237**, 793 (1980).

Born, R., Solar Phys. **38**, 127 (1974).

Brandenburg, A., Tuominen, I., Nordlund, A., Pulkkinen, P., and Stein, R.F., Astron. Astrophys. **232**, 277 (1990).

Brandenburg, A., in *The Cosmic Dynamo, IAU Sympo. No. 157*, eds. F. Krause *et al.*, p. 111 (1993).

Brandenburg, A., Nordlund, A., Stein, R.F., and Torkelsson, U., Astrophys. J. **446**, 741 (1995).

Brants, J.J., Solar Phys. **98**, 197 (1985).

Brants, J.J., and Steenbeck, J.C.M., Solar Phys. **96**, 229 (1985).

Brueckner, G., and Bartoe, J.D.-F., Astrophys. J. **272**, 329 (1983).

Brunel, F., Tajima, T., and Dawson, J.M., Phys. Rev. **49**, 323 (1982).

Bruzek, A., Solar Phys. **2**, 451 (1967).

Bruzek, A., Solar Phys. **8**, 29 (1969).

Büchner, J., and Zelenyï, L.M., J. Geophys. Res. **92**(13), 456 (1987).

Bumba, V., and Howard, R., Astrophys. J. **141**, 1492 (1965).

Cable, S.B., and Tajima, T., Phys. Rev. A **46**, 3413 (1992).

Canfield, R.C., *et al.*, Astrophys. J. **464**, 1016 (1996).

269

Cargil, P.J., and Priest, E.R., Astrophys. J. **266**, 383 (1983).

Carmichael, H., in *Proc. of AAS-NASA Symp. on the Physics of Solar Flares*, ed. Hess, W.N., (NASA-SP 50, 1964) p. 451.

Cattaneo, F., Hurlburt, N.E., and Toomre, J., Astrophys. J. 349, L63 (1990).

Cattaneo, F., Vainstein, S.I., Astrophys. J. **376**, L21 (1991).

Cattaneo, F., Astrophys. J. **434**, 200 (1994).

Cattaneo, F., Hughes, D., and Kim, Phys. Rev. Lett. **76**, 2057 (1996).

Cattaneo, F., and Hughes, D., Phys. Rev. E, in press (1996).

Chandrasekhar, S., *Hydrodynamic and Hydromagnetic Stability* (1961).

Chiba, M., and Tosa, M., Mon. Not. R. Astr. Soc. **41**, 241 (1989).

Choe, G.S., and Lee, L., Astrophys. J. **472**, 330 (1996).

Chou, D., and Wang, H., Solar Phys. **110**, 81 (1987).

Chou, D., and Zirin, H., Astrophys. J. **333**, 420 (1988).

Chou, W., Tajima, T., Matsumoto, R., and Shibata, K., *Proc. IAU Colloq. No. 153*, "Magnetodynamic Phenomena in the Solar Atmosphere," eds. Uchida, Y., *et al.* (Kluwer, Dordrecht, 1996), p. 613.

Cowling, T.G., Mon. Not. R. Astr. Soc. **94**, 39 (1934).

Culhane, L., *et al*, Solar Phys. **136**, 89 (1991).

D'Silva, S., Choudhuri, A.R., Solar Phys. **136**, 201 (1991).

Dawson, J.M., Phys. Rev. **113**, 383 (1959).

Davidson, R.C., *Methods in Nonlinear Plasma Theory*, (Academic, New York, 1972), p. 35.

Davies, R.M., and Taylor, G.I., Proc. Roy. Soc. Lond. A **200**, 375 (1950).

Defouw, R.J., Astrophys. J. **209**, 266 (1976).

Dere, K., Adv. Space Res. **14**, 13 (1994).

Doxas, I., Horton, W., Sandusky, K., Tajima, T., and Steinfolson, R., J. Geophys. Res. **95**, 12033, 1990.

Drake, J.F., and Lee, Y.C., Phys. Fluids **20**, 1341 (1977).

Dulk, G.A., McLean, D.J., and Nelson, G.J., in *Solar Radiophpysics*, eds. McLean, D.J., and Labrum, N.R., (Cambridge Univ. Press, 1985), p. 53.

Dryer, M., Wu, S.T., Steinolfson, R.S., and Wilson, R.M., Astrophys. J. **227**, 1059 (1979).

Eddy, J.A., Science **192**, 1189 (1976).

Fan, Y., Fisher, G.H., Deluca, E., Astrophys. J. **405**, 390 (1993).

Finn, J.M., and Ott, E., Phys. Fluids **31**, 2992 (1988).

Fisher, G.H., and Hawley, S.L., Astrophys. J. **367**, 243 (1990).

270

Fogglizo, T., and Tagger, M., Astron. Astrophys. **287**, 297 (1994).

Forbes, T.G., and Priest, E.R., Solar Phys. **84**, 169 (1983).

Forbes, T.G., and Priest, E.R., Solar Phys. **94**, 315 (1984).

Forbes, T.G., and Acton, L.W., Astrophys. J. **459**, 330 (1996).

Forbes, T.G., and Priest, E.R., Rev. Geophys. **25**, 1583 (1987).

Forbes, T.G., J. Geophys. Res. **95**, 11919 (1990).

Forbes, T.G., and Malherbe, J.M., Solar Phys. **135**, 361 (1991).

Fujimoto, M., and Sawa, T., Geophys. Astrophys. Fluid Dyn. **50**, 159 (1990).

Furth, H.P., Killeen, J., and Rosenbluth, M.N., Phys. Fluids **6**, 459 (1963).

Galeev, A.A., and ZelenyI, L.M., Sov. Phys. JETP **43**, 1113 (1976).

Galloway, D.J., and Moore, D.R., Geophys. Astrophys. Fluid Dyn. **12**, 73 (1979).

Galloway, D.J., and Weiss, N.O., Astrophys. J. **243**, 945 (1981).

Galloway, D.J., and Frisch, U., Geophys. Astrophys. Fluid Dyn. **36**, 53 (1986).

Gilman, P.A., and Glatzmaier, G.A., Astrophys. J. Suppl. **45**, 335 (1981).

Giz, A.T., and Shu, F.H., Astrophys. J. **404**, 185 (1993).

Glackin, D.L., Solar Phys. **43**, 317 (1975).

Glaztmaier, G.A., and Roberts, P.H., Nature **377**, 203 (1995).

Glaztmaier, G.A., and Roberts, P.H., Science **274**, 1887 (1996).

Gold, T., and Hoyle, F., Mon. Not. R. Astr. Soc. **120**, 89 (1960).

Golub, L., Krieger A.S., and Vaiana, G.S., Solar Phys. **49**, 79 (1976).

Gough, D.O., Kosovichev, A.G., Sekii, T. ,Libbrecht, K.G., and Woodard, M.F., in *Proc. Gong 1992: Seismic Investigation of the Sun and Stars*, ed. Brown, T.M., (ASP Conference Series, 1993) Vol. 42, pp. 213–216.

Hanaoka, Y., and Kurokawa, H., Solar Phys. **124**, 227 (1989).

Hanaoka, Y., *et al*, Publ. Astr. Soc. Jpn. **46**, 205 (1994).

Hanami, H., and Tajima, T., Astrophys. J. **377**, 694 (1991).

Hanawa, T., Matsumoto, R., and Shibata, K., Astrophys. J. Lett. **393**, L71 (1992a).

Hanawa, T., Nakamura, F., and Nakano, T., Publ. Astr. Soc. Jpn. **44**, 509 (1992b).

Harrison, R.A., Astron. Astrophys. **162**, 283 (1986).

Harvey, K.L., Thesis, Utrecht Univ. (1993).

Hasan, S.S., Astron. Astrophys. **143**, 39 (1985).

Hawley, J.F., Gammie, C.F., and Balbus, S.A., Astrophys. J. **440**, 742 (1995).

Hawley, J.F., Gammie, C.F., and Balbus, S.A., Astrophys. J. **464**, 690 (1996).

Heyvaerts, J., Priest, E.R., Rust, D.M., Astrophys. J. **216**, 123 (1977).

Hiei, E., *et al.*, Geophys. Res. Lett. **20**, 2785 (1993).

Hirayama, T., Solar Phys. **34**, 323 (1974).

Hirayama, T., in *Lecture Notes in Physics, No. 387*, "Flare Physics in Solar Activity Maximum 22," eds. Uchida, Y., *et al.* (New York, Springer, 1991) p. 197.

Horiuchi, T., Matsumoto, R., Hanawa, T., and Shibata, K., Publ. Astr. Soc. Jpn. **40**, 147 (1988).

Horton, W., and Tajima, T., J. Geophys. Res. **93**, 2741 (1988).

Horton, W., and Tajima, T., Geophys. Res. Lett. **17**, 123 (1990).

Horton, W., and Tajima, T., J. Geophys. Res. **96**, 15811 (1991a).

Horton, W., and Tajima, T., Geophys. Res. Lett. **18**, 1583 (1991b).

Horton, W., Hernandez, J., Kim, J-Y., and Tajima, T., in *Physics of Space Plasmas, 1991, SPI Conf. Proc. Rep. Ser., Vol. 27*, "Orbital stability, transport, and convective heating in the current sheet plasma," eds. Chang, T., Crew, G.B., and Jasperse, J.R., (Scientific Publishers, Cambridge, Mass., 1992).

Hudson, H.S., Solar Phys. **133**, 357 (1991).

Hudson, H.S., in *Proc. of Kofu Symp.* eds. Enome, S., and Hirayama, T., (Nobeyama Radio Observatory Report No. 360), p. 11 (1994).

Hughes, D.W., Geophys. Astrophys. Fluid Dyn. **32**, 273 (1988).

Hughes, D.W., and Procter, M.R.E., Ann. Rev. Fluid Mech. **20**, 187 (1988).

Hulburt, N.E., and Toomre, J., Astrophys. J. **327**, 920 (1988).

Kageyama, A., *et al.*, Phys. Plasmas **2**, 1421 (1995).

Kahler, S., Spicer, D., Uchida, Y., and Zirin, H., in *A Monograph from Skylab Workshop II*, "Solar Flares," ed. P.A. Sturrock, (Colorado Univ. Press, 1980). p. 83.

Kaisig, M., Tajima, T., Shibata, K., Nozawa, S., and Matsumoto, R., Astrophys. J. **358**, 698 (1990).

Kamaya, T., Mineshige, S., Shibata, K., and Matsumoto, R., Astrophys. J. Lett. **458**, L25 (1996).

Karpen, J.T., Antiochos, S.K., and Devore, C.R., Astrophys. J., in press (1997).

Kawaguchi, I., and Kitai, R., Solar Phys. **46**, 125 (1976).

Kawaguchi, I., Solar Phys. **65**, 207 (1980).

Knölker, M., Schüssler, M., and Weisshaar, E., Astron. Astrophys. **202**, 275 (1988).

Kopp, R.A., and Pneuman, G.W., Solar Phys. **50**, 85 (1976).

Kosugi, T., *et al.*, Solar Phys. **136**, 17 (1991).

Krause, F., Radler, K.H., Rudiger, G., *The Cosmic Dynamo*, Proc. IAU Symp. No. 157 (Kluwer Academic Pub., 1993).

Kronberg, P.P., Rep. Prog. Phys. **57**, 325 (1994).

272

Kubo, R., J. Phys. Soc. J. **12**, 570 (1957).

Kurokawa, H., Hanaoka, Y., Shibata, K., and Uchida, Y., Solar Phys. **108**, 251 (1988).

Kurokawa, H., Space Sci Rev. **51**, 49 (1989).

Kurokawa, H., Nakai, Y., Funakoshi, Y., and Kitai, R., Adv. Space Res. **11**, 233 (1991).

Kurokawa, H., and Kawai, G., *Proc. IAU Colloq. No. 141: The Magnetic and Velocity Fields of Solar Active Regions*, (ASP Conference Series 46, 1993) eds. Zirin, H., Ai, G., and Wang, H., pp. 507–510.

Kruskal, M.D., and Schwarzschild, Mon. Not. R. Astr. Soc. (London) **223A**, 348 (1954).

Kundu, R., and Woodgate, B., (ed.) *Proc. SMM Workshop, Energetic Phenomena on the Sun, NASA Publ. No. 2439.*

Kusano, K., Suzuki, Y., Kubo, H., Miyoshi, T., and Nishikawa, K., Astrophys. J. **433**, 361 (1994).

Kusano, K., Suzuki, Y., and Nishikawa, K., Astrophys. J. **441**, 942 (1995).

Landau, L.D., and Lifshitz, E.M., *Electrodynamics of Continuous Media* translated by Sykes, J.B., and Bell, J.S., (Pergamon, New York, 1960) Chapter 7.

Leka, K.D., Ph.D. Thesis, Hawaii Univ. (1994).

Leka, K.D., *et al.*, in *The X-ray Solar Physics from Yohkoh*, eds. Uchida, Y., *et al.*, (Universal Academy Press, Tokyo, 1994).

Libbrecht, K.G., in *Seismology of the Sun and Sun-like Stars* ed. E.J. Rolfe, ESA-SP286, p. 131 (1988).

Lin, R.P., *et al.*, Astrophys. J. **283**, 421 (1984).

Linker, J.R., and Mikic, Z., Astrophys. J. **438**, L45 (1995).

Low, B.C., Astrophys. J. **251**, 352 (1981).

Low, B.C., Astrophys. J. **254**, 796 (1982).

Lyons, L.R., and Speiser, T.W., J. Geophys. Res. **90**, 8543 (1985).

Magara, T., Mineshige, S., Yokoyama, T., and Shibata, K., Astrophys. J. **466**, 1054 (1996).

Mariska, J.T., *Solar Transition Regions*, (Springer-Verlag, 1992).

Masuda, S., Ph.D. Thesis, Univ. of Tokyo (1994).

Masuda, S., *et al.*, Nature **371**, 495 (1994).

Masuda, S., *et al.*, Publ. Astr. Soc. Jpn. **47**, 677 (1995).

Matsumoto, R., Horiuchi, T. Shibata, K., and Hanawa, T., Publ. Astr. Soc. Jpn. **40**, 171 (1988).

Matsumoto, R., Horiuchi, T., Hanawa, T., and Shibata, K., Astrophys. J. **356**, 259 (1990).

Matsumoto, R., and Shibata, K., in *Proc. IAU Symp. No. 144*, "The Interstellar Disk-Halo Connection in Galaxies," ed. J.B.G.M. Bloemen, (Kluwer Academic Pub., 1991) pp. 429–432.

Matsumoto, R., and Shibata, K., Publ. Astr. Soc. Jpn. **44**, 167 (1992).

Matsumoto, R., Tajima, T., Shibata, K., and Kaisig, M., Astrophys. J. **414**, 357 (1993).

273

Matsumoto, R., and Tajima, T., Astrophys. J. **445**, 767 (1995).

Matsumoto, R., Tajima, T., Chou, W., and Shibata, K., in *IAU Coll. 153*, eds. Y. Uchida *et al.*, (Kluwer, Dordrecht, 1996) p. 355.

Malagoli, A., Cattaneo, F., and Brummell, N.H., Astrophys. J. **316**, L33 (1990).

McAllister, A., *et al.*, Publ. Astr. Soc. Jpn. **44**, L205 (1992).

McAllister, A., Dryer, M., McIntosh, P., and Singer, H., J. Geophys. Res. **101**, 13497 (1996).

McAllister, A., Kurokawa, H., Shibata, K., and Nitta, N., Solar Phys. **169**, 123 (1996).

Mikic, Z., Barnes, D.C., and Schnack, D.D., Astrophys. J. **328**, 830 (1988).

Mikic, Z., and Linker, J.A. Astrophys. J. **430**, 898 (1994).

Mineshige, S., Kusunose, M., and Matsumoto, R., Astrophys. J. Lett. **445**, L43 (1995).

Miyamoto, M., Kitamoto, S., Hayashida, K., and Egoshi, W., Astrophys. J. Lett. **442**, L13 (1995).

Moffat, H.K., *Magnetic Field Generation in Electrically Conducting Fluids*, (Cambridge Univ. Press, 1978).

Moore, R.L., and Roumeliotis, G., in *Eruptive Solar Flares*, eds. Svestka, Z., Jackson, B.V., and Machado, M.E, (Lecture Notes in Physics, No. 399, Springer Verlag, 1992) p. 69.

Moreno-Insertis, F., Astron. Astrophys. **166**, 291 (1986).

Mouschovias, T.Ch., Astrophys. J. **192**, 37 (1974).

Muller, R., in *Solar and Stellar Granulation*, eds. R.J. Rutten and G. Severino, (Kluwer Academic Pub., 1989), p. 101.

Nakai, N., Kuno, N., Hanada, T., and Sofue, Y., Publ. Astr. Soc. Jpn. **46**, 527 (1994).

Nakajima, N., J. Phys. Soc. Jpn. **56**, 3911 (1987).

Nakajima, H., Kosugi, T., Kai, K., and Enome, S., Nature **305**, 292 (1983).

Nesis, A., Bogdan, T.J., Cattaneo, F., Hanslmeier, A. Knölker, M., and Malagoli, A. Astrophys. J. **407**, 316 (1992).

Newcomb, W.A., Phys. Fluids **4**, 391 (1969).

Nitta, N., in *Observation of Magnetic Reconnection in the Solar Atmosphere*, eds. Bentley, R., and Mariska, J., (ASP, 1997), in press.

Nozawa, S., Shibata, K., Matsumoto, R., Tajima, T., Sterling, A.C., Uchida, Y., Ferrari, A., and Rosner, R., Astrophys. J. Suppl. **78**, 267 (1992).

Ogawara, Y., *et al.*, Solar Phys. **136**, 1 (1991).

Ogawara, Y., *et al.*, Publ. Astr. Soc. Jpn. Lett. **44**, L41 (1992).

Ohyama, M., and Shibata, K., in *Proc. IAU Colloq., No. 153*, eds. Y. Uchida *et al.*, (Kluwer, Dordrecht, 1996), p. 525.

Ohyama, M., and Shibata, K., Publ. Astr. Soc. Jpn. **49**, in press (1997).

Okubo, A. *et al.*, in *Observations of Magnetic Reconnection in the Solar Atmosphere*, eds. Bentley, R., and Mariska, J., (ASP, 1997), in press.

Ono, Y., Yamada, M., Akao, T., Tajima, T., and Matsumoto, R., Phys. Rev. Lett. **76**, 3328 (1996).

Ossakow, S.L., and Chatuvedi, P.K., J. Geophys. Res. **83**, 2085 (1978).

Ott, E., Phys. Rev. Lett. **9**, 1429 (1972).

Parker, E.N., Astrophys. J. **121**, 491 (1955).

Parker, E.N., J. Geophys. Res. **62**, 509 (1957).

Parker, E.N., Astrophys. J. **145**, 811 (1966).

Parker, E.N., in *Nebulae and Insterstellar Matter*, vol. VII, "Stars and stellar systems" (eds. Middlehurst, B.M., and Aller, L.H.), (University of Chicago Press, 1968) chap. XIV.

Parker, E.N., Astrophys. J. **163**, 255 (1971).

Parker, E.N., Astrophys. J. **198**, 205 (1975).

Parker, E.N., Astrophys. J. **221**, 368 (1978).

Parker, E.N., *Cosmical Magnetic Field*, (Clarendon Press, Oxford, 1979) p. 314.

Parker, E.N., Astrophys. J. **330**, 474 (1988).

Petschek, H.E., in *Proc. AAS-NASA Symposium on Physics of Solar Flares*, ed. Hess, W.N. (NASA SP-50, Washington, D.C., 1964) p. 425.

Pevtsov, A.A., Canfield, R.C., and Metcalf, T.R., Astrophys. J. **425**, L117 (1994).

Priest, E.R., in *Solar Flare Magnetohydrodynamics*, ed. Priest, E.R., (Gordon and Breach Science Pub., 1981) p. 1.

Priest, E.R. *Solar Magnetohydrodynamics*, (Reidel, Dordrecht, 1982) p. 344.

Priest, E.R., and Forbes, T.G., J. Geophys. Res. **91**, 5579 (1987).

Proctor, M.R.E., and Weiss, N.O., Rep. Prog. Phy. **45**, 1317 (1982).

Proctor, M.R.E., and Gilbert, A.D. (eds.), *Lectures on Solar and Planetary Dynamo*, (Cambridge Univ. Press, 1994).

Roberts, P.H., Mathematika **19**, 169 (1972).

Rosner, R., Tucker, W.H., Vaiana, G.S., Astrophys. J. **222**, 317 (1978).

Rosner, R. and Vaiana, G.S., in *X-ray Astronomy*, eds. Giacconi, R., & Setti, G., (Dordrecht: Reidel, 1980), p. 129.

Rosner, R., and Weiss, N.O., in *The Solar Cycle*, (ASP Conference series, 1992) Vol. 27, p. 511.

Rust, D.M., Solar Phys. **25**, 141 (1972).

Rust, D.M., Geophys. Res. Lett. **21**, 241 (1994).

Ruzmaikin, A.A., Sokoloff, D.D., and Shukurov, A.M., *The Magnetic Fields of Galaxies* (Kluwer, Dordrecht, 1988).

275

Sakai, J.-I., and Washimi, H., Astrophys. J. **258**, 823 (1982).

Sakai, J.-I., and Tajima, T., in *Proc. of the Joint Varenna-Abatsumani Int. School and Workshop on Plasma Astrophysics*, (ESA SP–251, 1986), p. 77.

Sakai, J.-I., and Ohsawa, Y., Space Sci. Rev. **46**, 113 (1987).

Sakai, J.-I., and Koide, S., Solar Phys. **142**, 399 (1992).

Sakurai, T., Publ. Astr. Soc. Jpn. **28**, 177 (1976).

Sakurai, T., and Uchida, Y., Solar Phys. **52**, 397 (1977).

Sakurai, T., Solar Phys. **121**, 347 (1989).

Sato, T., and Hayashi, T., Phys. Fluids **22**, 1189 (1979).

Sawa, T., and Fujimoto, M., Publ. Astr. Soc. Jpn. **38**, 133 (1986).

Schmitt, D., in *Advances in Solar Physics, Lecture Notes in Physics No. 432*, (Springer-Verlag, 1993) p. 61.

Schmitt, J.H.M.M., and Rosner, R., Astrophys. J. 265, 901 (1983).

Schmitt, J.H.M.M., Rosner, R., and Bohn, H.U., Astrophys. J. *282*, 316 (1984).

Scholer, M., J. Geophys. Res. **94**, 8805 (1989).

Schüssler, M., Nature **288**, 150 (1980).

Schüssler, M., *in Solar and Stellar Magnetic Fields, IAU Symp. No. 102*, (1983) p. 213.

Sedov, L.I., *Similarity and Dimensional Methods in Mechanics*, (Academic, New York, 1959).

Sergeev, V.A., and Tsyganenko, N.A., Planet. Space Sci. 30, **999** (1982).

Sharp, D.H., Physica **12D**, 3 (1984).

Shibata, K., Solar Phys. **66**, 61 (1980).

Shibata, K., and Uchida, Y., Publ. Astr. Soc. Jpn. **37**, 31 (1985).

Shibata, K., and Uchida, Y., Solar Phys. **103**, 299 (1986a).

Shibata, K., and Uchida, Y., Publ. Astr. Soc. Jpn. **38**, 631 (1986b).

Shibata, K., Tajima, T., Matsumoto, R., Horiuchi, T., Hanawa, T., Rosner, R., and Uchida, Y., Astrophys. J. **338**, 471 (1989a).

Shibata, K., Tajima, T., Steinolfson, R.S., and Matsumoto, R., Astrophys. J. **345**, 584 (1989b).

Shibata, K., Tajima, T., and Matsumoto, R., Phys. Fluids B **2**, 1989 (1990a).

Shibata, K., Tajima, T., and Matsumoto, R., Astrophys. J. **350**, 295 (1990b).

Shibata, K., Tajima, T., Steinolfson, R.S., and Matsumoto, R., Astrophys. J. Lett. **351**, L25 (1990c).

Shibata, K., Nozawa, S., Matsumoto, R., Tajima, T., and Sterling, A.C., in *Proc. of Heidelberg Conference on Mechanisms of Chromospheric and Coronal Heating*, ed. P. Ulmschneider, (Springer Verlag, 1991), p. 609.

276

Shibata, K., and Matsumoto, R., Nature **353**, 633 (1991).

Shibata, K., in *Proc. of "Flare Physics in Solar Activity Maximum 22*, eds. Uchida, Y., Canfield, R., Watanabe, T., and Hiei, E., (Series of Lecture Notes in Physics, No. 387), (Springer Verlag, 1991) p. 205.

Shibata, K., Nozawa, S., and Matsumoto, R., Publ. Astr. Soc. Jpn. **44**, 265–272 (1992a).

Shibata, K. *et al.*, Publ. Astr. Soc. Jpn. **44**, L173 (1992b).

Shibata, K. *et al.*, *X-ray Solar Physics from Yohkoh*, eds. Uchida, Y., *et al.*, (Universal Academy Pub., 1993), p. 29.

Shibata, K., *et al.*, Astrophys. J. Lett. **431**, L51 (1994a).

Shibata, K., Yokoyama, T., and Shimojo, M., in *Proc. of Kofu Symp.*, eds. Enome, S., and Hirayama, T. Nobeyama Radio Observatory Report No. 360, (1994b) p. 75.

Shibata, K., in *Proc. The Sun as a Variable Star*, eds. J.M. Pap *et al.*, (Cambridge Univ. Press, 1994) p. 89.

Shibata, K., *et al.*, Astrophys. J. Lett. **451**, L83 (1995).

Shibata, K., Adv. Sp. Res. **17**, (4/5)9 (1996).

Shibata, K., Yokoyama, T., and Shimojo, M., J. Geomag. Geoelectr. **48**, 19 (1996).

Shimizu, T., *et al.*, Publ. Astr. Soc. Jpn. **44**, L147 (1992).

Shimizu, T., *et al.*, Astrophys. J. **422**, 906 (1994).

Shimizu, T., Publ. Astr. Soc. Jpn. **47**, 251 (1995).

Shimojo, M., *et al.*, Publ. Astr. Soc. Jpn. **48**, 123 (1996).

Shu, F.H., Astron. Astrophys. **33**, 55 (1974).

Sofue, Y., Publ. Astr. Soc. Jpn. **25**, 207 (1973).

Sofue, Y., Fujimoto, M., and Tosa, M., Publ. Astr. Soc. Jpn. **28**, 317 (1976).

Solanki, S., Space Sci. Rev. **63**, 1 (1994).

Sonnerup, B.V.Ö., J. Phys. Plasmas **4**, 161 (1970).

Spicer, D.S., Solar Phys. **53**, 305 (1977).

Spiegel, E.A., and Weiss, N.O., Nature **287**, 616 (1980).

Spitzer, L., *Physics of Fully Ionized Gases*, (Interscience, New York, 1962).

Spruit, H.C., Solar Phys. **61**, 363 (1979).

Spruit, H.C., and Zweibel, E.G., Solar Phys. **62**, 15 (1979).

Spruit, H.C., and van Ballooijen, A.A., Astron. Astrophys. **106**, 58 (1982).

Spruit, H.C., and Roberts, B., Nature **304**, 401 (1983).

Spruit, H.C., and Nordlund, A., and Title, A.M., Ann. Rev. Astron. Astrophys. **28**, 263 (1990).

277

Steenbeck, M., Krause, F., and Radler, K.H.Z., Naturforsch A **21**, 369 (1966).

Steinolfson, R.S., and van Hoven, G., Phys. Fluids **27**, 1207 (1984).

Steinolfson, R.S., and Tajima, T., Astrophys. J. **322**, 503 (1987).

Stenflo, J.O., Solar Phys. **32**, 41 (1973).

Strong, K.T., *et al.*, Publ. Astr. Soc. Jpn. **44**, L63 (1992).

Stix, M., in *Basic Mechanics of Solar Activity, IAU Symp.* **71**, eds. Bumba, V., and Kleczek, J., (Reidel, Dordrecht, 1976), p. 367.

Sturrock, P.A., in *Strucure and Development of Solar Active Regions, IAU Symp. No. 35*, ed. Kiepenheuer, K.O., (Reidel Pub., Dordrecht, 1968) p. 471.

Sturrock, P.A., Solar Phys. **121**, 387 (1989).

Sturrock, P.A., in *Lecture Note in Physics, No. 399*, "Eruptive Solar Flares" eds. Svestka, Z., Jackson, B.V., and Machado, M.E. (New York, Springer, 1992), p. 397,

Svestka, Z., *Solar Flares*, (Reidel, Dordrecht, 1976).

Svestka, Z., and Cliver, E.W., in *Lecture Note in Physics, No. 399*, "Eruptive Solar Flares" ed. Svestka, Z., Jackson, B.V., and Machado, M.E., (New York, Springer, 1992) p. 1..

Sweet, P.A., in *Electromagnetic Phenomena in Cosmic Physics, IAU Symp. No. 6*, ed. Lehnert, B., (Cambridge Univ. Press, 1958), p. 123.

Tajima, T., Brunel, F., and Sakai, J-I., Astrophys. J. **245**, L45 (1982).

Tajima, T., and Sakai, J-I., IEEE Trans. Plasma Sci. **PS14**, 926 (1986).

Tajima, T., and Sakai, J-I., Nakajima, H., Kosugi, T., Brunel, F., and Kundu, R., Astrophys. J. **321**, 1031 (1987).

Tajima, T., *Computational Plasma Physics With Applications to Fusion and Astrophysics*, (Addison Wesley, 1989).

Tajima, T. and Sakai, J-I., Sov. J. Plasma Phys. **15**, 519 (1989a).

Tajima, T., and Sakai, J-I., Sov. J. Plasma Phys. **15**, 606 (1989b).

Tajima, T., Benz, A.O., Thaker, M., and Leboeuf, J.-N., Astrophys. J. **353**, 666 (1990).

Takeuchi, A., Publ. Astr. Soc. Jpn. **45**, 811 (1993).

Tanaka, K., Publ. Astr. Soc. Jpn. **39**, 1 (1987).

Tanaka, M., and Sato, T., J. Geophys. Res. **86**, 5541 (1981).

Tarbel, T. *et al.*, in "Solar Photosphere: Structure, Convection, and Magnetic Fields," *Proc. IAU Symp.* ed. Stenflo, J.O., (Kluwer, 1990), p. 147.

Terasawa, T., and Scholer, M., Science, **244**, 1050 (1989).

Tosa, M., and Fujimoto, M., Publ. Astr. Soc. Jpn. **30**, 315 (1978).

Tsuneta, S., *et al.*, Solar Phys. **136**, 37–67 (1991).

Tsuneta, S., *et al.*, Publ. Astr. Soc. Jpn. **44**, L63 (1992a).

Tsuneta, S., *et al.*, Publ. Astr. Soc. Jpn. **44**, L211 (1992b).

Tsuneta, S., in *Proc. IAU Colloq. No. 141*, "Magnetic and Velocity Fields of Solar Active Regions" eds. Zirin, H. *et al.*, (Astr. Soc. Pacific, 1993) p. 239.

Tsuneta, S., Astrophys. J. **456**, 840 (1996).

Tsyganenko, N.A., Space Sci. **35**, 1347 (1987).

Uchida, Y., and Sakurai, T., Solar Phys. **51**, 413 (1977).

Uchida, Y., and Shibata, K., Publ. Astr. Soc. Jpn. **37**, 515 (1985).

Uchida, Y., and Shibata, K., Solar Phys. **116**, 291 (1988).

Uchida, Y., Mizuno, A., Nozawa, S., and Fukui, Y., Nature **42**, 69 (1990).

Ugai, M., and Tsuda, T., J. Phys. Plasmas **17**, 337 (1977).

Ugai, M., Geophys. Res. Lett. **14**, 103 (1987).

Ugai, M., Phys. Fluids B **4**, 2953 (1992).

Unno, W., and Audo, H., Geophys. Ap. Dyn. **12**, 107 (1979).

Vainshtein, S.I., and Zeldovich, Ya.B., Sov. Phys. Usp. **15**, 159 (1972).

Vainshtein, S.I., Parker, E.N., and Rosner, R., Astrophys. J. **408**, 678 (1993).

Vainshtein, S.I., and Rosner, R., Astrophys. J. **376**, 199 (1991).

Vainshtein, S.I., Sagdeev, R., and Rosner, R., Phys. Rev. E, **53**(5), 4729 (1996).

Watanabe, T., *et al.*, Publ. Astr. Soc. Jpn. **44**, L199 (1992).

Watanabe, T., in *Proc. of Kofu Symp.*, eds. Enome, S., and Hirayama, T., (Nobeyama Radio Observatory Report No. 360), p. 99 (1994).

Webb, A.R., and Roberts, B., Solar Phys. **59**, 249 (1978).

White, R.B., Monticello, D.A., Rosenbluth, M.N., and Waddell, B.V., Phys. Fluids **20**, 800 (1977).

White, R.B., in *Handbook of Plasma Physics*, ed. Rosenbluth, M.N., and Sagdeev, R.Z. (North-Holland, Amsterdam, 1983), vol. 1; White, R.B., Rev. Mod. Phys. **58**, 183 (1986).

Wu, S.T., Dryer, M., Nakagawa, Y., and Han, S.M., Astrophys. J. **219**, 324 (1978).

Yang, M., Otto, A., Muzzell, D., and Lee, L.C., J. Geophys. Res. **99**, 8657 (1994).

Yokoi, N., Astron. Astrophys. **311**, 731 (1996).

Yokoyama, T., and Shibata, K., Astrophys. J. Lett. **436**, L197 (1994).

Yokoyama, T., and Shibata, K., Nature **375**, 42 (1995).

Yokoyama, T., and Shibata, K., Publ. Astr. Soc. Jpn. **48**, 353 (1996).

Yokoyama, T., and Shibata, K., Astrophys. J. Lett. **474**, L61 (1997).

Yokoyama, T., Ph.D. Thesis, The Graduate University for the Advanced Studies, National Astronomical Observatory (1995).

Yoshimura, H., Astrophys. J. Supp. **29**, 467 (1975).

Yoshimura, H., Astrophys. J. **226**, 706 (1978).

Yoshimori, M., *et al.*, Solar Phys. **136**, 69 (1991).

Yoshizawa, A., and Yokoi, N., Astrophys. J. **407**, 540 (1993).

Zaidman, E.G., and Tajima, T., Astrophys. J. **338**, 1139–1147 (1989).

Zeldovich, Ya.B., and Raizer, Yu.P., *Physics of Shock Waves and High Temperature Hydrodynamic Phenomena*, (Academic, New York, 1967), Vol. 2.

Zirin, H., *Astrophysics of the Sun*, (Cambridge Univ. Press, Cambridge, 1988).

Zwaan, C., Solar Phys. **100**, 397 (1985).

Zwaan, C., Ann. Rev. Astron. Astrophys. **25**, 83 (1987).

Chapter 4

Fundamental Processes in Gravitational Objects

4.1 Gravitational Contraction

In this section we will discuss the fundamentals of the gravitational contraction of a self-gravitating cloud. This is because (1) gravitational physics has many physical laws similar to those in plasma physics, due to the similarity between the gravitational force and the Coulomb force, (2) the most energetic phenomena observed in the universe (i.e. jets ejected from nuclei of active galaxies and quasars) probably originate from the gravitational energy release in the gravitational contraction of the self-gravitating cloud or in the accretion of mass by massive objects. In later sections, we will study the mechanism to extract energy, mass, momentum, and magnetic flux from the gravitationally contracting cloud.

In order to demonstrate the enormity of this problem, we give some numerical examples in the case of a star formation. Stars are formed by the gravitational contraction of an interstellar cloud with density $\sim 10^{-21} \text{g/cm}^3$ and size $\sim 10^{18} \text{cm}$ (Fig. 4.1). The average density of stars is of the order of $\sim 1 \text{g/cm}^3$, and the radius of stars is $\sim 10^{11} \text{cm}$. Thus, the density changes by the order of 10^{21}, or equivalently, the size changes by the order of 10^7 when stars are born. These numbers are enormous, and this is why the star formation is one of the outstanding problems in plasma astrophysics.

4.1.1 Jeans Instability

The gravitational contraction starts as a gravitational instability, which is sometimes called *Jeans instability* (Jeans, 1928). Since the physics of this instability is similar to the plasma oscillation, we will study it in detail here.

Let us consider a uniform, isothermal medium with no rotation and no magnetic field, and consider only one-dimensional motion in the x-direction. Assuming the isothermal equation of state, $p = \rho C_s^2$, the basic equations are

$$\frac{\partial \rho}{\partial t} + \frac{\partial}{\partial x}(\rho v) = 0, \qquad (4.1.1)$$

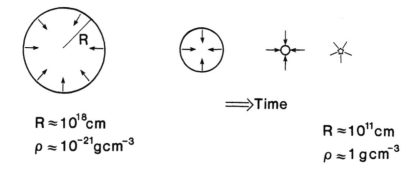

⟹Time

$$R \approx 10^{18} \text{cm}$$
$$\rho \approx 10^{-21} \text{gcm}^{-3}$$

$$R \approx 10^{11} \text{cm}$$
$$\rho \approx 1 \text{gcm}^{-3}$$

FIGURE 4.1 How are stars born?

$$\frac{\partial v}{\partial t} + v \frac{\partial v}{\partial x} + \frac{C_s^2}{\rho} \frac{\partial \rho}{\partial x} + \frac{\partial \Phi}{\partial x} = 0, \tag{4.1.2}$$

$$\frac{\partial^2 \Phi}{\partial x^2} = 4\pi \rho G, \tag{4.1.3}$$

where ρ is the mass density, v is the fluid velocity in x-direction, C_s is the isothermal sound speed, and Φ is the gravitational potential. Assuming that $\rho = \rho_0 + \rho_1, \rho_1 \ll \rho_0$, etc., we linearize Eqs. (4.1.1)–(4.1.3), where ρ_0 is the mass density in the unperturbed cloud and is taken to be constant here. Then, the linearized equations are

$$\frac{\partial \rho_1}{\partial t} + \rho_0 \frac{\partial v_1}{\partial x} = 0, \tag{4.1.4}$$

$$\frac{\partial v_1}{\partial t} + \frac{C_s^2}{\rho_0} \frac{\partial \rho_1}{\partial x} + \frac{\partial \Phi_1}{\partial x} = 0, \tag{4.1.5}$$

$$\frac{\partial^2 \Phi_1}{\partial x^2} = 4\pi \rho_1 G, \tag{4.1.6}$$

where the subscript $_1$ and $_0$ denote 1st and 0th order quantities. After algebraic manipulations, Eqs. (4.1.4)–(4.1.6) become

$$-\left(\frac{\partial^2 \rho_1}{\partial t^2} - C_s^2 \frac{\partial^2 \rho_1}{\partial x^2}\right) + 4\pi G \rho_0 \rho_1 = 0. \tag{4.1.7}$$

If we assume that $\rho_1 = \tilde{\rho}_1 \exp(i\omega t + ikx)$, Eq. (4.1.7) becomes

$$\omega^2 - k^2 C_s^2 + 4\pi G \rho_0 = 0. \tag{4.1.8}$$

This means that if

$$k^2 C_s^2 < 4\pi G \rho_0, \tag{4.1.9}$$

ω becomes pure imaginary ($i\omega$ is real) so that the positive $i\omega$ leads to the exponential growth of perturbation with time. That is, we have an instability (Jeans instability). On the other hand, if $k^2 C_s^2 > 4\pi G \rho_0$, ω is real, and the perturbation undergoes oscillation; the gas layer is stable. The physical meaning of Eq. (4.1.9) is that the instability occurs when the gas pressure force (the left-hand side) is smaller than the gravitational force (the right-hand side). Eq. (4.1.9) is rewritten as

$$\lambda > \lambda_J = C_s \left(\frac{\pi}{G\rho_0}\right)^{1/2}, \tag{4.1.10}$$

where $\lambda = 2\pi/k$ is the wavelength of the perturbation, and λ_J is called the *Jeans length*. The mass contained in a cube of one Jeans length ($M_J = \lambda_J^3 \rho_0$) is called the *Jeans mass*.

Because the gravitational force is similar to the coulomb force, the physics and mathematics of plasma oscillation are almost parallel to those of Jeans instability. The only difference is the sign of the force; the gravitational force is attraction, while the coulomb force between charges with the same sign is repulsion. Thus, we do not have the plasma "instability," but instead have plasma *oscillation*. The basic equation for one-dimensional plasma oscillation is

$$\frac{\partial n_e}{\partial t} + \frac{\partial}{\partial x}(n_e v) = 0, \tag{4.1.11}$$

$$\frac{\partial v}{\partial t} + v\frac{\partial v}{\partial x} + \frac{C_s^2}{n_e}\frac{\partial n_e}{\partial x} + \frac{e}{m_e}\frac{\partial \Psi}{\partial x} = 0, \tag{4.1.12}$$

$$\frac{\partial^2 \Psi}{\partial x^2} = -4\pi(n_e - n_i)e, \tag{4.1.13}$$

where n_e and n_i are electron and ion number densities, e is the electron charge, m_e is the electron mass, and Ψ is the electric potential. Thus, we get Eqs. (4.1.1)–(4.1.3) by taking

$$n_e \rightarrow \rho,$$

285

$$n_i \to 0,$$
$$-e^2/m_e \to G,$$
$$e\Psi/m_e \to \Phi$$

in Eqs. (4.1.11)–(4.1.12). Thus, the dispersion relation for the plasma oscillation is directly obtained from Eq. (4.1.8) as

$$\omega^2 - k^2 C_s^2 - 4\pi n_e e^2/m_e = 0. \tag{4.1.14}$$

In the cold limit ($C_s = 0$), we find

$$\omega = \left(\frac{4\pi n_e e^2}{m_e}\right)^{1/2},$$

which is called *plasma frequency* as described in Sec. 2.1.

The dispersion relation for the Jeans instability [Eq. (4.1.8)] shows that the growth rate increases with an increase in the wavelength. This means that only one object is created in the universe if the universe is initially uniform. But, we have many objects in our actual universe. Hence, something is wrong in the above analysis. What is it?

The answer is in Eqs. (4.1.4)–(4.1.6). In deriving these equations, we implicitly assumed that $\Phi_0 = $ const. and $\rho_0 = $ const. But these assumptions do not satisfy the original Poisson Eq. (4.1.3). This is why we have strange behavior in the growth rate when the wavelength tends toward infinity. In order to obtain the correct dispersion relation, we must first solve the equilibrium state for the self-gravitating system (nonuniform density and potential), and then solve the eigenvalue problem for the linearized equations of (4.1.1)–(4.1.3). An example of such a correct analysis is shown in Fig. 4.2 for the case of an infinite disk with a uniform surface density (Hayashi, 1987).

4.1.2   Gravitational Contraction of Spherical Clouds

4.1.2.1   Contraction of Spherical Clouds with Zero Pressure

Let us now consider the nonlinear evolution of gravitational contraction of a spherically symmetric (nonrotating and nonmagnetic) cloud. There is a standard initial condition for this problem (Larson, 1969), i.e. a static, isothermal, spherical cloud with uniform density. Note that this cloud does not satisfy the static equilibrium equation. Thus, strictly speaking, we do not have instability, but simply nonequilibrium. The basic equation for this problem is

$$\frac{\partial \rho}{\partial t} + \frac{1}{r^2}\frac{\partial}{\partial r}(\rho v r^2) = 0, \tag{4.1.15}$$

$$\frac{\partial v}{\partial t} + v\frac{\partial v}{\partial r} = -\frac{C_s^2}{\rho}\frac{\partial \rho}{\partial r} - \frac{GM(r)}{r^2}, \tag{4.1.16}$$

$$\frac{\partial M(r)}{\partial r} = 4\pi r^2 \rho. \tag{4.1.17}$$

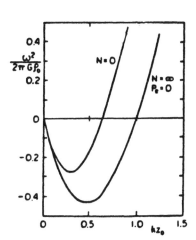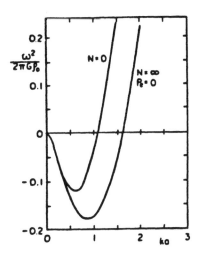

FIGURE 4.2 Dispersion relation for the gravitational instability in a cylinder and a sheet (Hayashi, 1987). Here, the polytropic equation of state $p \propto \rho^{1+\frac{1}{N}}$ is assumed, i.e., $N = 0$ corresponds to incompressible plasma, while $N = \infty$ corresponds to isothermal plasma.

Since this equation is nonlinear, it is difficult to solve analytically. Therefore, we will first consider the case with no pressure (Hunter, 1962). Although the equation with no pressure is still nonlinear, we can solve it because the Lagrangian equation of motion under the assumption with zero pressure is linear. From Eqs. (4.1.15) and (4.1.17), we have

$$\frac{\partial M}{\partial t} + v \frac{\partial M}{\partial r} = 0. \tag{4.1.18}$$

Assuming $C_s = 0$ (zero-pressure), we obtain

$$\frac{\partial v}{\partial t} + v \frac{\partial v}{\partial r} = -\frac{GM}{r^2}. \tag{4.1.19}$$

Now, we define the Lagrangian coordinates as follows,

$$\tau \equiv t, \tag{4.1.20}$$

$$R \equiv r - \int_0^\tau v(R, \tau') d\tau', \tag{4.1.21}$$

where (r, t) are the Eulerian coordinates. Noting that

$$v = \partial r(R, \tau)/\partial \tau, \tag{4.1.22}$$

Eq. (4.1.19) becomes

$$\frac{\partial^2 r(\tau)}{\partial \tau^2} = -\frac{GM}{r^2}. \tag{4.1.23}$$

This is the well-known equation for the orbit of a planet rotating around the Sun. The solution for this equation is

$$\left(\frac{8\pi}{3} G\rho_0\right)^{1/2} t = \sin^{-1}\left(1 - \frac{r}{r_0}\right)^{1/2} + \left(\frac{r}{r_0}\right)^{1/2}\left(1 - \frac{r}{r_0}\right)^{1/2}, \tag{4.1.24}$$

where r_0 is the initial radius of the cloud, and the time when $r \to 0$ is called *free fall time* and is written as

$$t_{ff} = \frac{\pi}{2}\left(\frac{8\pi G\rho_0}{3}\right)^{-1/2} = \left(\frac{32G\rho_0}{3\pi}\right)^{-1/2}. \tag{4.1.25}$$

The final solution for the density, mass, and velocity is

$$\rho = \rho_0 \left(\frac{r_0}{r_c(t)}\right)^3, \tag{4.1.26}$$

$$M = M_0 x^3, \tag{4.1.27}$$

$$v = x\dot{r}_c = -\left(\frac{8\pi}{3} G\rho_0\right)^{1/2}\left(\frac{r_0}{r_c(t)} - 1\right)^{1/2} x r_0, \tag{4.1.28}$$

where $x = r/r_c(t)$ and r_c is the solution of

$$\frac{dr_c}{dt} = -\left(\frac{8\pi}{3} G\rho_0\right)^{1/2}\left(\frac{r_0}{r_c} - 1\right)^{1/2} r_0. \tag{4.1.29}$$

The time evolution of the solution is illustrated in Fig. 4.3.

As in linear theory, we have a nonlinear theory for the plasma oscillation similar to above nonlinear theory of the gravitational contraction. Assume again the zero-pressure for one-dimensional slab plasma (Dawson, 1959; Davidson, 1972). Basic equations are

$$\frac{\partial n_e}{\partial t} + \frac{\partial}{\partial x}(n_e v_e) = 0, \tag{4.1.30}$$

$$\frac{\partial v_e}{\partial t} + v_e \frac{\partial v_e}{\partial x} = -\frac{e}{m_e} E, \tag{4.1.31}$$

$$\frac{\partial E}{\partial x} = -4\pi e(n_e - n_0). \tag{4.1.32}$$

288

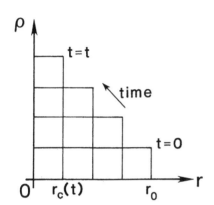
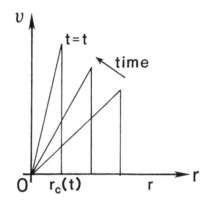

FIGURE 4.3 The time evolution of spherical collapse of zero pressure sphere.

From Eqs. (4.1.30) and (4.1.31), we have

$$\frac{\partial E}{\partial t} + v_e \frac{\partial E}{\partial x} = 4\pi e n_0 v_e. \tag{4.1.33}$$

This is essentially the same as Eq. (4.1.18) in the case of gravitational contraction. We again define the Lagrangian coordinates (x_0, τ), by noting that

$$R \rightarrow x_0$$

$$r \rightarrow x$$

in Eq. (4.1.21), where (x, t) are the Eulerian coordinates. Then, the equation of motion in a Lagrangian frame is

$$\frac{\partial^2 v_e}{\partial \tau^2} + \omega_{pe}^2 v_e = 0. \tag{4.1.34}$$

Since this is the linear equation, it is easily solved. The solution when the initial density has sinusoidal distribution is illusrated in Fig. 4.4.

289

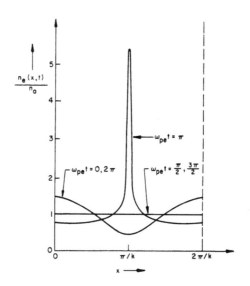

FIGURE 4.4 The solution of the nonlinear plasma oscillation (Davidson, 1972).

4.1.2.2 Contraction of Spherical Clouds with Finite Pressure

The contraction with a finite pressure is a much more difficult problem than that in the pressureless case, and hence was first studied by using numerical simulations (Bodenheimer and Sweigert, 1968; Penston, 1969; Larson, 1969).

Numerical results show the following characteristics (Fig. 4.5):

(1) In the initial stage of contraction, when the contracting gas is nearly isothermal, the density distribution tends to $\rho \propto r^{-2}$, which is the density distribution of hydrostatic isothermal sphere, and the (maximum) inflow velocity far from the center is nearly constant $\simeq 3.3 C_s$.

(2) In the later stage, after the central core becomes opaque and nearly adiabatic so that it stops to contract, the contraction of a gas near the central core is more like the free fall accretion, and the density distribution has $\rho \propto r^{-3/2}$, the same as that in a spherically symmetric accretion flow (Bondi, 1952). The inflow velocity increases as the gas contracts, and has the distribution $v \propto r^{-1/2}$, which again corresponds to that of the free-fall accretion

290

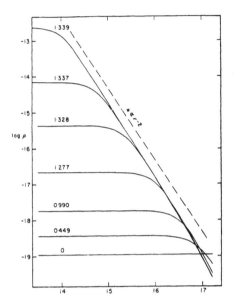

FIGURE 4.5 Results of Larson (1969)'s 1D hydrodynamic numerical simulations of the collapse of a spherical cloud.

flow.

On the basis of these numerical results, Larson (1969) and Penston (1969) studied the problem analytically, and found one self-similar solution[1] which explains the numerical results in the stage (1). They first define the self-similar variables as

$$\zeta = \frac{r}{c_s t},$$

(4.1.35)

$$\tau = t.$$

(4.1.36)

Furthermore, it is assumed that

$$\rho = \tilde{\rho}(\tau)\sigma(\zeta),$$

(4.1.37)

$$v = C_s u(\zeta),$$

(4.1.38)

---

[1]It is interesting to note that there exist self-similar solutions also for collapsing isothermal cylinders with self-gravity (e.g., see Inutsuka and Miyama, 1992).

291

$$M = \widetilde{M}(\tau)m(\zeta).$$

(4.1.39)

Substituting equations (4.1.37)–(4.1.39) into (4.15)–(4.17), we have

$$\frac{\dot{\widetilde{M}}}{\widetilde{M}} + \frac{m'}{m}(u - \zeta)\frac{1}{\tau} = 0,$$

(4.1.40)

$$\frac{\widetilde{M}}{\tilde{\rho}}m' = 4\pi\zeta^2\sigma C_s^3\tau^3,$$

(4.1.41)

$$u'(u - \zeta) = -\frac{\sigma'}{\sigma} - \frac{G\widetilde{M}m}{C_s^3\tau^3}$$

(4.1.42)

From the requirement of separation of variables, we find

$$\widetilde{M} = C_s^3\tau/G,$$

(4.1.43)

$$\tilde{\rho} = \frac{1}{4\pi G\tau^2}.$$

(4.1.44)

The substitution of equations (4.1.43)–(4.1.44) into equations (4.1.40)–(4.1.41) yields

$$m + (u - \zeta)m' = 0,$$

(4.1.45)

$$m' = \zeta^2\sigma.$$

(4.1.46)

The term m' can be eliminated from the above relations to give

$$m = \zeta^2\sigma(\zeta - u).$$

(4.1.47)

Finally, eliminating m, we have the following equations for σ and u.

$$\frac{\sigma'}{\sigma}\left[(\zeta - u)^2 - 1\right] = \left[\sigma - \frac{2}{\zeta}(\zeta - u)\right](\zeta - u),$$

(4.1.48)

$$u'\left[(\zeta - u)^2 - 1\right] = \left[\sigma(\zeta - u) - \frac{2}{\zeta}\right](\zeta - u).$$

(4.1.49)

where $' \equiv d/d\zeta$. For $\zeta = r/(C_st) \gg 1$ and $u' = 0$, we have $\sigma \simeq 2/\zeta^2$, which corresponds to $\rho \propto r^{-2}$. On the other hand, for $\zeta \ll 1$ and $\sigma' = 0$, we find $\zeta \propto u$, corresponding to $v \propto r/(C_st)$. The solution is shown in Fig. 4.6. It is interesting to note that the property $\rho \propto r^{-2}$ is similar to the density distribution in hydrostatic equilibrium of an isothermal gas sphere.

Shu (1977) found a self-similar solution which explains the later stage of the gravitational contraction (i.e. accretion stage), by using the same equations [Eqs. (4.1.48) and (4.1.49)]. The only difference is the boundary condition. Shu's boundary condition is

$$u = 0 \quad \text{at} \quad \zeta = 1,$$

(4.1.50)

$$u = -\infty \quad \text{at} \quad \zeta = 0.$$

(4.1.51)

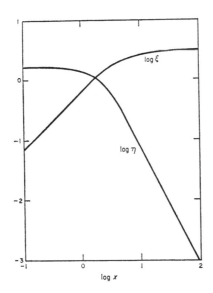

FIGURE 4.6 Larson-Penston's self-similar solution for the 1D collapse of a spherical cloud (Larson, 1969).

The results are illustrated in Fig. 4.7. For $\zeta \ll 1$, we have

$$m \to m_0,$$

$$\sigma \propto \zeta^{-3/2} \quad (\rho \propto r^{-3/2}),$$

$$u \propto \zeta^{-1/2} \quad (v \propto r^{-1/2}),$$

or using the relations (4.1.43) and (4.1.44)

$$M = m_0 C_s^3 \tau / G, \tag{4.1.52}$$

$$\rho = \frac{1}{4\pi G \tau^2} \left(\frac{m_0}{2}\right)^{1/2} \left(\frac{r}{C_s \tau}\right)^{-3/2}, \tag{4.1.53}$$

$$v = C_s \left(\frac{r}{2 m_0 C_s \tau}\right)^{-1/2}. \tag{4.1.54}$$

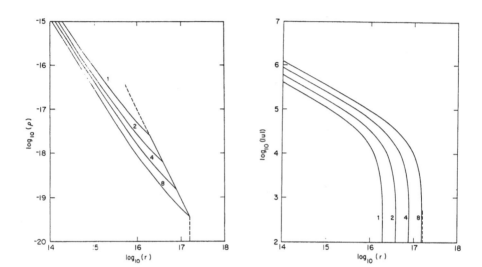

FIGURE 4.7 Shu's self-similar solution of a collapse of a spherical cloud (Shu, 1977).

in agreement with numerical results. Here $m_0 = 0.975$, when the initial condition is the singular (hydrostatic) isothermal sphere, $\rho = C_s^2/(2\pi G r^2)$. An interesting property of this solution is that the cloud contracts from inside to outside; the rarefaction wave propagates outward on the static isothermal gas sphere to initiate the gravitational contraction. Furthermore, the mass accretion rate $\dot{M} \propto C_s^3$ is independent of the scale (r).

4.1.2.3 Present Picture of Star Formation

Figure 4.8 illustrates the present picture of 1D star formation (Stahler *et al.*, 1980). The accretion rate is determined by the sound speed in the ambient cloud as described by Shu's similarity solution;

$$\dot{M} \simeq 0.975 \, C_s^3/G. \tag{4.1.55}$$

It is interesting to note that this equation does not include any characteristic mass. The collpase occurs from inside to outside and the front of accretion propagates outward as expansion wave at sound speed.

$$\dot{M} = 0.975 \frac{C_s^3}{G} \text{ (from Shu's self-similar solution)}$$

(there is no typical mass)

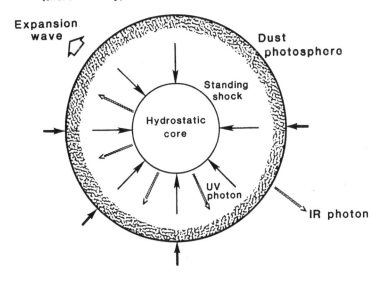

FIGURE 4.8 Present picture of 1D star formation (Stahler *et al.*, 1980)

The density of accreting gas increases as the gas falls deeper and deeper, and when the density of accreting gas exceeds $\rho \sim 10^{-13}$ g/cm^3, the gas becomes opaque and hence the isothermal condition breaks down. The gas now begins to undergo adiabatic accretion (or collapse). In this case, the ratio of the gas pressure to gravitational force,

$$\frac{\partial p}{\partial r} \bigg/ \left(\frac{GM\rho}{r^2} \right) \propto \frac{\rho^\gamma}{r} \bigg/ \frac{\rho}{r^2} = \rho^{\gamma-1} r \propto r^{-3(\gamma-1)+1} = r^{4-3\gamma}, \qquad (4.1.56)$$

which increases with decreasing r for $\gamma > 4/3$. Hence the accretion (or collapse) finally stops due to the enhanced gas pressure, and forms a *hydrostatic core* at the center. This corresponds to birth of a star.

When the central hydrostatic core is formed, the accretion flow well exceeds the sound speed, and collides with the central hydrostatic core, forming a standing *accretion shock*. The temperature just behind the accretion shock becomes very high ($\sim 10^6$ K), because the

295

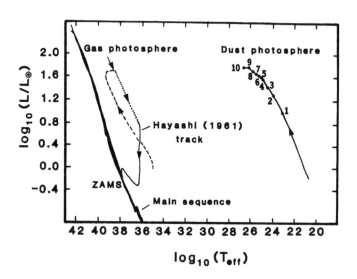

FIGURE 4.9 Hayashi track in HR diagram (Stahler *et al.*, 1980).

accretion flow speed is

$$v_{\text{accretion}} \sim (GM/r)^{1/2} \sim 100 \left(\frac{M}{1M_\odot}\right)^{1/2} \left(\frac{r}{10^{11}\text{cm}}\right) \text{ km/s}, \tag{4.1.57}$$

and the shock heated plasmas emit X-rays and UV radiation, ionizing the near-core region. However, the core cannot be seen from outside in X-rays, UV, nor in visible wavelengths, because the dense dust envelope surrounding the core absorbs these radiations. The absorbed radiation finally becomes IR photons which are emitted from the dust photosphere. Thus, in the accretion phase the evolutionary track in the HR diagram is in the very low temperature region. After almost all the envelope is accreted to the core, the core begins to be seen in visible wavelength as a newborn star. At this time, the star is on the *Hayashi track* (Fig. 4.9), where the star undergoes the quasi-static contraction (Hayashi, 1961).

Figure 4.10 shows the comparison between theory and observation for pre-main-sequence contraction of stars. The white circles show observed young stars called T-Tauri stars. The thick curve indicates the birth line of young stars predicted by the theory. The solid curve

296

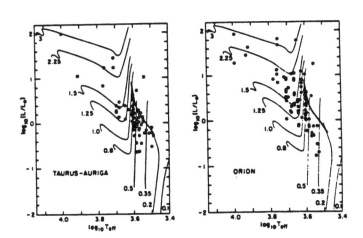

FIGURE 4.10 Comparison of theory and observations in HR diagram (Stahler *et al.*, 1980).

shows the quasi-static evolutionary track of pre-main-sequence stars. It is seen that the theory and observation are in qualitative agreement.

The accretion rate in the equation (46) is rewritten as

$$\dot{M} \simeq C_s^3/G \simeq 6 \times 10^{-5} \left(\frac{T}{10\text{K}}\right)^{3/2} \ M_\odot/\text{yr}. \tag{4.1.58}$$

In a molecular cloud with $T \sim 10$ K, the accretion rate becomes of order of $10^{-5} M_\odot/\text{yr}$, and hence a star is formed within $\sim 10^5$ years. Then the release rate of gravitational energy becomes

$$L = GM\dot{M}/r$$

$$\sim 100 \left(\frac{M}{1M_\odot}\right) \left(\frac{\dot{M}}{6 \times 10^{-5} M_\odot/\text{yr}}\right) \left(\frac{r}{10^{11}\text{cm}}\right) \ L_\odot \sim 10^{35}\text{erg/s}. \tag{4.1.59}$$

This is comparable to the observed maximum luminosity of protostars with $1M_\odot$ (Fig. 4.9).

On the other hand, if we apply the equation (4.1.58) and (4.1.59) to the galactic nucleus with the average temperature of $T \sim 10^4$ K, we find $\dot{M} \sim 2M_\odot/\text{yr}$ and $L \sim 10^{13}L_\odot \sim 10^{46}$ erg/s. Interestingly, this huge luminosity is comparable to the observed luminosity of active galactic nuclei.

4.1.3 Contraction of Rotating Clouds

Although 1D model seems to be successful, there are fundamental difficulties when developing the theory of gravitational contraction of interstellar clouds; one is the effect of rotation, and the other is the effect of magnetic field.

The effect of rotation is very important for the gravitational contraction of a cloud, because

$$\frac{\text{centrifugal force}}{\text{gravitational force}} \sim \frac{\rho v_\varphi^2/r}{\rho GM/r^2} = \frac{J^2/r^3}{GM/r^2} \propto \frac{1}{r} \tag{4.1.60}$$

increases with decreasing r, if angular momentum $J = rv_\varphi$ is conserved. Furthermore, the observed values of this ratio for molecular clouds are $0.1 \sim 1$, which means that a large fraction of angular momentum must be extracted from the cloud in order to make a star with radius $r_{\text{star}} \simeq 10^{-7}r_{\text{cloud}}$. In other words, the ratio of total angular momenta of clouds to stars are much larger than 1. This problem is known as the *angular momentum problem*.

Similarly, we can understand the importance of magnetic fields on the gravitational contraction of clouds;

$$\frac{\text{magnetic force}}{\text{gravitational force}} \sim \frac{B^2/(8\pi r)}{\rho GM/r^2} \propto \frac{rB^2}{\rho} = \text{constant}, \tag{4.1.61}$$

if $Br^2 = \text{constant}$, and $\rho r^3 = \text{constant}$. This means that if magnetic field is important in the initial interstellar cloud stage, it remains to be important up to the stellar stage. Actually, it is observed in some clouds that this ratio is of order of unity. Furthermore, the ratio of total magnetic flux of clouds to stars is much larger than 1. Thus, almost all magnetic flux must be removed from clouds during the gravitational contraction. This problem is known as the *magnetic flux problem*.

Both effects introduce multi-dimensionality in the problem, which makes the problem difficult. Consider again the large differences of density and size between stars and interstellar clouds. No one has succeeded to perform numerical simulation from an interstellar cloud to a star even for 2D problem. Therefore, in this sense, the theory of star formation as well as galaxy formation is in highly preliminary stage, and there are many difficult problems which are worth challenging.

From now on, we discuss fundamental points about the effect of rotation and magnetic field on the gravitational contraction of clouds.

4.1.3.1 Disk Formation and Runaway Collapse

Numerical simulation of gravitational contraction of a rotating, non-magnetic cloud was first performed by Larson (1972). The initial condition used in his simulation (sometimes

298

called the standard initial condition) is a static spherical cloud with uniform density, uniform temperature, and with uniform rotation (i.e. rigid rotation) with angular velocity Ω_0. The temperature is kept constant throughout the simulation. There are two fundamental parameters for this problem;

$$R_\alpha = \frac{\text{thermal energy}}{\text{gravitational energy}} = \frac{15C_s^2}{8\pi G \rho_0 R^2}, \tag{4.1.62}$$

$$R_\beta = \frac{\text{rotational energy}}{\text{gravitational energy}} = \frac{\Omega_0^2}{4\pi G \rho_0}. \tag{4.1.63}$$

It is to be noted that the product, $R_\alpha R_\beta$, is independent of the cloud size,

$$R_\alpha R_\beta = \frac{125}{24} \left(\frac{C_s J}{GM^2} \right)^2, \tag{4.1.64}$$

where M is the mass and J is the total angular momentum;

$$M = \left(\frac{4}{3} \right) \pi R^3 \rho_0, \tag{4.1.65}$$

$$J = \left(\frac{4\pi}{15} \right) \Omega_0 \rho_0 R^5. \tag{4.1.66}$$

This means that $R_\alpha R_\beta$ is a conserved quantity during the collapse if the transport of angular momentum is negligible.

Now consider 2D axisymmetric isothermal contraction. Many numerical simulations have been performed since the pioneering work by Larson (1972) on 2D axisymmetric gravitational contraction of a rotating cloud. Results were, however, not always in agreement with each other, even for exactly the same initial condition. For example, some reported the formation of a ring structure after the collapse, while some other found that the ring did not appear. The reason for these differences seems to be due to numerical errors, i.e. numerical transport of angular momentum.

Norman *et al.* (1980) carried out fine numerical simulation where the specific angular momentum is nearly conserved on a Lagrangian fluid element, and first found that the central part of the cloud undergoes self-similar type *runaway collapse*.

Narita *et al.* (1984) confirmed this runaway collapse and explained it semi-analytically. Figure 4.11 shows the typical example of the self-similar type runaway collapse of the rotating cloud. Figure 4.12 shows one-dimensional distribution of ρ and ω (angular velocity) on the equatorial plane, and shows

$$\rho \propto r^{-2}, \tag{4.1.67}$$

$$\omega \propto r^{-1}, \tag{4.1.68}$$

in the inner envelope. The former is similar to the hydrostatic density disribution of a spherical isothermal cloud, and the latter is essentially the same as the flat rotation observed

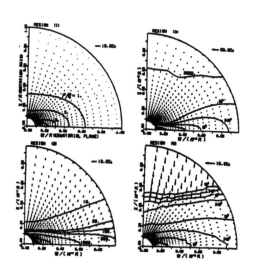

FIGURE 4.11 The self-similar type runaway collapse of the rotating cloud (Narita *et al.*, 1984).

in galaxies. According to the result of Narita *et al.* (1984), the runaway collapse occurs for

$$R_\alpha R_\beta \leq 0.2. \tag{4.1.69}$$

Otherwise ($R_\alpha R_\beta > 0.2$), the cloud undergoes an oscillation around the equilibrium state.

The reason why the runaway occurs is as follows (Narita *et al.*, 1984). The surface density of the disk is $\sigma \simeq \int_0^\infty \rho dz \propto \rho r \propto r^{-1}$, since $\rho \propto r^{-2}$. Hence the mass m inside r is in proportion to r, and the gravitational force $F_g \simeq Gm/r^2$ at r is $\propto r^{-1}$. On the other hand, the centrifugal force at r is given by $F_c \simeq r\omega^2 \propto r^{-1}$. Hence the ratio between centrifugal and gravitational forces F_c/F_g remains constant during the collapse, so that once the ratio becomes smaller than unity, the collapse cannot stop since the centrifugal bounce cannot occur.

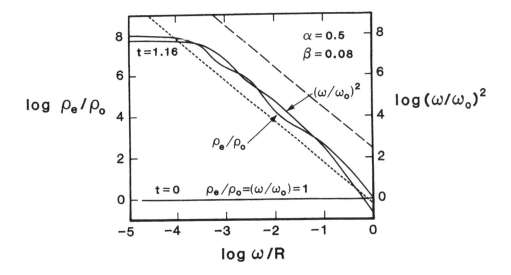

FIGURE 4.12 One dimensional distribution of ρ and ω (Narita *et al.*, 1984).

4.1.3.2 Equilibrium Solution of a Rotating Disk

The distribution of physical quantities obtained in Norman *et al.* (1980) and Narita *et al.* (1984) is similar to those in the analytical solution obtained for the equilibrium structure of a rotating cloud by Toomre (1982) and Hayashi *et al.* (1982). The basic equations for this problem (axisymmetric equilibrium state) are

$$C_s^2 \frac{\partial \ln \rho}{\partial r} - \frac{v_\varphi}{r} = F_r, \tag{4.1.70}$$

$$C_s^2 \frac{\partial \ln \rho}{\partial z} = F_z, \tag{4.1.71}$$

$$F_r = -\frac{\partial \Phi}{\partial r}, \tag{4.1.72}$$

301

$$F_z = -\frac{\partial \Phi}{\partial z}, \tag{4.1.73}$$

$$\frac{1}{r}\frac{\partial \Phi}{\partial r}\left(r\frac{\partial}{\partial r}\right) + \frac{\partial^2 \Phi}{\partial z^2} = 4\pi G\rho, \tag{4.1.74}$$

where (r, z) is the radial and vertical coordinates in the cylindrical coordinate. We now assume flat rotation $v_\varphi = $ constant, as found from numerical simulation. Taking

$$\rho(r, z) = g\frac{(r, z)}{r^2}, \tag{4.1.75}$$

we find

$$r\frac{\partial}{\partial r}r\frac{\partial \ln g}{\partial r} + r\frac{\partial}{\partial z}r\frac{\partial \ln g}{\partial z} = -4\pi Gg/C_s^2, \tag{4.1.76}$$

after some manipulation. This equation shows that g is a function of z/r alone, i.e. suggesting a self-similar character. In fact, assuming a new variable

$$z/r = \sinh\zeta, \tag{4.1.77}$$

we find that the equation (4.1.76) becomes

$$\frac{d^2 \ln g}{d\zeta^2} = -\left(\frac{4\pi G}{C_s^2}\right)g. \tag{4.1.78}$$

The solution of this simple equation is

$$g(\zeta) = \frac{\gamma^2 C_s^2}{2\pi G\cosh^2(\gamma\zeta)}, \tag{4.1.79}$$

where γ is an arbitrary constant which is to be determined by the boundary condition. In this case, the boundary condition is $rF_r \to 0$ as $r \to 0$, and we find

$$\gamma = 1 + \frac{v_\varphi^2}{2C_s^2}. \tag{4.1.80}$$

The solutions are illustrated in Fig. 4.13. The density is written as

$$\rho(r, z) = \frac{\gamma^2 C_s^2}{2\pi G\sinh^{-1}(z/r)}\frac{1}{r^2}. \tag{4.1.81}$$

Hence, the density on the equatorial plane ($z = 0$) becomes $\rho \propto r^{-2}$, explaining simulation results well. It is interesting to note that this density distribution shows neither a ring nor an ellipsoid.

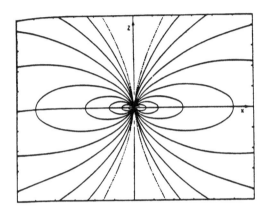

FIGURE 4.13 Toomre-Hayashi solution for equilibrium of rotating gas disk (Hayashi *et al.*, 1982).

4.1.3.3 3D Simulation

In three dimensional (3D) situation, a rotating disk (which is formed by a collapse of a rotating spherical cloud) can fragment into multiple clumps due to non-axisymmetric instabilities (Goldreich and Lynden-Bell 1965b) depending on the parameter values. This is a process creating binary stars, or multiple stellar systems.

Miyama *et al.* (1984) have carried out extensive 3D numerical simulations of 3D collapse of a (uniformly) rotating, isothermal cloud with various parameters R_α (= initial thermal energy/gravitational energy) and R_β (= initial rotational energy/gravitational energy). (See Fig. 4.14 for a typical case with parameters $R_\alpha = 0.4, R_\beta = 0.3$. As for references on earlier 3D numerical simulations, see Miyama *et al.*, 1984.) They found that the characteristics of collapse and fragmentation are determined essentially by the product of two parameters (see Fig. 4.15),

$$R_\alpha R_\beta = (125/24)(C_s J/GM^2)^2 = 5C_s^2 \Omega_0^2 R_0^4/(6G^2 M^2),$$

which is constant during the evolution, where R_0, Ω_0, M are the radius, angular rotational

303

velocity, and mass of a cloud at $t = 0$:

1. $R_\alpha R_\beta > 0.20$: a cloud does not shrink appreciably and oscillates about an equilibrium.

2. $0.20 > R_\alpha R_\beta > 0.12$: the inner part of a cloud contracts greatly to form a flattened disk. The flatness of the disk is nearly prportional to $1/R_\alpha R_\beta$. The disk does not fragment, but shows runaway collapse.

3. $0.12 > R_\alpha R_\beta$: a cloud forms a flattened disk, but it soon fragments and forms a multiple system, the multiplicity being larger for a cloud with smaller $R_\alpha R_\beta$.

What is the reason for the above results? This is explained as follows (Miyama *et al.*, 1984). Consider the stage when a cloud becomes nearly in equilibrium after a collapsing stage. In this stage, the z-component of gravity approximately balances with that of pressure gradient, so that the half thickness of the disk is given by

$$z \simeq C_s^2/(G\sigma) \simeq C_s^2 r^2/(Gm). \tag{4.1.82}$$

Here $\sigma \simeq \rho z$ is the column mass density, r is the radius of the disk, and $m \simeq \sigma r^2$ is the mass of the disk inside r, respectively. From the balance between centrifugal force and the r-component of gravity, we have

$$Gm/r^2 \simeq j^2/r^3, \tag{4.1.83}$$

where j is the specific angular momentum at r. Since initially we assumed a uniformly rotating spherical cloud, the specific angular momentum j for $m/M < 1/2$ is written as

$$j = R_0^2 \Omega_0 [1 - (1 - m/M)^{2/3}] \simeq (2/3) R_0^2 \Omega_0 m/M. \tag{4.1.84}$$

From equations (4.1.82)-(4.1.84), we finally find

$$z/r \simeq (8/15) R_\alpha R_\beta. \tag{4.1.85}$$

Thus, the flatness (r/z) of the disk formed by the collapse of a rotating cloud is determined by $1/R_\alpha R_\beta$. If it is smaller, the disk becomes thicker, and will be stable for fragmentation, whereas if it is larger, the disk will be thin and unstable for fragmentation. The number of fragments depends also on $R_\alpha R_\beta$, since the most unstable wavelength of the gravitational instability of a uniformly rotating disk is about

$$\lambda \simeq 2\pi z$$

(Goldreich and Lynden-Bell, 1965a, Miyama, 1992). It is also interesting to note that the parameter $R_\alpha R_\beta$ is related to $\gamma = 1 + v_\varphi^2/(2C_s^2)$ which is the key parameter determining the flatness of the disk (Hayashi *et al.*, 1982, see Fig. 4.13 and Sec. 4.1.3.2);

$$R_\alpha R_\beta = (15/4\pi^2)(\gamma - 1) \sin^2(\pi/2\gamma). \tag{4.1.86}$$

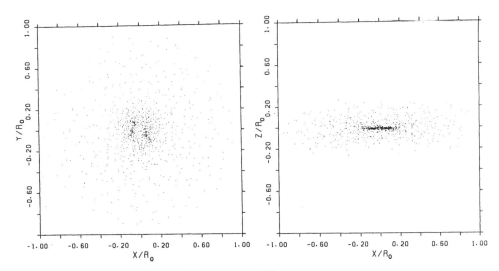

FIGURE 4.14   3D collapse of rotating cloud (Miyama *et al.*, 1984).

4.1.4   Contraction of Magnetic Clouds

As for this problem, not so many numerical simulations have been performed for both 2D and 3D case compared with the case of a rotating cloud. Main difficulties are; (1) Boundary conditions are difficult, i.e. Alfvén waves propagate along magnetic field lines to transfer angular momentum, but they are often reflected at the numerical boundary. (2) Non-ideal condition has to be considered to solve the magnetic flux problem.

4.1.4.1   Criteria for Collapse of Magnetic Clouds

Let's begin with a simple semi-analytical consideration on a condition of a magnetic cloud contraction. As we have seen before, the gravitational force must be larger than the magnetic force to initiate the gravitational contraction. That is,

$$\frac{GM\rho}{r^2} > \frac{B^2}{8\pi r}. \tag{4.1.87}$$

305

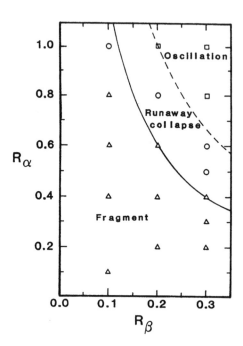

FIGURE 4.15 Parametric survey of 3D collapse of rotating cloud by Miyama *et al.* (1984).

Using the relation $M = (4\pi/3)r^3\rho$, this equation is written as

$$M > M_{cr} = \Phi(6G)^{-1/2}, \tag{4.1.88}$$

$$\text{or} \quad \sigma/B > (6G)^{-1/2}. \tag{4.1.89}$$

Here $\sigma = M/r^2$ is the surface (or column) density of the cloud, and M_{cr} is the critical mass above which the cloud can collapse,

$$M_{cr} \sim 10^3 \left(\frac{B}{30\mu G}\right)\left(\frac{r}{2\text{pc}}\right)^2 \ M_\odot. \tag{4.1.90}$$

In the case of the disk with magnetic field, the critical condition becomes (Nakano and Nakamura, 1978)

$$\sigma/B > (4\pi^2 G)^{-1/2}. \tag{4.1.91}$$

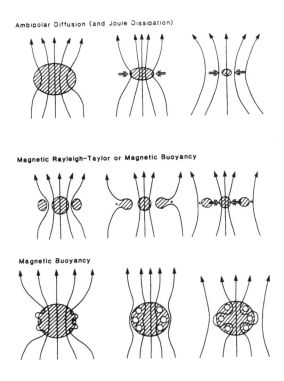

Ambipolar Diffusion (and Joule Dissipation)

Magnetic Rayleigh–Taylor or Magnetic Buoyancy

Magnetic Buoyancy

FIGURE 4.16 Various processes for extracting magnetic flux from a collapsing cloud or a star.

4.1.4.2 Ambipolar Diffusion

Usually the molecular cloud is in static equilibrium which satisfies force balance between magnetic, thermal, centrifugal, and gravitational forces. What triggers the collapse of the cloud? One solution is *ambipolar diffusion* (Mestel and Spitzer, 1956; Mouschovias and Paleologou, 1981; Nakano, 1982) (Fig. 4.16).

In a cold plasma with temperature as low as 100–1000 K, there are many neutral particles. (Note that there are some ions and free electrons even for such low temperatures, because ionization can occur due to cosmic rays.) Since neutral particles are free from Lorentz force, magnetic fields affect the motion of neutral particles only through collision between neutrals and charged particles. In that case, it can occur that the velocity of charged particles and neutrals are different. This process is called the *ambipolar diffusion.*

Consider the equation of motion for charged particles (ions, electrons, grains) and neutrals

(Zweibel, 1987),

$$\rho_c \frac{d\mathbf{v}_c}{dt} = \mathbf{F}_c + \frac{1}{c}\mathbf{J} \times \mathbf{B} - \rho_c \nu_{cn}(\mathbf{v}_c - \mathbf{v}_n), \tag{4.1.92}$$

$$\rho_n \frac{d\mathbf{v}_n}{dt} = \mathbf{F}_n - \rho_n \nu_{nc}(\mathbf{v}_n - \mathbf{v}_c), \tag{4.1.93}$$

where $\mathbf{v}_c$ is the velocity vector of charged particles, $\mathbf{v}_n$ is the velocity vector of neutral particles, and $\mathbf{F}_{c,n} = \rho_{c,n}\mathbf{g} - \nabla p_{c,n}$, and

$$\rho_c \nu_{cn} = \rho_n \nu_{nc} = \frac{m_c m_n n_c n_n}{m_c + m_n}\langle \sigma v \rangle, \tag{4.1.94}$$

and $\langle \sigma v \rangle$ is the rate coefficient for the collision between charged particles and neutrals, $\langle \sigma v \rangle \sim (3.3 - 13.6) \times 10^{-10} \text{cm}^3 \text{s}^{-1}$ for H-H collision for $10 \leq T \leq 1000 \text{ K}$ (Spitzer, 1978).

It follows from above equations that

$$\frac{\partial}{\partial t}(\mathbf{v}_c - \mathbf{v}_n) = -\frac{\nabla p_c}{\rho_c} + \frac{\nabla p_n}{\rho_n} + \frac{1}{\rho_c c}\mathbf{J} \times \mathbf{B} - (\nu_{cn} + \nu_{nc})(\mathbf{v}_c - \mathbf{v}_n). \tag{4.1.95}$$

We can assume steady state since the frictional time scale $(\nu_{cn} + \nu_{nc})^{-1}$ is typically short compared with the dynamical time scale,

$$t_{\text{fric}} \sim \frac{1}{\nu_{cn}} \sim 10^3 \left(\frac{n_n}{10^6 \text{cm}^{-3}}\right)^{-1} \text{sec}, \tag{4.1.96}$$

whereas, the dynamical time scale is

$$t_{\text{dyn}} \sim \frac{1}{(G\rho)^{1/2}} \sim 3 \times 10^{12} \left(\frac{n_n}{10^6 \text{cm}^{-3}}\right)^{-1} \text{sec}. \tag{4.1.97}$$

Further, since $p \ll B^2/8\pi$ in a molecular cloud, we can neglect the pressure terms, and get

$$\mathbf{v}_D = \mathbf{v}_c - \mathbf{v}_n = \frac{\mathbf{J} \times \mathbf{B}}{\rho_c(\nu_{cn} + \nu_{nc})c}. \tag{4.1.98}$$

Here $\mathbf{v}_D$ is called the ambipolar drift, and becomes

$$v_D \sim \frac{V_{AC}^2 t_{\text{fric}}}{L} \sim 10^4 \left(\frac{n_n}{10^6 \text{cm}^{-3}}\right)^{-2} \left(\frac{x}{10^{-7}}\right)^{-1} \left(\frac{B}{10^{-3}\text{G}}\right)^2 \left(\frac{L}{0.3\text{pc}}\right)^{-1} \text{cm/s}. \tag{4.1.99}$$

Consequently, the time scale of ambipolar diffusion is

$$t_D = \frac{L}{v_D} \sim \frac{\rho_c}{G\rho_n^2 t_{\text{fric}}} \sim 3 \times 10^6 \left(\frac{x}{10^{-7}}\right) \text{years}, \tag{4.1.100}$$

where we assumed the force balance between Lorentz and gravitational forces $\rho_c V_{AC}^2/(\rho_n L^2) \sim G\rho_n$, where n_n is assumed to be 10^6cm^{-3}, and $x = n_c/n_n$ is the ionization degree of plasma.

308

Note that when $n_n \sim 10^6 \mathrm{cm}^{-3}$, the free fall time is about 10^5 years. Hence the ambipolar diffusion is relatively slow process compared with the dynamical process.

Originally it has been thought that the ambipolar diffusion can solve the magnetic flux problem. However, Nakano (1984) showed that the magnetic flux cannot be escaped effectively during the quasi-static contraction. Since the ambipolar diffusion time is much longer than the free fall time scale, the magnetic flux is nearly frozen into the collapsing cloud. Nakano (1984) found that the ohmic diffusion finally solve the problem if the density exceed $10^{11} \mathrm{cm}^{-3}$.

4.1.4.3 Magnetic Braking

The most effective way to solve the angular momentum problem is *magnetic braking* (Mestel and Spitzer, 1956; Nakano, 1972; Mouschovias and Paleologou, 1980).

Consider a rotating cloud, penetrated by uniform magnetic field parallel to the rotation axis (Fig. 4.17). Magnetic field lines are twisted by a rotating disk, and hence the magnetic twist propagates toward both polar directions along magnetic field lines as torsional Alfvén waves. As the Alfvén wave propagates external medium, the gas in the external medium begins to rotates, i.e. it gains angular momentum. Since the source of angular momentum is the rotating cloud, it is equivalent to the angular momentum transfer. When the total angular momentum transferred to the external medium becomes comparable to that of the original rotating cloud, a significant braking takes place. Since the rotation velocity associated with the Alfvén wave is approximately the same as that of the disk (because the Alfvén velocity in the external medium is much larger than the rotation velocity of the cloud), a significant braking occurs when the *moment of inertia* of the gas through which the Alfvén waves have propagated becomes comparable to the moment of inertia of the initial cloud; i.e., the braking time $\tau_\parallel$ is estimated to be

$$\tau_\parallel = \int_{z_d}^{z_i} \frac{dz}{v_{A,ex}} \simeq (z_i - z_d)/v_{A,ex},$$
(4.1.101)

where $v_{A,ex}$ is the Alfvén velocity in the external medium, z_d is the half thickness of the cloud, and z_i is the distance where the moment of inertia in the external medium

$$I = \int \rho r^2 dV,$$
(4.1.102)

becomes comparable to that of the cloud, i.e.,

$$(z_i - z_d)\rho_{ex} = z_d \rho_d.$$
(4.1.103)

This is because in this case the radial (r) dependence is the same for external medium and cloud. Then we find that

$$\tau_\parallel \simeq \frac{z_d}{v_{A,ex}} \frac{\rho_d}{\rho_{ex}} \simeq 10^6 \left(\frac{z_d}{0.01 \text{ pc}} \right) \left(\frac{\rho_d/\rho_{ex}}{10^3} \right) \left(\frac{v_{A,ex}}{10 \text{ km/s}} \right)^{-1} \text{ years.}$$
(4.1.104)

309

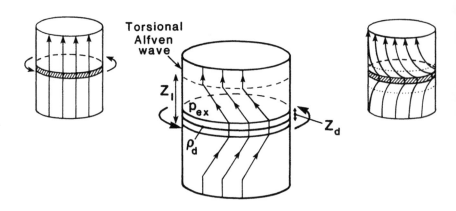

FIGURE 4.17 Magnetic braking in the cas of parallel field $(B\|\Omega)$. Note the generation and propagation of torsional Alfvén waves, which carries the angular momentum.

Note here the Alfvén velocity in the molecular cloud is about

$$V_A = \frac{B}{(4\pi\rho)^{1/2}} \simeq 10 \left(\frac{B}{10^{-4}\,\text{G}}\right) \left(\frac{n}{10^3\,\text{cm}^{-3}}\right)^{-1/2} \text{km/s}, \qquad (4.1.105)$$

where n is the hydrogen number density. On the other hand, the rotational velocity of molecular clouds is usually less than 1 km/s. Hence linear approximation is good for propagation of Alfvén waves.

Let us now consider the perpendicular case ($B \perp \Omega$; Fig. 4.18). In this case, the volume that the Alfvén wave propagates increases much faster than the parallel case ($B\|\Omega$), so that the braking time is shorter. The moment of inertia in this case is

$$I_\perp = \int_R^{R_I} \rho r^2 2\pi r\, dr = 2\pi\rho_{ex} \left(\frac{R_I^4}{4} - \frac{R^4}{4}\right). \qquad (4.1.106)$$

310

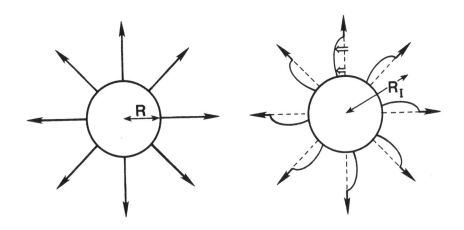

FIGURE 4.18 Magnetic braking in the case of perpendicular field $(\mathbf{B} \perp \mathbf{\Omega})$.

Hence we find

$$R_I = R \left(1 + \frac{\rho_d}{\rho_{ex}}\right)^{1/4}.$$ (4.1.107)

The braking time is estimated as

$$\tau_\perp = \int_R^{R_I} \frac{dr}{V_A(r)} = \frac{R}{2V_A(R)} \left[\left(1 + \frac{\rho_d}{\rho_{ex}}\right)^{1/2} - 1\right].$$ (4.1.108)

This time scale is much shorter than that for parallel rotator;

$$\tau_\perp \sim 3 \times 10^4 \left(\frac{R_d}{0.01 \text{ pc}}\right) \left(\frac{\rho_d/\rho_{ex}}{10^3}\right)^{1/2} \left(\frac{v_{A,ex}}{10 \text{ km/s}}\right)^{-1} \text{ years.}$$ (4.1.109)

Consequently, we now understand why rotating clouds tend to be *aligned rotators* $(\mathbf{B}\|\mathbf{\Omega})$.

311

4.1.4.4 Quasistatic Contraction

Consider a magnetic cloud which is in magnetostatic equilibrium and satisfies stability criteria [Eq. (4.1.89)]. If the ambipolar diffusion occurs, the mass in the cloud gradually diffuses (contracts) to the inner region, i.e. the cloud undergoes quasi-static contraction. As the clouds gradually contracts, the surface density increases, and hence the cloud would finally lose equilibrium and begins to collapse. This explains why stars are formed from stable molecular clouds.

Let us consider the equilibrium structure of a magnetized, non-rotating cloud (Mouschovias 1976; Nakano, 1984). Basic equations are;

$$-\rho\nabla\Psi - \nabla p + \frac{1}{c}\mathbf{J}\times\mathbf{B} = 0, \qquad (4.1.110)$$

$$\nabla^2\Psi = 4\pi G\rho, \qquad (4.1.111)$$

$$p = \rho C_s^2. \qquad (4.1.112)$$

We now assume axi-symmetry ($\partial/\partial\varphi = 0$ in cylindrical coordinate) and isothermality (C_s=constant), and introduce magnetic flux (or stream) function Φ,

$$\Phi(r, z) = rA_\varphi(r, z), \qquad (4.1.113)$$

$$\mathbf{B} = \left(\frac{1}{r}\frac{\partial\Phi}{\partial z}, -\frac{1}{r}\frac{\partial\Phi}{\partial r}\right), \qquad (4.1.114)$$

where A_φ is φ component of vector potential. It follows that

$$\mathbf{B}\cdot\nabla\Phi = 0, \qquad (4.1.115)$$

i.e. the flux function Φ is constant along magnetic field lines.

Multiplying $\mathbf{B}$ with equation (4.1.110), we get

$$-\rho\mathbf{B}\cdot\nabla\Psi - \mathbf{B}\cdot\nabla p = 0, \qquad (4.1.116)$$

which is rewritten as

$$\frac{p}{C_s^2}\frac{\partial\Psi}{\partial s} + \frac{\partial p}{\partial s} = 0, \qquad (4.1.117)$$

where s is the length measured along field lines (Φ = constant). This equation is easily integrated and we get

$$p = p_0(\Phi)\exp(-\Psi/C_s^2). \qquad (4.1.118)$$

Inserting this equation into (4.1.110), we find

$$\frac{dp_0(\Phi)}{d\Phi}\exp(-\Psi/C_s^2) + \frac{1}{4\pi}\frac{\Delta^*\Phi}{r^2} = 0, \qquad (4.1.119)$$

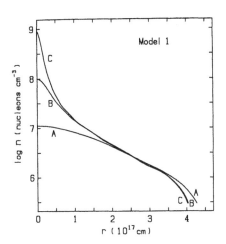

FIGURE 4.19 Quasistatic contraction of a magnetic cloud (Nakano, 1984)

where

$$\Delta^* = r \frac{\partial}{\partial r} \left(\frac{1}{r} \frac{\partial \Phi}{\partial r} \right) + \frac{\partial^2 \Phi}{\partial z^2}. \tag{4.1.120}$$

The equation (4.1.119) is a generalized *Grad-Shafranov equation* which is well known in fusion plasma physics. The Poisson equation (4.1.111) is now written as

$$\frac{1}{r} \frac{\partial}{\partial r} \left(r \frac{\partial \Psi}{\partial r} \right) + \frac{\partial^2 \Psi}{\partial z^2} = \frac{4\pi G}{C_s^2} p_0(\Phi) \exp \left(-\frac{\Psi}{C_s^2} \right). \tag{4.1.121}$$

Mouschovias (1976), Nakano (1984), and Tomisaka *et al.* (1988, 1989, 1990) solved these equations and followed the quasi-static contraction of a magnetized cloud. One example calculated by Nakano (1984) is shown in Fig. 4.19. It is seen that a central part gradually contracts even if the outer part is not changed much. After the final stage, there is no equilibrium solution, i.e. the cloud begins dynamical contraction.

313

4.1.4.5 Dynamical Collapse

If the cloud undergoes a spherical contraction, such as in the case of weak magnetic field, the magnetic field strength increases as

$$B \propto \rho^{2/3}, \tag{4.1.122}$$

since $Br^2 = $ constant, and $\rho r^3 = $ constant. On the other hand, if a cloud undergoes vertical collapse along vertical field line, the magnetic field strength remains constant.

Scott and Black (1980) first carried out numerical simulations of 2D dynamical collapse of a non-rotating magnetic cloud, and found, interestingly, intermediate behavior for the relationship between field strength and density;

$$B \propto \rho^{1/2}. \tag{4.1.123}$$

This has been confirmed by a number of following numerical simulations of dynamical collapse of a magnetic cloud (Phillips, 1986; Dorfi, 1982; Bentz, 1984; Mouschovias and Morton, 1991; Fielder and Mouschovias, 1992; Basu and Mouschovias, 1994; Tomisaka, 1995, 1996; Nakamura, Hanawa, Nakano, 1995; among them, Mouschovias and Morton, 1991, and Basu and Mouschovias, 1994 included an effect of ambipolar diffusion). The equation (4.1.123) suggests that approximate equipartition holds between magnetic energy and thermal energy, i.e., $B^2/8\pi \sim p = \rho C_s^2 \propto \rho$ for $C_s = $ constant, and is roughly consistent with the observed relationship between field strength and density (Troland and Heiles; 1986; see Fig. 1.16).

In the following, we will discuss the physics of dynamical collapse of a non-rotating, magnetized cloud to understand why above relation holds and how the colappose proceeds, on the basis of 2D simulations of Tomisaka (1995, 1996) and Nakamura $et\ al.$ (1995).

As an initial condition, Tomisaka (1995) and Nakamura $et\ al.$ (1995) assumed an isothermal cylindrical cloud penetrated by longitudinal magnetic fields with constant β in magneto-hydrostatic equilibrium. This magnetized cylinder is unstable to the gravitational instability, and the most unstable wavelength (Nakamura et al. 1993) becomes

$$\lambda_{\max} \simeq 20C_s/(G\rho_c)^{1/2}, \tag{4.1.124}$$

nearly independent of β when $\beta > 1$, where C_s is the isothermal sound speed, and ρ_c is the density on the axis of the cylindrical cloud.

They followed the nonlinear evolution of this gravitational instability, and found the following. When $\beta \sim 1$, the cloud contracts along the field lines to form a disk perpendicular to magnetic field lines. Supercritical cores $(M > M_{cr})$ are formed in a cylindrical cloud for a wide range of physical conditions. In such a case, the core undergoes indefinite collapse (i.e., $runaway\ collapse$) to form a self-similar structure (Nakamura $et\ al.$, 1995).

Figure 4.20 (Tomisaka, 1996) shows one example of such self-similar collapse. Fig. 4.20(a) is the stage when the central density reaches 10^5 times of the initial density, and 4.20(b) shows an enlargement of the central region (simulated by the nested grid scheme). Figure 4.20(c) further shows an inner region of the core with a size of 1/1000 of the initial radius

314

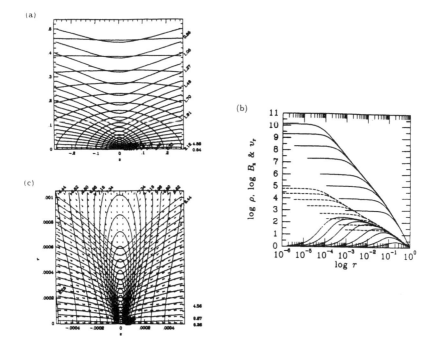

FIGURE 4.20 2D dynamical collapse of a non-rotating magnetic cloud (from Tomisaka, 1996).

of the cloud. At this stage, the central density reaches 10^{10} times of the initial density. Figures 4.20(d) and (e) show the radial distribution of the density and the magnetic field at $z = 0$, revealing the self-similar distribution of the density and the magnetic field strength;

$$\rho(r) \propto B_z(r)^2 \propto r^{-2}. \tag{4.1.125}$$

Interestingly, this density distribution is similar to that of the singular solution of the hydrostatic quilibrium for an isothermal gas sphere, and also that of the self-similar solution of the 1D collapse of a spherical isothermal cloud (Larson, 1969; Penston, 1969; see Fig. 4.5). This is also similar to the density distribution on the equatorial plane of a collapsing rotating isothermal cloud (e.g., Norman *et al.*, 1980; Narita *et al.*, 1984; Matsumoto *et al.*, 1997; Fig. 4.12).

What is the physical mechanism leading to $B_z(r) \propto \rho(r)^{1/2}$? The magnetic fields are frozen into plasmas, so that the column density, σ, integrated along the magnetic field line is proportional to the magnetic field strength;

$$\sigma \sim \rho_d z_d \propto B_z. \tag{4.1.126}$$

315

Here, z_d is the thickness of the disk, and ρ_d is the density of the disk at $z = 0$. If the disk is approximated to be an isothermal plane-parallel disk, the thickness of the disk is written as

$$z_d \simeq C_s^2/(G\sigma),\qquad(4.1.127)$$

(see Eq. 4.1.82). Combining these two equations, we have $z_d \propto \rho_d^{-1/2}$, so that $B_z \propto \rho_d^{1/2}$.

It is interesting to note that when the initial magnetic field is weak, the cloud collapse becomes similar to that of a non-magnetic spherical cloud so that the magnetic field strength increases like the equation (4.1.122). In this case, the plasma β decreases as the cloud contracts;

$$\beta \propto p/B^2 \propto r^{-3}/r^{-4} \propto r.\qquad(4.1.128)$$

Thus, even if the initial β is large (i.e., even if the initial magnetic field is weak), β decreases until the magnetic pressure becomes comparable to the gas pressure. This is the reason why the approximate equipartition holds between magnetic energy and thermal energy during the collapse of a magnetic cloud.

Nakamura et al. (1995) searched for a quasi-1D self-similar solution of this problem, assuming that the disk is infinitesmally thin. They indeed found an approximate similarity solution similar to the Larson-Penston solution, which explains above results very well. The reader should be referred to Nakamura et al. (1995) for a detailed derivation of this interesting similarity solution. (See also Shu and Li, 1997) for a related work.)

Finally, it should be noted that above runaway collapse is eventually followed by the *inside-out collapse* after the central core become opaque (Sec. 4.1.2.2). Saigo and Hanawa (1997) presented similarity solutions that describe the runway collapse and its subsequent inside-out collapse, and showed that in the inside-out collapse phase, the disk has two parts: an inner rotating disk in quasi-equilibrium and an outer dynamically infalling envelope.

4.2 Shear Flows

As the gravitational contraction of clouds proceeds, a gravitating object (e.g., a star) is formed at the center. After this stage, the contracting cloud evolves into an *accretion disk* rotating around a central object. The proto-solar nebula, where many planets are formed, is a kind of such accretion disk. The accretion disk is also formed in the late stage of stellar evolution in binary stars when the gas in a companion star fall into a primary star. Any kind of mass accretion to a star (either a black hole, a neutron star, a white dwarf, or a main sequence star) produces an accretion disk around the central object. Even a giant accretion disk is considered to be rotating around a supermassive black hole (with mass of $10^9 M_\odot$) in the nuclei of active galaxies. The accretion disk is one of the "stars" in modern astrophysics, showing vigorous activity such as bursts, flares, and jets.

When the self gravity of the disk is neglected, the rotation velocity of the disk is the *Keplerian speed*

$$V_k = \left(\frac{GM}{r}\right)^{1/2}\qquad(4.2.1)$$

or the angular velocity is

$$\Omega_k = \left(\frac{GM}{r^3}\right)^{1/2} \propto r^{-3/2}. \tag{4.2.2}$$

That is, the rotation in the disk is the differential rotation, which is a kind of *shear flow*. It is well known that the shear flow is unstable and produces turbulence when hydrodynamic Reynolds number is high. This instability is called the *Kelvin-Helmholtz instability* (K-H instability). People have long thought that the accretion disk is also very unstable for K-H instability and produces the turbulence which may be the source of anomalous viscosity needed for angular momentum transport in the accretion disk. Recently, however, an entirely different view has emerged; it is the magnetic field that makes accretion disks very unstable. This instability is called *magnetorotational instability* (or *Balbus-Hawley instability*).

Although the physics of the K-H instability is somewhat different from that of the magnetorotational instability, the formalism has many common points, and it is even instructive to learn what the K-H instability is and why the K-H instability does not work so well in accretion disks. For this reason, we will first discuss the K-H instability, in particular, its elementary physics, and then go to more fascinating subject, magnetorational instability and its role in producing anomalous magnetic viscosity in accretion disks.

4.2.1 Kelvin-Helmholtz Instability

Consider a shear flow with velocity V_1 in the upper half plane and V_2 in the lower half plane (Fig. 4.21). Since the kinetic energy associated with the velocity difference $(V_1 - V_2)$ is surplus to the system, this free kinetic energy can become the energy source of turbulence or ultimately the internal energy of plasmas. Naturally, we expect the instability would occur in a crossing time scale of the flow over the characteristic length scale.

Let us begin with the formal, but elementary treatment. The basic equation of the non-magnetized fluid is

$$\frac{\partial \rho}{pt} + \text{div}(\rho \mathbf{v}) = 0, \tag{4.2.3}$$

$$\rho \frac{d\mathbf{v}}{dt} = -\nabla p, \tag{4.2.4}$$

$$\frac{dp}{dt} = \frac{\gamma p}{\rho} \frac{d\rho}{dt}. \tag{4.2.5}$$

Here we assume general situation such that the unperturbed flow (in the x-direction) has the flow profile $\mathbf{V} = (U(y), 0, 0)$. Linearizing the quantities as $\rho = \rho_0 + \rho_1$ ($\rho_1 \ll \rho_0$), $\rho_1 \propto \exp(ikx + i\omega t)$, we have

$$\omega_d v_y = C_s^2 \left[\frac{-v'_y \omega_d + kU'v_y}{(\omega_d^2 - k^2 C_s^2)}\right]', \tag{4.2.6}$$

317

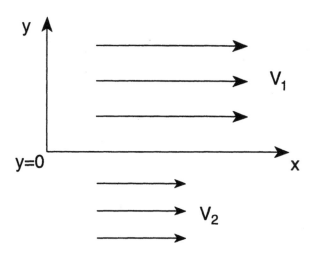

FIGURE 4.21 A Shear Flow.

where $' = d/dy$, v_y is the perturbed velocity in y-direction, and

$$\omega_d = \omega + kU.$$

In the incompressible plasmas $(C_s = \infty)$, the equation becomes

$$\omega_d v_y'' - (kU'' + k^2\omega_d)v_y = 0. \qquad (4.2.7)$$

This is the equation which we have to solve *non-locally* as an eigen value problem (see Sec. 4.2.1.2). However, if we consider the simplest model, the two streams separated by the discontinuity, we can solve the equation locally which will be treated in the next section to get the insight on the growth rate of the instability.

4.2.1.1 Growth Rate

The flow profile we assume is

$$\begin{aligned} U &= V_1 \quad \text{for} \quad y > 0, \\ U &= V_2 \quad \text{for} \quad y < 0, \end{aligned} \qquad (4.2.8)$$

where the flow is in x-direction. In each half space, the basic equation for eigen function v_y becomes

$$v_y'' - k^2 v_y = 0, \qquad (4.2.9)$$

so that we have a solution

$$\begin{aligned} v_y &\propto \exp(-ky) \text{ for } y > 0 \\ v_y &\propto \exp(ky) \quad \text{for } y < 0 \end{aligned} \qquad (4.2.10)$$

Since the displacement at the boundary between two streams must be continuous, we have the following condition,

$$\frac{d}{dt}\delta y = \left(\frac{\partial}{\partial t} + U\frac{\partial}{\partial x}\right)\delta y = v_y, \qquad (4.2.11)$$

leading to the condition;

$$v_y/(\omega + kU) \quad \text{must be continuous.} \qquad (4.2.12)$$

From this, our solution becomes

$$\begin{aligned} -v_y &= (\omega + kV_1)\exp(-ky) \text{ for } y > 0 \\ v_y &= (\omega + kV_2)\exp(ky) \quad \text{for } y < 0 \end{aligned} \qquad (4.2.13)$$

Here we have also the following boundary condition at the boundary between two streams (Chandrasekhar, 1961)

$$\Delta_s(-v_y\omega_d + kU'v_y) = 0. \qquad (4.2.14)$$

where

$$\Delta_s f = f_{y=+0} - f_{y=-0}. \qquad (4.2.15)$$

Applying this condition to the above solution, we finally get

$$(\omega + kV_1)^2 + (\omega + kV_2)^2 = 0. \qquad (4.2.16)$$

Consequently, we have the dispersion relation

$$\omega = -k\frac{V_1 + V_2}{2} + ik\frac{V_1 - V_2}{2}. \qquad (4.2.17)$$

The imaginary part of ω is the growth rate of the K-H instability, and we find that the time scale of the instability is the flow crossing time over one wavelength, as expected from the elementary consideration. [2]

---

[2] The K-H instability becomes important in various astrophysical and space plasma situations, such as astrophysical jets and magnetopause boundary layer. As for the former subject, the reader should refer to e.g., Blandford and Pringle (1976), Ferrari et al. (1981), and as for the latter subject, see e.g., Miura (1992).

4.2.1.2 Richardson Number

The Kelvin-Helmholtz instability is often discussed for the stratified gas layer, i.e., in the situation such that the fast flow is streaming horizontally above the slow dense gas layer stratified in a gravitational field. A good example is the wind above the water surface. In this case, the gravitational force acts as a restoring force. Hence only when the free kinetic energy of the flow exceeds the gravitational energy, the instability can occur.

Let us examine this problem in more detail. We assume the fast flow with velocity $U + \delta U$ and density ρ is streaming above (by δz) the slow dense gas layer with $\rho + \delta \rho$ and U. Then the excess kinetic energy is

$$\delta W_{\text{kin}} = \frac{1}{4} \rho (\delta U)^2 \tag{4.2.18}$$

while the work that must be done to effect the interchange of the slow dense flow and the fast light flow against the gravity is

$$\delta W_{\text{grav}} = -g\delta\rho\delta z. \tag{4.2.19}$$

Hence the condition for stability is that the excess kinetic energy must be smaller than the gravitational energy, i.e.,

$$\frac{1}{4} \rho (\delta U)^2 < -g\delta\rho\delta z. \tag{4.2.20}$$

This is rewritten as

$$\left(\frac{dU}{dz}\right)^2 < -4 \frac{g}{\rho} \frac{d\rho}{dz} \tag{4.2.21}$$

$$J = -\frac{g}{\rho} \frac{d\rho/dz}{(dU/dz)^2} > \frac{1}{4} \tag{4.2.22}$$

for stability. J is called the *Richardson number*.

4.2.1.3 Local vs. Nonlocal Analysis

In section 4.2.1.1, we solved the eigenvalue equation using the local approximation. This was possible because the problem treated was a very simple one, i.e., the two streams separated by the discontinuity. In general, the shear flow problem is not so simple, and must be solved *non-locally*. That is, *we have to solve the boundary value problem assuming proper boundary condition*. This is not only applied to the usual shear flow problem, but also to the differentially rotating flows such as accretion disks (and galactic disks). One reason of this complication results from the mathematical complexity of the fundamental eigenvalue equation (Eqs. 4.2.6 and 4.2.7) which has (apparent) singularity (or *corotation point*) at $\omega + kU = 0$. This sigularity is coupled with the term U'' (see Eq. 4.2.7),

$$v_y'' - \left(\frac{kU''}{(\omega + kU)} + k^2\right) v_y = 0. \tag{4.2.23}$$

Hence if $U'' = 0$, we may have different behavior. In fact, Dyson (1960) and Case (1960) showed that *if the flow profile is $U = U_0 y/d$ in a semi-infinite space, the flow becomes completely stable even if the Rechardson criterion (for stability) is violated* (Chandrasekhar 1961). On the other hand, the flow profile with tanh form is shown to be unstable (Drazin, 1958). Hence the condition $U'' \neq 0$ is essential for this problem, and this is equivalent to the existence of the boundary at some place which assures the problem is *non-local*.

Similar characteristic is also known in the shear flow instability in accretion disks, i.e., so called *Papaloizou-Pringle instability* (Papaloizou and Pringle, 1984). In order to have an instability, the existence of the boundary or the reflection of sound waves at some radius is essential.

Generally speaking, the eigen value equation for ideal (dissipationless) MHD problem without shear flow can be written in the *self-adjoint form*. In such case, eigenvalue is real and hence we have either pure oscillation or pure instability. However, in the problem including shear flow, the eigen value equation cannot be written in the self-adjoint form; eigen value is complex in spite of the fact that there is no dissipation term in the equation. This point is similar to the Landau damping in a sense that there is no dissipation in the collisionless wave-particle system but nevertheless we have (Landau) damping of a wave.

Here it should be remembered that there are two types of instability in this kind of shear flow problem. One is the *absolute instability* and the other is *convective instability* (Fig. 4.22).

The former occurs in a fixed position, while the other occurs on a moving frame. If we observe the convective instabilty in a fixed frame, we will not observe the exponential growth of the quantities. Only when we move on the frame propagating with some wave mode, the exponential growth is obseved. The well known example of the convective instability is the spiral mode of the gravitational instability in a rotating disk (Goldreich and Lynden-Bell, 1965b). This type of instability is analyzed using time dependent approach. The example of the absolute instability is the Papaloizou-Pringle instability. This can be analyzed by the normal mode method.

The non-local normal mode analysis of magnetorotational instability will be discussed in section 4.2.3.1.

4.2.1.4   Rotating Flow: Rayleigh's Criterion

We shall briefly discuss the stability of non-magnetized rotating flow, and heuristically derive the *Rayleigh's Criterion*. According to Chandrasekhar (1961, p. 273), Rayleigh stated "in the absense of viscosity, the necessary and sufficient condition for a distribution of angular velocity $\Omega(r)$ to be stable is

$$\frac{d}{dr}\left(r^2\Omega\right)^2 > 0 \qquad (4.2.24)$$

everywhere in the internal; and further, that the distribution is unstable if $(r^2\Omega)^2$ should decrease anywhere inside the internal."

Suppose the rotating fluid element with angular momentum L_1 at r_1 and the other fluid

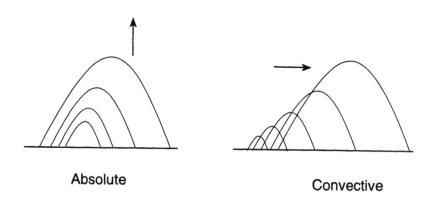

element with L_2 at r_2. If we exchange two elements, the change in the kinetic energy is

$$\delta W = \left(\frac{L_2^2}{r_1} + \frac{L_1^2}{r_2}\right) - \left(\frac{L_1^2}{r_1} + \frac{L_2^2}{r_2}\right) = \left(\frac{1}{r_1^2} - \frac{1}{r_2^2}\right)(L_2^2 - L_1^2). \qquad (4.2.25)$$

If $L_1 > L_2$ for $r_2 > r_1$, we find $\delta W < 0$, i.e., instability begins. In other words, if angular momentum increases outward, the stability is assured in a rotating fluid. This (stability) criterion is formally written as

$$\frac{dL}{dr} > 0 \qquad (4.2.26)$$

or

$$\frac{d}{dr}(r^2\Omega) > 0. \qquad (4.2.27)$$

In the case of accretion disks, the angular momentum distribution is

$$L \propto r^{1/2}. \qquad (4.2.28)$$

322

Hence from this point of view, the accretion disk is stable, unless there exists non-axisymmetric unstable Papaloizou-Pringle mode due to the presence of reflection boundary. (Recently, Balbus *et al.* (1996) confirmed that Rayleigh criterion is quite generally applicable not only for linear regime but also for nonlinear regime, by performing nonlinear simulations.)

This situation, however, becomes entirely different if there is magnetic field, even if it is very weak. In this case, the powerful instability occurs. This instability, which is now called *magnetorotational instability* or Balbus-Hawley (1991) instability, has already been noted by Velikhov (1959) and Chandrasekhar (1961). According to Chandrasekhar (1961, p. 389), *in the limit of zero magnetic field, a sufficient condition for stability is that the angular speed, Ω, is a monotonic increasing function of r. .... It is remarkable that we do not recover Rayleigh's criterion in the limit of zero magnetic field.* In the next section, we will study this remarkable instability in more detial.

4.2.2 Magnetorotational Instabililty

4.2.2.1 Motivation: Need for Anomalous Viscosity in Accretion Disks

When a plasma with angular momentum accretes to a gravitating object, it forms a differentially rotating disk called accretion disks. See Chap. 1 and Figs. 1.5, 1.8, and 1.9. In fast time scale matter rotates forming a disk. In a slower (longer) time scale this matter accretes toward the center by losing some of its angular momentum. It is this physics which has been studied extensively in recent years. In the standard theory of accretion disks (Shakura and Sunyaev, 1973), the α-prescription of viscosity is adopted, in which the $r\varphi$-component of stress tensor is assumed to be proportional to pressure as $\tau_{r\varphi} = \alpha P$. The pressure may include the magnetic pressure and radiation pressure. The $\tau_{r\varphi}$ component contributes to the radial angular momentum transport. Thus the phenomenological viscosity parameter α determines the time scale of the evolution of accretion disk.

In dwarf novae, by fitting the theoretical and observed durations of the quiescent phase and the bursting phase, it is suggested that α is of the order of 0.1 during the bursting phase; on the other hand, α is about 0.02 in the quiescent phase (Cannizo *et al.*, 1988). Since the molecular viscosity cannot provide such a high rate of angular momentum transport (by many orders of magnitude too small), various models of anomalous viscosity have been pushed forward; the convective turbulence (Lin and Papaloizou, 1980; Kley *et al.*, 1993), the global hydrodynamic shear flow instability (e.g. Papaloizou and Pringle, 1984; Drury, 1985; Goldreich *et al.*, 1986; Kato, 1987; Glatzel, 1987), and the hydromagnetic turbulence (e.g. Eardley and Lightman, 1975; Ichimaru, 1977; Pudritz, 1981; Kato, 1984; Kato and Horiuchi, 1985, 1986). Although the hydrodynamic shear flow instability has been studied extensively as a possible mechanism of generating turbulence in accretion disks, its contribution to α was shown to be only $O(10^{-3})$ in geometrically thin Keplerian disks (Kaisig, 1989) even if the effects of compressibility are taken into account.

In hydromagnetic models of anomalous viscosity, the angular momentum is transported inside the disk by the magnetic stress incurred by fluctuating magnetic fields. Balbus and Hawley (1991) pointed out the importance of the local MHD instability on the generation

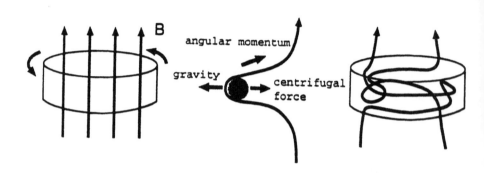

FIGURE 4.23 Mechanism of magneto-rotational instability

of fluctuating magnetic fields. Historically, this instability was found by Velikhov (1959) and described in the textbook of Chandrasekhar (1961). However, until Balbus and Hawley pointed out this mechanism, various candidate instabilities fell short of explaining the needed anomalous viscosity for the angular momentum transport in the disk. The mechanism of this instability is as follows (see Fig. 4.22):

When a differentially rotating cylindrical plasma is threaded by axial magnetic field, if we perturb a local part of the fluid inward, angular momentum is extracted from the fluid element because the perturbed part rotates faster than the outer part. Since the centrifugal force reduces due to the angular momentum loss, perturbation grows if the resulting radial force exceeds the restoring magnetic tension force. This mechanism may be more easily understood by mechanical analogies with two particles connected by a spring orbiting around a gravitating object (Balbus and Hawley, 1992). Since the particle in the inner orbit loses angular momentum, it falls down so long as the radial force (gravitational force–centrifugal force) exceeds the tension force. This is the mechanism of the *magneto-rotational instability* or the *Balbus-Hawley instability*.

324

4.2.2.2 Critical Wavelength and Growth Rate

Here we shall derive the condition for the occurrence of the magnetorotational (Balbus-Hawley) instability and its growth rate using simple physical argument.

Suppose a fluid element, which is attached to the veritcal magentic field lines in the accretion disk, is displaced inward by $\Delta r (> 0)$ (Fig. 4.23). Here we assume that the fluid element rotates at the same angular speed (Ω) as the original one (rigid rotation) because of angular momentum transfer due to magnetic field. Then

$$\text{centrifugal force} = (r - \Delta r)\rho\Omega^2 \tag{4.2.29}$$

$$\text{gravitational force} = \frac{\rho GM}{r^2}\left(1 + \frac{2\Delta r}{r}\right) \tag{4.2.30}$$

$$\text{magnetic tension force} = \frac{B^2}{4\pi}\frac{\Delta r}{(\lambda/2\pi)^2}. \tag{4.2.31}$$

Noting that $\Omega^2 = GM/r^3$, we find that the net force in the radial direction is

$$F = \left(\frac{B^2}{4\pi(\lambda/2\pi)^2} - \rho\Omega^2\right)\Delta r. \tag{4.2.32}$$

If this force is negative, the displaced fluid element moves further to the inward radial direction; i.e., the instability sets in. Thus the condition for the instability is

$$-\rho\Omega^2 + \frac{B^2}{4\pi(\lambda/2\pi)^2} < 0 \tag{4.2.33}$$

or in other words,

$$\lambda/2\pi > V_A/\Omega. \tag{4.2.34}$$

From this, we can understand that this instabilty occur for the long wavelength, and there is a critical wavelength, $\lambda_c = 2\pi V_A/\Omega$. These features are similar to those of the Parker instability (Sec. 3.2.2). The reason for this similarity comes from that both instabilities are stabilized by the magnetic tension force.

Next, we shall calculate the approximate growth rate of this instability. The equation of motion of the displaced fluid element is

$$\rho\frac{d^2\Delta r}{dt^2} = F = \left(-\frac{B^2}{4\pi(\lambda/2\pi)^2} + \rho\Omega^2\right)\Delta r. \tag{4.2.35}$$

If we replace d^2/dt^2 with $-\omega^2$ (ω is the growth rate) and write $k = 2\pi/\lambda$, then we find

$$\omega^2 = \Omega^2 - k^2 V_A^2. \tag{4.2.36}$$

In the limit of the long wavelength $k \to 0$, the growth rate becomes Ω, i.e., this instability is fairly fast instability occurring at the rotation time scale of the disk. Also important is that this instability occurs even if the magnetic field is very weak (V_A or $B \to 0$). This is a very important point because people often neglected the effect of the magnetic field in accretion disk simply because they believe magnetic field is very weak in the disk. However, above discussion shows that at least the linear instability can occur even if magnetic field is very weak. If this instability grows to the nonlinear level which affect the dynamics of accretion disk, then we can say that we cannot neglect magnetic fields any more. In fact, this instability transfers angular momentum in the radial direction, so that it can give a mechanism to create the source of the anomalous viscosity in accretion disk as we will see in more detail later (Sec. 4.2.3).

On the other hand, if magnetic field is strong, this instability is stabilized. The condition of such stability is that the critical wavelength λ_c exceeds the thickness of the disk H. In this case, the critical field strength for stability is

$$B > B_c \sim H\Omega(4\pi\rho)^{1/2}/2\pi \tag{4.2.37}$$

or

$$\beta = p_{\text{gas}}/p_{\text{mag}} < \beta_c \sim (kH)^2 \sim 10. \tag{4.2.38}$$

Thus, this instability can occur only in high beta disks when the magnetic field is vertical to the disk.

4.2.2.3 Local Linear Stability Analysis for Vertical Magnetic Field

In order to see more detailed (and exact) behavior of the magnetorotational instability, we follow the pioneering paper by Balbus and Hawley (1991) for the linear stability analysis and Hawley and Balbus (1991) for nonlinear evolution.

We consider a case with no vertical gravity and assume that the unperturbed disk is uniform in the vertical direction and is threaded by uniform vertical magnetic field. Since the compressibility is not important for the instability, we assume an incompressible fluid for simplicity. The basic MHD equations in the frame rotating with angular velocity Ω are

$$\frac{\partial \mathbf{v}}{\partial t} + (\mathbf{v} \cdot \nabla)\mathbf{v} = -\frac{1}{\rho}\nabla p + \frac{1}{4\pi\rho}(\nabla \times \mathbf{B}) \times \mathbf{B} + g$$

$$+ 2\mathbf{v} \times \mathbf{\Omega} + (\mathbf{\Omega} \times \mathbf{r}) \times \mathbf{\Omega}, \tag{4.2.39a}$$

$$\frac{\partial \mathbf{B}}{\partial t} = \nabla \times (\mathbf{v} \times \mathbf{B}) + \eta\nabla^2\mathbf{B}, \tag{4.2.39b}$$

and

$$\nabla \cdot \mathbf{v} = 0,$$

where g is the gravitational acceleration and $\mathbf{r}$ is the position vector. Other symbols have their usual meanings.

We use the local Cartesian coordinate (x, y, z) in the rotating frame where x-axis is in the radial direction, the y-axis in the azimuthal direction, and the z-axis is parallel to

Ω. We consider axisymmetric Eulerian linear perturbations with space-time dependence $\exp(ik_r r + ik_z z - i\omega t)$. After some manipulations, we get the following dispersion relation,

$$\frac{k^2}{k_z^2}\tilde{\omega}'^4 - \kappa\omega'^2 - 4\Omega^2 k_z^2 v_A^2 = 0, \tag{4.2.40}$$

where $k^2 = k_r^2 + k_z^2$, $\omega'^2 = \omega^2 - k_z^2 v_A^2$ and

$$\kappa = \frac{2\Omega}{r}\frac{d(r^2\Omega)}{dr}, \tag{4.2.41}$$

is the epicyclic frequency. From this, we find that the necessary condition for the instability is

$$k^2 v_A^2 + \frac{d\Omega^2}{d\ln r} < 0, \tag{4.2.42}$$

and hence

$$\frac{d\Omega^2}{dr} < 0. \tag{4.2.43}$$

We also find that there is a critical wavenumber k_c and the instability occurs for

$$k < k_c = \left|\frac{d\Omega^2}{d\ln r}\right|/v_A = 3^{1/2}\Omega/v_A, \tag{4.2.44}$$

for Keplerian disk. Figure 4.24 shows the growth rate as a function of k_z ($q_z = k_z v_A/\Omega$) for different k_r.

Note the stabilizing effect by nonzero k_r. It is also important to note that the field strength enters the equation always with wavenumber, so that the growth rate is independent of the field strength by choosing the wavenumber appropriately.

4.2.2.4 Nonlinear Evolution for Vertical Magnetic Field*

Hawley and Balbus (1991) then studied the nonlinear evolution of the instability, using 2.5D MHD numerical simulations for a local part of an accretion disk penetrated by vertical magnetic field, and confirmed the linear analysis results by Balbus and Hawley (1991). Especially they confirmed that the maximum growth rate is independent of magnetic field strength. Figure 4.25 shows typical results of the nonlinear evolution of the instability, and reveals that the nonlinear evolution results in the growth of structure on large scales even for very weak fields ($\beta = 1000 - 16000$). The most important dynamic effect is the redistribution of angular momentum. Subsequent simulations by Hawley et al. (1995), Brandenburg et al. (1995), Matsumoto and Tajima (1995) all confirm that the effective viscosity parameter becomes of the order of 0.01–0.1 by this process (Sec. 4.2.3). It is also found that magnetic reconnection is important to determine the saturation level.

Although the above discussion was restricted to the case of vertical magnetic fields, this magnetorotational instability can occur also for pure toroidal magnetic fields. The instability

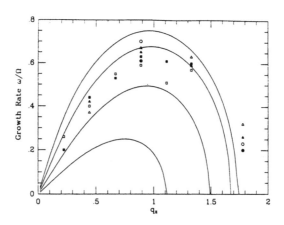

FIGURE 4.24 Growth rate of the magnetorational instability as a function of $q_z = k_z v_A / \Omega$. Curves represent values from linear theory for $q_r = k_r v_A / \Omega = 0$, 0.44, 0.89, and 1.33 (from Hawley and Bablus, 1991).

is not sensitive to disk boundary condition. The dissipation such as resistivity sets the lower limit for the magnetic field strength for the instability (Matsumoto and Tajima, 1985, Sec. 4.2.3.3), though this lower limit is very small for usual condition in accretion disk (except for proto-solar nebula) if the resistivity is due to coulomb collision.

Finally, it is also to be noted that in addition to the magnetic viscosity, the 3D nonlinear simulations of the magnetorotational instability have revealed that the weak-field magnetorotational instability consititutes a hydromagnetic dynamo (Brandenburg *et al.*, 1995; Hawley *et al.*, 1995).

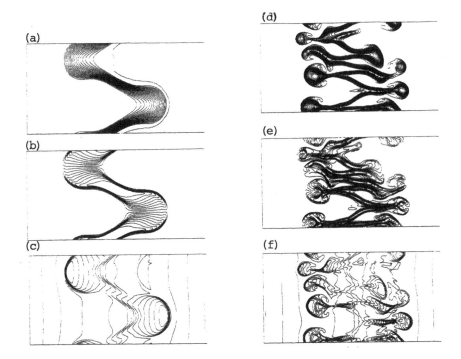

FIGURE 4.25 Typical nonlinear evolution of the magnetorotational instability (from Hawley and Balbus, 1991). Contour plots of (a) the initial poloidal magnetic field lines, and (b) the poloidal magnetic field lines, (c) the toroidal field at 3.3 orbits in the case of $\beta = 1000$. (d)-(f) are for the case of $\beta = 16000$ and (d) poloidal field lines, (e) toroidal field, and (f) angular momentum at 3.2 orbit.

4.2.3 Magnetic Viscosity in Accretion Disks

4.2.3.1 Nonlocal Linear Stability Analysis for General Magnetic Shearing Field

As we discussed in Sec. 4.2.2, differentially rotating disks are subject to the axisymmetric instability for perfectly conducting plasma in the presence of poloidal magnetic fields (Balbus and Hawley, 1991). For nonaxisymmetric perturbations, we find localized unstable eigenmodes whose eigenfunction is confined between two Alfvén singularities at $\omega_d = \pm\omega_A$ where ω_d is the Doppler-shifted wave frequency, and $\omega_A = k_\parallel v_A$ is the Alfvén frequency Here, $k_\parallel$ is the wavenumber parallel to magnetic fields. The radial width of the unstable eigenfunction is $\Delta x \sim \omega_A/(Ak_y)$, where A is the *Oort's constant*, and k_y is the azimutha wave number. The growth rate of the fundamental mode is larger for smaller value of k_y/k_z The maximum growth rate when $k_y/k_z \sim 0.1$ is $\sim 0.2\Omega$ for the Keplerian disk with loca

329

angular velocity Ω. It is found that the purely growing mode disappears when $k_y/k_z > 0.12$. In a perfectly conducting disk, the instability grows even when the seed magnetic field is infinitesimal. Inclusion of the resistivity, however, leads to the appearance of an instability threshold. When the resistivity η depends on the instability-induced turbulent magnetic fields $\delta\mathbf{B}$ as $\eta(\langle\delta B^2\rangle)$, the marginal stability condition self-consistently determines the α parameter of the angular momentum transport due to the magnetic stress. For fully ionized disks, the magnetic viscosity parameter α_B is between 0.001 and 1.

A theoretical framework for modes in the presence of shear flows is developed (Matsumoto and Tajima, 1995). We show that nonaxisymmetric unstable eigenmodes exist in differentially rotating magnetized disks. The origin of these eigenmodes is the trapping of Alfvén waves between two Alfvén singularities where the Doppler shifted wave frequency equals to the Alfvén frequency. These modes are distinct from those by Balbus and Hawley (1992), which were global noneigenmodes, as mentioned. The linear theory finds that the instability now has the threshold with respect to the magnetic field strength in the presence of a small but finite resistivity. On the other hand the presence of finite viscosity does not change the qualitative behavior of the instability.

Theoretical approaches to the mode growth in a system with a velocity shear depend on the ratio of the shear scale-length L_s and the relevant wavelength L_n. For the global spiral pattern in galaxies whose wavelength is much longer than the shear scale length ($s = L_n/L_s \gg 1$), the *WKB method* is not appropriate. When $s = L_n/L_s \ll 1$, the WKB method may be used to determine the response to a localized fixed-frequency excitation of a shearing plasma (e.g. Tajima *et al.*, 1991; Waelbroeck *et al.*, 1994). It should be emphasized that unstable eigenmodes can exist even if the velocity shear is significant. One such example is the drift acoustic wave instabilities in the presence of sheared flows where exact eigenfunctions have been obtained (Waelbroeck *et al.*, 1992). Another example is the nonaxisymmetric hydrodynamical *shear flow instability* in accretion disks (e.g. Papaloizou and Pringle, 1984; Drury, 1985). In nonmagnetized, nonself-gravitating differentially rotating disks, spontaneous growth of nonaxisymmetric modes occurs by the *overreflection* of waves at the *corotation resonance* where $\omega/k_y = v_y$ (Drury, 1985; Goldreich *et al.*, 1986; Kato, 1987). When the disk has a reflecting edge on the same side where overreflection occurs, the disk becomes unstable. By imposing the appropriate boundary conditions, several authors have obtained standing, growing modes as solutions of the eigenvalue problem (e.g. Goldreich *et al.*, 1986; Kato, 1987; Hanawa, 1987; Glatzel, 1987). When unstable eigenmodes exist, they will eventually dominate over transiently amplified waves.

However, we are interested in localized instabilities and their subsequent turbulence that are not influenced by the disk edge boundaries (Matsumoto and Tajima, 1995). These determine the local property of plasma transport such as the angular momentum transport, i.e, the problem of the viscosity of differentially rotating disks. Thus we adopt the *eigenmode analysis formulation* suitable for localized instabilities. Note that in the presence of shear flow localized modes show strong wave absorption characteristics and deviate sharply from sinusoidal behaviors and that thus a simpler and more naive local Fourier analysis is not adequate or at least only approximate. On the other hand, such problems as the global pattern of spiral arm magnetic fields need the *temporal domain formulation* mentioned earlier.

The stability of rotating disks can be studied by linearizing the basic equations around the equilibrium state and looking for solutions of the form $\phi(x, t)\exp[i(k_y y + k_z z)]$. The Laplace transform of the perturbation, $\bar{\phi}(x, \omega)$, is employed

$$\bar{\phi}(x, \omega) = \int_0^\infty dt \tilde{\phi} e^{i\omega t} \qquad (4.2.45)$$

for $\mathrm{Im}(\omega) > 0$. Let the basic equation take the form

$$\frac{d^2\phi}{dx^2} + L(x, \omega)\bar{\phi} = \Gamma(x, \omega). \qquad (4.2.46)$$

The initial condition enters through the function $\Gamma(x, \omega)$. This equation may be solved in terms of Green's function,

$$G(x, \hat{x}, \omega) = [\bar{\phi}_+(x, \omega)\bar{\phi}_-(\hat{x}, \omega)H(x - \hat{x}) + \bar{\phi}_+(\hat{x}, \omega)\bar{\phi}_-(x, \omega)H(\hat{x} - x)]/D(\omega). \qquad (4.2.47)$$

Here $\bar{\phi}_+$ and $\bar{\phi}_-$ are the solutions of the homogeneous equation vanishing at $+\infty$ and $-\infty$ respectively, and $H(x)$ is the Heaviside step function. These $\bar{\phi}_+$ and $\bar{\phi}_-$ are the standard eigenfunctions in the standard eigenmode analysis (Waelbroeck *et al.*, 1994). $D(\omega)$ is the *Wronskian* of these solutions, $D(\omega) = \bar{\phi}'_+(x, \omega)\bar{\phi}_-(x, \omega) - \bar{\phi}_+(x, \omega)\bar{\phi}'_-(x, \omega)$. Having introduced Green's function in Eq. (4.2.47), we can show that this analysis can be useful even for the cases where no eigenmode exists. The solution is

$$\bar{\phi}(x, \omega) = \int_{-\infty}^\infty d\hat{x} G(x, \hat{x}, \omega)\Gamma(\hat{x}, \omega),$$

and the response as a function of time t is then

$$\hat{\phi}(x, t) = \frac{1}{2\pi} \int_{-\infty + ic}^{\infty + ic} d\omega e^{-i\omega t}\bar{\phi}(x, \omega), \qquad (4.2.48)$$

where the Laplace inversion integral extends over the contour lying above all the singularities in the integrand. Equation (4.2.48) can have temporally growing solutions even if there is no eigenvalue and eigenfunction to Eq. (4.2.46). In fact the *global spiral modes* of Goldreich and Lynden-Bell are such a case. Still this formal solution of the initial value problem through $\bar{\phi}_+$ and $\bar{\phi}_-$ is equivalent to the temporal formulation by Goldreich and Lynden-Bell (1965) if we assume the functional form $\bar{\phi}(x, t) = \hat{\phi}(t)\exp[i(k_x(0) + 2Ak_y t)x]$.

Unstable eigenmodes occur when the Wronskian has a zero in the upper complex ω plane. It means that the solution vanishing as $\xi \to -\infty$ is proportional to that vanishing as $\xi \to \infty$, or that there exists a globally well-behaved solution. The latter constitutes the standard eigenmode problem. In the below we consider the localized eigenmode problem, as discussed here and will find exponentially growing localized eigenfunctions. Such modes are distinct from the temporal domain modes discussed by Balbus and Hawley (1992). On the other hand Balbus and Hawley (1991) did not treat radial eigenfunction structures.

4.2.3.2 Alfvén Singularities and Eigenmodes*

In this subsection we derive the wave equation in differentially rotating magnetized disks and solve the eigenvalue problem. In the unperturbed state, the density, pressure, and magnetic fields are assumed to be uniform. By assuming that $v_x = v_z = 0$ in the unperturbed state, the unperturbed momentum equation is

$$\mathbf{g} + 2\mathbf{v}_0 \times \mathbf{\Omega} + (\mathbf{\Omega} \times \mathbf{r}) \times \mathbf{\Omega} = 0. \tag{4.2.49}$$

We further assume that $B_z = 0$ in the unperturbed state. We linearize the basic equations and look for eigenmode solutions of the form $\tilde{\phi}(x,t)\exp[i(k_y y + k_z z)]$ according to the above discussion. The Laplace transform of the momentum equation and the induction equation are

$$-i\omega_d \overline{\mathbf{v}} - 2A\overline{v}_x \hat{y} - 2\overline{\mathbf{v}} \times \mathbf{\Omega} - \frac{(\mathbf{B}\nabla)\overline{\mathbf{b}}}{4\pi\rho} + \nabla\left(\frac{\mathbf{B}\overline{\mathbf{b}}}{4\pi\rho} + \frac{\delta\overline{p}}{\rho}\right) - \nu\nabla^2\overline{\mathbf{v}} = \tilde{\mathbf{v}}(x,0), \tag{4.2.50}$$

and

$$-i\omega_d \overline{\mathbf{b}} + 2A\overline{b}_x \hat{y} - (\mathbf{B}\nabla)\overline{\mathbf{v}} - \eta\nabla^2\overline{\mathbf{b}} = \tilde{\mathbf{b}}(x,0), \tag{4.2.51}$$

where

$$\omega_d = \omega + 2Ak_y x \tag{4.2.52}$$

is the Doppler shifted frequency, $\overline{\mathbf{b}}$, $\overline{\mathbf{v}}$, and $\delta\overline{p}$ are Laplace transform of the magnetic field, velocity, and the pressure perturbations, respectively.

Substituting these results into the Laplace transform of the continuity equation $\nabla \cdot \mathbf{v} = 0$ yields the initial value equation

$$\omega_\eta^2 \omega_{\eta\nu}^4 \frac{d^2\overline{v}_x}{dx^2} + 4A\omega_A^2 k_y \omega_\eta \omega_{\eta\nu}^2 \frac{d\overline{v}_x}{dx}$$
$$+ \left[-(k_y^2 + k_z^2)\omega_\eta^2\omega_{\eta\nu}^4 - 8A^2 k_y^2 \omega_A^2 \omega_{\eta\nu}^2 + \kappa^2 k_z^2 \omega_\eta^2(\omega_\eta^2 + A\omega_A^4/\Omega_B)\right]\overline{v}_x = \Gamma(x,\omega) \tag{4.2.53}$$

where

$$\omega_\eta = \omega_d + i\eta(k_y^2 + k_z^2 - \frac{d^2}{dx^2}), \tag{4.2.54}$$

$$\omega_\nu = \omega_d + i\nu(k_y^2 + k_z^2 - \frac{d^2}{dx^2}), \tag{4.2.55}$$

and

$$\omega_{\eta\nu}^2 = \omega_\eta\omega_\nu - \omega_A^2. \tag{4.2.56}$$

Here $\Omega_B = \Omega - A$, $\kappa = 2(\Omega\Omega_B)^{1/2}$ is the *epicyclic frequency*, and ω_A is the Alfvén frequency

$$\omega_A^2 = \frac{(\mathbf{k} \cdot \mathbf{B})^2}{4\pi\rho} = k_\parallel^2 v_A^2. \tag{4.2.57}$$

332

The initial condition enters through the right-hand side of Eq. (4.2.53).

First, we consider the case when there is no dissipation ($\eta = \nu = 0$). The homogeneous part of Eq. (4.2.53) reduces to the second order differential equation as

$$\frac{d^2 \bar{v}_x}{dx^2} + \frac{4A\omega_A^2 k_y}{\omega_d(\omega_d^2 - \omega_A^2)} \frac{d\bar{v}_x}{dx} + \left[-(k_y^2 + k_z^2) - \frac{8A^2 k_y^2 \omega_A^2}{\omega_d^2(\omega_d^2 - \omega_A^2)} + \kappa^2 k_z^2 \frac{\omega_d^2 + A\omega_A^2/\Omega_B}{(\omega_d^2 - \omega_A^2)^2} \right] \bar{v}_x = 0.$$

(4.2.58)

It is noted that in the presence of shear flows the eigenmode equation such as Eq. (4.2.58) does not take the self-adjoint form anymore. This is in contrast to the standard MHD problem without shear flows, which has the *self-adjoint form* (Bernstein et al., 1958). Note that although this can be formally changed to self-adjoint form (4.2.58) must not be converted into the self-adjoint form since it includes singular term (Arfken and Weber, 1995, p. 539). Thus the eigenvalues ω are not guaranteed to be real or pure imaginary. We express Eq. (4.2.58) in terms of

$$\xi = \frac{2Ak_y x}{\omega_A}$$

(4.2.59)

as

$$\frac{d^2 \bar{v}_x}{d\xi^2} + \frac{2\omega_A^3}{\omega_d(\omega_d^2 - \omega_A^2)} \frac{d\bar{v}_x}{d\xi}$$
$$+ \left[-\left(1 + \frac{1}{q}\right)\left(\frac{\omega_A}{2A}\right)^2 - \frac{2\omega_A^4}{\omega_d^2(\omega_d^2 - \omega_A^2)} + \left(\frac{\kappa}{2A}\right)^2 \left(\frac{\omega_A^2}{q}\right) \frac{\omega_d^2 + A\omega_A^2/\Omega_B}{(\omega_d^2 - \omega_A^2)^2} \right] \bar{v}_x = 0, \quad (4.2.60)$$

where

$$q = \frac{k_y^2}{k_z^2}.$$

(4.2.61)

This differential Eq. (4.2.60) has two singularities at $\omega_d = \pm\omega_A$. These are the shear *Alfvén singularities* where the absorption and mode conversion of Alfvén waves take place (e.g. Ross et al., 1982). The locations of Alfvén singularities in the complex plane are

$$\xi_A = \pm 1 - \frac{\omega}{\omega_A}.$$

(4.2.62)

By applying the *Frobenius method* around $\omega_d = \pm\omega_A$ (or $\xi = \xi_A$), and solving the *indicial equation*, we find that the exponent s in the series expansion

$$\bar{v}_x(\xi) = \sum_{n=0}^{\infty} a_n(\xi - \xi_A)^{n+s}$$

becomes complex. Thus the solutions which pass through these points are singular (called *regular singular points*). The corotation point $\omega_d = 0$ (or $\xi_c = -\omega/\omega_A$) appears to be also singular in Eq. (4.2.60). However, by solving the indicial equation, we find that the solution is regular at the corotation point.

The solutions of Eq. (4.2.60) which vanish as $\xi \to \pm\infty$ have an asymptotic form

$$\bar{v}_x \propto \exp[\mp(1 + 1/q)^{1/2} \omega_A \xi/(2A)].$$

(4.2.63)

333

We now numerically look for eigenmodes whose eigenfunction connects these asymptotic solutions. At the numerical boundaries at $\xi = \pm 5$, we impose a boundary condition $\bar{v}'_x/\bar{v}_x = k_{\mp}$, where $k_{\mp}$ are negative and positive solutions of the quadratic equation given by inserting the functional form $\bar{v}_x \propto \exp(k_{\pm}x)$ into Eq. (4.2.60). The eigenvalues and eigenfunctions which satisfy these boundary conditions are obtained through the shooting method.

Figure 4.26 shows examples of eigenfunctions $\bar{v}_x$ for the Keplerian disk ($A/\Omega = 3/4$) when $\omega_A = 0.1\Omega$ and $q = 0.01$. The solid curve and the dashed curve represent the real part and the imaginary part of the eigenfunction, respectively. Figure 4.26a is for the fundamental mode, and Figure 4.26b is for the next nodal mode. Since the eigenvalues are pure imaginary ($\omega = 0.069\Omega i$ for the fundamental mode, and $\omega = 0.036\Omega i$ for the next nodal mode) in these cases, the eigenfunctions are symmetric (or antisymmetric) with respect to $\xi = 0$. The eigenfunctions are confined between two Alfvén singularities located at $\xi = \pm 1$ where the Alfvén waves vanish. The standing growing mode appears between these singularities (Matsumoto and Tajima, 1995). The distance between these resonance points is $\Delta x \sim \omega_A/(Ak_y)$. When the unperturbed field is purely toroidal, we find $\Delta x \sim v_A/A \sim (v_A/C_s)(\Omega/A)H$, where H is the thickness of the disk and C_s is the sound speed. Thus when $v_A \ll C_s$, the mode is localized in the radial direction with the mode width small compared to the disk thickness. When the unperturbed magnetic fields have poloidal components, Δx is proportional to $k_{\parallel}/k_y$. In such a case, the standing wave can have a large radial extent for nearly axisymmetric perturbations ($k_y \ll k_{\parallel}$). The purely toroidal field is a special case in which Δx is independent of $q = k_y^2/k_z^2$.

The maximum growth occurs when $\omega_A \sim \Omega$. When $q = 0.01$, the maximum growth rate of the fundamental mode is $\sim 0.2\Omega$. This value is less than a half the maximum growth rate of the axisymmetric magneto-rotational instability. The reduction of the maximum growth rate for nonaxisymmetric perturbations is interpreted to arise from the radial confinement of the mode. The growth rate of the modes decreases with increasing q. The purely growing mode disappears when $q > 0.015$. Although unstable modes with complex eigenvalues still survive, their significance will be less than the purely growing mode partly because their growth rate is lower. Thus we expect that unstable eigenmodes whose azimuthal wavelength is much longer than the vertical wavelength will become dominant.

In the regime where $\omega_A \ll \Omega$, the growth rate of the fundamental mode is roughly proportional to ω_A and we write the growth rate as

$$\gamma^2 = f(q)\omega_A^2. \tag{4.2.64}$$

The proportionality coefficient $f(q)$ is a decreasing function of q and $f(q) \sim 0.5$ when $q \sim 0.01$. With dissipation, corresponding expressions with η and ν can be derived.

Here we compare the shooting code results of eigenvalues with the local (Fourier) dispersion relation. By replacing d/dx in Eq. (4.2.53) with ik_x in a region around $x = 0$, the local dispersion relation is reduced to

$$\omega_\eta^2(\omega_\eta\omega_\nu - \omega_A^2)^2 - \omega_f^2\omega_\eta^2(\omega_\eta^2 + A\omega_A^2/\Omega_B)$$
$$+ (8A^2q_y - 4Aq_{xy}i\omega_\eta)\omega_A^2(\omega_\eta\omega_\nu - \omega_A^2) = 0, \tag{4.2.65}$$

where $\omega_I^2 = \kappa^2 k_z^2/k^2$, $k^2 = k_x^2 + k_y^2 + k_z^2$, $q_y = k_y^2/k^2$, and $q_{xy} = k_x k_y/k^2$. For axisymmetric perturbations ($k_y = 0$) in a nondissipative disk ($\eta = \nu = 0$), Eq. (4.2.65) reproduces the local dispersion relation derived by Balbus and Hawley (1991). In such a perturbation, unstable modes appear when $\omega_A^2 < (A/\Omega_B)\omega_I^2$.

When the unperturbed magnetic field is toroidal ($B_x = B_z = 0$), the approximate solutions of Eq. (4.2.65) in the regime $|\gamma| \sim \omega_A \ll \omega_I$ yields the growth rate for a disk with $\eta = \nu = 0$ as

$$\gamma^2 = \frac{A}{2\Omega_B}\left[1 - \frac{2A}{\Omega}q \pm \sqrt{1 - 4\left(1 + \frac{\Omega_B}{\Omega}\right)q + \left(\frac{2A}{\Omega}\right)^2 q^2}\right]\omega_A^2. \qquad (4.2.66)$$

In the Keplerian disk, purely growing modes appear only when $q < 2/9$.

4.2.3.3 Effect of Resistivity

First, we consider a case where the unperturbed magnetic fields are purely poloidal ($B_x = B_y = 0$). The criterion for the instability for axisymmetric ($k_y = 0$) perturbations is

$$(\omega_A^2 + \eta\nu k^4)^2 + \omega_I^2(\eta^2 k^4 - \frac{A}{\Omega_B}\omega_A^2) < 0. \qquad (4.2.67)$$

When $\eta = 0$, we have seen that unstable modes appear for weak magnetic fields which satisfy $B_z^2 < 4\pi\rho(A/\Omega_B)(\kappa^2/k^2)$. There is no instability threshold for B_z from below, as $B_z = 0+$ satisfies this condition (Balbus and Hawley, 1991). Note, however, that when $B_z = 0$ there is no instability.

When $\eta \neq 0$, however, there is a critical value for B_z

$$B_{zc}^2 = 4\pi\rho\left(\frac{\Omega_B}{A}\right)\frac{\eta^2 k^4}{k_z^2}, \qquad (4.2.68)$$

below which the second term in the left-hand side of Eq. (4.2.73) is positive. We now find that the Balbus-Hawley mode is stabilized for sufficiently weak seed magnetic fields or sufficiently large resistivity. Such a threshold does not appear when $\eta = 0$ even if $\nu \neq 0$, and thus we conclude that the kinematic viscosity does not play an essential role for the stabilization of the Balbus-Hawley mode, but the *resistivity does*.

4.2.3.4 Anomalous Resistivity and Magnetic Viscosity*

When $\eta \neq 0$, unstable modes disappear when ω_A is below a threshold. The threshold for the upper branch is $\omega_A \sim 0.05\Omega$ when $q = 0.01$. This threshold is close to that determined by equating the nonresistive growth rate ($\gamma \sim k_y v_A$) with the resistive damping rate (ηk^2).

When charged particles, electrons in particular, move in *stochastic magnetic fields*, their orbits (or streamlines) in such fields become diffusive. This diffusiveness may arise both from the field aligned directional motion and from that perpendicular to the field. Such diffusive

335

orbits may give rise to effective scattering of electrons and thus to effective resistivity. In general such fluctuating magnetic field effects can be incorporated by a quasilinear theory. Thus the resistivity becomes a function of fluctuations $\langle \delta B^2 \rangle$ as $\eta = \eta(\langle \delta B^2 \rangle)$ in a generic expression. As concrete examples of such an expression, we employ one by Ichimaru (1975) and one by Horton et al. (1984) in the following. Although details of their theories and thus their formalisms are different, a common nonlinear feature exists, as both theories have the quadratically nonlinear δB dependence of η.

Although the collisional resistivity η_0 of the disk is small to allow the instability, the enhanced *(anomalous) resistivity* η due to magnetic fluctuations becomes large enough to saturate the instability when the turbulence sets in. Thus, the strength of magnetic fluctuations is determined near or at the marginal stability dictated by this anomalous resistivity when the unstable hydromagnetic modes are robust. The concept of the marginal stability has been invoked in fusion plasma problems (e.g. Manheimer and Boris, 1977; Kishimoto et al., 1996) and magnetic diffusion problems during the star formation (Norman and Heyvaerts, 1985) (see Sec. 2.6). When the governing instability is (magneto)hydrodynamic and encompasses many mode surfaces, the system can reach a quasisteady state by making it close to but slightly above the *marginal stability*. (To be precise, the small deviation away from the marginal is determined by the input rate of energy to the disk such as the binary star mass injection etc.) Otherwise, large scale "vortices" exerting strong modulating influence on the background shear profile can be quickly transported away (intermittent burst and returns to a quasi-steady state). In this way the level of magnetic fluctuations is self-consistently determined once the global conditions of the disk are given.

First, we adopt the resistivity expression by Ichimaru (1975),

$$\eta = \frac{c^2}{4\pi\sigma} = \left(\frac{\pi}{2}\right)^{1/2} \frac{1}{4\pi n_e m C_s} \sum_k \int d\omega \frac{k_J^2}{k^3} \langle |b^2|(\mathbf{k}, \omega) \rangle, \qquad (4.2.69)$$

where σ is the electrical conductivity, m the mass of ions, n_e the number density of electrons, $\langle |b^2|(\mathbf{k}, \omega) \rangle$ is the spectral intensity of the *magnetic fluctuations* (see also Sec. 2.3 and Sec. 5.3), and k_J is the component of wave vector parallel to the mean electric current. This expression has been used to model the magnetic turbulence in accretion disks by Ichimaru (1977) and Kato (1984). The three-dimensional computation of Hawley et al. (1995) and Matsumoto and Tajima (1995) indicate numerous x-point like structures in the nonlinear stage of the magnetorotational instability. When the plasma is fully ionized ($n = n_e$), we set $\rho = n_e m$. The magnetic fluctuations generated by the magnetorotational instability are highly anisotropic such that $k_y^2 \ll k_z^2 \sim k_x^2$. The mean electric current is predominantly in the x-z plane because we find the toroidal component of magnetic fields dominate over the poloidal component once the instability sets in.

By assuming that the magnetic fluctuations have a spectrum peaked around $\mathbf{k}_{\max}$, and by denoting

$$\langle \delta B^2 \rangle = \int d\omega \langle |b^2|(\mathbf{k}_{\max}, \omega) \rangle, \qquad (4.2.70)$$

we obtain

$$\eta = \left(\frac{\pi}{2}\right)^{1/2} \frac{1}{4\pi n_e m C_s} \left(\frac{k_J^2}{k^3}\right)_{\text{max}} \langle \delta B^2 \rangle,$$ (4.2.71)

where $(k_J^2/k^3)_{\text{max}}$ is evaluated at $\mathbf{k} = \mathbf{k}_{\text{max}}$.

Since magnetic fields induced by the magnetorotational instability eventually dominate over seed magnetic fields, we equate B^2 to $\langle \delta B^2 \rangle$ to evaluate the saturation level. The saturation level of magnetic fluctuations can be determined by equating the growth rate γ of the instability with the anomalous resistivity damping ηk^2. By using Eq. (4.2.64) for γ and Eq. (4.2.71) for η, we obtain

$$\frac{\langle \delta B^2 \rangle}{4\pi n_e m C_s^2} = \chi_e \left(\frac{2}{\pi}\right) \left(\frac{k^2 k_z^2}{k_J^4}\right) \left(\frac{k_\parallel^2}{k_z^2}\right) f(q),$$ (4.2.72)

where $\chi_e = n_e/n$ is the ionization rate. The factor $k_\parallel^2/k_z^2$ is unity for purely poloidal field. For cases with a purely toroidal field $k_\parallel^2/k_z^2 = q$. In the marginally stable state $k_\parallel^2/k_z^2$ is between q and 1 because the magnetic fields are already perturbed.

As the disk plasma is close to the marginality, the *magnetic viscosity parameter* $\alpha_B = -\langle \delta B_x \delta B_y \rangle/(4\pi \rho C_s^2)$ may be approximately written (Matsumoto and Tajima, 1995) by using the linearized momentum equation [Eq. (4.2.50)] and linearized induction equation [Eq. (4.2.51)] with $\eta = \nu = 0$ as

$$\alpha_B = \chi_e \frac{\langle \delta B^2 \rangle}{4\pi n_e m C_s^2} \frac{-\left\langle \frac{\delta B_y}{\delta B_x} \right\rangle}{\left\langle \left(\frac{\delta B_y}{\delta B_x}\right)^2 \right\rangle + \left\langle \left(\frac{\delta B_z}{\delta B_x}\right)^2 \right\rangle + 1},$$ (4.2.73)

where

$$\left\langle \frac{\delta B_y}{\delta B_x} \right\rangle = \frac{-2\Omega\gamma + (\gamma^2 + \omega_A^2 - 4A\Omega)(k_y/k_x)}{\gamma^2 + \omega_A^2 + 2\Omega\gamma(k_y/k_x)},$$ (4.2.74)

and

$$\left\langle \frac{\delta B_z}{\delta B_x} \right\rangle = -q^{1/2} \frac{(\gamma^2 + \omega_A^2)(k_x/k_y + k_y/k_x + 2A/\gamma)}{\gamma^2 + \omega_A^2 + 2\Omega\gamma(k_y/k_x)}.$$ (4.2.75)

When deriving these equations, we replaced d/dx by ik_x. The notations $\langle \delta B_x \delta B_y \rangle$ etc. denote the spatial average.

The instability-induced velocity fields also contribute to the radial *angular momentum transport*. The viscosity parameter corresponding to the *Reynolds* stress $\rho \langle v_x v_y \rangle$ due to this instability is expressed (Matsumoto and Tajima, 1995) as

$$\alpha_v = \frac{\langle v_x v_y \rangle}{C_s^2} = -\alpha_B \left(\frac{\gamma^2}{\omega_A^2}\right) \frac{\langle v_y/v_x \rangle}{\langle \delta B_y/\delta B_x \rangle},$$ (4.2.76)

where

$$\left\langle \frac{v_y}{v_x} \right\rangle = \left\langle \frac{\delta B_y}{\delta B_x} \right\rangle + \frac{2A}{\gamma}.$$ (4.2.77)

Since $k_y \ll k_x$, we can evaluate as

$$\left\langle \frac{\delta B_y}{\delta B_x} \right\rangle \sim \frac{-2\Omega\gamma}{\gamma^2 + \omega_A^2} \sim \frac{-2f(q)^{1/2}}{f(q) + 1} \left(\frac{\Omega}{\omega_A} \right),$$

and

$$\left\langle \frac{\delta B_z}{\delta B_x} \right\rangle \sim -\frac{k_x}{k_z}.$$

When $f(q) \sim 1$, $k_x \sim k_z$, $\omega_A \sim \Omega$, and $\chi_e = 1$, the *magnetic viscosity* α_B is a third of the magnetic fluctuation $\langle \delta B^2 \rangle / (4\pi n_e m C_s^2)$, and $\alpha_v \sim (2A/\Omega - 1)\alpha_B$.

When the poloidal field is dominant ($k_\parallel^2/k_z^2 \sim 1$), the magnetic fluctuation [Eq. (4.2.72)] is maximized when $q = 0$. The functional value $f(q)$ is approximately given by Eq. (4.2.65) as $f(q = 0) = A/(\Omega - A)$. The magnetic fluctuation level is determined by assuming that $k_z^2 \sim k_x^2 \sim k^2$ and $\chi_e = 1$. The maximum growth rate $\gamma = A$ and the corresponding ω_A $[\omega_A^2 = A(\Omega + \Omega_B)]$ are substituted into Eqs. (4.2.74), (4.2.75), and (4.2.77) to determine α_B and α_v. When the toroidal field is dominant ($k_\parallel^2/k_z^2 \sim q$), the magnetic fluctuation level is low when $q \ll 1$. By eigenmode analysis, we find that for a wide range of the shear parameter A, the function $qf(q)$ is maximized when $q = q_{max} \sim 0.01$. The magnetic fluctuation level for initially toroidal field case is determined by assuming $k_J^2 \sim k_x^2 \sim k_z^2 \sim k^2/2$ and $\chi_e = 1$. The maximum growth rate of the purely growing eigenmode and corresponding ω_A are used to determine α_B and α_v. The ratio of the wavenumbers k_y/k_x is determined from the width of unstable eigenfunction as $k_x \sim 2\pi A k_y/\omega_A$. The detailed values of α_B and α_v are given in Matsumoto and Tajima (1995).

One important implication of the present theory is that α_B depends on the temperature of accretion disks. Theory indicates that in the innermost region of accretion disks around stellar mass black holes or neutron stars where $T \sim 10^7 K$, the magnetic viscosity parameter is $\alpha_B = O(0.1)$ where the disk is fully ionized. In lower temperature ($T < 10^4 K$) disks whose ionization level is low, we expect that the saturation level of magnetic fluctuations is lower because the electron density is proportional to χ_e. Since α_B determined from Eq. (4.2.79) is proportional to the ionization fraction χ_e, the magnetic viscosity drastically decreases when the hydrogen recombination occurs.

Let us discuss in star forming regions whether large resistivity can stabilize the magnetorotational instability. Since the temperature of protostellar disks is so low ($T \sim 100K$) that even the original *collisional resistivity* is substantial. We assume that the thickness of the *protostellar disk* is $H \sim 10^{14}$ cm and the number density is $n \sim 10^{10}$ cm^{-3}. Using the resistivity expression in a partially ionized medium (e.g. Norman and Heyvaerts, 1985), we obtain $\eta \sim 3 \times 10^{-5} H C_s$. By applying Eq. (4.2.68), we find the critical field strength for poloidal magnetorotational instability is $B_{zc} \sim 10^{-6} G$, which is much smaller than the strength of large-scale poloidal magnetic field in protostellar disks. Thus, the magnetorotational instability can grow even in protostellar disks. The saturation level of the instability, however, will be low due to the low ionization level. The observational constraint on the value of α in protostellar disks comes from the excess infrared spectra around *T Tauri stars*

338

(Adams *et al.*, 1987), which suggests the presence of a viscous disk. The infrared excess vanishes on timescales of several 10^6 yr (Strom *et al.*, 1989). Since the viscous evolution time scale when $T = 100\,\mathrm{K}$, $R = 10^{15}\mathrm{cm}$, and $H/R = 0.1$ is $t_v \sim (\alpha C_s)^{-1}(R^2/H) \sim (3/\alpha) \times 10^3$ yr, the viscosity parameter α is the order of 10^{-3}. Such a smaller value of α is consistent with the above discussion that the magnetic viscosity is turned off in protostellar disks.

The amplification of a seed magnetic field by the magnetorotational instability also occurs in spiral galaxies. In the region where the rotation speed decreases with radius ($A/\Omega > 0.5$), the magnetic field can be amplified up to $1/\beta > 0.01$ by this mechanism. The resulting predominantly toroidal magnetic field ($q = k_y^2/k_z^2 \sim 0.01$) is consistent with observations of magnetic fields in spiral galaxies (e.g. Sofue *et al.*, 1986; Tajima *et al.*, 1987). The strength of magnetic fields ($\beta \sim 1$) in spiral galaxies, however, suggests that other amplification mechanisms such as the resonance with the spiral arm (Chiba and Tosa, 1990) are important, or the saturation level of the instability is higher.

When the magnetorotational instability grows in spiral galaxies, the interstellar gas will fall toward the center by magnetic viscosity. The time scale of infall is roughly $t_v \sim 10^9(1/\alpha)(R/10\mathrm{kpc})(C_s/10\mathrm{kms}^{-1})^{-1}(H/R)^{-1}$ yr. This time scale is longer than the age of universe in the outer part where the rotation curve is flat ($A/\Omega \sim 0.5$). In the middle part ($R = 1-3\,\mathrm{kpc}$ in our galaxy) where $A/\Omega > 0.5$, it is possible that viscous infall occurs within 10^{10} yr. The accretion rate $\dot{M} \sim 10^2\alpha(R/10\,\mathrm{kpc})(C_s/10\mathrm{kms}^{-1})(n/1\mathrm{cm}^{-3})(H/R)^2 M_\odot/\mathrm{yr}$ can exceed $0.1 M_\odot/\mathrm{yr}$, required to explain the activity of *Seyfert galaxies*.

4.2.3.5   3D Numerical Simulation of the Magneto-Rotational Instability*

Three-dimensional numerical simulations of the magnetorotational instability have been carried out by Hawley, Gammie, and Balbus (1995), Matsumoto and Tajima (1995), and Brandenburg *et al.* (1995). Here we present the result of Matsumoto and Tajima (1995).

The basic equations in corotating frame are:

- Equation of Continuity

$$\frac{\partial \rho}{\partial t} + \nabla(\rho \mathbf{v}) = 0, \tag{4.2.78}$$

- Equation of Motion: same as Eq. (4.2.39a)

- Induction Equation: same as Eq. (4.2.39b)

- Energy Equation

$$\frac{\partial \rho \epsilon}{\partial t} + \nabla(\rho \epsilon \mathbf{v}) + p\nabla \mathbf{v} = \eta \mathbf{J}^2. \tag{4.2.79}$$

$$\mathbf{J} = \frac{\nabla \times \mathbf{B}}{4\pi}. \tag{4.2.80}$$

- Equation of State

$$P = \rho \epsilon (\gamma_a - 1), \tag{4.2.81}$$

339

where γ_a is the adiabatic index. We use $\gamma_a = 5/3$.

For simplicity, the effect of gravitational stratification is neglected and $g_z = 0$ is assumed. In the unperturbed state, the density, temperature, and magnetic field is assumed to be uniform. The unperturbed velocity is $(v_x, v_y, v_z) = (0, -1.5\Omega x, 0)$. Here Ω is the rotational angular velocity at the particular radius $(x = 0)$ of the disk. We use the units $C_s = \Omega = \rho_0 = 1$ where C_s is the sound speed and ρ_0 is the initial density. The unit of the length equals to the scale height H of the disk.

We adopt the shearing box model (Hawley et al., 1995) and impose the following boundary condition of the plane $x = 0$ and $x = L_x$ where L_x is the size of the simulation box in x-direction (radial direction)

$$f(x, y, z) = f(x + L_x, y - 2AL_x t, z).$$

This boundary condition was implemented by Wisdom and Tremaine (1988) to simulate the planetary rings, and later used by Hawley and Balbus (1992), Hawley et al. (1995) and Brandenburg et al. (1995). The boundaries at $y = 0$, $y = L_y$, $z = 0$ and $z = L_z$ are periodic boundaries.

To initiate the simulation, random perturbations for v_y and P at $t = 0$ are imposed. The amplitudes of perturbations are $\delta v_y = 10^{-3}C_s$ and $\delta P = 10^{-3}\rho_0 C_s^2$. We dropped the artificial diffusion term from the induction equation and set $\eta = 0$ except the region where the current density is larger than a critical value $J_c = 100\rho^{1/2}\Omega$. The effective magnetic Reynolds number by numerical resistivity is the order of 10^5. The magnetic Reynolds number is artificially reduced to $O(10^2)$ in the current sheet where $|J| > J_c$.

In the following, we show the numerical results for a typical model starting with B_y alone with the Keplerian velocity shear $(A = 3\Omega/4)$. The initial magnetic field is uniform and purely toroidal. The initial plasma β $(= 8\pi P_0/B_0^2)$ is 100. For these parameters, the characteristic size of waves expected to be excited is $\delta x \sim v_A/A \sim 0.2H$ (see Sec. 4.2.3.4). Thus the size of the simulation box and the number of grid points are chosen such that we can numerically resolve this mode. The size of the simulation box is $(L_x, L_y, L_z) = (0.5H, 2H, 0.5H)$. The number of grid points for this run is $(N_x, N_y, N_z) = (41, 41, 41)$.

Figure 4.27 shows time sequence of magnetic field lines. Magnetic field lines are highly tangled up. This justified our physical picture (Sec. 4.2.3.4) of turbulent and chaotic magnetic field lines anticipated from linear theory and adopted for nonlinear theory to evaluate the saturation by enhanced resistivity.

Figure 4.28 shows isocontours of the azimuthal velocity perturbation $\delta v_y = v_y + 2Ax$ in the $z = 0.25H$ plane.

After waves with very short wavelength appear $(t = 8.4/\Omega)$, the mode whose radial wavelength is about $0.2H$ grows exponentially $(t = 11.2 - 16.4/\Omega)$. The preferential growth of the perturbation with $\lambda_x = 0.2H$ is also seen in Fig. 4.29 in isocontours of δv_y in the $x - z$ plane.

The azimuthal wavelength of the exponentially growing mode is nearly the box size $(\lambda_y = 2H)$, while the vertical wavelength is $\lambda_z \sim 0.3H$. The growth of the mode around

340

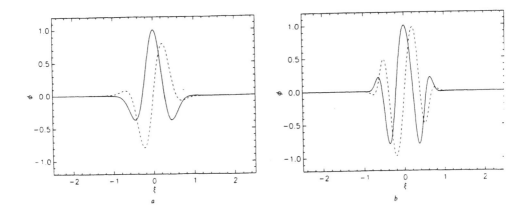

FIGURE 4.26 Examples of the eigenfunctions of the nonaxisymmetric magnetic shearing instability in a disk with the Keplerian velocity shear $(A/\Omega = 3/4)$. *Solid curve*, real part of eigenfunction; *dashed curve*, imaginary part of eigenfunction. The model parameters are $\omega_A = k_\parallel v_A = 0.1\Omega$ and $q = k_y^2/k_x^2 = 0.01$. (a) Fundamental mode (eigenvalue is $\omega = 0.692\Omega_i$); (b) next nodal mode (eigenvalue is $\omega = 0.0357\Omega_i$). The eigenfunctions are localized between $\xi = 2Ak_yx/\omega_A = \pm1$ (Matsumoto and Tajima, 1995).

$x = -0.1H$ can be seen. It should be noted that since we are using the shearing box periodic boundary condition, and allowable mode numbers do not restrict the mode rational surface (the center of the eigenmode x-position), there is no unique surface position x. Thus such unstable waves excited at various x-position will overlap with each other. This is in fact seen in Fig. 4.28.

In the later stage $(t > 20/\Omega)$, the growth of the eigenmode nearly saturates and the system shows more complex behavior. Figure 4.30 shows the time history of $\alpha_B = -\langle B_x B_y \rangle/(4\pi P_0)$, $\alpha_v = \rho_0 \langle v_x v_y \rangle/P_0$, and $(\langle B^2 \rangle)/(8\pi P_0)$. These quantities increase exponentially during $t < 20/\Omega$. Experimentally obtained growth rate in this stage is 0.23Ω. The magnetic viscosity α_B increases up to 0.02 while the kinematic viscosity α_v is about 0.005.

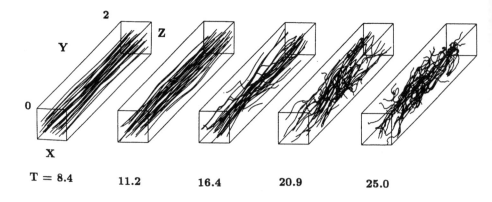

FIGURE 4.27 Magnetic field lines for model T in the eigenmode growth state $[t = (8.4 - 16.4)/\Omega]$ and the nearly saturation stage $[t = (20.9 - 25.0/\Omega]$ (Matsumoto and Tajima, 1995).

4.2.3.6   Effects of the Parker Instability*

When the vertical gravity is included, magnetic field escapes from the disk to the corona due to the Parker instability (the magneto-buoyancy instability; see Sec. 3.2). Since the growth rate of the Parker instability is $2 - 5H/v_A$, the growth rate of the Parker instability becomes comparable to that of the magnetic shearing instability as β approaches unity. The linear growth rates of the Parker instability in disks have been obtained by Horiuchi *et al.* (1988) by taking into account the spatial variation of vertical gravity. Nonlinear evolution of the Parker instability in disks have been studied by Matsumoto *et al.* (1988, 1990) by using two-dimensional MHD code. The effects of rotation and shear flow, however, were neglected in their model. The stabilizing effects of rotation on the Parker instability were studied by Zweibel and Kulsrud (1975), Shu (1974), and Hanawa *et al.* (1992). More recently, Chou *et al.* (1996) carried out three-dimensional magnetohydrodynamic simulations of the Parker instability of a flux tube imbedded in the isothermal atmosphere. In the linear stage,

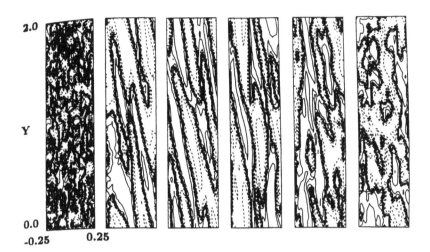

FIGURE 4.28 Result of three-dimensional MHD simulation of the disk with the Keplerian velocity shear (model T). The unperturbed magnetic field is uniform and purely toroidal. The initial plasma β is 100. The isocontours of the azimuthal velocity perturbation δv_y in the $z = 0.25H$ plane are shown. The contour step width is 0.5 in logarithmic scale. Solid curves and dashed curves represent the positive velocity and the negative velocity, respectively (from Matsumoto and Tajima, 1985).

the growth rate of the Parker instability is reduced by rotation. In the nonlinear stage, the magnetic flux tube is deformed into a helical shape. Fogglizo and Tagger (1994) solved the linearized equations time dependently by using the swinging coordinates: the growth rate of the Parker instability is reduced. The Parker instability grows for long wavelength perturbations ($\lambda > 10H$) along horizontal magnetic fields. Due to this instability magnetic fields, created in the magneto-rotational instability (or any other mechanism such as dynamo) in the accretion disk, escape into the corona, which sets the lower limit for the plasma β in the disk. The time scale of escape is of the order of $H/v_A \gg 1/\Omega$ when $\beta \gg 1$. Inclusion of the rotation has stabilizing influence on the Parker instability.

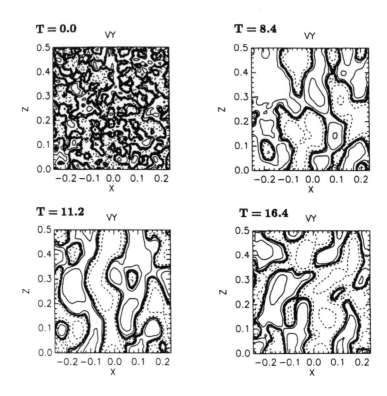

FIGURE 4.29   Same as in Fig. 4.28, but for δv_y in the x-y-plane $(y = H)$ (Matsumoto and Tajima, 1995).

4.2.3.7   Effects of Large-Scale Vertical Field*

So far, we mainly studied the growth of magnetic shearing instabilities and the Parker instabilities for a disk with purely toroidal magnetic field. Uchida and Shibata (1985) and Shibata and Uchida (1986) studied the interaction between a differentially rotating disk and a large scale vertical magnetic field (see Fig. 1.8). In such a configuration, angular momentum of the disk can be extracted through the outflow along the twisted magnetic fields, forming a bipolar jet (Sec. 4.3). Accompanied by the jet is avalanche of matter accretion toward the central compact object. Matsumoto *et al.* (1996b) studied a global 3D simulation of an accretion disk threaded by a large scale vertical field. Their model consisted of a rotating torus, a spherical non-rotating halo, and a uniform vertical magnetic field at $t = 0$. The rotating torus is characterized by the angular distribution

$$L = L_0 r^a \tag{4.2.82}$$

344

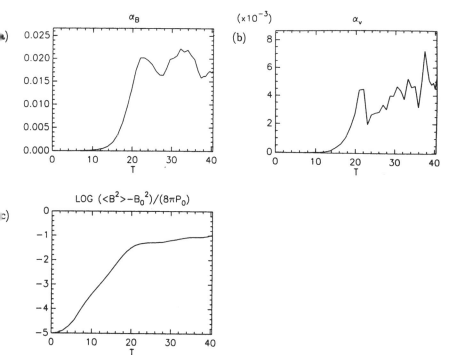

FIGURE 4.30 Time history of (a) $\alpha_B = -\langle B_x B_y \rangle /(4\pi P_0)$, (b) $\rho_0 \langle v_x, v_y \rangle /P_0$, and (c) the magnetic fluctuation $(\langle B^2 \rangle - B_0^2 /(8\pi P_0))$ for toroidal field model (model T) (Matsumoto and Tajima, 1995).

the polytropic equation of state

$$P = K_p^{1+\frac{1}{n}} \tag{4.2.83}$$

and the dynamical equilibrium

$$-\frac{GM}{(r^2 + z^2)^{1/2}} + \frac{1}{2(1-a)} L_0^2 r^{2a-2} + (n+1)\frac{P}{\rho} = \text{const.} \tag{4.2.84}$$

The spherical halo is given as

$$\rho = \rho_h \exp\left[\alpha\left(\frac{r_0}{\sqrt{r^2 + z^2}} - 1\right)\right], \tag{4.2.85}$$

where $\alpha = (GM/RT_{\text{halo}})/r_0$.

Matsumoto et al. (1996b)'s simulation found that (i) the avalanche in accretion proceeds along spiral channels; (ii) the growth of global non-axisymmetric modes in differentially

345

FIGURE 4.31 (a) Magnetic field lines and equatorial density; (b) Projection of magnetic field lines (Matsumoto *et al.*, 1996b).

rotating magnetized disks (magnetic Papaloizou-Pringle instability) is observed; (iii) a helical structure in jets is formed; (iv) the jet speed is approximately of the Kepler rotation speed; (v) the magnetic fields in the disk are such that the toroidal field dominates in the innermost region of the disk while a spiral low-β region is formed; (vi) growth of the Parker instability toward the halo is observed. See Figs. 4.31 and 4.32.

4.2.3.8 Two States of Accretion Disks*

In standard accretion disk models (Shakura and Sunyaev, 1973), an accretion disk is assumed to be optically thick, geometrically thin, and rotating with Keplerian angular speed. The gravitational energy released by viscosity as heat is assumed to be radiated locally from the surface of the disk. The standard model can explain many features of X-ray emission from X-ray binaries especially the black body soft X-ray component, and UV emission from dwarf nova disks, and UV black body component in Seyfert galaxies. By comparing the

346

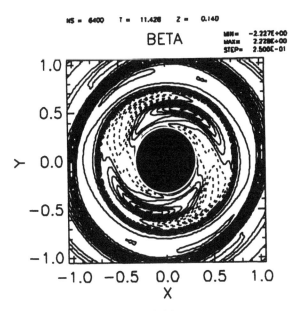

FIGURE 4.32 Isocontours of plasma β in the disk at $z = 0.14r_0$. The dashed curves show the region where $\beta < 1$ (Matsumoto *et al.*, 1996b).

theory and observation of dwarf novae outbursts, the viscosity parameter α in the quiescent phase of the dwarf novae is the order of 0.01, consistent with the results of three-dimensional MHD simulations of the magnetorotational instability and its saturation in Keplerian disks. The standard theory, however, cannot explain the power law component of X-ray emission in galactic black hole candidates and active galactic nuclei. In order to explain the hard power law component of the spectrum, Shapiro, Lightman, and Eardley (1976) proposed a optically thin, two temperature accretion disk model. Their disk model, however, is thermally unstable. Recently, a new branch of optically thin disk solutions are extensively studied (e.g. Abramowicz *et al.*, 1995; Narayan and Yi, 1995). In this model, the local heat generation balances with the radial advection. This branch is thus called *advection dominated disk model*. Since the radial advection plays an essential role for energy transport in advection dominated disks, the magnetic structure and the magnetic viscosity will also be affected. In this section, we review the theory and observations of black hole disks and their relation to the magnetic structure and turbulence in accretion disks.

347

● **High State (Soft State)**

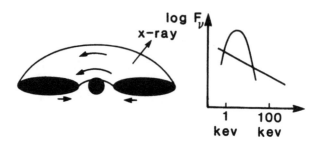

● **Low State (Hard State)**

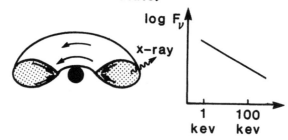

FIGURE 4.33 Two states of accretion disks: "High state" (soft state) vs. "Low state" (hard state): the disk dynamics and X-ray spectra

It is known that accretion disks in black hole candidates have two spectral states (e.g. Liang and Nolan, 1984). One is the high state and the other is the low state. In the high (or soft) state, the spectra has blackbody component which can be explained by emission from optically thick accretion disks. On the other hand, in the low (or hard) state, the spectra obey a power law which may come from optically thin accretion disk (Fig. 4.33).

A well-known object which shows the bimodal behavior is Cygnus X-1. Other black hole candidates such as A0420-00 and Nova Muscae also showed the bimodal transitions between the low state and the high state.

The spectrum state seems to be correlated with the luminosity of the disk. The high-low transition occurs in the decay phase of the outburst when the luminosity is about 1% of the maximum (Lund, 1993; Miyamoto et al., 1995). Low mass X-ray binaries containing neutron stars also exhibit high-low transitions when their absolute luminosity is as low as 10^{36} erg/s (Mitsuda et al., 1989). The origin of the low-high state transition can be explained naturally by using a thermal equilibrium curves of accretion disks in the diagram of surface density

348

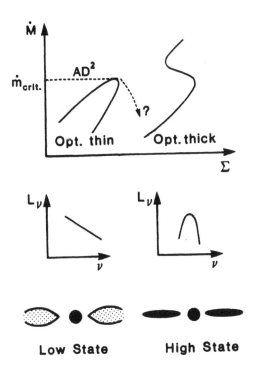

FIGURE 4.34 Thermal transition between the two states of an accretion disk. The optical porperties and the viscosity.

versus mass flow rate. Figure 4.34 shows the thermal equilibrium curve when we adopt the α viscosity (e.g. Narayan and Yi, 1995).

The effects of radial advection have been taken into account. The equilibrium curves have two branches. The left branch is the optically thin disk and the right branch is the optically thick disk. The S-shaped curve in the optically thick branch corresponds to the gas pressure dominated disk, radiation pressure dominated disk, and the advection dominated disk, respectively from bottom to top (Abramowicz *et al.*, 1988). The optically thin branch consists of the cooling dominated branch (lower) and the advection dominated branch (upper). In the cooling dominated branch, the heat generated by the viscosity is radiated immediately from that place; the local heat generation balances with the local radiation loss. On the other hand, in the advection dominated branch, most of the heat is advected inward by radial accretion. Thus the advection dominated, optically thin disk is underluminous. The disk luminosity is $L \sim (\dot{m}^2/\dot{m}_{\rm crit})L_{\rm edd}$ where $\dot{m}$ is the accretion rate in the unit of the Eddington accretion rate. The optically thin, advection dominated branch was proposed and studied by Ichimaru (1977) and Matsumoto *et al.* (1985). The existence of the upper limit

349

of the mass accretion rate for this branch was pointed out by Abramowicz et al. (1995) and Nayayan and Yi (1995). The maximum accretion rate for the advection dominated disk is $\dot{m}_{crit} \sim 0.3\alpha^2$ (Narayan and Yi, 1995). Thus, if we can determine the critical mass accretion rate $\dot{m}_{crit}$ from observations of the low-state high-state transition in black hole disks, we can estimate the value of α. Since the cooling dominated optically thin branch is thermally unstable, optically thin disks will stay in the advection dominated branch. The optically thin, advection dominated disk corresponds to the low-state disk, and the optically thick disk corresponds to the high-state disk.

Now we have the following scenario regarding the evolution of transient black hole candidates. Suppose that the disk is on the optically thin advection-dominated branch, thereby emitting hard power-law X-rays (the low state). A low-to-high transition is excited when the accretion rate increases and the advection dominated branch disappears. (See Fig. 4.34.) When accretion rate is higher, the disk bifurcates to the optically thick branch. Most of the power then begins to be emitted in soft X-rays. After reaching the maximum luminosity, the mass flow rate decreases. When most of the disk mass accretes and the disk becomes effectively optically thin, the disk undergoes the high-to-low transition. Some black hole candidates such as V404 Cygni have not exhibited a low-to-high transition probably because the mass accretion rate is too low. The disks in Sgr A at the center of our galaxy and many of active galactic nuclei also seem to be staying always in the low-state.

Black hole candidates show different time variability characteristics in the low state and in high state (Miyamoto et al., 1995). In the low state, the disk exhibits aperiodic X-ray fluctuations (or flickerings). A notable feature of the time variability in a low-state disk is the $1/f$ noise like spectrum of fluctuations (Sec. 2.2 and Sec. 2.1.7) and the large amplitude fluctuations up to 40% rms of the total luminosity (van der Klis, 1994). In the high state, the amplitude of fluctuation is small (see Fig. 4.35).

Mineshige et al. (1995) proposed a model of $1/f$-like fluctuations based on the notion of the self-organized criticality (SOC); see Sec. 2.2 and Sec. 2.6. According to this model the inner portions of black hole disks are composed of numerous reservoirs. When a reservoir is filled with accreting gas, something critical occurs, and stored mass and energy in that reservoir are quickly released, emitting X-rays. A similar model has been applied to the solar magnetic fields and had some success in explaining the power-law peak intensity distribution of flares (Lu and Hamilton, 1991). This suggests that the critical phenomenon producing X-ray shots in black hole candidates are intermittent magnetic reconnection similar to solar flares. In solar flares, the contribution of fluctuating X-rays to the total luminosity of the Sun is only 0.01% at most. The magnetic activities of the Sun are restricted to the coronal region. The small amplitude X-ray time variabilities in the high state of black hole disks can be explained by such coronal activities. Magnetic reconnection inside the disk may also contribute to the X-ray variability in high state disks. The amplitude of variabilities, however, will be small compared to the disk total luminosity because the gas pressure dominates the magnetic pressure in standard type disks. Note that in three-dimensional simulation of Matsumoto et al. (1996b), high-β disk tends to the magnetically self-sustained state with $\beta \sim 10$ (Stone et al., 1996; Matsumoto et al., 1996b). The time variabilities of high state

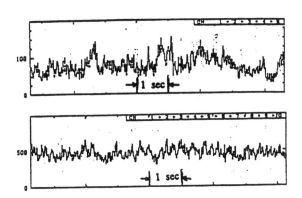

FIGURE 4.35 X-ray time variation of Cyg $X - 1$ (Miyamoto *et al.*, 1993).

with our model that high state disks are standard type high-β disks. (See also Sec. 4.3.4.4.)

The magnetic structure of low state disks seems to be quite different from high state disks. In order to explain the sporadic large amplitude fluctuations in low-state disks by a magnetic mechanism, the magnetically active region should extend over the entire disk. Thus a low state disk is a low-β disk, in which the magnetic pressure is large and even dominate over the gas pressure. The possible existence of such a low-β disk was proposed by Shibata *et al.* (1990) and confirmed by the three-dimensional MHD simulation as we have shown by Matsumoto *et al.* (1996b) (Fig. 4.30).

A low β disk is like a corona itself. It is made up of magnetically constricted blobs. The blobs may be optically thick, thus emitting some amount of soft X-rays. However, since the volume-filling factor of blobs is small, the majority of power is emitted as hard radiation. [This situation is similar to the one considered for the Cosmic X-ray Background Radiation (Sec. 5.2)]. A sequence of magnetic reconnection or magnetic disruption will yield sporadic, large-amplitude X-ray time variabilities. According to the results of 3D MHD simulation (see Fig. 4.32), the magnetic viscosity parameter in low-β disks is higher than the high-β

Parameter	Disk Type	
	High-β Disk	Low-β Disk
$\beta = P_{gas}/P_{mag}$	$\beta > 1$ (P_{gas} supported)	$\beta \lesssim 1$ (P_{mag} supported)
Configuration	Optically thick disk (cooling-dominated) + corona	Optically thin disk (advection-dominated) consistin
Dissipation of magnetic fields	Escape via buoyancy and reconnection	Reconnection
Dissipation of energy	Continuous	Sporadic
Compactness	$l_{soft}/l > 1$	$l_{soft}/l < 1$
Spectrum	Soft + hard tail	Hard power law
Power-law index	~2–3	~1.5–2
Fluctuations	Small	Large

TABLE 4.1 High-β disk and low-β disk (Mineshige *et al.*, 1995)

disk and will be $\alpha_B > 0.1$ (Matsuzaki *et al.*, 1997).

The picture for the bimodal state of a black hole disk is summarized in Table 4.1 and Fig. 4.36 (Mineshige *et al.*, 1995).

A high-state disk is a high-β, optically thick, standard-type disk. In quiescent disks with no net global vertical magnetic fields, the disk tends to be a magnetically self-sustained state with $\beta \sim 10$. [It is of interest to observe that magnetic turbulence evolution precipitates toward two and only two final states $\beta < 1$ and $\beta \sim 10$, regardless of the initial condition of β, as we surveyed in Sec. 2.3 (Kinney *et al.*, 1993, 1995)]. The effective magnetic viscosity α_B is around 0.01. This is the high-state disk. The high-β disk in galactic black hole candidates or around neutron star produces predominantly soft X-rays. As shown by 3D MHD simulation, some fraction of the magnetic energy emerges via buoyancy in the form of loop-like structures, emitting hard radiation. The contribution of the hard radiation to the total luminosity is small. A low-β disk can be created when the disk matter dynamically accretes to the central object. For example, when the disk is threaded by large-scale magnetic field, the avalanche flow due to magnetic braking generates low-β region inside the disk as

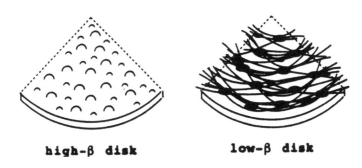

high-β disk　　　　**low-β disk**

FIGURE 4.36　Schematic views of a high-β disk (*left*) and a low-β disk (*right*). Magnetic fields emerge via buoyancy in the form of looplike structures in the high-β disk, whereas the low-β disk is composed of blobs threaded by mainly toroidal fields. (Mineshige *et al.*, 1995)

shown in Fig. 4.32. Although more detailed study is necessary, advection dominated disks will tend to be low-β because radial advection continuously generates radial magnetic field. Once a low-β disk is formed, it can stay in the low-β state for time-scale much longer than the dynamical time.

4.3　Jet

As the cloud gas falls to the inner and inner portion of the accretion disk, the magnetic fields originally penetrating the contracting cloud are convected towards the central object (either a star or a black hole), and twisted more and more by the rotation of the disk. When the twisted field gets sufficiently strong, disk plasmas start to be accelerated to both polar directions to form *bipolar jets* via magnetic force (i.e. centrifugal force on corotating field lines and magnetic pressure force due to toroidal fields). The jets are results of *release of gravitational energy due to accretion* as well as of *angular momentum extraction from the disk*. If the original global magnetic field is very weak, the dynamo mechanism can work to develop a global field to launch bipolar jets.

Since this is a simple, basic process independent of detailed physical conditions, it can occur everywhere in the universe if the contracting (or accreting) plasmas have angular momentum and (even weak) magnetic fields.

In fact, the recent development of radio and X-ray astronomy have revealed many similar jets everywhere in the universe, such as star forming regions, cataclysmic variables, X-ray binaries, and active galactic nuclei. This is probably a manifestation of above simple physical rule.

In this section, starting from the quick survey of observations of astrophysical jets, we will discuss a basic theory on the formation of MHD jets.

4.3.1 Observations of Astrophysical Jets

Jets ejected from active galactic nuclei (AGN) are one of the most specacular and enigmatic phenomena in our universe (Fig. 4.37). It has been thought that mass accretion in (or from) an accretion disk rotating around a supermassive ($\sim 10^8 M_\odot$) black hole at the center of AGN would explain their vigorous energy release. However, no one has yet observed an evidence of supermassive black hole, and even an accretion disk has not yet been spatially resolved well, since the expected size of accretion disk ($1 - 100\,\mathrm{AU}$) around the BH is too small to be resolved well. Note the very long distance from us. Hence the formation mechanism of AGN jet is still a puzzle.

Interestingly, however, it has been found that there are many jet phenomena in the universe, which are similar to radio jets ejected from AGN. These are jets ejected from young stellar objects (YSO), X-ray binary (XRB), and cataclysmic variable (CV) (Table 4.2). The central objects observed in these jets are not necessarily black holes, but are protostars for YSO, white dwarfs for CV, and neutron stars or black holes for XRB. Therefore, the acceleration mechanism itself is not related to general relativistic effects, though the general relativistic MHD around black holes is one of the central subjects to understand the physics of AGN jets (Sec. 2.7).

Hereafter, we will quickly survey what are the jets ejected from various central objects, AGN, YSO, XRB, and CV. For more details of observations and models of astrophysical jets, the reader should be referred to the following books or reviews; Ferrari and Pacholczyk (1983); Begelman *et al.* (1984); Margon (1984); Hughes (1991); Lada (1985); Burgarella *et al.* (1993); Tsinganos (1996); Bachiller (1996).

TABLE 4.2 Astrophysical Jets (Blandford, 1993)

physical quantities	AGN	YSO	XRB	CV
central objects	BH	protostar	BH/NS	WD
V_j/c	1	10^{-3}	0.3	10^{-2}
V_{esc}/c	1	10^{-3}	0.3	10^{-2}
L_j (erg/s)	10^{44}	10^{35}	10^{40}	10^{33}
$M/M_\odot$	10^{8}	1	10	1
R (cm)	10^{14}	10^{11}	10^{6}	10^{9}
R_{coll} (cm)	10^{17}	10^{16}	10^{11}	10^{11}
B(Gauss)	10^{4}	10^{3}	10^{9}	10^{6}

Note: AGN=Active Galactic Nuclei, YSO=Young Stellar Objects,
XRB=X-ray Binaries, CV=Cataclysmic Variables,
BH=Black Holes, NS =Neutron Stars, WD=White Dwarfs

4.3.1.1 AGN (Active Galactic Nuclei) Jets

Figure 4.37 shows a typical example of AGN jets, a radio jet ejected from nuclei of radio galaxies, NGC6251 (Bridle and Perley, 1984). It is remarkable that the jet is well collimated in the same direction from very small parsec scale ($\sim 10^{19}$ cm) (near the nucleus) to largest scale of more than 100 kpc scale ($\sim 10^{24}$ cm). The total energy and the power of an AGN jet plus its nuclei is estimated to be about 10^{61} erg and 10^{46} erg/s, and it is certainly one of the most violent phenomena in the universe. The jets terminate at the so called *lobes* which have been observed as *double radio sources* for many years. It has now been believed that these lobes are energized by the jets.

Though the direct observation of the velocity of the jet with doppler shift measurement has not yet been succeeded, the proper motion of the parsec-scale jets show that the apparent velocities are larger than the speed of light! This is called the *superluminal motion*, and of course it is simply an apparent effect due to special relativity, though this effect suggests that the real velocity is close to the speed of light. Figure 4.38 shows one of famous examples of the superluminal motion found in a parsec-scale jet of 3C273 observed with VLBI at 1.3 and 8 cm. In this case, the apparent speed of the jet is about 6c, and thus suggest 0.99 c if the angle (θ) between the jet direction and us is 10°, since the apparent speed is related to the real speed by the following formula,

$$V_{\text{apparent}} = v \sin \theta / [1 - (v/c) \cos \theta]. \qquad (4.3.1)$$

Since the radio emission from these extragalactic jets is nonthermal emission (synchrotron emission), we do not have any information on thermal plasmas in jets. We know neither

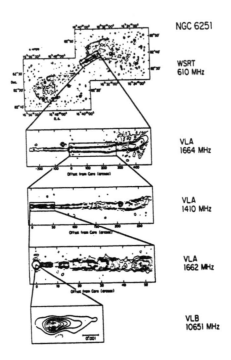

FIGURE 4.37 Radio images of jets ejected from nucleus of radio galaxy, NGC6251 (from Bridle and Perley, 1984).

true speed of the jets, nor temperature and density of the jets. However, since the emission from the jets is synchrotron, there is no doubt that there are magnetic fields, though it has not yet been made clear whether these magnetic fields are dynamically important or not.

On the other hand, our Galactic center shows various evidence of strong magnetic fields. Yusef-Zadeh *et al.* (1984) found a beautiful filamentary structure in so called "radio arc" (Fig. 1.19) which is very similar to filamentary structures in prominences observed on the Sun. (See Sec. 2.3.) The radio arc is located about 50 pc from the Galactic center, and has similar dimension. Inoue *et al.* (1984) and Tsuboi *et al.* (1985, 1986) found strong polarization and estimated the field strength as order of 10–100 μG around the radio arc, and they called the structure the "polarization lobes." Even a stronger magnetic field ($\sim$ 1 mG) has been suggested to the structure in the radio arc (Morris, 1989). This strong magnetic fields have energy density comparable to the gravitational and rotational energies of the rotating disk, and will affect significantly the evolution of the disk. Even the smaller magnetic field can affect the dynamics of outflow from that region. Sofue and Handa (1984) found the "Galactic center radio lobes" (Fig. 4.39) with a scale of 100 pc, and discussed the possibility that this

356

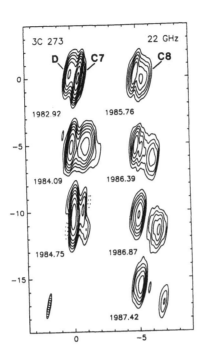

FIGURE 4.38 Radio images of superluminal jets ejected from 3C273 (from Unwin, 1990).

phenomeon may be a prototype of more energetic AGN jet and proposed a model of a jet accelerated by the magnetic force (Uchida *et al.*, 1985; Shibata and Uchida, 1987; Shibata, 1989).

4.3.1.2 YSO (Young Stellar Objects) jets

Recent high resolution optical images taken by Hubble Space Telescope have revealed the remarkable morphology of jets ejected from protostars; the shapes of the protostellar *optical jets* or *ionized jets* are very similar to those found in active galaxies. These jets show well collimated linear structure from very small scale (~ 10 AU ~ 0.0001 pc) to much larger scale of 0.1 pc (Fig. 4.40) (Snell *et al.*, 1985). There is an increasing evidence of accretion disks (or rotating disks) at the footpoint of the jets (Fig. 4.41), (Kaifu *et al.*, 1984; Kawabe *et al.*, 1993). The jets terminate at the bow-shock-like structure which resembles the *lobes* observed in extragalactic radio jets (Fig. 4.42). The jets show the knots or wiggle structures which are also similar to those found in AGN jets. Some of these features (including jets themselves)

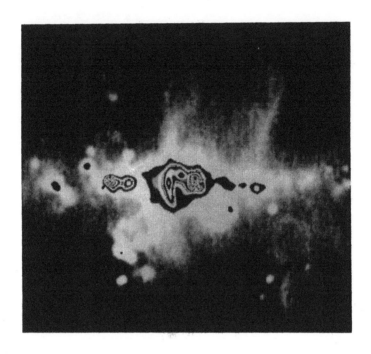

FIGURE 4.39 Galactic center radio lobe observed in our Galactic center (Sofue and Handa, 1984).

have been known as *Herbig Haro objects (HH objects)*. [3] The velocity of the *optical jets* is of order of a few 100 km/s, and mass loss rate is estimated to be about $10^{-8} - 10^{-9} M_\odot/year$.

In addition to optical jets, *bipolar molecular flows* with much larger scale (a few pc) have been observed in star forming regions (Snell *et al.*, 1985; Lada, 1985; Fukui, *et al.*, 1993). The molecular flows show weak collimation (Fig. 4.41), and the velocity of the flow is $10 - 30$ km/s, much slower than that of the optical jet, while the mass flow rate is much larger, $\sim 10^{-5} - 10^{-6} M_\odot/year$.

Recently, it has been found that there are one more componet of high velocity jet, *high velocity neutral wind*, surrounding the optical jets. The velocity of this jet is comparable to that of the optical jet, but the mass loss rate is an order of magnitude higher than that of the optical jets.

---

[3]The term HH objects are simply a historical term, and we had better use jets, bow shocks, knots, etc. to call structures of the phenomena.

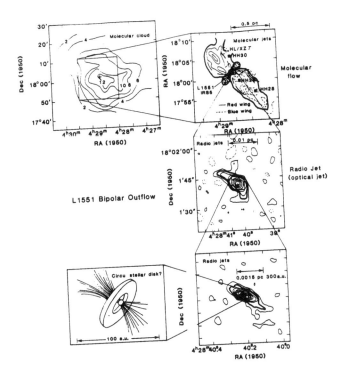

FIGURE 4.40 A typical example of jets ejected from YSO (young stellar objects), L1551 jets (Snell *et al.*, 1985).

4.3.1.3 Jets ejected from XRB (X-ray Binaries) and CV (Cataclysmic Variables)

The most famous jet in this category is the *SS433* jet (Margon, 1984), and still has its uniqueness because similar jets have not yet been observed. The SS433 jet show remarkably stable velocity at 0.26c both at Hα and X-ray wavelengths. The jet extends from $< 10^{12}$cm to 10^{17}cm as observed by radio waves (Fig. 4.43). The SS433 is a close binary system in which accretion disks are formed around the central objects, though it has not yet been established whether the central object is a black hole or a neutron star.

Recently, superluminal jets similar to AGN superluminal jets were discovered in our Galaxy (Mirable and Rodriguez, 1994). The central source is the X-ray binary source. It is expected that more and more jets similar to AGN jets will be found in our Galaxy.

If the central object is not a neutron star nor a black hole, what would happen? If the central object is a white dwarf, we observe *cataclysmic variables* which show various evidence of accretion disks and jets (Kafatos *et al.*, 1989). When the central star is a normal main

359

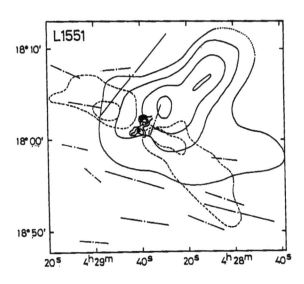

FIGURE 4.41   Bipolar flow and rotating disk in L1551 (Kaifu *et al.*, 1984).

sequence star, we find *symbiotic stars*. Observations show some evidence of high speed jets (at 1000 km/s) ejected from such symbiotic stars (Iijima, 1994).

4.3.2   Thermally Driven Wind/Jet

Let's begin with the simplest wind model, i.e. one dimensional spherical, steady, isothermal wind (without magnetic field) which was first developed by Parker (1958) for solar wind. Basic equations are

$$\frac{d}{dr}(\rho v_r r^2) = 0, \tag{4.3.2}$$

$$v_r \frac{dv_r}{dr} + \frac{1}{\rho}\frac{dp}{dr} + \frac{GM}{r^2} = 0. \tag{4.3.3}$$

Noting that

$$p = \rho C_s^2, \tag{4.3.4}$$

360

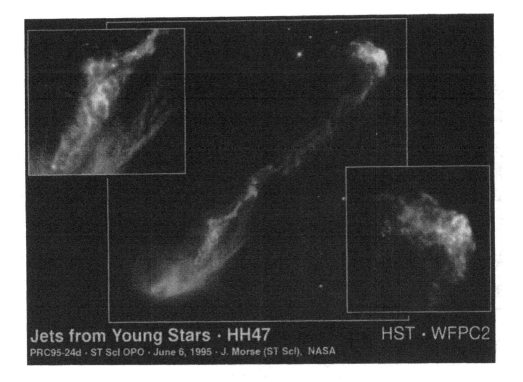

FIGURE 4.42 YSO optical jets observed with HST.

and C_s (= isothermal sound speed)=constant, and using (4.3.2), we find the equation (4.3.3) can be rewritten as

$$v_r \frac{dv_r}{dr} = \left(\frac{2C_s^2}{r} - \frac{GM}{r^2} \right) \bigg/ \left(1 - \frac{C_s^2}{v_r^2} \right).$$
(4.3.5)

This equation has a singularity at the point $v_r = C_s$. This point is called *critical point* or *sonic point*. In order to avoid the unphysical divergence at the critical point, the quantity $\frac{2C_s^2}{r} - \frac{GM}{r^2}$ must also become zero. This determines the radius of the critical point;

$$r = r_c = \frac{GM}{2C_s^2}.$$
(4.3.6)

This becomes

$$r_c \simeq 6\, R_\odot \left(\frac{T}{10^6 \text{K}} \right)^{-1} \left(\frac{M}{1 M_\odot} \right).$$
(4.3.7)

Hence the critical point is located at about $3 - 6 R_\odot$ for the case of the solar wind with $T = (1 - 2) \times 10^6 \text{K}$.

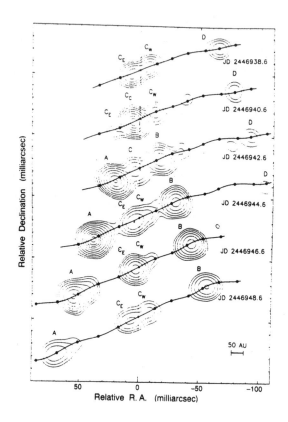

FIGURE 4.43 The radio images of the SS433 jet (from Vermeulen *et al.*, 1993).

Consider then the behavior of thermally driven wind (e.g. solar wind). Near the Sun $r < r_c$, the wind velocity is smaller than the sound speed, $v_r < C_s$. In this case, the right-hand side of equation (4.3.5) is positive, so that the velocity of the wind increases. At the critical point, we can calculate dv_r/dr using the perturbation method $v_r \simeq C_s + \delta v, r \simeq r_c + \delta r$. Then we find

$$\left(\frac{dv_r}{dr}\right)_c = \pm C_s/r_c. \tag{4.3.8}$$

So there are two branches, positive (acceleration) and negative (deceleration). In order to get continuous solution, we have to adopt positive solution in this case. Beyond the critical point, the wind velocity exceeds the sound speed and both signs of numerator and denominator become negative so that the left hand is still positive and the acceleration continues to infinity. Note that in the case of isothermal wind, the velocity becomes infinite at infinity, because the energy is not conserved but is added everywhere. The solution family is illustrated in Fig. 4.44.

362

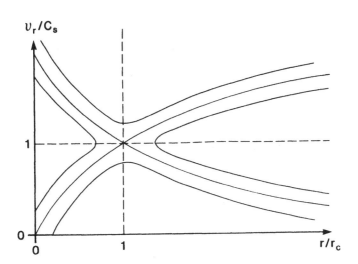

FIGURE 4.44 Solution curves of the thermally driven wind.

4.3.3 Magnetically Driven Wind/Jet

Although AGN jets are still one of the most enigmatic phenomena in our universe, it is considered that the magnetic mechanism is one of the promising mechanism to accelerate these jets. On the other hand, many researchers have recently begun to consider that YSO jets might be magnetically driven, partly because strong magnetic fields are observed in star forming regions, and partly because magnetic fields are considered to be very important to form stars as discussed in Sec. 4.1. Since the observed morphologies of AGN jets and YSO jets are remarkably similar, it is very likely that both jets are magnetically driven. For these reasons, there are increasing number of studies of magnetically driven jets (mostly steady models; e.g., Blandford and Payne, 1982; Hartman and McGregor, 1982; Pudritz and Norman, 1986; Camenzind, 1986; Sakurai, 1987; Kaburaki and Itoh, 1987; Lovelace *et al.*, 1986, 1991; Heyvaerts and Norman, 1989; Takahashi *et al.*, 1990; Chiueh *et al.*, 1991; Fukue *et al.*, 1991; Trussoni and Tsinganos, 1993; Königl and Ruden, 1993; Cao and Spruit, 1994; Begelman and Li, 1994; Tomimatsu, 1994; Sauty and Tsinganos, 1994; Shu *et al.*, 1994;

363

Contopoulos, 1995; Kudoh and Shibata, 1995,1997a; see Tsinganos, 1996 for many other references).

In the following, we will discuss the basic formalism of the steady theory of magnetically driven jets and winds. The nonsteady MHD jet models will be discussed in Sec. 4.3.4.

4.3.3.1 1D Steady Magnetically Driven (Centrifugal) Wind

A one-dimensional, steady, centrifugal wind theory was developed by Weber and Davis (1967) to model a solar wind on the equatorial plane. Taking spherical coordinate (r, φ, θ), we assume steady state $\partial/\partial t = 0$, axisymmetry $\partial/\partial \varphi = 0$, no non-plane magnetic and velocity fields (i.e. $\mathbf{B} = (B_r, B_\varphi, 0)$, $\mathbf{v} = (v_r, v_\varphi, 0)$), ideal (adiabatic) MHD, and 1D ($\partial/\partial \theta = 0$) on the equatorial plane ($\theta = \pi/2$). Then, basic MHD equations are integrated into the following six conservation equations;

$$\rho v_r r^2 = f, \tag{4.3.9}$$

$$r^2 B_r = \Phi, \tag{4.3.10}$$

$$r \left(v_\varphi - \frac{B_r B_\varphi}{4\pi \rho v_r} \right) = \Omega r_A^2, \tag{4.3.11}$$

$$r(v_r B_\varphi - v_\varphi B_r) = -\Omega r^2 B_r, \tag{4.3.12}$$

$$p = K\rho^\gamma, \tag{4.3.13}$$

$$\frac{1}{2}v_r^2 + \frac{1}{2}(v_\varphi - \Omega r)^2 + \frac{\gamma}{\gamma - 1}\frac{p}{\rho} - \frac{GM}{r} - \frac{\Omega^2 r^2}{2} = E. \tag{4.3.14}$$

Here f is the mass flux, Φ is the magnetic flux, Ω is the angular velocity of the Sun (or star or any rotating body) from which wind or jet comes out, r_A is the Alfvén radius (discussed later), K is a constant depending only on entropy, and E is the total energy of the wind. These six parameters f, Φ, Ω, r_A^2, K, E are integral constants. The unknown variables are also six, ρ, v_r, B_r, v_φ, B_φ, and p. Hence, if these six constants are given, the equations are solved so that six unknown physical quantities are determined at each r.

Eliminating v_φ in equations (4.3.11) and (4.3.12), we find

$$\frac{B_\varphi}{B_r} = -\frac{r\Omega}{v_r}\frac{(1 - r_A^2/r^2)}{(1 - v_{Ar}^2/v_r^2)}. \tag{4.3.15}$$

It follows from this equation that r must be equal to r_A when v_r is equal to v_{Ar}. Here $v_{Ar} = B_r/(4\pi\rho)^{1/2}$ is the Alfvén velocity due to the radial component of magnetic field. r_A is called *Alfvén radius* or *Alfvén point.*

Before solving these equations, it will be useful to calculate the asymptotic behavior of the physical quantities in this wind. As r increases to ∞, we find

$$B_r \propto r^{-2}. \tag{4.3.16}$$

Since in adiabatic wind the wind velocity v_r should tend to be constant terminal velocity v_∞ from energy conservation, i.e.,

$$v_r \to v_\infty, \tag{4.3.17}$$

we obtain

$$\rho \propto r^{-2}, \tag{4.3.18}$$

$$v_{Ar} \propto B_r/\rho^{1/2} \propto r^{-1}, \tag{4.3.19}$$

$$B_\varphi/B_r \propto r, \tag{4.3.20}$$

$$B_\varphi \propto r^{-1}. \tag{4.3.21}$$

Hence the degree of magnetic twist or wrapping B_φ/B_r increases with distance r.

Let us now calculate singular points in this wind following the notation of Sakurai (1985). Eliminating variables $v_r, B_r, B_\varphi, v_\varphi, p$ in the energy equation (Bernoulli equation) (4.3.14) using other equations, we finally get the following equation;

$$H(r, \rho) = \frac{f^2}{2} \frac{1}{\rho^2 r^4} \frac{\gamma K}{\gamma - 1} \rho^{\gamma-1} - \frac{GM}{r} + \frac{\Omega^2 r^2}{2} \left(\frac{(1 - r_A^2/r^2)^2}{(1 - \rho/\rho_A)^2} - 1 \right), \tag{4.3.22}$$

where

$$\frac{\rho_A}{\rho} = \frac{v_r r^2}{v_{Ar} r_A^2}. \tag{4.3.23}$$

Since the equation (4.3.9) is written as

$$\frac{1}{v_r} \frac{dv_r}{dr} = -\frac{1}{\rho} \frac{d\rho}{dr} - \frac{2}{r} = -\frac{\frac{\partial H}{\partial r} + \frac{2\rho}{r} \frac{\partial H}{\partial \rho}}{\rho \frac{\partial H}{\partial \rho}}. \tag{4.3.24}$$

Hence the point where $\partial H/\partial \rho = 0$ becomes the singular point. From equation (4.3.22), we obtain

$$\rho \frac{\partial H}{\partial \rho} = -\frac{(v_r^2 - v_{sr}^2)(v_r^2 - v_{fr}^2)}{v_r^2 - v_{Ar}^2}. \tag{4.3.25}$$

Here

$$v_{sr}^2 = \frac{1}{2} \left[C_s^2 + v_{Ar}^2 + v_{A\varphi}^2 - \left((C_s^2 + v_{Ar}^2 + v_{A\varphi}^2)^2 - 4C_s^2 v_{Ar}^2 \right)^{1/2} \right], \tag{4.3.26}$$

$$v_{fr}^2 = \frac{1}{2} \left[C_s^2 + v_{Ar}^2 + v_{A\varphi}^2 + \left((C_s^2 + v_{Ar}^2 + v_{A\varphi}^2)^2 - 4C_s^2 v_{Ar}^2 \right)^{1/2} \right]. \tag{4.3.27}$$

Similarly,

$$\rho \frac{\partial H}{\partial r} = -\frac{2v_r^2}{r} + \frac{GM}{r^2} - r\Omega^2 \left[1 - \frac{v_r^4 (1 - r_A^4/r^4)}{(v_r^2 - v_{Ar}^2)} \right]. \tag{4.3.28}$$

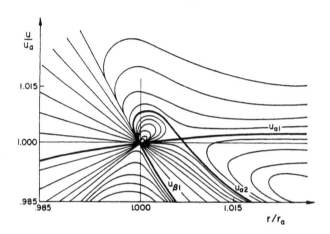

FIGURE 4.45 The solution curves of 1D magneto-centrifugal wind (Weber and Davis, 1967).

From these equations (4.3.26)–(4.3.30), we find when $\partial H/\partial \rho = 0$ (i.e. $v_r = v_{sr}$ or $v_r = v_{fr}$), $\partial H/\partial r$ must be equal to zero. The points where $\partial H/\partial r = 0$ are called *slow point* $(r = r_{sr})$ and *fast point* $(r = r_{fr})$. Both critical points and family of solutions are shown in Fig. 4.45 (taken from Weber and Davis (1967)). Figure 4.46 shows the classification of the solution family as a function of the centrifugal force and thermal force.

If the centrifugal force is dominant compared with thermal force, that kind of wind is called *fast magnetic rotator*. In this case, the terminal velocity becomes comparable to the rotation velocity at the Alfvén radius;

$$v_\infty \sim r_A \Omega. \tag{4.3.29}$$

That is, the magnetic field corotates with the rotating body up to the Alfvén surface. Let us estimate the terminal velocity in this case. The Bernoulli's equation (4.3.16) can be written as

$$E = \frac{1}{2}v_r^2 + \frac{1}{2}v_\infty^2 + \frac{\gamma}{\gamma-1}\frac{p}{\rho} - \frac{GM}{r} - \frac{rB_r B_\varphi}{4\pi\rho}\frac{\Omega}{v_r}, \tag{4.3.30}$$

366

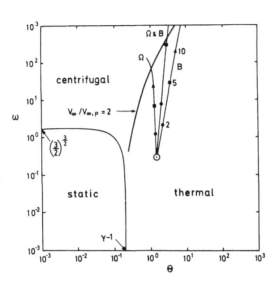

FIGURE 4.46 The classification of the solution for 1D mangneto-thermal-centrifugal wind (Sakurai, 1985).

if we add a constant $-r_A^2 \Omega^2$.

When $r \to \infty$, the Bernoulli's law (4.3.32) becomes

$$E \simeq \frac{1}{2} v_\infty^2 - \left(\frac{r B_r B_\varphi}{4 \pi \rho} \right) \frac{\Omega}{v_\infty}. \tag{4.3.31}$$

Using the relation [see (4.3.17)], $B_\varphi \simeq -\Omega r / v_\infty$, we have

$$E \simeq \frac{1}{2} v_\infty^2 + \left(\frac{\Omega^2 \Phi^2}{4 \pi \dot{M}} \right) \frac{1}{v_\infty}, \tag{4.3.32}$$

where $\dot{M} = 4 \pi \rho v_r r^2$. The energy becomes minimum if

$$v_\infty \sim \left(\frac{\Phi^2 \Omega^2}{\dot{M} v} \right)^{1/3} \propto B^{2/3} \propto \beta^{-1/3}. \tag{4.3.33}$$

367

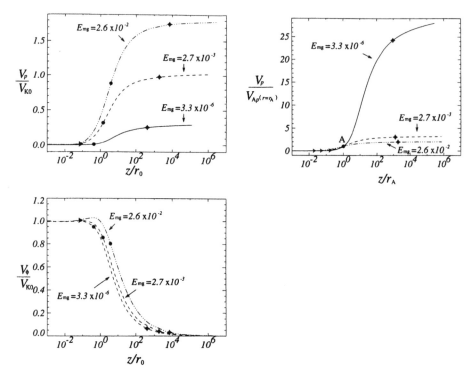

FIGURE 4.47 The typical solutions with the positions of slow (triangle), Alfvén (circle) and fast points (diamond) in 1D magnetically driven jet from a Keplerian accretion disk (Kudoh and Shibata, 1997a). Note that in the case of weak magnetic field $E_{mg} = (v_A/v_k)^2 = 3.3 \times 10^{-6}$), main acceleration of the jet occurs after passing Alfvén point, whereas in the strong field case $(E_{mg} = 2.6 \times 10^{-2})$, main acceleration occurs before/around the Alfvén point (see also Sec. 4.3.4.6).

This is called Michel's *minimum energy solution* (Michel, 1969). This behaviors were confirmed by Belcher and McGregor (1976) and Kudoh and Shibata (1995, 1997a) by examining the 1D solution for various parameter values in the case of fast rotators. Figure 4.47 shows several typical solutions with their positions of slow, Alfvén and fast points.

4.3.3.2 2D Steady Magnetically Driven Wind Theory*

We now extend Weber-Davis 1D theory to 2D theory (Okamoto, 1975; Heineman and Olbert, 1978; Sakurai, 1985). In 2D case, we have a new equation describing the force balance perpendicular to poloidal field lines, called *Grad Shafranov equation*. This equation was originally developed in fusion plasma physics, since it describes the equilibrium of magnetically confined fusion plasmas. Since the case without gravity and flow is very simple and educational, we will discuss such case at first, and then move gradually to more complicated cases.

(a) Grad-Shafranov equation (no flow, no gravity)

The equation for force balance perpendicular to magnetic field lines (i.e., trans-field equation) can be simplified to what is called the Grad-Shafranov equation. Let's derive this GS equation.

The momentum equation is

$$-\nabla p + \frac{1}{4\pi}\mathrm{rot}\mathbf{B}\times\mathbf{B} = 0. \tag{4.3.34}$$

Assuming axisymmetry, we have

$$\mathbf{B} = \nabla\times(A_\varphi\mathbf{e}_\varphi) + B_\varphi\mathbf{e}_\varphi, \tag{4.3.35}$$

$$\Phi(r,z) = rA_\varphi(r,z), \tag{4.3.36}$$

$$\mathbf{B}_p = \nabla\times(A_\varphi\mathbf{e}_\varphi) = \left(\frac{1}{r}\frac{\partial\Phi}{\partial z} - \frac{1}{r}\frac{\partial\Phi}{\partial r}\right), \tag{4.3.37}$$

where Φ is a magnetic flux function. It follows from equation (4.3.37) that

$$\mathbf{B}\cdot\nabla\Phi = 0, \tag{4.3.38}$$

i.e., the flux function Φ is constant along magnetic field lines. From Eq. (4.3.35), we find

$$\mathbf{B}\cdot\nabla p = 0, \tag{4.3.39}$$

i.e.,

$$p = p(\Phi), \tag{4.3.40}$$

showing that the gas pressure is constant along magnetic field lines. From φ component of Eq. (4.3.35), we have

$$\left[\mathrm{rot}\mathbf{B}\times\mathbf{B}\right]_\varphi = \frac{1}{r}\mathbf{B}\cdot\nabla(rB_\varphi) = 0. \tag{4.3.41}$$

This leads to

$$rB_\varphi = I(\Phi). \tag{4.3.42}$$

From equations (4.3.35), (4.3.37), (4.3.40), (4.3.42), we find

$$\nabla p + \frac{1}{4\pi}\frac{\Delta^*\Phi}{r^2}\nabla\Phi + \frac{rB_\varphi}{r^2}\nabla(rB_\varphi) = 0. \tag{4.3.43}$$

369

Here

$$\Delta^*\Phi \equiv r\frac{\partial}{\partial r}\left(\frac{1}{r}\frac{\partial \Phi}{\partial r}\right) + \frac{\partial^2 \Phi}{\partial z^2}. \tag{4.3.44}$$

Equation (4.3.43) is rewritten as

$$\frac{dp(\Phi)}{d\Phi} + \frac{1}{4\pi}\frac{\Delta^*\Phi}{r^2} + \frac{I(\Phi)}{4\pi r^2}\frac{dI(\Phi)}{d\Phi} = 0. \tag{4.3.45}$$

This is the well-known Grad-Shafranov equation.

(b) Generalized Grad-Shafranov equation for rotating, self-gravitating cloud without flow along B

In this case (e.g., Tomisaka *et al.*, 1988), the basic equations are

$$-\nabla p + \frac{1}{4\pi}\text{rot}\mathbf{B} \times \mathbf{B} - \rho\nabla\Psi - \rho\mathbf{\Omega} \times (\mathbf{\Omega} \times \mathbf{r}) = 0, \tag{4.3.46}$$

$$\nabla^2\Psi = 4\pi G\rho, \tag{4.3.47}$$

$$p = \rho C_s^2, \tag{4.3.48}$$

$$\nabla \times (\mathbf{V} \times \mathbf{B}) = \nabla \times (\Omega r\mathbf{e}_\varphi \times \mathbf{B}) = 0, \tag{4.3.49}$$

where Ψ is the gravitational potential. From Eq. (4.3.49) we find

$$\nabla \times (\Omega r\mathbf{e}_\varphi \times \mathbf{B}) = r\mathbf{B} \cdot \nabla\Omega = 0, \tag{4.3.50}$$

i.e.,

$$\Omega = \Omega(\Phi). \tag{4.3.51}$$

This shows that the angular speed is constant along each field line, i.e., each field line rotates rigidly. This is called *Ferraro's law of isorotation*. Then the equation of motion is integrated along the field line, yielding

$$p = q(\Phi)\exp[-(\Psi - r^2\Omega^2/2)/C_s^2]. \tag{4.3.52}$$

From (4.3.46)–(4.3.52), we get the generalized Grad-Shafranov equation;

$$\left(\frac{dq(\Phi)}{d\Phi} + \frac{q(\Phi)}{C_s^2}r^2\Omega\frac{d\Omega(\Phi)}{d\Phi}\right)\exp[-(\Psi - r^2\Omega^2/2)/C_s^2] + \frac{1}{4\pi}\frac{\Delta^*\Phi}{r^2} + \frac{I}{4\pi r^2}\frac{dI(\Phi)}{d\Phi} = 0. \tag{4.3.53}$$

Here the gravitational potential is obtained from the following equation;

$$\Delta\Psi = 4\pi G\frac{q(\Phi)}{C_s^2}\exp[-(\Psi - r^2\Omega^2/2)/C_s^2]. \tag{4.3.54}$$

The case of $\Omega = 0$ has been solved by Mouschovias (1976) and Nakano (1984).

(c) Generalized Grad-Shafranov equation for 2D MHD wind

370

The basic equations for 2D MHD wind (e.g., Heineman and Olbert , 1978, Sakurai, 1985; Lovelace *et al.*, 1986) are written as follows;

$$\rho(\mathbf{V} \cdot \nabla)\mathbf{V} = -\nabla p + \frac{1}{4\pi}\mathrm{rot}\mathbf{B} \times \mathbf{B} - \rho\nabla\Psi, \qquad (4.3.55)$$

$$\nabla \cdot (\rho\mathbf{V}) = 0, \qquad (4.3.56)$$

$$\nabla \times (\mathbf{V} \times \mathbf{B}) = 0, \qquad (4.3.57)$$

$$\mathbf{V} \cdot \nabla(p\rho^{-\gamma}) = 0, \qquad (4.3.58)$$

where

$$\mathbf{B} = \left(\frac{1}{r}\frac{\partial\Phi}{\partial z}, B_\varphi, -\frac{1}{r}\frac{\partial\Phi}{\partial r}\right). \qquad (4.3.59)$$

Equation (4.3.56) yields

$$\mathbf{V} = \alpha\frac{\mathbf{B}}{\rho} + r\Omega\mathbf{e}_\varphi. \qquad (4.3.60)$$

This means that the poloidal velocity vector $\mathbf{V}_p = (V_r, 0, V_z)$ is parallel to the poloidal magnetic field vector $\mathbf{B}_p = (B_r, 0, B_z)$.

Inserting (4.3.59) into (4.3.56), we have

$$(\mathbf{B} \cdot \nabla)\alpha = 0, \qquad (4.3.61)$$

i.e.,

$$\alpha = \alpha(\Phi). \qquad (4.3.62)$$

From equation (4.3.57) we find

$$[\nabla \times (\mathbf{V} \times \mathbf{B})]_\varphi = r\mathbf{B} \cdot \nabla\Omega = 0, \qquad (4.3.63)$$

i.e.,

$$\Omega = \Omega(\Phi). \qquad (4.3.64)$$

We find *Ferraro's law of isorotation* also for the MHD wind; the angular speed is constant along each field line, i.e., each field line rotates rigidly.

From equation (4.3.58), we have

$$p = K(\Phi)\rho^{-\gamma}. \qquad (4.3.65)$$

On the other hand, φ component of equation (4.3.55) yields

$$\mathbf{V} \cdot \nabla(rV_\varphi) = \frac{1}{4\pi\rho}\mathbf{B} \cdot \nabla(rB_\varphi). \qquad (4.3.66)$$

371

Using (4.3.60), this equation becomes

$$r\left(V_\varphi - \frac{B_\varphi}{4\pi\alpha(\Phi)}\right) = \ell(\Phi) = \Omega(\Phi)r_A(\Phi)^2, \tag{4.3.67}$$

which shows angular momentum conservation, where ℓ is a specific angular momentum carried by the centrifugal wind along each field line.

The equation of motion along $\mathbf{B}$ can be integrated to

$$\frac{1}{2}V_p^2 + \frac{1}{2}(V_\varphi - r\Omega)^2 + \frac{\gamma}{\gamma-1}\frac{p}{\rho} + \Psi - \frac{1}{2}\Omega^2 r^2 = E(\Phi). \tag{4.3.68}$$

This is generalized *Bernoulli's equation.* Using (4.3.60), (4.3.63), (4.3.65), (4.3.67), the equation (4.3.68) is rewritten as

$$\frac{1}{2}\frac{\alpha^2}{\rho^2}\left(\frac{\nabla\Phi}{r}\right)^2 + \frac{1}{2}\left[\frac{4\pi\alpha^2\Omega(r_A^2 - r^2)}{r(4\pi\alpha^2 - \rho)}\right] + \frac{\gamma K}{\gamma-1}\rho^{\gamma-1} - \frac{GM}{r} - \frac{\Omega^2 r^2}{2} = E(\Phi). \tag{4.3.69}$$

This equation gives $\rho = \rho(\Phi, r)$.

From equations (4.3.60) and (4.3.67), we have

$$V_\varphi - r\Omega = \frac{4\pi\alpha^2\Omega}{r}\frac{(r_A^2 - r^2)}{(4\pi\alpha^2 - \rho)}. \tag{4.3.70}$$

Therefore, $r = r_A$ at the point where

$$\rho = \rho_A \equiv 4\pi\alpha^2. \tag{4.3.71}$$

The latter condition is rewritten as

$$V_p^2 = \frac{\alpha^2 B_p^2}{\rho^2} = \frac{B_p^2}{4\pi\rho} = V_{Ap}^2. \tag{4.3.72}$$

Hence we find that V_p becomes equal to V_{Ap} when r is equal to the Alfvén raidus r_A.

Let us now consider the force balance perpendicular to magnetic field. After some manipulation, the equation of motion (4.3.55) perpendicular to $\mathbf{B}$ becomes

$$\nabla\cdot\left[\left(\frac{\alpha^2}{\rho} - \frac{1}{4\pi}\right)\frac{\nabla\Phi}{r^2}\right] = \left(\frac{\alpha^2}{\rho} - \frac{1}{4\pi}\right)\frac{\Delta^*\Phi}{r^2} + \nabla\left(\frac{\alpha^2}{\rho}\right)\frac{\nabla\Phi}{r^2}$$

$$= \rho\left(E' - \frac{1}{\gamma-1}\frac{K'}{K}\frac{p}{\rho} + r^2\Omega\Omega'\right) + \frac{B_p^2}{\rho}\alpha\alpha' + Q\left[\frac{Q}{4\pi}\Omega^2 r^2\alpha\alpha' - \alpha^2\Omega^2(r_A^2)' - \alpha^2\Omega\Omega'(r_A^2 - r^2)\right], \tag{4.3.73}$$

where

$$Q = \frac{4\pi\rho(r_A^2 - r^2)}{r^2(4\pi\alpha^2 - \rho)}, \tag{4.3.74}$$

and $' = d/d\Phi$. This is the *generalized Grad-Shafranov equation* for the 2D centrifugal wind problem, and is sometimes called *trans-field equation*.

The equation (4.3.73) is a second order quasi-linear partial differential equation for Φ. The main operators in this equation are

$$\frac{1}{4\pi r^2}\left(\frac{V_p^2}{V_{Ap}^2} - 1\right)\left(A_{rr}\frac{\partial^2 \Phi}{\partial r^2} + A_{rz}\frac{\partial^2 \Phi}{\partial r\partial z} + A_{zz}\frac{\partial^2 \Phi}{\partial z^2}\right), \tag{4.3.75}$$

where

$$A_{rr} = 1 - V_p^2 V_z^2 / R, \tag{4.3.76}$$

$$A_{rz} = 2V_p^2 V_r V_z / R, \tag{4.3.77}$$

$$A_{zz} = 1 - V_p^2 V_r^2 / R, \tag{4.3.78}$$

$$R = V_p^4 - V_p^2(C_s^2 + V_{Ap}^2 + V_{A\varphi}^2) + C_s^2 V_{Ap}^2. \tag{4.3.79}$$

The first bracket in equation (4.3.75) shows that this equation has a singularity at $V_p = V_{Ap}$ (i.e., at the Alfvén radius). The second bracket, on the other hand, shows that if

$$D \equiv A_{rz}^2 - 4A_{rr}A_{zz} = \left[V_p^2(C_s^2 + V_{Ap}^2 + V_{A\varphi}^2) - C_s^2 V_{Ap}^2\right]/R^2 \tag{4.3.80}$$

is less than zero, the equation is an elliptic type, whereas if $D > 0$, it is a hyperbolic type. Here the equation is rewritten as

$$D = (C_s^2 + V_{Ap}^2 + V_{A\varphi}^2)\frac{(V_p^2 - V_{cp}^2)}{(V_p^2 - V_{sp}^2)(V_p^2 - V_{fp}^2)}, \tag{4.3.81}$$

where

$$V_{cp} \equiv \frac{C_s V_{Ap}}{\left(C_s^2 + V_{Ap}^2 + V_{A\varphi}^2\right)^{1/2}}, \tag{4.3.82}$$

and V_{sp} and V_{fp} are solutions of the equation (4.3.79) $= 0$. Hence the equation becomes hyperbolic for $V_p > V_{fp}$ and $V_{cp} < V_p < V_{sp}$ and elliptic for $V_p < V_{cp}$ and $V_{sp} < V_p < V_{fp}$. Blandford and Payne (1982) found a self-similar solution for steady 2D magneto-centrifugally driven jets ejected from the accretion disk, using cold approximation (Fig. 4.48). They noticed that if the polar angle of the poloidal field line become less than 60 degree, the centrifugal force becomes larger than the gravitational force on the field line rotating at the Kepler speed of the footpoint of the field line (Fig. 4.49).

Sakurai (1985) first numerically solved the generalized GS equation for the magneto-thermal-centrifugally driven wind (slow rotator), and found that the wind is *collimated* towards the rotation axis by the *pinch* of toroidal fields. Fig. 4.50 shows one example of his solution. Sakurai (1987) then applied his method to obtain 2D steady solution of the jet ejected from the accretion disk. Heyvaerts and Norman (1989) analytically established that any stationary axisymmetric magnetized wind will collimate along the symmetry axis at large distances from the source. (See also Tomimatsu, 1994, for the collimation of a special *relativistic MHD wind*.)

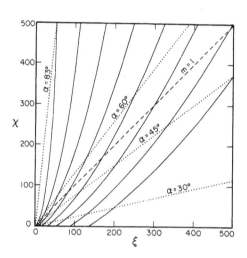

FIGURE 4.48 A self-similar solution of 2D steady magneto-centrifugally driven jet from accretion disk (Blandford and Payne, 1982).

4.3.4   Nonsteady MHD Jets

If the accretion disk (or simply the rotating disk) is penetrated by global poloidal magnetic field, the rotating disk helically twists ploidal field, generating torsional Alfvén wave propagating into both polar directions. When the amplitude of the Alfvén wave is sufficiently large ($v_\varphi \sim V_A$), the magnetic pressure due to $\nabla B_\varphi^2/8\pi$ associated with Alfvén wave can no longer be neglected. Similarly, the centrifugal force associated with the rotation of the disk can not be neglected if the polar angle of the field lines becomes sufficiently large (Blandford and Payne, 1982). In these cases, the mangnetically acceleleted plasma jet appears along the poloidal mangetic field. If the pinch force due to $B_\varphi^2/4\pi r$ becomes sufficiently strong, the flow can be collimated by the pinch force around the polar axis, and it would be seen as a jet. When the acceleration process stably continues for many disk orbital periods, the resulting jet would become a steady jet/wind as discussed in previous sections. However, it is also possible that the intrinsic acceleration mechanism is nonsteady, or *intermittent*, near

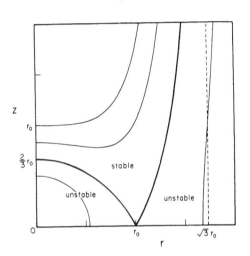

FIGURE 4.49 The effective potential around the particular point of accretion disk (Blandford and Payne 1982). If the polar angle is smaller than 60 degrees, the plasmas on the field lines become unstable and start to flow outwards.

the boundary between the central object (black hole or star) and the accretion disk. (See also Sec. 2.8 and Sec. 4.3.5.) Recent high resolution observations of superluminal jets in the Galaxy (Mirabel and Rodriguez, 1994) and YSO jets (e.g. Ray and Mundt, 1993) show, in fact, evidence of intermittent ejection of jets. Thus, it is very important to study nonsteady MHD jets ejected from accretion disks (Uchida and Shibata, 1985; Shibata and Uchida, 1985, 1986a,b, 1987, 1990; Stone and Norman, 1994; Ustyugova *et al.*, 1995; Matsumoto *et al.*, 1996a; Hayashi *et al.*, 1996; Ouyed *et al.*, 1997; Kudoh and Shibata, 1997b; Hirose*et al.*, 1997).

It will also be interesting to study the long term behavior of nonsteady jets to see whether they tend to a quasi-steady state predicted by steady theory or not.

Top-View   Side-View

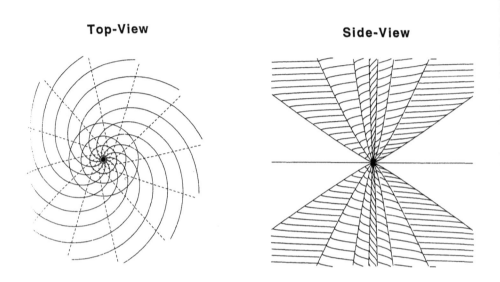

FIGURE 4.50 A numerical solution for 2D magneto-thermal-centrifugally driven wind (slow rotator) (Sakurai, 1985).

4.3.4.1 Magnetic Twist Jets Produced by the Interaction between a Rotating Disk and a Poloidal Magnetic Field

Consider a rotating disk penetrated by uniform poloidal (vertical) magnetic field at a rotational velocity smaller than the Keplerian speed. In this case, the rotating disk will helically twist magnetic field lines as well as contract toward the inner radius, generating strong torsional Alfvén wave. This *nonlinear torsional Alfvén wave* exert magnetic pressure gradient force and centrifugal force to accelerate surface plasma in the disk to both bipolar directions. This process produces bipolar jets. Similar processes can occur even if initially the centrifugal force balances with the gravitational force (i.e. Keplerian case), because the *magnetic braking* of the disk eventually enables the disk contract. Shibata and Uchida (1985a, 1986a,b, 1990a) studied this problem in detail, by performing 2.5D nonsteady MHD numerical simulations, and Uchida and Shibata (1985) applied the result to the formation of bipolar molecular flows in star forming regions (Fig. 4.51; see also related theoretical models e.g., Pudritz and Norman, 1986; Lovelace *et al.*, 1991; Königl and Ruden, 1993). On the other hand, Uchida

376

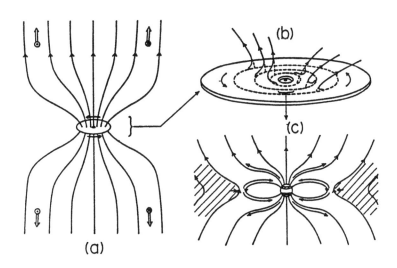

FIGURE 4.51   The situation of the bipolar flow model of Uchida and Shibata (1985).

et al. (1985) and Shibata and Uchida (1987) applied this mechanism to the Galactic center radio lobe (Sofue and Handa, 1984).

Figure 4.52 illustrates typical example of the nonsteady MHD jet model simulated by Shibata and Uchida (1986a,b). Initially ($t = 0$), a cold dense disk is rotating around a point mass at the origin ($x = 0, z = 0$), at sub-Keplerian rotation velocity. Around the disk, there is a non-rotating hot corona in hydrostatic equilibrium. Pressure balance is assumed between the corona and the disk. A uniform magnetic field penetrates both the disk and the corona. There is no equilibrium in vertical direction in the disk; i.e., the disk mass freely collapse to the equatorial plane, but does not much affect the overall dynamics. Further assumptions are, (1) no self-gravity (only external gravity produced by a point mass), (2) ideal MHD, (3) 2.5D axisymmetry, (4) Newtonian. There are fundamentally four free parameters for the initial condition of this problem;

$$R_1 = (C_s/V_k)^2 = \frac{\text{thermal energy}}{\text{gravitational energy}}, \qquad (4.3.83)$$

$$R_2 = (V_A/V_k)^2 = \frac{\text{magnetic energy}}{\text{gravitational energy}} \tag{4.3.84}$$

$$R_3 = (V_\varphi/V_k)^2 = \frac{\text{rotational energy}}{\text{gravitational energy}} \tag{4.3.85}$$

$$R_4 = T_{\text{corona}}/T_{\text{disk}} = \rho_{\text{disk}}/\rho_{\text{corona}}, \tag{4.3.86}$$

where V_k is the Keplerian velocity $(= (GM/r)^{1/2})$, V_A and C_s are the Alfvén speed and the sound speed in the disk, V_φ is the rotation velocity, T_{corona} and T_{disk} are temperatures of the corona and the disk, respectively. The boundary conditions are as follows: Outer boundaries are free boundaries through which plasmas, waves, magnetic flux pass freely. There is an inner boundary around the origin (a point mass), which is also assumed to be free. The unit of length, velocity, time are $r_0, C_{s,c0}, t_0$, where

$$t_0 = r_0/C_{s,c0} = t_{ff}(R_4 R_1)^{-1/2}, \tag{4.3.87}$$

and $C_{s,c0}$ is the sound speed in the initial corona at $(r^2 + z^2)^{1/2} = r_0$. Here r_0 is the inner radius of the initial disk, $t_{ff} = r_0/V_k$. Sometimes, we use the parameters β and α, which are related to other parameters as

$$\beta = 8\pi p_{\text{gas},d0}/B_0^2 = \frac{2C_{s,d0}^2}{\gamma V_{A,d0}^2} = \frac{2}{\gamma}\frac{R_1}{R_2}, \tag{4.3.88}$$

$$\alpha = \frac{V_\varphi}{V_k} = R_3^{1/2}. \tag{4.3.89}$$

Figures 4.52 and 4.53 shows typical examples of the formation of a magnetic twist jet by this mechanism (Shibata and Uchida, 1986a,b). The parameters are $(R_1, R_2, R_3, R_4) =$ (0.03, 0.0072, 0.64, 400), $(\beta = 0.5)$. The grid size is 0.02, and the total grid numbers are 80×80. The resulting time evolution is divided into the following three stages.

1) stage 1 $(0 < t/t_{ff} < 1$, linear stage): In this stage, we see generation and propagation of torsional Alfvén waves through the corona. The disk begins to contract toward the inner region, which produces also disturbance (fast mode MHD wave). It is interesting to note that the contraction velocity of the disk surface gas is slightly larger than that of the disk inner gas. This is clearly seen as the bending of magnetic field lines at the disk surface. This is because the braking effect (angular momentum loss rate per unit mass) is largest at the disk surface, and essentially the same as Balbus-Hawley mechanism (Sec. 4.2.2). In this case the sub-Keplerian velocity $(V_\varphi = R_3^{1/2}V_k = 0.8V_k)$ is assumed, but the magnetic braking still plays an important role.

2) stage 2 $(1 < t/t_{ff} < 1.7$, polar coronal flow stage): As the disk contracts toward the inner region, the magnetic field lines are pulled also toward the central axis, compressing the coronal plasmas. The enhanced gas pressure in the polar coronal tube drives the weak

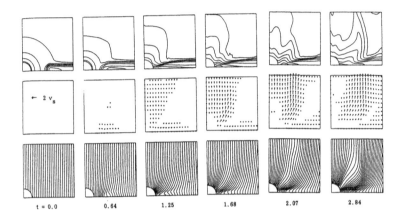

FIGURE 4.52 2.5D MHD sim. of magnetic twist jet (Shibata and Uchida, 1986a): ρ, v, B.

plasma outflow from the corona along the polar coronal flux tube.

3) stage 3 ($1.7 < t/t_{ff} < 3.2$, cold jet stage): As the disk contracts toward the inner region, the magnetic twist is further generated near the surface of the disk. Finally, at $t/t_{ff} \simeq 1.5 - 1.7$, the gas in the surface layer of the disk begins to be accelerated due to $\mathbf{J} \times \mathbf{B}$ force (magnetic pressure grandient force and centrifugal force), and forms a hollow cylindrical bopolar jet. The final velocity of the jet is about $1 - 2V_k \sim V_A(\text{disk} - \text{surface})$. Note also that there is a tenuous hot magnetic twist jet ahead of the cold dense jet, It is important to note that the rotational velocity V_φ is seen just on the jet. This may be seen as helical velocity.

From Fig. 4.54, we find gas in the surface layer of the disk is ejected upward, while gas inside the disk continues to infall. The angular momentum loss is fast near the disk surface and slow inside the disk (Fig. 4.54c), while the coronal plasmas quickly gain angular momentum from the disk. After the jet is ejected, the angular momentum loss and gain become much faster, because of the inclined magnetic field line configuration and the ejection

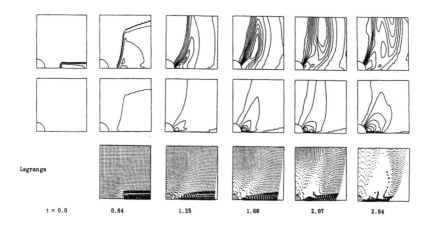

Lagrange

t = 0.0   0.64   1.25   1.68   2.07   2.84

FIGURE 4.53   2.5D MHD sim. of magnetic twist jet (Shibata and Uchida, 1986a): v_φ, B_φ, Lagrange.

of jet.

Figure 4.55 shows the dependence on the plasma β. It is seen that the magnetic field is more rigid in low $\beta(= 0.3)$ case, while it is more undulating in high $\beta(= 5)$ case. It is also found that the velocity of jet is higher in low β case than in high β case. Empirically, this is written as

$$V_{\text{jet}} \sim \beta^{-0.3 \sim -0.4} \sim B^{0.5 \sim 0.7}. \tag{4.3.90}$$

Interestingly, this relation is roughly in agreement with the relation of Michel's minimum energy solution for a fast rotator (equation (4.3.35) in previous subsection). According to Iki and Shibata (1986, unpublished), the low β jet is accelerated mainly by the centrifugal force, whereas the high β jet is by the magnetic pressure gradient force. The dependence on the initial rotational velocity is shown in Figure 4.56. From this, we find that even the initially Keplerian case leads to contraction due to magnetic braking and hence eventually the jet, though the velocity of the jet is slower than the sub-Keplerian case. More detailed comparison between the jet formation and the Balbus-Hawley mchanism (or Velikhov-Chandrasekhar

380

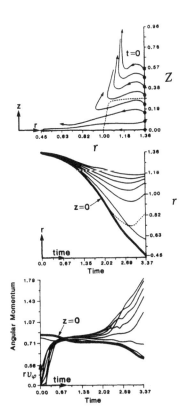

FIGURE 4.54 2.5D MHD sim. of magnetic twist jet (Shibata and Uchida, 1986b): Trajectories of test particles, angular momentum.

instability; see Sec. 4.2) is discussed in Stone and Norman (1994).

Finally, a comment is given on the mass loss and velocity. The mass loss rate due to the jet is generally of order of 0.1 or less of the disk mass accretion rate. The velocity of the jet is comaparable to the Keplerian velocity at the inner edge of the initial disk, though it increases with increasing field strength.

4.3.4.2 Interaction of Magnetic Twist Jets with Interstellar Clouds*

The previous magnetic twist jet model shows some key physical processes of the acceleration of the jet. However the region of simulation is limited near the inner edge of the initial disk, so that it was not possible to see large scale propagation.

Motivated by the observations which show some dense blob structure inside the bipolar flow L1551 (Uchida *et al.*, 1987), Shibata and Uchida (1990) have studied the interaction between the cloud and the jet, by developing the code to inlude non-uniform meshes. Then

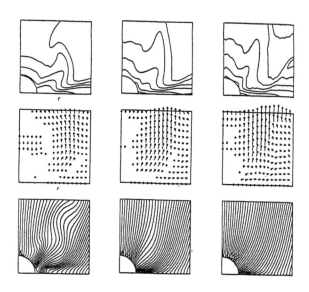

FIGURE 4.55 2.5D MHD sim. of magnetic twist jet (Shibata and Uchida, 1986b): Dependence on β.

they succeeded in obtaining a larger scale simulation, as shown in Fig. 4.57. Figure 4.57 shows three dimensional magnetic field configuration of a magnetic twist jet a in much larger scale (Shibata and Uchida, 1990). The jet corresponds to a tightly twisted part emerging after $t = 2.37$. It is clearly shown that the torsinal Alfvén wave (mild twist) propagates ahead of the magnetic twist jet.

In the above model, the collimation is mainly due to the initial poloidal magnetic field lines near the disk. However, if the initial magnetic field is diverging, the jet becomes collimated by the pinch of toroidal fields as shown in Fig. 4.58 (Shibata, 1996). This is the same as in the collimation of a steady centrifugally driven jet/wind (Sakurai, 1987; Heyvaerts and Norman, 1989; Sauty and Tsinganos, 1994).

4.3.4.3 Relation to Magnetorotational (Balbus-Hawley) Instability

Stone and Norman (1994) studied the same problem as that of Shibata and Uchida (1986a,b), and confirmed the basic points on the production of jets. The new feature they found is the

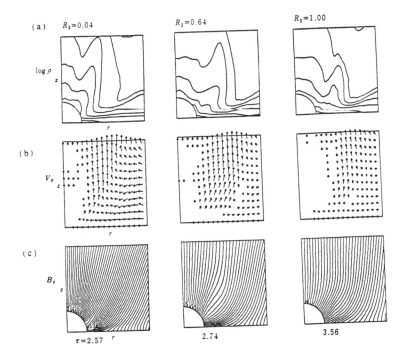

(a) $R_1=0.04$ $R_1=0.64$ $R_1=1.00$

$\log \rho$
z

r

(b)

V_p
z

r

(c)

B_p
z

$\tau=2.57$ r 2.74   3.56

FIGURE 4.56   2.5D MHD sim. of magnetic twist jet (Shibata and Uchida, 1986b): Dependence on $\alpha = v_\varphi/v_k$.

magneto-rotational instability (Balbus and Hawley, 1991) when the initial magnetic field is weak.[4]

It is interesting to note that some of the key feature of this magneto-rotational instability, such as avalanche (accretion) flow at the surface of the disk, have already been seen in some results of Shibata and Uchida's simulations (Shibata and Uchida, 1986a,b, 1987, 1990) as well as in early results of Matsumoto *et al.*'s simulations of a magnetized thick disks before 1990 (see below). Although this feature was physically understood at that time, it has never been explicitly emphasized nor been studied in detail. This is because the spatial resolution of these simulations was not good enough. (From this history, we can learn that we have to consider numerical results more seriously even if the spatial resolution is not so good. There are a lot of hints in numerical simulation results from which we can develop important theories.)

---

[4]It may be said that even the basic physics of nonsteady MHD jet-accretion (Shibata and Uchida, 1986a,b; Stone and Norman, 1994) is the same as that of the Balbus-Hawley (1991) instability.

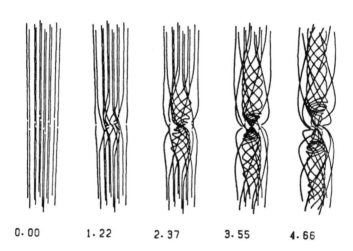

| 0.00 | 1.22 | 2.37 | 3.55 | 4.66 |

FIGURE 4.57 3D magnetic field configuration of *magnetic-twist* jet developed by Shibata and Uchida (1990).

4.3.4.4 Nonsteady MHD Jets from Thick Disks*

In order to develop an AGN jet model, Matsumoto *et al.* (1996a) studied the case of thick disks penetrated by the poloidal field, by performing nonsteady 2.5D MHD simulations similar to Shibata and Uchida (1986a,b). (See Sec. 4.2.2.5.) Their results (see Fig. 4.59) show; (1) Avalanche (accretion) flow occurs along the surface of thick disks. (2) Magneto-rotational instability (Balbus and Hawley, 1991) occurs inside thick disks. (3) The velocity of jets is again comparable to the Keplerian velocity, $V_{\text{jet}} \sim 0.6 - 2.3 V_k \propto B^{0.15-0.25}$ if initially $(V_A/V_\varphi)^2 < 10^{-3}$. (4) The mass loss rate by the jet $\dot{M}$ is in proportion to magnetic field strength; $\dot{M} \propto B$. For the three-dimensional cases see Fig. 4.60.

4.3.4.5 Interaction between Stellar Magnetosphere and Accretion Disk Magnetic Field*

Hirose (1994); (Hirose *et al.*, 1997) studied the reconnection between (non-rotating) protostellar dipole magnetic field and poloidal magnetic fields carried by an accretion disk, as an

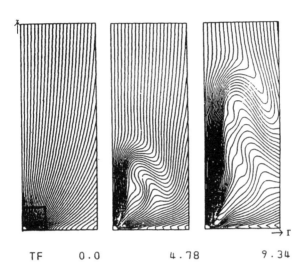

TF   0.0   4.78   9.34

FIGURE 4.58 Initially diverging field case (Shibata, 1996).

extension of the previous Uchida–Shibata (1984, 1985) model of protostellar jets. In their case, a neutral line (or a neutral ring in 3D space) is formed between stellar magnetic fields and disk magnetic field. (This is, in some sense, similar to dayside reconnection of the terrestrial magnetosphere.) They found that the reconnection itself accelerates plasma directly from the disk, in addition to the general acceleration of disk plasma via the $J \times B$ force (the centrifugal force and magnetic pressure gradient force of toroidal fields), and suggested that optical jets may be generated by this reconnection acceleration. This reconneciton process is important to understand the evolution of protostellar rotation, and hence in future the interaction between a protostellar magnetosphere and a magnetized accretion disk will have to be studied in detail.

Hayashi *et al.* (1996), on the other hand, studied the interaction of a stellar dipole field with an accretion disk, assuming that the dipole field initially threads the accretion disk. This situation is similar to that of Shu *et al.* (1994)'s X-wind model. They found the following: As the closed magnetic loops connecting the central star and the disk are twisted by the rotation of the disk more and more, magnetic loops expand and finally trigger the magnetic

385

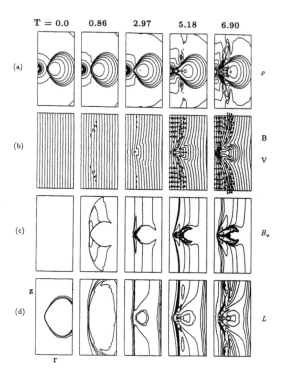

$$\mathbf{T} = 0.0 \quad 0.86 \quad 2.97 \quad 5.18 \quad 6.90$$

FIGURE 4.59   2.5D MHD simulation of a magnetic-twist jet ejected from a thick disk (Matsumoto *et al.*, 1996a).

reconnection inside the expanding loop to eject a detached magnetic island (plasmoid) similar to solar coronal mass ejection. They proposed that this process explains both hard X-ray flares observed by the ASCA satellite (Koyama *et al.*, 1994) and optical jets.

4.3.4.6   Relation to Steady Wind/Jet and Remaining Questions

What is the relation between these nonsteady MHD jets and steady magnetically driven jets (Blandford and Payne, 1982)? In order to answer this question, Kudoh and Shibata (1995, 1997a) studied one dimensional (1.5D) steady magnetically driven jets along a fixed poloidal field line for a wide range of parameters, assuming the shape of the poloidal magnetic field line. There are two free parameters in their problem;

$$E_{mg} = \left(\frac{V_A}{V_k}\right)^2 \simeq 3.8 \times 10^{-4} \left(\frac{B}{1\mathrm{G}}\right)^2 \left(\frac{M}{M_\odot}\right)^{-1} \left(\frac{n}{10^{12}\mathrm{cm}^{-3}}\right)^{-1} \left(\frac{R}{15R_\odot}\right),$$

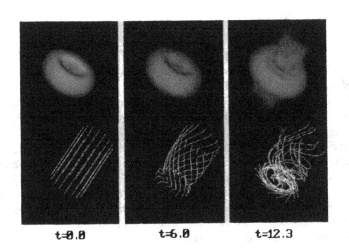

t=0.0   t=6.0   t=12.3

FIGURE 4.60   3D MHD simulation of a magnetic-twist jet ejected from a thick disk (from Matsumoto *et al.*, 1996b).

$$E_{th} = \frac{1}{\gamma}\left(\frac{C_s}{V_k}\right)^2 \simeq 6.5 \times 10^{-3}\left(\frac{T_d}{10^4 \text{K}}\right)\left(\frac{M}{M_\odot}\right)^{-1}\left(\frac{R}{15 R_\odot}\right),$$

where V_A is the Alfvén speed based on the poloidal field, and the physical values in these equations are suitable for protostellar jets. They found that the inclination angle of poloidal magnetic field lines at the surface of accretion disks is important to determine the properties of the jet as first noted by Blandford and Payne (1982) and later stressed by Cao and Spruit (1994). As the angle between the poloidal field and the disk surface decreases, mass flux of the jets increases. If the angle becomes less than 60 degree, a high mass flux jet with strong toroidal fields can arise, depending on the poloidal field strength. Namely, in such a low angle case, the solution can be generally classified into two branches,

(1) *centrifugally driven jet*,

(2) *magnetic pressure driven jet*.

The former arises when the poloidal field is strong, i.e., the poloidal field is dominant near the surface of the disk, whereas the latter arises when the poloidal field is weak, i.e., the

387

toroidal field is dominant near the disk (Fig. 4.61). In the former case, the main acceleration occurs below the Alfvén point by the centrifugal force, whereas in the latter the acceleration occurs above the Alfvén point by the magnetic pressure force of the toroidal field. The latter branch corresponds to the steady MHD jet model developed by Lovelace *et al.* (1991); see also related 2.5D work by Ustyugova *et al.* (1995). In the case of *magnetic pressure driven jet*, (i.e. $E_{mg} < 0.01$), Kudoh and Shibata (1995, 1997a) further found;

(1) the mass loss rate is in proportion to magnetic field strength, i.e. $\dot{M} \propto E_{mg}^{0.5} \propto B$,

(2) the terminal velocity of the jet becomes Michel's minimum energy solution, though the dependence on B becomes weaker than had been thought for centrifugally driven jet because of $\dot{M} \propto B$;

$$V_\infty \propto (E_{mg}/\dot{M})^{1/3} \propto B^{1/3},$$

(3) the terminal velocity is of order of Keplerian velocity at the footpoint of the jet for a wide range of parameters; $V_\infty \propto 0.5 - 4.0 V_k$ for $10^{-5} < E_{mg} < 0.1$.

These results explain simulation results of Matsumoto *et al.* (1996a) very well, and explain also previous 2.5D nonsteady MHD simulations (Shibata and Uchida, 1986a,b, 1987, 1990; Stone and Norman, 1994) which showed that the velocity of a nonsteady jet is an order of the Keplerian velocity and the toroidal magnetic field is dominant near the disk.[5] On the basis of these results, Kudoh and Shibata (1995, 1997a) predicted the physical condition of the disk from which YSO optical jets and/or high velocity neutral winds are ejected. Namely, the jet production radius must be less than 20 $R_\odot$ (for one solar mass protostar) to explain the velocity of optical jets and/or high velocity neutral wind, and the magnetic field strength $> 10\,G$, and the temperature must be larger than 1000 K at $r \simeq 15\ R_\odot$.

Using the same poloidal field line configuration, Kudoh and Shibata (1997b) further studied nonsteady 1.5D MHD simulations assuming the initial and boundary conditions similar to those of Shibata and Uchida (1986a) to see the relation nonsteady jets and steady jets. Note that the boundary condition of this problem is not suitable to obtain a steady solution. Nevertheless, such a boundary condition is adopted to see the long term behavior of Shibata-Uchida simulation model. (Note also that still at present it is not easy to perform long term 2.5D MHD simulation of this model.) Although the boundary condition is not suitable to obtain a steady solution, Kudoh and Shibata found that the solution shows some characteristics of the steady solution in the early stage of evolution, around 10 orbital periods. When magnetic field is weak, the positions of Alfvén and slow points agree well with those of the steady solution. Even in the cases where these positions do not agree with those of the steady solution, the dependences of the maximum velocity and mass loss rate of jets upon E_{mg} agree with those found in the steady solution (Fig. 4.61). Since the dynamics of 1.5D MHD model are common with that of 2.5D MHD model except for accretion, we can say that the characteristics of nonsteady jets found in 2.5D MHD model are essentially similar to those of the steady model. For example, the maximum velocity of jets found in 2.5D MHD model ($V_{jet} \sim 1 - 2 V_k$) is explained by the steady model well.

---

[5]Strictly speaking, the results of Shibata and Uchida (1986a,b) are intermediate case between *centrifugally driven jet* end *magnetic pressure driven jet*, and hence the jet velocity showed stronger dependence on B than the pure magnetic pressure driven jet.

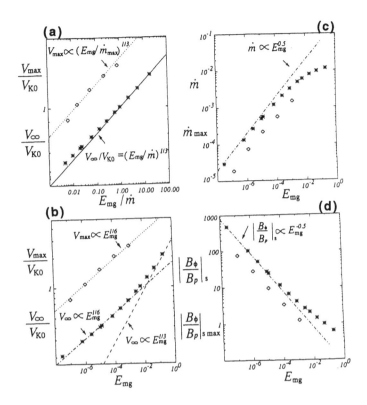

FIGURE 4.61 1.5D steady and nonsteady magnetically driven jet model developed by Kudoh and Shibata (1995, 1997a,b). (a) Terminal velocity for steady case (asterisk) and maximum velocity for nonsteady case (diamond) as a function of $E_{mg}/\dot{m}$, where $\dot{m}$ is normalized by $\rho_d V_k R^2$. The solid line shows michel's minimum energy solution. (b) The normalized mass flux versus E_{mg}. (c) Terminal (or maximum) velocity as a function of E_{mg}. (d) The ratio of the toroidal field to the poloidal field at slow point, B_φ/B_p, versus E_{mg}. In these figures, the steady solution is denoted by asterisk, and the nonsteady solution is shown by diamond.

4.3.4.7 Three-Dimensional Propagation of MHD Jets*

Todo *et al.* (1993) have studied the stability of MHD jets by performing three dimensional MHD simulations. In their model, a jet was assumed initially to be propagating along a helically twisted flux tube. The initial helical field is force free and stable for the kink instability. A dense region is located somewhere ahead of the jet, and the subsequent interaction between the helical jet and the dense region was studied in detail. [See Todo *et al.* (1992) for 2D version, and Uchida *et al.* (1992) for 1D version.] It was found that the collision of the *magnetic-twist-jet* with the dense region produces strong magnetic twist ($B_\varphi \gg B_z$) (Shibata, 1990) which eventually becomes unstable for the kink instability in three dimensional space. The resulting magnetic field and velocity field configurations are similar to those observed around the filamentary structures seen in some HH objects in star forming regions (Todo, 1993).

4.3.5 High Energy Particle Acceleration

4.3.5.1 γ ray Bursts with Alfvén Waves

It is widely believed that some γ-ray bursts particularly those of the galactic origin come from neutron stars. A disturbance near the surface of a magnetized neutron star, whether due to a "starquake," an instability in an accretion disk, or some other phenomenon, can launch strong compressional Alfvén waves into the atmosphere. Such waves can accelerate particles by magnetic trapping; as the density of the atmosphere decreases away from the surface of the star, the condition for trapping can be satisfied for an extended time, resulting in acceleration of some particles to high energies. The mechanism (Holcomb and Tajima, 1991) of *phase-locked acceleration* naturally produces a fairly hard, *power-law spectrum* (see Sec. 2.6) when particle energy is converted to photons by bremsstrahlung. The photon spectra from the simple model can qualitatively reproduce several features of observed spectra. The mechanism can also account for the paucity of X-ray energy, relative to the energy in γ-rays, that is observed in γ-ray bursts. It is not sensitive to the underlying model and can account for the features of bursters for both the "starquake" and the "disk" models; in the case of the "disk" model, we can also explain the observation in temporally-resolved data of a burst of gamma rays, followed by X-rays.

Gamma-ray bursts have remained unexplained since their discovery (Klebesadel *et al.*, 1973). Their spatial distribution is at least consistent with isotropy, to observed distances (Hartmann, 1989, 1990), so either they are of extragalactic origin, or else they arise in an isotropic component of the Galaxy. Some observations of the bursts show lines which have been associated with cyclotron absorption (Mazets *et al.*, 1981; Murakami, 1988); this implies the presence of strong magnetic fields, which suggests that in some cases magnetized neutron stars are the source. Theoretical modeling of line formation in neutron-star atmospheres (Fenimore *et al.*, 1988; Alexander and Mészáros, 1989; Wang *et al.*, 1989) provides further support for this interpretation. Other evidence, such as the possible coincidence of a burst with a supernova remnant in the LMC (Cline *et al.*, 1980), points toward neutron stars. Many theoretical models of gamma-ray bursters have relied on neutron stars as the source, with the energy for the gamma rays provided by such mechanisms as a "starquake" (Pacini and Ruderman, 1974; Tsyganenko, 1975; Fabian *et al.*, 1976; Mitrofanov, 1984; Muslinov and Tsygan, 1986; Epstein, 1988; Blaes *et al.*, 1989), nuclear burning of accreted matter (Hameury *et al.*, 1982; Woosley and Wallace, 1982), and instabilities in an accretion disk around an old neutron star (Michel, 1985; Epstein, 1985; Melia *et al.*, 1986). The paucity of low-energy X-ray emission from the bursters implies that the bursts form in a region of fairly low density (Ichimaru, 1987); other models (Ho and Epstein, 1989; Dermer, 1990) do not require this criterion, but must generally demand high electron currents in the emission region. [γ-ray bursts from blackhole candidates (Tanaka, 1989) have been mentioned in

It is believed that young neutron stars are surrounded by $e^+ - e^-$ plasma (Goldreich and Julian, 1969). The gravitational field of the neutron star will result in a density gradient in the atmosphere, with high-density plasma close to the star and lower densities farther out. A disturbance in the atmosphere could launch a torsional or compressional wave, such as an Alfvén wave, into the relatively dense electron-positron atmosphere of the neutron star. In a hot plasma, a compressional Alfvén wave, for example, can have an associated electric field with a component parallel to the propagation. A longitudinal electric field could also be generated by nonlinear or nonuniform effects. The presence of a longitudinal electric field as well as an ambient magnetic field can allow an Alfvén wave to accelerate particles, in both the transverse and the longitudinal directions, by the magnetic trapping mechanism (see Sakai and Ohsawa, 1987a,b) in which particles whose velocity matches the phase velocity of the wave absorb energy from the wave. It has been shown (Meerson et al., 1990) that in the case of a wave with a nonuniform phase velocity, detrapping can be delayed, and particles can be accelerated to high energies, if the phase velocity increases appropriately as the acceleration proceeds. In our model, the decreasing density produces continuous acceleration, under appropriate conditions, as the wave propagates into an atmosphere in which the Alfvén velocity is increasing. Moreover, under certain circumstances, a pulse of accelerated particles propagates rapidly into the low-density region of the atmosphere.

4.3.5.2 Explosive Magnetic Reconnection in a Disk and Associated Acceleration*

Electromagnetic field evolution occurring in an accretion disk around a compact object can be explosive and gives rise to an explosive mechanism of particle acceleration (Haswell and Tajima, 1992). Flux-freezing in the differentially rotating disk causes the seed and/or generated magnetic field to wrap up tightly (see Sec. 4.2.3), becoming highly sheared and locally predominantly azimuthal in orientation. We show how asymptotically nonlinear solutions for the electromagnetic fields may arise in isolated plasma blobs. These fields are capable of rapidly accelerating charged particles from the disk. Our results have implications for the hard component of astrophysical jets, for accretion disk coronae, and for cosmic rays. In particular, acceleration through the present mechanism from active galactic nuclei (AGN) can give rise to energies beyond 10^{20} eV. Such a mechanism may present an explanation for the extragalactic origin of the most energetic observed cosmic rays.

We consider field evolution driven by the (MHD) fluid dynamics in the disk. The particle acceleration in our model arises as a result of electromagnetic field evolution. Magnetic field compression and reconnection take place periodically as a result of the differentially rotating fluid and the highly sheared, predominantly azimuthal magnetic field (Tajima and Gilden, 1987). The regeneration or sustenance of the magnetic field due to the dynamo effect occurs continuously. We focus on the behavior of elements of plasma at or near the inner edge of the accretion disk, where the accretion of matter permeated with magnetic field onto the compact object helps to drive the enhancement of the magnetic field. A cartoon of the resulting magnetic collapse and reconnection events, and the consequent particle acceleration,

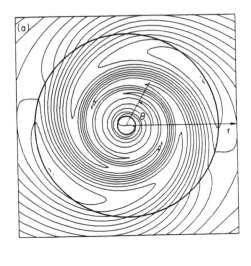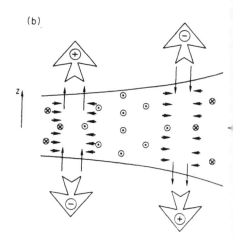

FIGURE 4.62 Cartoon depictions of the magnetic field topology. (a) The structure found in the Tajima-Gilden ▮ simulations. The magnetic field is predominantly azimuthal, exhibiting a banded structure with many reversals in mag▮ field direction. (b) A cross-section of a portion of the inner disk showing two adjacent magnetic reversals. This car▮ illustrates the magnetic collapse process explored in this paper. The magnetic field direction is indicated with the star▮ symbols. The small, dark arrows show the direction of the bulk MHD fluid flow; the long thin arrows show the ele▮ field vectors, and the larger, unfilled arrows indicate the direction of ejection of charged particles which decouple ▮ the MHD flow (Haswell *et al.*, 1992).

is presented in Fig. 4.62.

In order to separate the effects due to the steady-state rotational velocity, we transf▮ to a rotating frame of reference, chosen to be rotating with the steady angular velocity the blob of plasma under scrutiny.

The explosive evolution of the electromagnetic fields is heuristically derived. The te▮ nical definition of the explosive growth is that the quantity grows faster than exponenti▮ in time so that it could increase toward infinity in a finite time: an example of an *explo▮ instability* is the expression $(t_0 - t)^{-\alpha}$ with $\alpha > 0$. (See Sec. 2.2.) Here, and subsequently▮ is a constant. The purpose of this discussion is to illuminate the basic physical processe▮

work here; the results of a more complete treatment will be given at the end of this section.

We are interested in the azimuthal component of the magnetic field. This is coupled with the radial component of the velocity. Therefore we examine the radial component of momentum equation and the azimuthal component of the induction equation:

$$\frac{\partial u_r}{\partial t} + \frac{u_r}{r}\frac{\partial}{\partial r}(r u_r) - 2\Omega u_\theta = \frac{c}{4\pi}\frac{1}{r}\frac{\partial}{\partial r}(r B_\theta) B_\theta + \eta_{\text{vis}}\left(\frac{1}{r^2}\frac{\partial}{\partial r}r^2\frac{\partial u_r}{\partial r} - \frac{2u_r}{r^2}\right), \qquad (4.3.91)$$

$$\frac{\partial B_\theta}{\partial t} + \Omega B_r - \frac{\partial}{\partial r}(u_r B_\theta - u_\theta B_r) = -\frac{c^2\eta}{4\pi}\frac{\partial}{\partial r}\left(\frac{1}{r}\frac{\partial}{\partial r}(r B_\theta)\right). \qquad (4.3.92)$$

If we assume that $u_\theta = B_r = 0$ at $t = 0$, then the above two equations are closed by themselves. In general, the linear terms may be important and the azimuthal velocity and radial magnetic field are coupled into the evolution of the terms of interest; however we contend that the azimuthal magnetic field is likely to be dominant in equations (4.3.91) and (4.3.92), since this component of the magnetic field is stretched and enhanced by the differential rotation in the disk (the dynamo effect; Sec. 3.1). To satisfy the prerequisites of an asymptotically nonlinear treatment, we require:

$$2\Omega u_\theta \ll \frac{u_r}{r}\frac{\partial}{\partial r}(r u_r), \qquad (4.3.93)$$

and

$$\Omega B_r \ll u_r k_r B_\theta, \qquad (4.3.94)$$

where k_r is the scale length of radial variations of $u_r B_\theta$. Whenever these inequalities are satisfied as a result of a fluctuation in the velocity and magnetic field structure of a plasma element, we may simplify Eqs. (4.3.91) and (4.3.92) to:

$$\frac{\partial u_r}{\partial t} = -\frac{u_r}{r}\frac{\partial}{\partial r}(r u_r) + \frac{c}{4\pi}\frac{1}{r}\frac{\partial}{\partial r}(r B_\theta) B_\theta, \qquad (4.3.95)$$

$$\frac{\partial B_\theta}{\partial t} = \frac{\partial}{\partial r}(u_r B_\theta - u_\theta B_r). \qquad (4.3.96)$$

These equations, along with the continuity equation (Tajima et al., 1987), have the structure of quadratic nonlinearity on the right-hand side, equated to the time evolution terms on the left-hand side. For illustration let us consider Eqs. (4.3.95) and (4.3.96) alone, which admit solutions of the form

$$B_\theta \sim \frac{B_0(\mathbf{x})}{(t_0 - t)}, \qquad (4.3.97)$$

$$u_r \sim \frac{u_0(\mathbf{x})}{(t_0 - t)}, \qquad (4.3.98)$$

where B_0 and u_0 are constants. It goes without saying that such solutions are only valid when the nonlinearities are sufficiently strong to justify the treatment outlined above, and when t

393

is not too close to t_0. If the continuity equation is taken into account, it can play a role in stopping the explosive process from diverging to infinity. Furthermore, as we shall see at the end of this section, if one includes the continuity equation in one dimensional fashion, one can derive explosive solutions similar to Eqs. (4.3.97) and (4.3.98) with a slightly modified exponent to $(t_0 - t)$ (Tajima and Gilden, 1987). It could be argued that magnetic buoyancy effects will tend to disperse the growing field represented by equation (4.3.98). As shown by Shibata $et\ al.$ (1990) (see also Sec. 3.2), when the plasma β of the disk is low, the Parker instability (Sec. 3.2.2) may be stabilized.

Now we will examine the consequences of these highly nonlinear solutions for the magnetic field evolution. Firstly we consider the response of the electric field E_z to the growth in B_θ. The azimuthal component of the induction equation is:

$$\frac{\partial B_\theta}{\partial t} = c\frac{\partial E_z}{\partial r};$$

(4.3.99)

hence substituting from Eq. (4.3.97) for B_θ, $\frac{B_0}{(t_0-t)^2} = c\frac{\partial E_z}{\partial r}$. From this we can see that E_z consists of a curl-free part and a part which varies temporally as

$$\frac{B_0}{(t_0 - t)^2}.$$

(4.3.100)

Away from the magnetic field null point (the null point is at the reversal), the induced electric field is nothing but the manifestation of the radial flow of the fluid, which in the final analysis is driven by the gravitational force; locally it could be said to be driven by the fluid velocity described in Eq. (4.3.98). Exact exponents to Eqs. (4.3.97)–(4.3.98) are subject to change upon more precise mathematical treatments (Tajima and Sakai, 1989) and examined in Sec. 3.3.5. However, the qualitative features are captured here.

The explosive electromagnetic field gives rise to the MHD $\mathbf{E} \times \mathbf{B}$ motion (bulk fluid motion) causing plasma on both sides of the reversal to flow towards the field null point. The physical phenomenon described in the field solutions is self-driving. The current sheet at the field reversal induces the growing magnetic fields; the growth of the fields causes compression which pinches the current sheet. Since the driving force that leads to this explosive (or impulsive) collapse of the plasma is the magnetic force. This process is called the *magnetic collapse*.

The growing fields are dictated by the form of Eqs. (4.3.97) and (4.3.98). The solutions are scale-free: neither the temporal nor the spatial scale has been quantitatively fixed by our arguments. We require merely that the spatial scale be small compared with the field reversal scale itself, and that the characteristic timescale of the explosive processes be sufficiently short to prevent buoyancy forces from removing the generated field. The radial size of the present magnetic structure, L_B, is related to the spatial scale of the current sheet, λ, through the magnetic Reynolds number, R_m. The magnetic Reynolds number is defined as the ratio of the resistive timescale, τ_{res}, to the dynamical timescale, τ_{dyn}, $i.e.$

$$R_m \sim \frac{|\nabla \times (\mathbf{u} \times \mathbf{B})|}{|\eta\nabla^2\mathbf{B}|};$$

(4.3.101)

R_m is a fundamental dimensionless plasma parameter. The expression in the numerator of Eq. (4.3.101) consists of four terms:

$$B_\theta \frac{1}{r}\frac{\partial \mathbf{u}}{\partial \theta}; \qquad B_r \frac{\partial \mathbf{u}}{\partial r}; \qquad u_r \frac{\partial \mathbf{B}}{\partial r}; \qquad \text{and} \qquad u_\theta \frac{1}{r}\frac{\partial \mathbf{B}}{\partial \theta}. \qquad (4.3.102)$$

Since $\mathbf{B}$ and $\mathbf{u}$ are both predominantly azimuthal, and both have larger radial gradients than azimuthal gradients, all these terms are probably of approximately the same order. Choosing the last of these terms as representative and performing dimensional analysis yields:

$$R_m \sim \frac{u_\theta L_B^2}{\eta R}. \qquad (4.3.103)$$

Here R is the scale of variations of $\mathbf{B}$ in the azimuthal direction, which could be visualized as the "wrap length" between magnetic field reversals. The sheet current width, λ, is determined by the balance of diffusion of the magnetic field and rotational regeneration. Thus the spatial scale, λ, is obtained from dimensional analysis of

$$\nabla \times (\mathbf{u} \times \mathbf{B}) \sim \eta \nabla^2 \mathbf{B}, \qquad (4.3.104)$$

where the spatial derivative on the left-hand side is over wrap length, while the one on the right-hand side is over the current sheet. From Eq. (4.3.104) we obtain:

$$\lambda \sim \sqrt{\frac{\eta R}{u_\theta}}. \qquad (4.3.105)$$

Hence with equation (4.3.103)

$$\lambda \sim \frac{L_B}{\sqrt{R_m}}. \qquad (4.3.106)$$

Note that the form of the dependence on R_m in Eq. (4.3.59) is similar to that found in Perkins and Zweibel (1987).

The stage at which our approximation breaks down can depend on both the temporal and the spatial scale. For relatively slow, large-scale bursts, the energy embodied in the field may be limited by the gravitational energy, the limiting magnetic field, B_{collapse}, being given by:

$$\frac{B_{\text{collapse}}^2}{8\pi} \sim \frac{GM\rho L_B}{R_0^2}. \qquad (4.3.107)$$

Here M is the mass of the compact object and ρ is the ambient mass density in the disk. This limit arises in the following way: once the magnetic field strength becomes strong enough to compress the disk plasma sufficiently that regions of reversed magnetic field are brought into contact, then reconnection will occur. In order to compress the plasma, the magnetic field will have to do work against the gravitational force. Rearranging equation (4.3.107) we obtain

$$B_{\text{collapse}} \sim c\sqrt{4\pi\rho} \left(\frac{r_S}{R_0}\right)\left(\frac{L_B}{r_S}\right)^{\frac{1}{2}}, \qquad (4.3.108)$$

where r_S is the Schwarzschild radius.

Obviously the explosive growth eventually leads to a breakdown of the assumptions used to derive Eqs. (4.3.95) and (4.3.96). Compression of the plasma will cause increased pressure; the pressure could then inhibit the pinching caused by the growing magnetic field, and our approximation breaks down. The explosive evolution then ceases. Another possibility is a fast instability in which the field topology changes: for instance magnetic field may peel off into the coronal region.

The explosive process recurs repetitively: the magnetic field is being continually wound up, leading to the regions of "runaway" growth in which the fields grow according to Eqs. (4.3.97)–(4.3.98); each of these bursts is ultimately terminated by magnetic reconnection across the field reversal (Sec. 3.3.4), leading to a locally more relaxed magnetic topology. Interesting phenomena such as particle acceleration take place in plasma elements undergoing this electromagnetic evolution. In the remainder of this section we will examine the consequences of such cataclysmic field evolution.

When magnetic reconnection occurs across the field reversal, the energy built up in the electromagnetic fields will be released into the disk/boundary layer plasma (only a small part of this energy is expended in accelerating charged particles, since a relatively small number of particles is involved.) This energy is entrained in the kinetic energy in the compressive bulk flow towards the reversal at the moment of reconnection, and ultimately causes heating of the local disk plasma. The release and subsequent radiation of this energy may explain the flickering observed in the light curves of interacting binary systems containing accretion disks. For an example of this behavior in the *cataclysmic variable* HT Cas, see Horne (1991).

The model discussed produces a hard, or high energy outflow from an accretion disk as an indirect result of the liberation of gravitational energy in the region of the compact object through kinetic and electromagnetic processes. We emphasize that the acceleration is essentially prompt and impulsive and that the exploding plasma element is tiny compared to the dimensions of the disk.

It is reasonable to surmise that the characteristic spatial scale of the bursting regions is proportional to the Schwarzschild radius of the central compact object, r_S. In this case the energy of the emitted particles is proportional to the mass of the central object. Thus the model predicts particles of far higher energy (and presumably more of them) from disks around *AGN* than from disks around stellar mass compact objects. This has interesting implications for the production of cosmic rays. It is accepted that the bulk of cosmic rays, those with energies $\lesssim 10^{17}$ eV, have Galactic origins, and are contained by the Galactic magnetic field. A mass of 3 $M_\odot$ for the compact object leads to particles accelerated to energies of $\sim 10^{13}$ eV (Haswell and Tajima, 1992). Thus our mechanism applied to magnetically cataclysmic accretion disks in interacting binary star systems produces particles of energies consistent with the observed *Galactic cosmic rays*. On the other hand, ultra-high energy cosmic rays with energies $\gtrsim 10^{17}$ eV apparently have extra-galactic origins within the local supercluster (Strong *et al.*, 1974). These *ultra-high energy cosmic rays* may be identified with particles accelerated from AGN systems with compact object masses $\geq 10^5 M_\odot$.

For ultra-high energy cosmic rays (Linsley, 1981) discusses energies up to 10^{20} eV, and concludes that those with energies greater than 10^{19} eV are of extragalactic origin. Rochester and Turver (1981) conclude that cosmic rays of energies $10^{17} - 10^{20}$ eV are extragalactic in origin. The latter paper discusses acceleration mechanisms: the *Fermi mechanism* (Fermi, 1949) fails to explain the observed heavy element component of cosmic rays, while shock waves from supernovae can provide such acceleration. Shock wave acceleration is, however, probably deficient when high energy cosmic rays are considered; though it is, however, a suitable candidate, for example, for gamma ray ($\lesssim 10^9$ eV) bursts (Holcomb et al., 1991).

The interaction of high γ particles with the cosmic background radiation leading to pair production of electron-positrons and pions has been predicted to cause the *Greisen-Zatsepin cutoff* in the predicted energy spectrum at $\sim 10^{19}$ eV (Strong et al., 1974). In fact, the observed ultrahigh energy spectrum of cosmic rays does not show this sharp cutoff predicted (Greisen, 1966; Zatsepin and Kuzwin, 1966), but extends beyond 10^{20} eV. This can be explained if only the more energetic cosmic rays are extragalactic. Strong et al. (1974) and Linsley (1981) suggested that (i) the apparent isotropy of cosmic rays beyond 10^{18} eV indicates the extragalactic origin and (ii) some directiveness of $\sim 10^{20}$ eV components from the local supercluster direction indicates an origin within the local supercluster for this range of energies. See also Watson (1980), Stecker (1968, 1989) discusses steady state cosmic ray theory, addressing the local supercluster origin. For more discussion there are many recent papers in the cosmic ray physics conference proceedings (Chudakov, 1987). Haswell and Tajima (1992)'s mechanism for the AGN with waves $10^7 M_\odot$ yields energies in excess of 10^{20} eV, though other effects such as collisions, have to be considered.

Problem 4–1: Show that the following equation

$$\rho = \frac{C_s^2}{2\pi G r^2}$$

satisfies the exact hydrostatic equilibrium for the isothermal gas sphere.

Problem 4–2: Prove equation 4.3.1

Problem 4–3: When the magnetic fields have purely toroidal fields without poloidal field lines, discuss the acceleration of jets, assuming 1D geometry (i.e. magnetic pressure driven jet/wind (Maruyama and Fujimoto, 1987; Fukue et al., 1991).

Problem 4–4: Prove Eq. (4.3.60).

Problem 4–5: Prove that

$$\alpha = \frac{\rho V_r}{B_r} = \frac{\rho V_p}{B_p} = \frac{\dot{M}}{\Phi_p} \tag{4.3.109}$$

where $\dot{M}$ and Φ_p are mass flux and poloidal magnetic flux, respectively, which are constant along magnetic line of force.

Problem 4–6: Construct adiabatic, thermal wind model in which the total energy is conserved;

$$\frac{1}{2}v_r^2 + \frac{\gamma}{\gamma-1}\frac{p}{\rho} - \frac{GM}{r} = E. \qquad (4.3.110)$$

$$p = K\rho^\gamma, \qquad (4.3.111)$$

Calculate the velocity at infinity (terminal velocity) and the position of critical point.

Problem 4–7: Prove Eq. (4.3.30).

Problem 4–8: Develop a formulation for special and general relativistic 1D steady MHD wind theories. (Camenzind, 1986; Takahashi *et al.*, 1990). Discuss a role of a light cylinder.

Problem 4–9: Steady accretion flows can be analyzed by the same formalism as the steady wind theory. Develop a spherical non-magnetic accretion theory (Bondi, 1952). Derive the mass accretion rate onto a star with mass M when the density and the sound speed of the interstellar matter are ρ_0 and c_s. What is the difference from the mass accretion rate derived from Shu's self-similar solution for 1D collapse of a spherical cloud?

Problem 4–10: Although the Alfvén waves are said to be transverse waves, why can they (compressional Alfvén waves) accelerate particles?

Problem 4–11: Develop a theory of general relativistic MHD accretion flow (Hirotani *et al.*, 1992).

Problem 4–12: Derive the approximate growth rate of the Jeans (gravitational) instability in a rigidly rotating disk at angular velocity Ω. What is a role of the rotation? (See, e.g., Goldreich and Lynden-Bell 1965a.)

Problem 4–13: Derive the magnetic braking time of the Sun's rotation using the formulae developed for 1D magneto-centrifugally driven wind (see Sec. 4.3.3 and also Weber and Davis, 1967; Washimi and Shibata, S., 1993). On the basis of this calculation, discuss the early evolutional history of the solar rotation.

Problem 4–14: Prove the following relation between accretion rate $\dot{M}_{acc}$ and mass loss rate $\dot{M}_{jet}$ by a magneto-centrifugally driven jet from an accretion disk;

$$\frac{\dot{M}_{jet}}{\dot{M}_{acc}} \sim \left(\frac{r_d}{r_A}\right)^2,$$

where r_d and r_A are the radius of the accretion disk and the Alfvén point. (Hint: Use the approximate relation $V_{jet} \sim r_A\Omega$, where Ω is the angular velocity of the accretion disk. See, e.g., Pudritz and Norman, 1986.) From this, discuss the role of the MHD jet on the formation of stars and AGNs.

Problem 4–15: What are the differences of non-relativisitc jets and (special) relativistic jets? (See, e.g., Yokosawa *et al.*, 1982, Koide *et al.*, 1996).

Problem 4–16: Discuss the consequences of the Kelvin-Helmholtz (K-H) instability in astrophysical jets. Can magnetic field stabilize the K-H instability? (See, e.g. Tajima and Leboeuf, 1980; Ferrari and Trussoni, 1983).

Problem 4–17: Derive the pulsar wind equation, and discuss the difference and similarity between pulsar winds and astrophysical jets. (e.g., Shibata, 1991).

Abramowicz, M.A., et al., Astrophys. J. **332**, 646 (1988).

Abramowicz, M.A., et al., Astrophys. J. **438**, L37 (1995).

Adams, F.C., Lada, C., and Shu, F.H., Astrophys. J. **312**, 788 (1987).

Alexander, S.G., and Mészáros, P., Astrophys. J. **344**, L1 (1989).

Balbus, S.A., and Hawley, J.F., Astrophys. J. **376**, 214 (1991).

Balbus, S.A., and Hawley, J.F., Astrophys. J. **400**, 610 (1992).

Balbus, S.A., Hawley, J.F., and Stone, J.M., Astrophys. J. **467**, 76 (1996).

Basu, S., Mouschovias, T.Ch., Astrophys. J. **444**, 770 (1994).

Begelman, M.C., and Li, Z.-Y., Astrophys. J. **426**, 269 (1994).

Belcher, S.A., and MacGregor, K.B., Astrophys. J. **210**, 498 (1976).

Bentz, W., Astron. Astrophys. **139**, 378 (1984).

Bernstein, I.B., Frieman, E.A., Kruskal, M.D., and Kulsrud, R.M., *Proceedings of the Royal Society*, vol. A244, p. 17 (1958).

Blaes, O., Blandford, R.D., Goldreich, P., and Madau, P., Astrophys. J. **343**, 839 (1989).

Blandford, R.D., and Pringle, J.E., Mon. Not. R. Astr. Soc. **176**, 443 (1976).

Blandford, R.D., and Payne, D.G., Mon. Not. R. Astr. Soc. **199**, 883 (1982).

Blandford, R.D., in *Astrophysical Jets*, Burgarella, D. *et al.*, eds. (Cambridge Univ. Press, 1993), p. 15.

Bodenheimer, P., and Sweigert, A., Astrophys. J. **152**, 515 (1968).

Bondi, H., Mon. Not. R. Astr. Soc. **112**, 195 (1952).

Bridle, A.H., and Perley, R.A., Ann. Rev. Astron. Astrophys. **22**, 319 (1984).

Brandenburg, A., Nordlund, A., Stein, R.F., and Torkelsson, U., Astrophys. J. **446**, 741 (1995).

Burgarella, D., Livio, M., and O'Dea, C.P., (eds), *Astrophysical Jets*, (Cambridge Univ. Press, 1993).

Camenzind, M., Astron. Astrophys. **162**, 32 (1986).

Cannizo, J.K., Shafer, A.W., and Wheeler, J.C., Astrophys. J. **333**, 227 (1988).

Cao, X., and Spruit, H.C., Astron. Astrophys. **287**, 80 (1994).

Case, K.M., Phys. Fluids **3**, 149 (1960).

Chandrasekhar, S., *Hydrodynamic and Hydromagnetic Stability* (Oxford: Clarendon Press, 1961), p. 384.

Chiba, M., and Tosa, M., Mon. Not. R. Astr. Soc. **244**, 714 (1990).

Chieuh, T., Li, Z.-Y., and Begelman, M.C., Astrophys. J. **377**, 462 (1991).

Chou, W., Tajima, T., Matsumoto, R., Shibata, K., *Proc. IAU Colloq. No. 153*, "Magnetodynamic Phenomena in the Solar Atmosphere," eds. Y. Uchida *et al.*, (Kluer, Dordrecht, 1996), p. 613.

400

Chudakov, A.E., in *Proceedings 20th Inter. Cosmic Ray Conf.*, Nauka, Moskow **6**, 494 (1987).

Cline, T.L., *et al.*, Astrophys. J. Lett. **237**, L1 (1980).

Contopoulos, J., Astrophys. J. **450**, 616 (1995).

Davidson, R.C., *Methods in Nonlinear Plasma Theory* (Academic, New York, 1972), p. 39.

Dawson, J.M., Phys. Rev. **113**, 383 (1959).

Dermer, C.D., Astrophys. J. Lett. **347**, L13 (1990).

Dorfi, E., Astron. Astrophys. **114**, 151 (1982).

Drazin, P.G., J. Fluid Mech. **4**, 214 (1958).

Drury, L.O'C., Mon. Not. R. Astr. Soc. **217**, 821 (1985).

Dyson, F.J., Phys. Fluids **3**, 155 (1960).

Eardley, D.M., and Lightman, A.P., Astrophys. J. **200**, 187 (1975).

Epstein, R.I., Astrophys. J. **291**, 822 (1985).

Epstein, R.I., Phys. Rep. **163**, 155 (1988).

Fabian, A.C., Icke, V., and Pringle, J.E., Ap. Sp. Sci. **42**, 77 (1976).

Fenimore, E.E., *et al.*, Astrophys. J. Lett. **335**, L71 (1988).

Fermi, E., Phys. Rev. **75**, 1169 (1949).

Ferrari, A., Trussoni, E., and Zaninetti, L., Mon. Not. R. Astr. Soc. **193**, 469 (1981).

Ferrari, A., and Trussoni, E., Mon. Not. R. Astr. Soc. **205**, 515 (1983).

Fielder, R.A., Mouschovias, T.Ch., Astrophys. J. **391**, 199 (1992).

Fogglizzo, T., and Tagger, M., Astron. Astrophys. **287**, 297 (1994).

Fukue, J., Shibata, K., and Okada, R., Publ. Astr. Soc. Jpn. **43**, 131 (1991).

Fukui, Y., *et al.*, in *Protostars and Planet III*, (1993), p. 603.

Glatzel, W., Mon. Not. R. Astr. Soc. **225**, 227 (1987).

Goldreich, P., and Lynden-Bell, D., Mon. Not. R. Astr. Soc. **130**, 125 (1965).

Goldreich, P., and Julian, W.H., Astrophys. J. **157**, 869 (1969).

Goldreich, P., Goodman, J., and Narayan, R., Mon. Not. R. Astr. Soc. **221**, 339 (1986).

Goldreich, P., and Lynden-Bell, D., Mon. Not. R. Astr. Soc. **130**, 97 (1965a).

Goldreich, P., and Lynden-Bell, D., Mon. Not. R. Astr. Soc. **130**, 125 (1965b).

Greisen, K., Phys. Rev. Lett. **16**, 748 (1966).

Hameury, J.M., Bonazzola, S., Heyvaerts, J., and Ventura, J., Astron. Astrophys. **111**, 242 (1982).

Hanawa, T., Astron. Astrophys. **185**, 160 (1987).

Hanawa, T., Matsumoto, R., and Shibata, K., Astrophys. J. Lett. **393**, L71 (1992a).

Hanawa, T., Nakamura, and T., Nakano, T., Publ. Astr. Soc. Jpn. **44**, 509 (1992b).

Hartman, L. and MacGregor, K.B., Astrophys. J. **259**, 180 (1982).

Hartmann, D., and Epstein, R.I., Astrophys. J. **346**, 960 (1989).

Hartmann, D., Epstein, R.I., and Woosley, S.E., Astrophys. J. **348**, 625 (1990).

Hasegawa, J., and Wakatani, M., Phys. Fluids **26**, 2770 (1983).

Haswell, C., Tajima, T., and Sakai, J.I., Astrophys. J. **401**, 495 (1992).

Hawley, J.F., and Balbus, S.A., Astrophys. J. **376**, 223 (1991).

Hawley, J.F., and Balbus, S.A., Astrophys. J. **400**, 595 (1992).

Hawley, J.F., Gammie, C.F., and Balbus, S.A., Astrophys. J. **440**, 742 (1995).

Hayashi, C., Publ. Astr. Soc. Jpn. **13**, 450 (1961).

Hayashi, C., Narita, S., and Miyama, S., Prog. Theor. Phys. **68**, 1949 (1982).

Hayashi, C., IAU Symp. No. **115**, 403 (1987).

Hayashi, M.R., Shibata, K., Matsumoto, R., Astrophys. J. **468**, L37 (1996).

Heineman, M., and Olbert, S., J. Geophys. Res. **83**, 2457 (1978).

Heyvaerts, J., and Norman, C., Astrophys. J. **347**, 1055 (1989).

Hirose, S., Ph.D. Thesis, University of Tokyo (1994).

Hirose, S., Uchida, Y., Shibata, K., Matsumoto, R., Publ. Astr. Soc. Jpn. **49**, in press (1997).

Hirotani, K., Takahashi, M., Nitta, S., Tomimatsu, A., Astrophys. J. **386**, 455 (1992).

Ho, C., and Epstein, R.I., Astrophys. J. **343**, 277 (1989).

Holcomb, K., and Tajima, T., Astrophys. J., **378**, 682 (1991).

Horiuchi, T., Matsumoto, R., Hanawa, T., and Shibata, K., Publ. Astr. Soc. Jpn. **40**, 147 (1988).

Horne, K., Wood, J.H., and Stiening, R.F., Astrophys. J. **378**, 271 (1991).

Horton, W., Tajima, T., and R. Galvao, in *Magnetic Reconnection in Space and Laboratory Plasmas*, Ed. Hones, and Edward, W., (American Geophysical Union, 1984), p. 45.

Horton, W., Tajima, T., and Kamimura, T., Phys. Fluids **30**, 3485 (1987).

Hughes, P.A. (ed.), *Beams and Jets in Astrophysics* (Cambridge Univ. Press, 1991).

Hunter, C., Astrophys. J. **136**, 594 (1962).

Ichimaru, S., Astrophys. J. **202**, 528 (1975).

Ichimaru, S., Astrophys. J. **208**, 701 (1977).

Iijima, T., *et al.*, Astron. Astrophys. **283**, 919 (1994).

Inoue, M., *et al.*, Publ. Astr. Soc. Jpn. **36**, 633 (1984).

Inutsuka, S., and Miyama, S.M., Astrophys. J. **388**, 392 (1992).

Jeans, J.H., *Astronomy and Cosgomony*, (Dover, 1928).

Kaburaki, O., and Itoh, M., Astron. Astrophys. **172**, 191 (1987).

Kafatos, M., *et al.*, Astrophys. J. **346**, 991 (1989).

Kaifu, N., Suzuki, S. Hasegawa, T. Morimoto, M. Inatani, J. Nagane, K., Miyazawa, K., Chikada, Y., Kanzawa, T., and Akabane, K., Astron. Astrophys. **134**, 7 (1984).

Kawabe, R., *et al.*, Astrophys. J. Lett. **404**, 63 (1993).

Kaisig, M., Astron. Astrophys. **218**, 89 (1989).

Kato, S., Publ. Astr. Soc. Jpn. **36**, 55 (1984).

Kato, S., and Horiuchi, T., Publ. Astr. Soc. Jpn. **37**, 399 (1985).

Kato, S., and Horiuchi, T., Publ. Astr. Soc. Jpn. **38**, 313 (1986).

Kato, S., Publ. Astr. Soc. Jpn. **39**, 645 (1987).

Kishimoto, Y.,*et al.*, Phys. Plasmas **3**, 1289 (1996).

Klebesadel, R.W., Strong, I. and Olson, R., Astrophys. J. Lett. **182**, L85 (1973).

Kley, W., Papaloizou, J.C.B., and Lin, D.N.C., Astrophys. J. **416**, 679 (1993).

Koide, S., Nishikawa, K.-I., Mutel, L., Astrophys. J. **463**, L71 (1996).

Koyama, K., *et al.*, Publ. Astr. Soc. Jpn. **46**, L125 (1994).

Königl, A., and Ruden, S.P., in *Protostars and Planets III*, (A93-42937 17-90), p. 641-687 (1993).

Kudoh, T., and Shibata, K., Astrophys. J. Lett. **452**, L41 (1995).

Kudoh, T., and Shibata, K., Astrophys. J. **474**, 362 (1997a).

Kudoh, T., and Shibata, K., Astrophys. J. **476**, 632 (1997b).

Lada, C.J., Ann. Rev. Astron. Astrophys. **23**, 267 (1985).

Larson, R.B., Mon. Not. R. Astr. Soc. **145**, 271 (1969).

Larson, R.B., Mon. Not. R. Astr. Soc. **156**, 437 (1972).

Li, Z.-Y., Astrophys. J. **444**, 848 (1995).

Liang, E.P., and Nolan, P.L., Sp. Sci. Rev. **38**, 353 (1984).

Lin, D.N.C., and Papaloizou, J.C.B., Mon. Not. R. Astr. Soc. **191**, 37 (1980).

Linsley, J., in *Origin of Cosmic Rays* eds. G. Setti, G. Spada, and A.W. Wolfendale (Reidel, Dordrecht, 1981), p. 53.

Liu, J., Horton, W., and Sedlak, J.E., Phys. Fluids **30**, 467 (1987).

Lovelace, R.V.E., Wang, J.C.L., and Sulkanen, M.E., Astrophys. J. **315**, 504 (1986).

Lovelace, R.V.E., Berk, H.L., and Contopoulos, J., Astrophys. J. **379**, 696 (1991).

Lu, E.T., and Hamilton, R.J., Astrophys. J. **380**, L89 (1991).

Lund, N., Astron. Astrophys. Suppl. **97**, 289

Manheimer, W., and Boris, J.P., Comments Plasma Phys. Contr. Fusion **3**, 15 (1977).

Margon, B., Ann. Rev. Astron. Astrophys. **22**, 507 (1984).

Maruyama, T., and Fujimoto, M., *IAU Symp. No. 115*, p. 381 (1987).

Matsumoto, R., Kato, S., and Fukue, J., in *Theoretical Aspects on Structure, Activity, and Evolution of Galaxies, III*, ed. S. Aoki *et al.*, (Tokyo Astr. Obs., 1985), p. 102.

Matsumoto, R., Horiuchi, T., Shibata, K., and Hanawa, T., Publ. Astr. Soc. Jpn. **40**, 171 (1988).

Matsumoto, R., Horiuchi, T., Hanawa, T., and Shibata, K., Astrophys. J. **356**, 259 (1990).

Matsumoto, R. and Shibata, K., Publ. Astr. Soc. Jpn. **44**, 167 (1992).

Matsumoto, R., Tajima, T., Shibata, K., and Kaisig, M., Astrophys. J. **414**, 357 (1993).

Matsumoto, R., Tajima, T., and Shibata, K., (1997), in preparation.

Matsumoto, R., and Tajima, T., Astrophys. J. **445**, 767 (1995).

Matsumoto, R., Uchida, Y., Hirose, S., Shibata, K., Hayashi, M.R., Ferrari, A., Bodo, G., and C. Norman, Astrophys. J. **461**, 115 (1996a).

Matsumoto, R. *et al.*, in *Basic Physics of Accretion Disks*, eds. S. Kato and S. Inagaki (1996b).

Matsumoto, T., Hanawa, T., and Nakamura, T., Astrophys. J. **478**, 569 (1997).

Matsuzaki, T., Matsumoto, R., and Shibata, K. in preparation (1997).

Mazets, E.P., Golenetskii, S.V., Aptekar', R.L., Gur'yan, Yu.A., and Il'inskii, V.N., Nature **290**, 378 (1981).

Meerson, B., Phys. Lett. A **150**, 290 (1990).

Melia, F., Rappaport, S., and Joss, P.C., Astrophys. J. Lett. **305**, L51 (1986).

Mestel, L., and Spitzer, L., Jr., Mon. Not. R. Astr. Soc. **116**, 503 (1956).

Michel, F.C., Astrophys. J. **158**, 727 (1969).

Michel, F.C., Astrophys. J. **290**, 721 (1985).

Mikhailovski, A.B. *Theory of Plasma Instabilities*, (Consultants Bureau, New York, 1974), vol. 1, p. 160.

Miller, R.L., Waelbroeck, F.L., Hassam, A.B., and Waltz, R.E., preprint (1994).

Mineshige, S., and Osaki, Y., Publ. Astr. Soc. Jpn. **35**, 377 (1983).

Mineshige, S., Kusunose, M., and Matsumoto, R., Astrophys. J. Lett. **445**, L43 (1995).

Mirable, I.F., and Rodriguez, L.F., Nature **371**, 46 (1994).

Mitrofanov, I.G., Ap. Sp. Sci. **205**, 245 (1984).

Mitsuda, K., *et al.*, Publ. Astr. Soc. Jpn. **41**, 97.

Miura, A., J. Geophys. Res. **97**, 10655 (1992).

Miyama, S.M., Hayashi, C., Narita, S., Astrophys. J. **279**, 621 (1984).

Miyama, S.M., Publ. Astr. Soc. Jpn. **44**, 193 (1992).

Miyamoto, S., in *Proc. IIAS Workshop on Mathematical Approach to Fluctuations II*, **2**, ed. T. Hida (Inter. Inst. Adv. Stud., Kyoto, 1993).

Miyamoto, M., Kitamoto, S., Hayashida, K., and Egoshi, W., Astrophys. J. Lett. **442**, L13 (1995).

Morris, M., (ed.), "Galactic Center, "in *Proc. IAU Symp. No. 136*, (1989).

Mouschovias, T.Ch., Astrophys. J. **206**, 753 (1976).

Mouschovias, T.Ch., and Paleologou, E.V., Astrophys. J. **237**, 877 (1980).

Mouschovias, T.Ch., and Paleologou, E.V., Astrophys. J. **246**, 48 (1981).

Mouschovias, T.Ch., and Morton, S.A., Astrophys. J. **371**, 296 (1991).

Muslinov, A.G., and Tsygan, A.I., Ap. Sp. Sci **120**, 27 (1986).

Murakami, T., 1988, *Physics of Neutron Stars and Black Holes*, Ed. Y. Tanaka (Universal Academy Press, Tokyo, 1988), p. 405.

Nakamura, T., Hanawa, T., and Nakano, T., Publ. Astr. Soc. Jpn. **45**, 551 (1993).

Nakamura, T., Hanawa, T., and Nakano, T., Astrophys. J. **444**, 770 (1995).

Nakano, T., Ann. Phys. (N.Y.) **73**, 326 (1972).

Nakano, T., and Nakamura, T., Publ. Astr. Soc. Jpn. **30**, 671 (1978).

Nakano, T., Publ. Astr. Soc. Jpn. **34**, 337 (1982).

Nakano, T., Fund. Cosmic Phys. **9**, 139 (1984).

Narayan, R., and Yi, I., Astrophys. J. **452**, 710 (1995).

Narita, S., McNally, D., Pearce, G.L., and Sorenson, S.A., Mon. Not. R. Astr. Soc. **203**, 491 (1983).

Narita, S., Hayashi, C., and Miyama, S.M., Prog. Theor. Phys. **72**, 1118 (1984).

Norman, M.L., Wilson, J.R., and Barton, R.T., Astrophys. J. **239**, 968 (1980).

Norman, C., and Heyvaerts, J., Astron. Astrophys. **147**, 247 (1985).

Norman, M.L., Smarr, L., and Winkler, K.H.A., *Numerical Astrophysics* eds. Centrella, J.M., LeBlanc, J.M., and Bowers, R.L. (Jone and Bartlett Publ., Boston, 1985) p. 88.

Okamoto, I., Mon. Not. R. Astr. Soc. **173**, 357 (1975).

Ouyed, R., Pudritz, R.E. and Stone, J.M., Nature **385**, 409 (1996).

Pacini, F., and Ruderman, M., Nature **251**, 399 (1974).

Papaloizou, J.C.B., and Pringle, J.E., Mon. Not. R. Astr. Soc. **208**, 721 (1984).

Parker, E.N., Astrophys. J. **128**, 664 (1958).

Parker, E.N., *Interplanetary Dynamical Processes*, (Interscience, New York, 1963).

Parker, E.N., Astrophys. J. **145**, 811 (1966).

Parker, E.N., Astrophys. J. **162**, 665 (1970).

Parker, E.N., *Cosmical Magnetic Field* (Clarendon Press, Oxford, 1979), p. 314.

Penston, M.V., Mon. Not. R. Astr. Soc. **144**, 425 (1969).

Perkins, F.W., and Zweibel, E., Phys. Fluids **30**, 1079 (1987).

Phillips, G.L., Mon. Not. R. Astr. Soc. **221**, 571 (1986).

Pudritz, R.E., Mon. Not. R. Astr. Soc. **195**, 881 (1981).

Pudritz, R.E., and Norman, C., Astrophys. J. **301**, 571 (1986).

Rochester, G.D., and Turver, K.E., Contemp. Phys. **22**, 425 (1981).

Ross, D.W., Chen, G.L., and Mahajan, S.M., Phys. Fluids **25**, 652 (1982).

Saigo, K., and Hanawa, T., submitted to Astrophys. J. (1997).

Sakai, J-I., and Ohsawa, Y., Space Sci. Rev. **46**, 113 (1987).

Sakurai, T., Astron. Astrophys. **152**, 121 (1985).

Sakurai, T., Publ. Astr. Soc. Jpn. **39**, 821 (1987).

Sauty, C., and Tsinganos, K., Astron. Astrophys. **287**, 893 (1994).

Scott, E.H. and Black, D.C., Astrophys. J. **239**, 166 (1980).

Shaing, K.C., Crume, E.C., and Houlberg, W.A., Phys. Fluids B, Part 2, **2**, 1492 (1990).

Shakura, N.I., and Sunyaev, R.A., Astron. Astrophys. **24**, 337 (1973).

Shapiro, S.L., Lightman, A.P., Eardley, D.M. Astrophys. J. **204**, 187.

Shibata, K., Publ. Astr. Soc. Jpn. **35**, 263 (1983).

Shibata, K., and Uchida, Y., Publ. Astr. Soc. Jpn. **37**, 31 (1985).

Shibata, K., and Uchida, Y., Ap. Sp. Sci. **118**, 443 (1986a).

Shibata, K., and Uchida, Y., Publ. Astr. Soc. Jpn. **38**, 631 (1986b).

Shibata, K., and Uchida, Y., Publ. Astr. Soc. Jpn. **39**, 559 (1987).

Shibata, K., Tajima, T., Matsumoto, R., Horiuchi, T., Hanawa, T., Rosner, R., and Uchida, Y., Astrophys. J. **338**, 471 (1989a).

Shibata, K., Tajima, T., Steinolfson, R., and Matsumoto, R., Astrophys. J. **345**, 584 (1989b).

Shibata, K., in *Proc. IAU Symp. No. 136*, "The Center of the Galaxy," eds. Morris, M. (Reidel, 1989), pp. 313.

Shibata, K., and Uchida, Y. in *Proc. of Nato Advanced Study*, "Theory of Accretion Disks," ed. Meyer, F. *et al.* (1989) pp. 65.

Shibata, K., Tajima, T., and Matsumoto, R., Astrophys. J. **350**, 295 (1990a).

Shibata, K., Tajima, T., and Matsumoto, R., Phys. Fluids B **2**, 1989 (1990b).

Shibata, K., and Uchida, Y., Publ. Astr. Soc. Jpn. **42**, 39 (1990).

Shibata, S., Astrophys. J. **378**, 239 (1991).

Shibata, K., and Matsumoto, R., Nature **353**, 633 (1991).

Shibata, K., Nozawa, S., and Matsumoto, R., Publ. Astr. Soc. Jpn. **44**, 265 (1992).

Shibata, K., in *Proc. "Solar and Astrophysical Magnetohydrodynamic Flows"*, ed. K. Tsinganos (Kluwer Academic Pub., 1996), p. 217.

Shu, F.H., Astrophys. J. **214**, 214 (1977).

Shu, F.H., Najita, J., Ostriker, E., Wilkin, F., Ruden, S. and Lizano, S. Astrophys. J. **429**, 781 (1994).

Shu, F.H., and Li, Z. Astrophys. J. **475**, 251 (1997).

Snell, R.L., *et al.*, Astrophys. J. **290**, 587 (1985).

Sofue, Y., and Hanada, T., Nature **310**, 568 (1984).

Sofue, Y., Fujimoto, M., and Wielebinski, R., Ann. Rev. Astron. Astrophys. **24**, 459 (1986).

Spitzer, L., Jr., *Physical Processes in Interstellar Medium*, (John Wiley & Sons, 1978).

Stahler, S.W., Shu, F.H., and Taam, R.E., Astrophys. J. **241**, 637 (1980).

Stecker, F.W., Phys. Rev. Lett. **21**, 1016 (1968).

Stecker, F.W., Nature **342**, 401 (1989).

Stone, J.M., and Norman, M.L., Astrophys. J. **433**, 746 (1994).

Stone, J.M., *et al.*, Astrophys. J..

Strom, S., Edwards, S., and Strom, K.M., *Formation and Evolution of Planetary Systems*, eds. Weaver, H.A., and Danly, L., (Cambridge University Press, 1989), p. 91.

Strong, A.W., Wdowczyk, J., and Wolfendale, A.W., J. Phys. A **7**, 1767 (1974).

Sturrock, P.A., Phys. Rev. **112**, 1488 (1958).

Tajima, T., and Leboeuf, J-N., Phys. Fluids **23**, 884 (1980).

Tajima, T., Sakai, J-I., Nakajima, H., Kosugi, T., Brunel, F., and Kundu, M.R., Astrophys. J. **321**, 1031 (1987).

Tajima, T., and Gilden, D., Astrophys. J. **320**, 741 (1987).

Tajima, T., and Sakai, J-I., Sov. J. Plasma Phys. **15**, 519 (1989).

Tajima, T., Horton, W., Morrison, P.J., Schutkeker, J., Kamimura, T., and Mima, K., Phys. Fluids B **3**, 938 (1991).

Takabe, H., Montierth, L., and Morse, R.L., Phys. Fluids **26**, 2299 (1983).

Takabe, H., Mima, K., Montierth, L., and Morse, R.L., Phys. Fluids **28**, 3676 (1985).

Takahashi, M., Nitta, S., Tatematsu, Y., and Tomimatsu, A., Astrophys. J. **363**, 206 (1990).

Tanaka, Y., "Blackholes in X-ray binaries: X-ray properties of the galactic blackhole candidates," in *Proc. 23rd ESLAB Symposium—X-ray Astronomy*, eds. Hung and Battrick (ESA, SP-296, Paris, 1989), p. 3.

Terry, P.W., and Horton, W., Phys. Fluids **25**, 491 (1982).

Todo, Y., Uchida, Y., Sato, T., and Rosner, R., Publ. Astr. Soc. Jpn. **44**, 245 (1992).

407

Todo, Y., Uchida, Y., Sato, T., and Rosner, R., Astrophys. J. **403**, 164 (1993).

Tomisaka, K., Ikeuchi, S., and Nakamura, T., Astrophys. J. **326**, 202 (1988).

Tomisaka, K., Ikeuchi, S., and Nakamura, T., Astrophys. J. **341**, 220 (1989).

Tomisaka, K., Ikeuchi, S., and Nakamura, T., Astrophys. J. **362**, 202 (1990).

Tomimatsu, A., Publ. Astr. Soc. Jpn. **46**, 123 (1994).

Tomisaka, K., Astrophys. J. **438**, 226 (1995).

Tomisaka, K., Publ. Astr. Soc. Jpn. **48**, 701 (1996).

Toomre, A., Astrophys. J. **259**, 535 (1982).

Trussoni, E., and Tsinganos, K., Astron. Astrophys. 269, 589 (1993).

Tout, C.A., and Pringle, J.E., Mon. Not. R. Astr. Soc. **259**, 604 (1992).

Tsuboi, M., et al., Astron. J. **92**, 818 (1986).

Tsinganos, K., (ed.) *Solar and Astrophysical MHD Flows*, (Kluwer Academic Publishers, 1996).

Tsyganenko, A.I., Astron. Astrophys. **44**, 21 (1975).

Uchida, Y., and Shibata, K., Publ. Astr. Soc. Jpn. **36**, 105 (1984).

Uchida, Y., and Shibata, K., Publ. Astr. Soc. Jpn. **37**, 515 (1985).

Uchida, Y., Shibata, K., and Sofue, Y., Nature **317**, 699 (1985).

Uchida, Y., Kaifu, N., Shibata, K., Hayashi, S.S., Hasegawa, T., and Hamatake, H., Nature Publ. Astr. Soc. Jpn. **39**, 907 (1987).

Uchida, Y., Mizuno, A., Nozawa, S., and Fukui, Y., Nature **42**, 69 (1990).

Uchida, Y., Todo, Y., Rosner, R., and Shibata, K., Publ. Astr. Soc. Jpn. **44**, 227 (1992).

Uchida, Y., Matsumoto, R., Hirose, S., and Shibata, K., in *Primordial Nucleosynthesis and Evolution of Early Universe*, eds. K. Sato, and J. Audouze (Kluwer Academic Publishers, 1991), p. 409.

Unwin, S.C., *Proc. Parsec-scale Radio Jets*, (Cambridge Univ. Press, 1990), p. 13.

Ustyugova, G.V., Koldoba, A.V., Romanova, M.N., Chechetkin, V.M., and Lovelace, R.V.E., Astrophys. J. Lett. **439**, L39 (1995).

Van der Klis, M., Astrophys. J. Suppl. **92**, 511 (1994).

Velikhov, E.P., Soviet JETP **35**, 995 (1959).

Vermeulen, R.C., et al., Astron. Astrophys. **270**, 177 (1993).

Waelbroeck, F.L., et al., Phys. Fluids B **4**, 2441 (1992).

Waelbroeck, F.L., Dong, J.Q., Horton, W., and Yushmanov, P.N., Phys. Plasmas 1, 3742 (1994).

Wagner, J.S., Lee, L.C., Wu, C.S., and Tajima, T., Geophys. Res. Lett. **10**, 483 (1983).

Wang, J.C.L., et al., Phys. Rev. Lett. **63**, 1550 (1989).

Washimi, H., and Shibata, S., Mon. Not. R. Astr. Soc. **262**, 936 (1993).

Watson, A.A., Quart. J. Roy. Astr. Soc. **21**, 1 (1980).

Weber, E.J., and Davis, L., Astrophys. J. **148**, 217 (1967).

Wisdom, J., and Tremaine, S., Astrophys. J. **95**, 925 (1988).

Woosley, S.E., and Wallace, R.K., Astrophys. J. **258**, 716 (1982).

Yokosawa, M., Ikeuchi, S., Sakashita, S., Publ. Astr. Soc. Jpn. **34**, 461 (1982).

Yusef-Zadeh, F., Chance, D., and Morris, M., Nature **310**, 557 (1984).

Zatsepin, G.T., and Kuzmin, V.A., JETP Lett. **4**, 78 (1966).

Zweibel, E.G., in *Interstellar Processes,* eds. D.J. Hollenback and H.A. Thronson, Jr. (Reidel, Dordrecht, 1987).

Zweibel, E.G., and Kulsrud, R.M., Astrophys. J. **201**, 63 (1975).

Chapter 5

Cosmological Plasma Astrophysics

5.1 Cosmology and the Plasma Epoch

The Big Bang cosmology has been suggested by Gamow and firmly supported by increasing accumulation of evidence, starting from the *Hubble expansion* of galaxies to the 3K black-body microwave background to the *primordial abundance* of nuclear elements. Some of the recent measurements of *COBE* (Cosmic Background Explorer) show highly uniform (with some faint characteristic large structures) and isotropic distribution of intensity and fitting closely to a black-body distribution. From the general theory of relativity the expansion rate $a(t)$ of the Universe after matter and radiation decoupled (the recombination epoch) is proportional to $t^{2/3}$, where t is the cosmic time since the Big Bang, while before the decoupling it is $t^{1/2}$ (Misner *et al.*, 1970). Accordingly, the temperature of the photons T is proportional to $t^{-2/3}$ after the recombination and $t^{-1/2}$ before that, as T decays inversely proportional to the scale length a. Thus prior to the recombination of electrons and protons into neutral hydrogen atoms we surmise that the constituent matter was that of plasma made up of electrons and protons (and a fraction of helium nuclei). Photons existed before and after in a large number, some $10^8 - 10^4$ times that of electrons. This epoch prior to the recombination is usually called the radiation epoch, while that after recombination is the gravitational epoch. If we pick the constituent matter, plasma, the epoch prior to the recombination may be called the plasma epoch. As we trace further back in time, when the plasma temperature reaches the energy of electron rest mass (twice) about MeV, electrons were accompanied by nearly as many positrons. The amount of electrons (and thus positrons) reaches that of photons in this epoch. The tiny difference between the electron and positron numbers is the present electron number. If we go back further in time and the temperature reaches the energy of mesons (about a few 100 MeV), the number of hadrons also becomes as large as that of photons or leptons. The epoch demarcated between the appearance of positrons and that of antihadrons mainly consists of photons, electrons, and positrons. As such, this is also a plasma. The epoch thus may be called the first phase of the plasma epoch. We may call the first and second (electron + proton plasma epoch) phases of the plasma epoch together as simply the plasma epoch. Prior to the plasma epoch we have epochs in which interaction other than the electromagnetic one come into important play, such as the strong interaction and weak interaction. In the present book we will not touch upon these epochs. We concentrate on the evolution of the Universe during the plasma epoch. This chapter handles the scope of plasma astrophysics that has least observational evidence (though mounting) and also least theoretical investigations. Accordingly, the description of this remains a rough *dessin* of our current (rather poor) understanding and hopefully this constitutes a sketch for future investigations.

5.1.1 Cosmological Constraints and Challenges

Relevant observational constraints on cosmology are summarized as follows:

(i) Highly uniform and isotropic *Cosmic Microwave Background* Radiation with the frequency fluctuation from the mean $\delta\omega/\omega \lesssim 5 \times 10^{-6}$ is observed (COBE).

This is proposed to be related to the plasma temperature fluctuation at the time of recombination to be as small $\delta T/T \sim \delta\omega/\omega \lesssim 5 \times 10^{-6}$.

(ii) Galaxies were formed with the redshift parameter $z \sim 3$ or before, as evidenced from quasars. Together with nonlinear N-body simulations of density perturbations to form clusters of galaxies, it is inferred that the density fluctuations of baryons at the time of recombination should be as large as $\delta n/n \sim 10^{-3}$ (at $t \sim 10^{13}$ sec after the Big Bang). This relatively large value of $\delta n/n$ needed for galaxy formation is due in part to the cosmic expansion turning the Jeans instability (exponential in time) into an algebraic growth in time, and in part to the hugeness of the intercluster (or intersupercluster) space distance covered by the time since the recombination till today.

(iii) Large structures much greater than clusters and even superclusters have been discovered. It seems that if these had to evolve from "scratch" at the recombination time, that would have been too short a time.

(iv) The observed galactic disk rotational velocity seems to decay too slowly for the central bulge's gravitational force to be responsible. Thus *dark matter* in the bulge as well as in the halo has been introduced.

(v) Intense *Cosmic X-ray Background Radiation* has been observed, which is uniform and isotropic. It is surmised that the X-ray spectra best fit with the bremsstrahlung of electrons around $z \sim 3$ with temperature $T_e \sim 300$ keV. In order to explain the high brightness of the X-rays, $\mathcal{L}_X$, we need a large baryon (and electron) mass density. However, the lack of the observed (so far) *Compton y-parameter* means that it is less than $10^{-4} - 10^{-5}$. The Compton y-parameter is the measure of how much Compton scattering happens and thus is proportional to the line integral of the electron density over the X-ray path length.

(vi) Evidence of large magnetic fields in galaxies and *intergalactic magnetic* fields $B \sim 10^{-9} - 10^{-6}$ G. Kronberg (1994) even suggested possible observational evidence of large intersupercluster magnetic fields. If these are the case, intergalactic magnetic fields appear to be too large and ubiquitous to arise from cosmic dynamo generation (Sec. 3.1) of magnetic fields during the gravitational epoch alone.

(vii) Lack of observed *Sachs-Wolfe effect* (1967) of the general relativistic red (or blue) shift of photon spectrum due to cosmological matter fluctuations in the CMBR remains a puzzle. The patchwork of various mechanisms to change photon spectrum is unlikely to give rise to clean cancellation.

(viii) Nearly scaleless cellular intermittent *galaxy correlation*. Galactic correlation shows highly interrelated structures and at the same time includes nearly all scales.

(ix) Natural abundance of light elements, such as He, Li, etc.

Some of these cosmological observations are difficult to explain by a single cute theory and one such feature often apparently seems to contradict with others. For example, N-body nonlinear simulation of the gravitational epoch indicates that the necessary amount of density fluctuation $\delta n/n$ at the dawn of the epoch (the recombination) is $\sim 10^{-3}$. This value is much greater than the recent observed value of $\delta T/T \sim 5 \times 10^{-6}$ by COBE, if $\delta n/n$ is related to $\delta T/T$ (adiabatic). Many cosmologists, therefore, feel obliged to introduce dark matter which has not been observed but is assumed to have enough density fluctuations to explain the galaxy and large structure formation. Introduction of dark matter alone, however, will not resolve other problems such as the lack of observed Sachs-Wolfe effect, the bright CXBR etc. Nevertheless, along with the differential rotation of our galaxy away from the Keplerian, this discrepancy between needed $\delta n/n$ and observed $\delta n/n$ could be due to the presence of some kind of dark matter. So far, however, nobody has observed evidence of any sufficient amount of dark matter nor is there any consensus on what it is. The "conventional wisdom" is that the plasma is opaque and tends to form the conventional uniform and isotropic thermal equilibrium state (Sec. 2.1.3) during the plasma epoch. Thus, it has been argued, there is no interesting happening such as structure formation during this epoch. We will examine the plasma epoch to see if it was uninteresting and structureless one or not in the subsequent sections.

5.1.2   Observations on Cosmological Magnetic Fields

The general trend of recent results indicates that, wherever we detect intergalactic hot gas and galaxies, we also find magnetic fields at levels of $\approx 10^{-7}$ G, or higher (Kronberg, 1997). The hitherto undetected, weaker fields in the general intergalactic medium outside of clusters and in large intergalactic voids might, in future, be measurable through observations of γ-rays and/or cosmic ray nuclei.

Observations over the last ten years have produced magnetic field detections not only in galaxy disks, but also in galaxy halos, clusters of galaxies, and in some very distant galaxy systems which product both *absorption lines* and Faraday rotation of the radiation from background quasars. Generally, the more we look for extragalactic magnetic fields, the more ubiquitous we find them to be.

Diffuse *synchrotron radiation* and *Faraday rotation* are the most practical methods for detecting magnetic fields in most galaxy systems and intergalactic space, although the Zeeman split as well as the induced polarization of optical starlight due to the intervening dust grains that may line up due to interstellar magnetic fields.

Intergalactic gas in galaxy clusters has a typical electron density of $10^{-4} - 10^{-2} \mathrm{cm}^{-3}$, temperature in the range 10^7 to 10^8 K, and an extent of 1 Mpc for the cluster core. At this temperature range, which makes clusters significant X-ray sources ($L_x \approx 10^{43} - 10^{45}$ erg/s), the ion sound speed is comparable to the galaxies' velocity dispersion in the cluster, which is 400-1200 km s^{-1}. A minority of clusters, such as the Coma cluster, contain an enhanced population of cosmic ray electrons which, over dimensions comparable to that of the hot gas, emit a diffuse "halo" of synchrotron radiation, thus revealing an intracluster magnetic field. For a more extensive review of magnetic fields in galaxy clusters, the reader is referred to Kronberg (1994).

Cosmic rays serve as particularly effective "illuminators" of intergalactic magnetic fields at lower radio frequencies. This is because of the relatively high spectral densit of synchrotron radiation at low frequencies, reflecting the high value of γ, typically 2.4-3, which defines the power law distribution of cosmic ray electron energies: $N(\varepsilon) \propto \varepsilon^{-\gamma}$. The critical frequency (near which most of the synchrotron radiation is emitted) is related to ε by $\nu_c = (3eB \sin\varphi/4\pi m^3 c^5)\varepsilon^2$ (Pacholczyk, 1970). An advantage of observing at the lowest possible radio frequencies is that we preferentially detect the lowest energy CR electrons, which survive the longest to "keep illuminating" the associated magnetic field. To calculate the longest possible loss time for a CR electron, we define a "cosmic background-equivalent" magnetic field, B_{bge}, for which a CR electron's energy loss rate by synchrotron radiation ($d\varepsilon/dt \approx |\mathbf{B}|^2\varepsilon^2$) equals that due to inverse Compton scattering off the microwave background radiation ($d\varepsilon/dt$) $\propto \varepsilon_{bg}\varepsilon^2$, where $\varepsilon_{bg} = 4.8 \cdot 10^{-13}(1+z)^4$ergcm^{-3}.

Figure 5.1 shows an image published by Kim et $al.$, (1989) of the 326 MHz intergalactic synchrotron emission arising from cosmic rays surrounding the Coma cluster of galaxies, which may be prototypical of the type of radiation that could be imaged in future at much lower frequencies. They found emission extending beyond the Coma cluster, indicating an extended, magnetic region on a supracluster scale.

5.2   Intergalactic Plasma

As X-ray observation of astrophysical objects progresses, X-ray emitting large scale hot plasmas have been found increasingly frequently. The greater the spatial scales (intergalactic to intercluster plasmas), the more time it takes to heat it. Thus the heating of cosmological scale plasmas takes cosmological time scale and it is not an easy task to come up with a plausible explanation for such a mechanism. In this section we look into a plausible theory to account for such with the presence of cosmological magnetic fields. The existence and origin of cosmological magnetic fields will be discussed in Sec. 5.3 and further in Sec. 5.4. In the present section with the presence of such fields we consider how we can explain the intergalactic intercluster plasma states as observed.

5.2.1   Cosmic X-ray Background Radiation

Network magnetic fields that weave through clusters of galaxies can be strongly constricted clusters violently relax gravitationally (Sec. 4.1) in the supercluster potential. These intercluster magnetic fields tend to constrict the trapped plasma, driving them to high densities

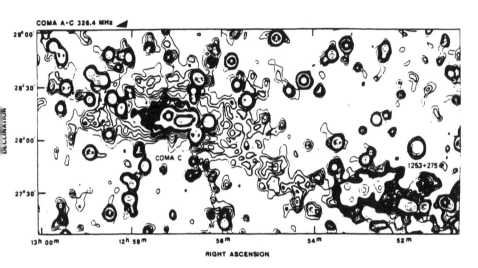

FIGURE 5.1 Radio image showing 326 MHz emission along an intergalactic "bridge," which appears to connect the Coma cluster of galaxies (Coma C) with the Coma A complex of radio emission elsewhere in the Coma supercluster (Kim *et al.*, 1989)

and high temperatures. These hot ($T \geq 10^8$ K) and dense plasmas are magnetically insulated from colder ($T \leq 10^4$ K) surrounding gases, forming intermittent intercluster medium. The dynamical processes (see Sec. 2.5) of these fields involve rapid (Sec. 4.3) magnetic relaxation toward the nearly force-free state by involving reconnection of field lines and rapid heating of plasmas by being continuously fed energy from the violent gravitational relaxation (cf. Sec. 4.1). The fundamental physical processes of magnetic constriction and subsequent plasma heating by the violent motions of compact objects that trap the magnetic fields are considered below. Brightening regions of such magnetically constricted plasmas have typical dimensions of order the size of clusters or even less, thus they will be seen as a diffuse X-ray source. This model could explain the large amount of necessary thermal energy that results in the cosmic X-ray background radiation in a large supercluster spatial scale, the rapid heating, the small amount of deviation of the cosmic microwave background radiation due to the Comptonization, and how to keep colder gases from evaporating. Compatible

415

with this model is the primordial origin of magnetic fields (Sec. 5.3): the primordial plasma could sustain a large amount of spontaneously generated magnetic fields and thus isothermal density fluctuations with little temperature signatures. We further consider the evolution of such generated magnetic fields by a dynamo in the epoch following this and preceding the above X-ray forming epoch.

The origin of the cosmic X-ray background radiation still remains a puzzle. This can be fitted by the bremsstrahlung from optically thin plasma with $T_X \sim 40\,$keV. There are two possible candidates for this; a hot, *diffuse intergalactic medium* (IGM), or sum of unresolved, discrete sources, such as *quasars* and *Seyfert galaxies*. At present observational data has not sufficient resolution to discriminate between these to theories.

In this section, we consider diffuse intergalactic media as a possible cause of the X-ray background. Especially we assume the presence of global cosmic magnetic fields and discuss if such intergalactic magnetic fields could remove the difficulties that existing non-magnetized IGM models are confronted with. [However, this mechanism does not exclude the presence of more compact X-ray sources]. There have been many works so far done direct to the field of the intergalactic medium, but less attention has been paid to the consequence of the violent activity of the magnetic fields.

Guilbert and Fabian (1986) concluded that the X-ray background can be explained if the IGM was heated up to $T_X \sim 400\,$keV at $z = 3.6$ due to some unknown mechanism (Field and Perrenod, 1977 and Taylor and Wright, 1989). There are, however, stringent constraints for possible nature of IGM: According to the recent COBE data of the cosmic microwave background radiation (CMBR), the allowed value for the Compton y parameter is at most 0.001 (Mather *et al.*, 1990). The presence of hot radiation inevitably causes, however, the distortion of CMBR via inverse-Compton scattering (Sunyaev and Zeldovich, 1972) and Lahav *et al.*, (1990). The lack of the Comptonization has a severe constraint on the non-magnetic IGM model for X-ray background. A hot IGM also requires huge mass in the universe, $\Omega_B > 0.2$, whereas the baryon densities, estimated from the theory of primordial nucleosynthesis, give $\Omega_B \sim 0.1$ or less, in apparent contradiction with the results of the IGM hypothesis.

To sum up, a huge amount of energy needed to heat the IGM and the little distortion observed in CMBR seems to be a paradox for the (uniform, non-magnetized) IGM model as an origin of X-ray background radiation. There is a possibility for IGM in a two-phase medium; namely an X-ray emitting, hot tenuous region is surrounded by low temperature, dense region. However, to get the huge X-ray emissivity we need a certain value of density for a hot medium, and, to make the pressure balance, the density of the cool medium should be even larger, requiring large Ω_B. It is also unclear how to stop evaporation of the cold gas. As mentioned above, because of these, many investigators have considered the X-ray mechanism from compact objects. We reconsider the X-ray mechanism from the IGM with magnetic field effects into consideration. Basic physical processes of *flux tube constriction* that may underlie the phenomenon are surveyed in Sec. 5.2.2. We discuss this X-ray mechanism in Sec. 5.2.4. Cosmological fields existed primordially for fundamental physical reasons (Tajima *et al.*, 1992) See Sec. 5.3. After such magnetic fields are spontaneously created, they need

416

to grow in spatial size and also to fight against the dilution due to cosmic expansion. This may be realized by the dynamo action. We note that observational evidence is increasing for large-scale magnetic fields, such as intercluster magnetic fields (Kim *et al.*, 1991 and Tribble, 1991).

5.2.2 Basic Physical Processes of Flux Tube Constriction

Before we enter a more realistic model of a network of magnetic fields anchored in violently moving clusters, we isolate two specific physical processes of flux tube dynamics in this section. The clusters may be moving violently relative to each other, both in terms of the mutual distance and in terms of the mutual orientation. The former will stretch or shorten the tubes, while the latter will twist and kink the tubes. In general, the web of networked magnetic fields undergo the combination of these motions (and more complicated ones). During such actions the complex geometry of networked fields is often forced to undergo reconnection. This involves the resistive process, accompanied by heating and acceleration of the plasma trapped in the magnetic field. This results in a change of topology of magnetic fields. In the present section we pick up the twisting motion of the flux tube and its resultant kink instability as a first example and examine the interaction of two flux tubes and their reconnection as a second.

Twist and kink of flux tubes

When a flux tube is mechanically twisted at two ends that rotate in the opposite senses, the field lines in the tube get wound up and as a result the field-aligned current ($\mathbf{j} \times \mathbf{B} \approx 0$) is induced. As long as $\mathbf{j} \times \mathbf{B} \approx 0$ (or $\mathbf{j} \times \mathbf{B} = \nabla p$), the plasma stays in equilibrium. However, if the current buildup is above a certain threshold, the entire flux tube now becomes unstable and exhibits a *kink instability*. In Fig. 5.2 we show an example (Zaidman and Tajima, 1989) of such a twisted flux tube undergoing the kink instability. Also in this simulation the twisting azimuthal velocity is sheared, i.e. $v_\theta(r)$, so that the twisting magnetic fields are now sheared as well, $B_\theta(r)$.

The sheared magnetic field structure may be best illustrated by the analysis of the magnetic fields in terms of the local rotational transform $\iota(r, z)$ (Shafranov, 1970) and its associated so-called *safety factor* $q(r, z) = 2\pi/\iota = \iota^{-1}$ locally defined as

$$q(r, z) = \frac{r B_z}{R B_\theta(\mathbf{x})},$$

where $R = L_z/2\pi$. Since the twist is a function of z, the *"rotational transform"* and the safety factor are functions of z and are thus local (z) quantities, in contrast to the original Shafranov's case. When q is, for example, 3 at $z = z_0$, the magnetic field is spiraling in the azimuthal direction with a pitch of $3L_z$. This would amount to a winding in the θ direction of the particular field line once while winding three times in the ζ direction (in the periodicity of z) if this local $q = 3$ was held for all z. Such a local q is depicted in Fig. 5.2(a). From Shafranov's theory a strong kink instability is expected when q becomes

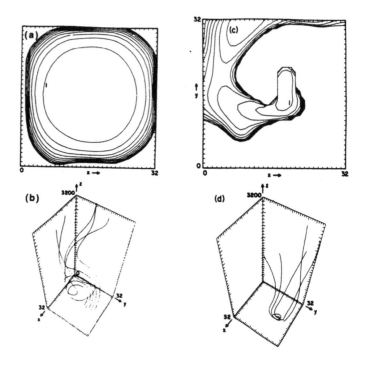

FIGURE 5.2 Twisted flux tube and its construction and heating. (a) and (b) at early twisting stage, while (c) and (d) at a later twist stage. (a) and (c) show the parameter q that characterizes the magnetic helicity. (b) and (d) show the field lines that are twisted (Zaidman and Tajima, 1989).

less than unity. In the present case the local q is just an approximate guideline. The result is in Fig. 5.2(b) for the magnetic line. As the twisting continues, the magnetic field lines become more wrapped showing a wider area with $q < 1$ [Fig. 5.2(c)]. Figure 5.2(d) shows an distortion with azimuthal mode number $m = 1$ as exemplified by a crescent-shaped island and by a dipole structure.

Instead of causing the twisted field lines by twisting, one can study the isolated effects of the kink instability by starting a force-free equilibrium with field-aligned current ($j \times B = 0$) being sufficiently strong to begin with. Shown in Fig. 5.3 are two cases of such initializations, one with the current profile $j_z(r) \propto [1 + (kr)^2]^{-1}$, and the other with the Bessel functional current profile (Matsumoto *et al.*, 1994). In these particular cases the flux tube was immersed under the influence of a uniform gravity in the negative z-direction. We observe that the flux tube becomes unstable against the kink mode and the resultant field lines form a supercoil structure. Some portion of the supercoil rises due both to the kink

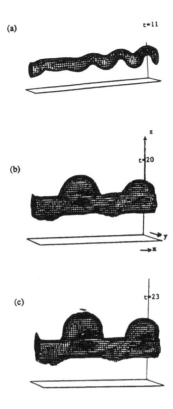

FIGURE 5.3 Flux tubes undergo the helical transition (a) the current-carrying flux tube undergoes the helical transition due to the kink instability; (b) and (c) the flux tube undergoes the transition due to the buoyancy instability (Matsumoto *et al.*, 1996).

and the magnetic buoyancy.

Shown in Fig. 5.4 is a case where we put in two disk-like gravitational attractors. The angular rotations of two disks are opposite and the magnetic field lines that penetrate through the disks are twisted as a result, as in the case of Fig. 5.2. As the twist gets stronger, the kink instability sets in as well as the evidence of the axial jet flows. This configuration and evolution may be thought of as a simplified version with only two clusters interacting through one flux tube between them for the more complex N clusters interacting with a web of intercluster magnetic fields.

Reconnection of field lines and flux tubes

When shear flux surfaces are moved by the kink instability and one flux surface is squeezed against the other, magnetic field lines are pinched and reconnected around the X-point.

419

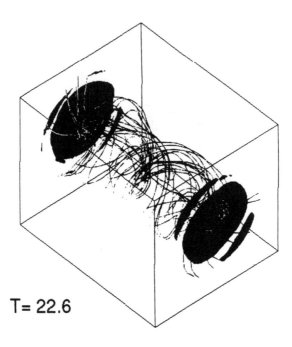

T= 22.6

FIGURE 5.4 Magnetic field line threading the differentially rotating disks after many disk rotations (from Matsumoto *et al.*, 1997)

This in fact happens in the kink mode associated with the action in the previous subsection. When the magnetic surface is totally reconnected and torn apart, a portion of plasma can be expelled out into the exterior as seen in Fig. 5.2. In order to isolate the reconnection process of flux tubes or magnetic field lines, we illustrate the interaction of two flux tubes that are originally in force-free equilibrium similar to the one in Fig. 5.3(a). See Fig. 5.5(a). If two flux tubes carry currents that are parallel, while the axial magnetic fields are antiparallel, the magnetic helicity $\int_i \mathbf{A} \cdot \mathbf{B} \, dV$ of two tubes are anti in the same sign. Since two parallel currents are attractive and susceptible to the coalescence instability (Bhattacharjee *et al.*, 1983). These two loops approach, as seen in Fig. 5.5(b). Similarly one can look at the case where the magnetic helicity of two loops are parallel. Although the classical tearing instability theory (White, 1983) predicts no difference in reconnecting rates in these two cases, Figs. 5.5(a)–(f) along with Figs. 5.5(g)–(h) show that the antiparallel helicity case is much faster (and explosive) than the parallel helicity case (Ono *et al.*, 1996). This effect

420

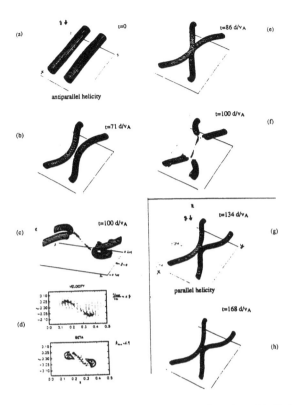

FIGURE 5.5 Two distinct ways of coalescence of two current-carrying flux tubes. (a)–(f), the case with antiparallel helicity for the two flux tubes. For this case *very* rapid reconnection [see Sec. 3.3] takes place around frame (e). The flow velocity and plasma β as a result of this rapid reconnection are shown in (c) and (d). On the other hand, (g)–(h) show the case with parallel helicity reconnection. This case proceeds rapidly, but not with rapid reconnection [Sweet-Parker process, Sec. 3.3].

has been predicted by Tajima (1982) and also seen in simulations (Leboeuf *et al.*, 1982). Details have been analyzed by Tajima and Sakai (1989a). The strong jet flows along the flux tubes after the reconnection are clearly seen in Fig. 5.5(d) for the antiparallel helicity reconnection. This antiparallel helicity reconnection can be many orders of magnitude faster than the tearing rate or the Sweet-Parker (1957) rate, and can be in the ball park of the Alfvén time scale; i.e. the Petschek mode. An experimental verification of such a phenomenon has recently been found (Ono *et al.*, 1993, 1996). Such a possibility of fast rate of reconnection is very significant, as astrophysical plasmas (particularly those of cosmology) are vast and the Lundquist number (the magnetic Reynolds number) tends to be huge and any intermediate time scale between the resistive and Alfvén time scales often tends to be too long for the age of the Universe or for its evolution.

5.2.3 Magnetically Constricted IGM Plasmas

We assume that global 'primordial' magnetic fields weave through clusters of galaxies. These magnetic fields are rapidly stretched, twisted, and braided by the *violent relaxation* of clusters of galaxies (Lynden-Bell, 1969) in the supercluster gravitational potential. It is thus natural that a substantial amount of the gravitational energy can be supplied to the magnetic energy at this stage. These intercluster fields tend to constrict the trapped gas, driving it to high densities and high temperatures. According to recent calculations of fast amplification of magnetic fields (see Sec. 3.1) by a dynamo such as the ABC dynamo (Galloway and Frisch, 1986), the essence of the process of plasma motion of the violent relaxation is captured by the ABC dynamo, leading to chaotic flows, which in turn drive chaotic magnetic field lines and thus amplify them. The generated magnetic fields may exhibit chaotic and intermittent properties and attain filamentary and/or cellular structures.

As rapid relaxation (Taylor, 1986) proceeds, we may apply a nearly magnetostatic equilibrium condition for such a magnetically constricted plasma;

$$\mathbf{J} \times \mathbf{B} + \nabla P \cong \text{const.} \tag{5.2.1}$$

Here P denotes the thermal pressure of the plasma. The kinetic temperature of a constricted plasma is hot enough to emit X-rays. The *galactic ridge brightening* of X-ray (Koyama *et al.*, 1986; Tanuma *et al.*, 1997) in the galactic disk may be related to this phenomenon, where the magnetic constriction may take place in a similar fashion. If we assumed a two-phase medium under pressure balance (Guilbert and Fabian, 1986), the space outside X-ray emitting hot plasmas should have been surrounded by a substantial amount of colder plasmas, requiring a huge mass in the colder component, or non-negligible distortion in CMBR by the hotter component.

It should be noted, however, that the (nearly) magnetostatic condition does not necessary entail that the pressure balance holds in the present model (see Sec. 2.5); i.e.

$$\left(\frac{B^2}{8\pi}\right) + P \neq \text{const.} \tag{5.2.2}$$

We now introduce a distribution function, $f(x, y, z)$, which is the probability to find a dense plasma at a certain point, (x, y, z). If we assume the completely random distribution of plasma inhomogeneities, then $f(x, y, z) = f(x)f(y)f(z)$. The average of the probability function over a unit volume may be written as

$$\langle f(x, y, z) \rangle dV = \langle f(x) \rangle \langle f(y) \rangle \langle f(z) \rangle dV \equiv f \, dV. \tag{5.2.3}$$

Here brackets denote a volume average of physical quantities. This averaged probability function, f, is the volume-filling factor, and is expected to be much smaller than unity in our case, as we shall see below. In the following, we let $\mathcal{L}_x$ the X-ray and luminosity density we consider the situations where $f \ll 1$ keeping the total X-ray luminosity constant,

$$\int \mathcal{L}_X dV^p = \int \mathcal{L}_X^{\text{uni}} \, dV^{\text{uni}}, \tag{5.2.4}$$

where the superscript, uni, represents the value under the uniform assumption, and dV^p is the volume occupied by hot plasmas. Since $\mathcal{L}_X \propto n^2\sqrt{T}$ and $dV^p = f dV^{uni}$, we have for the same temperature T, the density of X-ray emitting plasma is much higher than the uniform value:

$$n = f^{-1/2} n^{uni}. \tag{5.2.5}$$

The Compton y parameter, which is proportional to $nT\,dz$, where dz is a length along the direction of the sight, is then reduced to

$$y = f^{1/2} y^{uni}, \tag{5.2.6}$$

by a factor $f^{-1/2}(\gg 1)$ from the value obtained under the uniform assumption, because

$$dz^p = \langle f(x)f(y)\,f(z)dz \rangle = f\,dz. \tag{5.2.7}$$

Likewise, the total baryon density, Ω_B, is

$$\Omega_B \equiv \int n\,dV = f^{1/2} \Omega_B^{uni}, \tag{5.2.8}$$

and becomes by a factor $f^{-1/2}$ smaller. It is suggested from COBE results that "the limits on y would limit the X-ray background to only 1/36 of the observed value," (Mather, 1990) indicating that $f \leq 10^{-3}$ in the present framework.

We know from the absence of absorption lines in QSO spectra that cold gas cannot be neutral (Gunn and Peterson, 1965). The present model naturally satisfies this Gunn-Peterson test, because the present model need no cold dense gas for confinement of hot plasmas. It is, on the other hand, entirely possible that pockets of cold gas regions exist, as the thermal conduction is substantially reduced by the presence of magnetic fields, but still enough to ionize hydrogens in the cold gas.

5.2.4 Heating of IGM Plasmas by Magnetic Fields*

By the presence of magnetic fields, an efficient heating over a very large volume by a large amount may become possible. This is because, in the present model, no time is needed to form the network structure. For magnetic constriction to be possible, magnetic field strength is required to be at least

$$\frac{B^2}{4\pi} \geq n\,kT, \tag{5.2.9}$$

$$B \geq 2.5 \times 10^{-6} \text{ (Gauss)} \left(\frac{n}{10^{-6}\text{cm}^{-3}}\right)^{1/2} \left(\frac{T}{300\,\text{keV}}\right)^{1/2}. \tag{5.2.10}$$

Note that this field strength is the local (to the hot plasma region) one, and is

$$B \sim f_B^{-1/2} B^{uni}, \tag{5.2.11}$$

where f_B is the volume-filling factor for magnetic fields which is not necessarily the same as f for a plasma. For f_B of order 10^{-3}, the value obtained in Eq. (5.2.9) is consistent with the values of mean field strength of order 10^{-9} G derived by the measurements of the rotation measure (Fujimoto et al., 1971).

The heating provided by magnetic energy is

$$\mathcal{L}_X \sim \langle JE \rangle = \frac{1}{c} \frac{\partial B^2}{\partial t},$$ (5.2.12)

$$\frac{\partial B}{\partial t} = \nabla \times (v \times B) + \eta \nabla^2 B.$$ (5.2.13)

The heating rate occurring in Eq. (5.2.12) is related to magnetic reconnection

$$\tau_{\text{rec}} \sim R_m^{\epsilon-1} \frac{\ell^2}{\eta_{\text{SH}}},$$ (5.2.14)

where η_{SH} is the Spitzer-Härm resistivity and ℓ is the characteristic spatial scale of the plasma, ϵ is determined by the dynamics of reconnection and relaxation. As we discussed in Sec. 3.3, the parameter ϵ is generally $0 \le \epsilon \le 3/5$; $\epsilon = 0$ corresponds to the case of explosive reconnection (Petschek, 1965; Tajima et al., 1989b), whereas $\epsilon = 1/2$ is the case of Sweet-Parker-type (Parker, 1957, Sweet, 1958). The magnetic Reynolds number is

$$R_m \equiv \frac{1}{\nu_e} \left(\frac{v_A}{\ell} \right),$$ (5.2.15)

where v_A is Alfvén velocity,

$$v_A = \frac{B}{\sqrt{4\pi n m_p}},$$ (5.2.16)

ν_e is the collision frequency (T in eV),

$$\nu_e = 3 \times 10^{-6} n (\ln \Lambda) T^{-3/2},$$ (5.2.17)

and the Spitzer-Härm conductivity is

$$\eta_{\text{SH}} = 7 \times 10^6 \frac{\ln \Lambda}{T^{3/2}}.$$ (5.2.18)

For relevant values and $\epsilon = 0$, for example, we find

$$\int dV \, \mathcal{L}_X \sim 3 \times 10^{55} \frac{1}{\tau_{\text{rec}}} \left(\frac{B}{10^{-6}\,G} \right)^2 \left(\frac{L}{10^{25}\,\text{cm}} \right)^3.$$ (5.2.19)

where L is the supercluster size as we integrated over the volume. From the observed energy requirement (Guilbert and Fabian, 1986) $\mathcal{L}_X \sim 10^{62}/\tau_{\text{cool}}$, where γ_{cool} is the radiative cooling rate. We therefore get

$$\frac{\tau_{\text{rec}}}{\tau_{\text{cool}}} \simeq 10^{-6} \left(\frac{B}{10^{-6}\,G} \right)^2 \left(\frac{L}{10^{25}\,\text{cm}} \right)^3.$$ (5.2.20)

424

Equations (5.2.14) together with (5.2.15) determine the length scale on which reconnection takes place:

$$\ell = \left[10^{14}\, \frac{\eta_{\rm SH}}{\nu_e^{1-\epsilon}} \left(\frac{B}{10^{-6}\,G} \right)^2 v_A^{1-\epsilon} \right]^{1/3-\epsilon}. \qquad (5.2.21)$$

This reduces, for example for $\epsilon = 0$, to $\ell \propto n^{-1/6}\,B$, and takes the value of $\ell \sim 10^{11-14}$ cm for $\frac{1}{2} > \epsilon > 0$ with the cooling time of 10^{20} s with B(local) $\sim 1\mu\,G$. This is not unreasonable if B is highly intermittent and turbulent as widely suspected of chaotic field lines (Sec. 5.2.5). These fields are anchored in clusters or even galaxies.

This new picture for the IGM that removes the difficulties of the existing models facing the observed X-ray emissivities, the distortion in CMBR, and the total baryon number. The cosmic X-ray background is accounted for by hot plasmas constricted by weaving magnetic fields, driven by the violent relaxation of clusters of galaxies. This mechanism naturally explains the epoch of the cosmic X-ray background, as it is related to the epoch of violent gravitational relaxation; otherwise the X-ray spectrum would be much wider. The effects of such magnetic fields are that, with the known X-ray emissivity, (i) the Compton influence on the CMBR (Compton y parameter) and the total baryon density (Ω_B) are both reduced by a factor $f^{1/2}(\ll 1)$; (ii) there exists enough magnetic energies to account for X-ray emission and the necessary heating rate is related to the spatial scales of the fine structure of the chaotic magnetic fields. Observational confirmation (or rebuttal) of such a model in the future is highly desired.

5.2.5 The Intergalactic Magnetic Dynamo*

We now return to the problem of how the primordial large-scale magnetic fields and their structure may have been formed, which have contributed to the kind of fields needed for the cosmic X-ray background radiation considered in previous sections. Recently there has been a renewed effort to explain the existence of the large-scale structures in the universe. New observations indicate structures on the largest scales (Saunder et al., 1991; Broadhurst et al., 1990; De Lapparent, 1989), as well as an ever-decreasing limit on the nonuniformity of the cosmic microwave background at the time that matter and radiation decoupled (Mather et al., 1990; Smoot et al., 1992). It is well known that for given matter fluctuations compatible with this low level of electromagnetic fluctuations, there has been too little time since the recombination to allow for the formation of observed galactic and other structures. The current "cold dark matter" theory tries to answer this difficulty. It may be possible to introduce additional elements to consider that (i) the radiation (plasma) epoch in the primordial universe was crucial in preparing for the formation of large structures via magnetic interactions, and (ii) the magnetic structures involved were isothermal in nature and so did not leave an imprint on the high-frequency blackbody spectrum (i.e. the current 3K microwave spectrum).

During the period from 10^{-2} to 10^{13} second after the big bang, the universe consisted primarily of an expanding electron-(positron)-proton plasma, and the electromagnetic interaction was the dominant force. If fluctuations led to the growth of seed magnetic fields during this time, then the resulting fields may have had a significant effect on the distribution of matter. There have been several works on the magnetic field evolution in the early universe (Harrison, 1970; Sato *et al.*, 1971; Baierlein, 1978).

Investigations such as Harrison's have assumed primordial turbulence with nonzero vorticity. However, the assumption of the presence of turbulent flows seems to find less supporters these days because the presence of flow and the observed homogeneity in the 3K microwave background are believed to be incompatible (Rees, 1987). On the other hand, it is unclear whether the incompressible (or vortical) flow motion of low- or zero-frequency would leave an observable imprint on the cosmic background. The coupling of photons with adiabatic perturbations such as sound motions and that with nonadiabatic vortical motions are quite different. One of the major differences is that the latter need not incur density perturbations, and as such it is more difficult for nonadiabatic vortical motions to couple to photons (Tajima *et al.*, 1990). It is in fact quite natural to take the incompressible mode for the very large-scale slow motion that we are interested in. Furthermore, it has recently been shown (Tajima *et al.*, 1992a) that in the early epoch of the radiation era it is possible that the nearly zero-frequency magnetic fluctuations associated with a plasma in an (even perfect) thermal equilibrium can be substantial based on the theory of the fluctuation-dissipation theorem (Sitenko, 1967). The nearly static magnetic fluctuations can couple with the plasma, thus creating density and velocity fluctuations, while the high frequency photons couple more weakly with the plasma. The presence of such magnetic, density and velocity fluctuations, albeit with small-scale seed fields, could influence the subsequent evolution of spatially larger scales of fluctuations. Furthermore fluctuations with size $\lambda > ct$ (the horizon) are certainly not in thermal equilibrium and thus could give rise to non-equilibrium noise. These magnetic fluctuations decayed only slowly in time after they entered the horizon, with the diffusive relaxation time scale $t_{\text{dif}} \approx \omega_p^2 \lambda^2 / \nu c^2$, where ν is the collision frequency. If t_{dif} is longer than the interval between the entrance time into the horizon and the exit time (i.e., the recombination time), such magnetic fluctuations were unable to reach thermal equilibrium; see Fig. 5.6. (The velocity fluctuations decay faster due to viscosity than magnetic fluctuations.) This might provide large-scale seed fields. It is tempting to consider (Fujimoto, 1990) such seed fields that were amplified as a candidate for creating recently discovered large-scale structures.

Here we investigate, through a very simple model, the characteristic morphology and strength of the magnetic fields that may result from dynamo action in an expanding highly conducting medium. This will provide some qualitative understanding of the interplay between the exponentiation of the field that can result from an incompressible chaotic flow, and the effects of the expansion of the medium. Here we are concerned with large-scale magnetic fields.

In addition to the epoch prior to recombination, the epoch after recombination may also be relevant. After the recombination most constituent matter becomes charge neutral and the main interactive force becomes gravitational rather than electromagnetic. It is known, however, that there exists very hot tenuous plasma in the intergalactic (or intersupercluster) space (Giovannini *et al.*, 1990). Such hot plasma may be of recent creation; on the other hand,

426

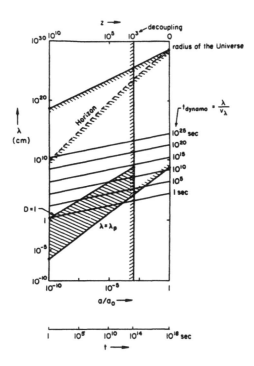

FIGURE 5.6 The size of relevant sizes and times in cosmological evolution from $t = 1$ sec after the Big Bang to the present. The typical size of magnetic bubbles due to the fluctuations is shown by the line $\lambda_p = c/\omega_p$.

it may well be of primordial nature (i.e., around the time of recombination). For example, the post-recombination violent relaxation of gravitationally unstable neutral matter can twist and stretch pre-existing magnetic fields that have been created prior to recombination. Matter may be heated by this stretching and twisting, thus creating and sustaining hot tenuous plasmas on the surface of denser matter and beyond. In addition, even without such, the ionization immediately after the recombination did not go below 10^{-6}. Such plasmas provide an alternative medium for the dynamo action of cosmological magnetic fields as well as galactic magnetic fields. Such magnetic fields may provide the necessary energy for X-ray emissions known as the cosmic X-ray background (Guilbert ₋nd Fabian, 1986). The dynamo action after the recombination was investigated by many others, including Ruzmaikin and Sokoloff (1977) and Zweibel (1988), and Kulsrud and Anderson (1992).

There do appear to exist cosmic magnetic fields of order 10^{-6} G (Norman, 1990). These recent works tend to support the view that the galactic dynamo has not had time to generate the observed fields. Such findings seem to suggest that they evolved either during the plasma epoch or after the recombination but before the formation of galaxies as described above.

In what follows we are interested in the cosmological dynamo as opposed to the galactic dynamo. Kueny and Tajima (1994) have examined this problem, using the "ABC flow" (named after Arnold, Beltrami, and Childress) and the associated fast dynamo, the mechanism examined by Finn and Ott (1988). The model indicates cellular morphology of enhanced magnetic fields by dynamo and the rest of the background diluted cosmological expansion which makes $|\mathbf{B}| \propto a^{-2}$. However, this model assumes an *ad hoc* flows and remains merely a model. Recently, Kulsrud *et al.* (1997) carried out a far more realistic calculation, whose essence is now reviewed below.

Prior to the formation of galaxies, the universe is by no means uniform. Indeed, it is now known that small relative density perturbations, $\delta\rho/\rho$, present at the time of recombination, grow to finite amplitude and form the present galaxies, clusters of galaxies, and other structures, at a variety of epochs. When $\delta\rho/\rho \approx 0.1$, shocks form and the resulting heated electrically conducting fluid produces electric currents that generate magnetic fields even from field-free initial conditions. Such fields, which are quite weak ($\approx 10^{-21}$ G), would, after compression into the galactic disk, produce the required seed field for the galactic dynamo.

However, this is not the end of the story. The shocks also generate *vorticity* on all scales. This vorticity is strong enough that its vortex cells turn over and generate turbulence. The energy spectra of this turbulence has the *Kolmogoroff power law*, and extends down to the viscous scale (see Sec. 2.3). Although the medium is compressible, the shear motions in the turbulence are themselves incompressible. This is because the difference in frequencies between compressible and shear motions do not permit the sound waves to interact strongly with the incompressible shear motions. The turbulence can be regarded as incompressible.

The eddies in the turbulence turn over at a rate that is fastest for the smallest eddy. The existence of a Kolmogoroff cascade from large eddies to small eddies is crucial because it enables the slowly rotating large eddies to drive rapidly rotating small eddies, that rapidly amplify the magnetic field.

Kulsrud *et al.* (1997) contend that conditions are always such that during the collapse of a protogalaxy the smallest eddy turns over several hundred times. This would lead to an increase of the magnetic field energy by a corresponding number of powers of e, the base of natural logarithms, if saturation processes did not set in. Instead, the magnetic field saturates and comes into equipartition with the pregalactic turbulence. Such a saturated field is of sufficient strength to provide a primordial origin for galactic magnetic fields.

If the dynamics of the magnetic field played no role, then the magnetic field would be very chaotic on small scales and the magnetic energy could be concentrated on scales smaller than the smallest eddy. However, as the field strengthens, the magnetic field on these smallest scales resists amplification, and the total magnetic field continues to strengthen only on scales comparable with that of the smallest hydrodynamic turbulent eddy. But later, when the field becomes even stronger, the smallest hydrodynamic eddies become suppressed due to the increasing drain of energy to the magnetic field. As a result, the spectrum of the turbulence becomes truncated at larger scales, and the scale of the smallest hydrodynamic

428

eddy increases. Eventually, only the largest eddies survive. During these later stages of saturation, only the magnetic energy at the largest scales is amplified, and the magnetic field eventually becomes coherent on these largest scales.

The first phase of the evolution of the protogalactic magnetic field can be numerically simulated by modifying a cosmological hydrodynamic code normally used to simulate structure formation (e.g. Ryu *et al.*, 1993). During this phase, the magnetic field is too weak to be of dynamic significance during the simulation. Thus, the magnetic field can be followed because it does not affect the motions of the plasma

$$\frac{\partial \mathbf{B}}{\partial t} = \nabla \times (\mathbf{v} \times \mathbf{B}), \tag{5.2.22}$$

where $\mathbf{v}$ is obtained from the hydrodynamic evolution without any magnetic forces.

The question arises concerning the initial value to take for the magnetic field. According to Eq. (5.2.22), it must be initially nonzero if $\mathbf{B}$ is to be nonzero in time. A number of proposals have been made for the generation of a small initial magnetic field in the early universe. One is the Biermann *battery mechanism*, which makes use of an extra pressure gradient term in Ohm's law (Bierman, 1950). The other is thermal fluctuations to be discussed in the next section (Sec. 5.3). The Biermann term arises from the nonvanishing pressure gradient in $\mathbf{E} = -\frac{1}{c}\mathbf{v} \times \mathbf{B} - \frac{1}{en_e}\nabla P_e$, which gives rise to the following modified magnetic induction equation

$$\frac{\partial \mathbf{B}}{\partial t} = \nabla \times (\mathbf{v} \times \mathbf{B}) + \frac{c\nabla p_e \times \nabla n_e}{n_e^2 e}, \tag{5.2.23}$$

where n_e is the electron density, and p_e is the electron pressure.

If the ionization fraction, X, is taken constant in space, and the electron temperature is taken equal to the neutral temperature, then $n_e/n_B(1+X) = p_e/p$. Here, n_B is the baryonic number density. From this result, Eq. (5.2.23) reduces to

$$\frac{\partial \mathbf{B}}{\partial t} = \nabla \times (\mathbf{v} \times \mathbf{B}) + \frac{\nabla p \times \nabla \rho}{\rho^2} \frac{cm_H}{e} \frac{1}{1+X}, \tag{5.2.24}$$

where m_H is the hydrogen mass. $\rho = n_B M/(1+X)$ for a hydrogen gas. For any barotropic flow ($p \equiv p(\rho)$), the last source term is zero because ∇p is parallel to ∇p. However, in general, for a real fluid in which curved shocks and photoheating can occur, $\nabla p \times \nabla p \neq 0$.

Multiplying Eq. (5.2.24) by $e/m_H c$, we get the equation for the cyclotron frequency $\omega_{\text{cyc}} = eB/m_H c$

$$\frac{\partial \omega_{\text{cyc}}}{\partial t} = \nabla \times (\mathbf{v} \times \omega_{\text{cyc}}) + \frac{\nabla p \times \nabla p}{\rho^2} \frac{1}{1+X} + \frac{\eta c}{4\pi} \nabla^2 \omega_{\text{cyc}}. \tag{5.2.25}$$

A term has been added to represent any resistive diffusion that may be present.

The ionization fraction enters into Eq. (5.2.25) through $1+X$, so even a very low ionization fraction is enough to generate magnetic fields. It is only necessary that there be enough electrons to carry the required current with a drift velocity relative to the ions v_D less than their thermal velocity v_e, i.e. $B/4\pi L = n_e e v_D/c < n_e e v_e/c$ where L is the scale size of variation of the magnetic field.

Kulsrud *et al.* incorporated Eq. (5.2.25) into a numerical simulation for the large-scale structure formation in a standard cold dark matter (CDM) model universe with total mass $\Omega = 1$. [The simulation (Kulsrud *et al.*, 1997) was done in a periodic box with $(32h^{-1} \text{Mpc})^3$ volume using 128^3 cells and 64^3 particles from $z_i = 20$ to $z_f = 0$. The values of other parameters used are baryon mass $\Omega_b = 0.06$, $h = 1/2$]. For the initial condition, the standard CDM power spectrum with $n = 1$ was adopted, which is modified by the transfer function given by Bardeen *et al.* (1986). Thus, the simulation is basically the same as that reported in Kang *et al.* (1994). The actual equation for the magnetic field in comoving coordinates is

$$\frac{\partial \mathbf{B}}{\partial t} = \frac{1}{a} \nabla \times (\mathbf{v} \times \mathbf{B}) - 2\frac{\dot{a}}{a} \mathbf{B} - \frac{1}{B_0 t_G} \frac{m_H c}{e} \frac{1}{1+\chi} \frac{\nabla \rho \times \nabla p}{\rho^2}, \tag{5.2.26}$$

where B_0 and t_G are the normalization constants for the magnetic field and time.

In Fig. 5.7, the temporal evolution of the resulting magnetic field is plotted. The upper panel shows the volume-averaged (solid line) and mass-averaged (dotted line) magnetic energy density $(B^2/8\pi)$ as a function of z. The lower panel shows the volume-averaged (solid line) and mass-averaged (dotted line) magnetic field strength (B). Note that $B \propto h$, and $h = 1/2$ was used. The magnetic field strength at first grows monotonically to the mass averaged value of order 10^{20} G by $z \sim 3$. After this time, the value of the averaged field strength leveled off without further increase. It is believed that the saturation is due to the finite numerical resistivity inherent in the numerical scheme used to solve Eq. (5.2.26).

The contours of the resulting baryonic density (ρ) and magnetic field strength (B) at $z = 2$ are shown in Fig. 5.8 and Fig. 5.9. The slice shown has a thickness of $2h^{-1}$ Mpc (or 8 cells). The upper panels show the whole region of $32 \times 32 h^{-1}$ Mpc, while the lower panels show a magnified region of $10 \times 10 h^{-1}$ Mpc. In Fig. 5.3, the regions with density higher than the volume averaged value $(0.06\bar{\rho})$ are contoured with contour levels $(0.06\bar{\rho}) \times 10^k$ and $k = 0, 0.1, 0.2, ..., 2$. Similarly in Fig. 5.9, the regions with magnetic field strength higher than the volume averaged value $(8 \times 10^{-23}$ G) are contoured with contour levels $(8 \times 10^{-23}$ G$) \times 10^k$ and $k = 0, 0.1, 0.2, ..., 2$. As expected, they are very well correlated. Since the magnetic field was mostly generated in the accretion shocks around the clusters, the high density core regions of the clusters have the strongest magnetic fields.

It is interesting to compare the maximum rms value of B, or alternatively $\omega_{cyc} = 10^4 B$, with the same rms mean for the vorticity, $\omega = \nabla \times \mathbf{v}$ where $\mathbf{v}$ is the fluid velocity. The maximum rms value for ω is $\sim 10^{-16}$ s^{-1} around the clusters in the numerical simulation. This is equal to the cyclotron frequency of an ion in a magnetic field of $\sim 10^{-20}$ G. In other words $eB/m_H c \approx \omega$ in the clusters. This is not surprising since the equation for the evolution of $-\omega$ is identical to that for $\omega_{cyc} = eB/m_H c$, except for dissipative terms.

By taking the curl of the equation of motion in the form

$$\frac{\partial \mathbf{v}}{\partial t} - \mathbf{v} \times (\nabla \times \mathbf{v}) + \frac{1}{2}\nabla v^2 = -\frac{\nabla p}{\rho} + \nu \nabla^2 \mathbf{v} \tag{5.2.27}$$

430

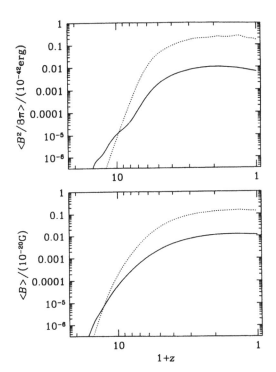

FIGURE 5.7 Temporal evolution of the magnetic field. The upper panel shows the volume-averaged (solid line) and mass-averaged (dotted line) magnetic energy density $(B^2/8\pi)$ as a function of z. The lower panel shows the volume-averaged (solid line) and mass-averaged (dotted line) magnetic field strength (B) (Kulsrud *et al.*, 1997).

where ν is the kinematic viscosity, one gets

$$\frac{\partial \boldsymbol{\omega}}{\partial t} = \nabla \times (\mathbf{v} \times \boldsymbol{\omega}) - \frac{\nabla p \times \nabla \rho}{\rho^2} + \nu \nabla^2 \boldsymbol{\omega}. \tag{5.2.28}$$

Now we see, on comparing Eq. (5.2.28) with Eq. (5.2.25), that if dissipative processes are ignored (conditions well satisfied except during the later stages of the simulation), and if we assume that both $\underset{\sim}{\omega}_{\text{cyc}}$ and $\underset{\sim}{\omega}$ are initially zero, then we should have

$$\underset{\sim}{\omega}_{\text{cyc}} = -\frac{\underset{\sim}{\omega}}{(1 \times \chi)}. \tag{5.2.29}$$

It must be appreciated that the $\nabla p \times \nabla \rho$ term is zero until some pressure is generated, since usually p is very small initially in the simulation. The generation of p generally happens

431

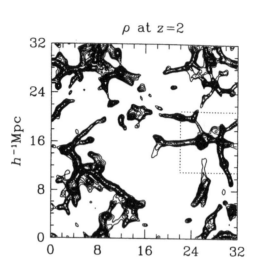

ρ at $z=2$

FIGURE 5.8 Density contours of a slice with a thickness of $2h^{-1}$ Mpc (or 8 cells) at $x = 2$. The contour lines with density higher than 0.06β are shown with levels 0.06×10^4 and $k = 0, 0.1, 0.2, ..., 2$ for the whole region of $32 \times 32h^{-1}$ Mpc (Kulsrud *et al.*, 1997).

in shocks where viscosity is certainly important. It can be argued that the jump in ω_{cyc} and $-\omega/(1+\chi)$ across a shock should be equal since, if we could treat Eq. (5.2.28) as valid through the shock, the integral of $\nu\nabla^2\omega$ is probably small. Thus, ω_{cyc} and ω satisfy essentially the same equation even in the shock. A check of the above relation has been carried out and found to be quite good (Kulsrud *et al.*, 1997).

Eventually viscosity does become important and ω tends to saturate in mean square average. However, since the twisting of the magnetic field by the $\nabla \times (\mathbf{v} \times \mathbf{B})$ term persists, one expect that $\mathbf{B}$ will continue to grow. This fact is supported by Batchelor's discussion in his early paper (Batchelor, 1950).

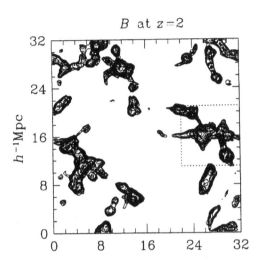

B at $z=2$

FIGURE 5.9 Magnetic field strength contours of a slice with a thickness of $2h^{-1}$ Mpc (or 8 cells) at $z = 2$. The contour lines with magnetic field strength higher than 8×10^{-23} G are shown with levels $8 \times 10h^{-23} \times 10^4$ and $k = 0, 0.1, 0.2, ..., 2$. Same region as in Fig. 5.8 (Kulsrud et al., 1997).

5.3 On the Origin of Cosmological Magnetic Fields

In the previous section 5.2 we discussed the dynamo mechanism operative prior to the galaxy formation epoch, where the violent motion of pregalactic epoch has been shown to multiply the seed magnetic fields that may be present at the epoch. Kulsrud et al. invoked the Biermann battery effect which acts to create seed magnetic fields when the pressure gradient is not entirely parallel to the electron density gradient. In addition to this mechanism, it may be possible that the ability of the primordial plasma to sustain magnetic fields out of their own fluctuations. A plasma with temperature T (energy unit) sustains fluctuations of electromagnetic fields and particle density even if it is assumed to be in a thermal equilibrium. The level of fluctuations in the plasma for a given wavelength and frequency of electromagnetic fields can be computed by the *fluctuation-dissipation theorem*. A large zero frequency peak of electromagnetic fluctuations is discovered by Tajima *et al.* (1992a). We show that the energy contained in this peak is complementary to the energy "lost" by the plasma cutoff effect. The level of the zero (or nearly zero) frequency magnetic fields is computed as $\langle B^2 \rangle^0 / 8\pi = 1/2\pi^3 \, T(\omega_p/c)^3$, where T and ω_p are the temperature and plasma frequency. This is the theoretical minimum magnetic field strength spontaneously generated, as no turbulence is assumed. The size of the fluctuations is $\lambda \sim (c/\omega_p)(\eta/\omega)^{1/2}$, where η and ω are the collision frequency and the (nearly zero) frequency of magnetic fields oscillations. These results are not in contradiction with the conventional black-body radiation spectra but its extension, and as such, do not contradict the observed lack of structure in the cosmic microwave background. The level of magnetic fields is significant at the early radiation (plasma) epoch of the Universe. Such magnetic fluctuations provide seed fields for later evolution, as discussed in Sec. 5.2. In the following we look at this mechanism for primordial magnetic fields and their properties.

Around the time of 1 second after the big bang, the standard cosmological theory says (Weinberg, 1972) that the weak (neutrino) interaction detached from reacting with the rest of the radiation and matter which are in thermal equilibrium. Around the time of 10 seconds ($z \sim 10^9$) to 10^3 seconds ($z \sim 10^8$) the strong (nuclear) interaction ceased to play a role in the evolution of the Universe. According to the standard theory "The Universe will go on expanding and cooling, but not much of interest will occur for 700,000 years ($\sim 10^{13}$ sec; $z \sim 10^3$). At that time the temperature will be at the point where electrons and nuclei can form stable atoms" (recombination) (Weinberg, 1972). During this epoch from $z = 10^{10}$ to 10^3 the radiation couples strongly with matter and thus has been called the *radiation epoch*. The main constituent of matter of this period is a plasma and the main interaction of this period is that of plasma dynamics, including that of radiation-plasma coupling. We first assume that at the dawn of the radiation (plasma) epoch ($t \sim 10^0$ sec; $z \sim 10^{10}$) photons and charged particles were in thermal equilibrium. In th next section (Sec. 5.4) we look into this point further in detail by considering the epoch prior to this. It will be learned that the presence of plasma plays an important role in shaping the radiation spectrum. This may be important, as it has been recognized that there appears to exist cosmic magnetic field of order 10^{-6} G which may or may not have had enough time to evolve in the gravitational epoch of the Universe (Norman, 1990). Furthermore, if the recent observation by Giovannini *et al.* (1990) proves to be the case, magnetic fields or their signature of the intersupercluster scale may be present, which hints at the primordial concoction of magnetic fields.

For the primordial Universe to be treated as a (gaseous) plasma, the collection of charged particles have to satisfy a certain condition. In the epoch of $t = 10^{-2} - 10^0$ sec the typical plasma density is $10^{28} - 10^{34} \mathrm{cm}^{-3}$. The plasma parameter $g = 1/(n \, \lambda_{De}^3)$ is much less than unity (Ichimaru, 1973) and is approximately 10^{-3} in the epoch of $10^0 - 10^{13}$ sec it is about 10^{-7}, where n is the density of electrons, λ_{De} is the electron Debye length. The Debye length is equal to c/ω_{pe} in the relativistic plasma in the first epoch ($z = 10^{11} - 10^{10}$), where c is the speed of light and ω_{pe} is the plasma frequency and the quantity c/ω_{pe} is usually referred to the *collisionless skin depth*. In both of the above epochs the mean distance between particles is much smaller than the typical collective length (the Debye length = the collisionless skin depth), which in turn is much smaller than the mean free path of electrons colliding with photons

$$\frac{1}{n^{1/3}} < \frac{c}{\omega_{pe}} < (n\,\sigma_{KN})^{-1}, \qquad (5.3.1)$$

where σ_{KN} is the Klein-Nishina cross-section of electron-photon collisions. When $T \gg mc^2$, the cross-section should be, instead of the Thompson cross-section, the Klein-Nishina formula:

$$\sigma_{KN} = \frac{3}{8}\left(\frac{mc^2}{\hbar\omega}\right)\sigma_T \qquad \text{(for } \hbar\omega \gg mc^2), \qquad (5.3.2)$$

while

$$\sigma_{KN} = \sigma_T \qquad \text{(for } \hbar\omega \ll mc^2), \qquad (5.3.3)$$

where the Thompson cross-section $\sigma_T = \frac{8\pi}{3}(e^2/mc^2)^2$.

In this plasma of $t = 10^{-2} - 10^0$ sec the (average) photon energy is $\hbar\omega \sim T \gg mc^2$ and we have

$$T \sim \hbar\omega \gg mc^2 > \hbar\omega_p. \qquad (5.3.4)$$

In the description of fluid behavior the Reynolds number sometimes plays an important role (Sec. 2.1). For wavelengths much larger than $(n\sigma_{KN})^{-1}$, the plasma behaves like a usual fluid and the Reynolds number may be expressed as Re $= \lambda^2\eta/\mu$, which can be much larger than unity, where η is the collision frequency or the effective collision frequency replaced by the Landau damping rate or other collisionless mechanisms such as the chaotic orbit effect. On the other hand, for wavelengths $\lambda \sim c/\omega_{pe} \ll (n\sigma_{KN})^{-1}$, the plasma is collisionless and nearly dissipationless. The list of plasma parameters in the early plasma epoch ($z = 10^{11} - 10^3$) are summarized in Table 5.1, in which our conclusions are also listed that are to be obtained in the following discussion.

We note that the past investigations of cosmological magnetic fields such as Harrison (1970, 1973) assumed primordial turbulence with nonzero vorticity and obtained magnetic fields of $\sim 10^{-18}$ Gauss for galactic scales. Kajantie and Kurki-Suonio (1986) discussed phase transition incurred fluctuations. In contrast to these works Tajima et al. (1992a) resort to no assumption as to the primordial condition but for the thermal equilibrium. This treatment is based on the theory of fluctuation-dissipation theorem. The calculation of magnetic fields is undertaken in Sec. 5.3.1. Some interesting physical properties are discussed in Sec. 5.3.2 and 5.3.3. In Sec. 5.3.4 we discuss cosmological implications of these magnetic fields.

5.3.1 The Fluctuation-Dissipation Theorem and Magnetic Fields*

In or near thermal equilibrium the plasma has thermal fluctuations whose level is related to the medium's dissipative characteristics and the temperature T, as formulated in the fluctuation-dissipation theorem (Kubo, 1957). More treatments for a plasma may be found in Rostoker, et al. (1965) and in Dawson (1968). We find an expression for the fluctuation spectrum of the magnetic field in an equilibrium plasma as a function of frequency.

		$t = 10^{-2}$	$t = 1$	$t = 10^{13}$	$t = 3 \times 10^{17}\,\text{sec}$
T	eV	10^7	10^6	0.4	$T_\gamma = 0.0003$
n	cm^{-3}	5×10^{34}	4×10^{31}	10^3	10^{-6}
L_{hor}	cm	10^8	10^{10}	10^{23}	10^{28}
B	Gauss	10^{16}	10^{13}	10^{-12*}	**
$\dfrac{\langle B^2 \rangle^0}{\langle B^2 \rangle^{bb}}$		0.1-1	10^{-2}	10^{-25*}	**
β		1	$10 - 10^2$	10^{15*}	**
Re		10^{17}	10^{18}	10^{15}	

*: instantaneous fluctuations assumed. Other evaluations are possible. See the text for detail.
**: no thermal equilibrium plasma is plausible.

TABLE 5.1 Primordial Magnetic Fluctuations with $\omega \sim 0$: The zero frequency magnetic fluctuations in early Universe. ($t = 10^{-2}$, 1, and 10^{13} sec after the big bang). The temperature T, density of the plasma electrons, the horizon size L_{hor}, the zero frequency magnetic fluctuations B, the ratio of the zero frequency magnetic fields $\langle B^2 \rangle^0/8\pi$ to the blackbody component $\langle B^2 \rangle_{\text{bb}}/8\pi$, the plasma beta β, and the (maximum) Reynolds number Re are tabulated. B and $\langle B^2 \rangle^0/\langle B^2 \rangle_{\text{bb}}$ are from Eq. (5.3.26) and β from Eq. (5.3.27) (Tajima et al., 1992a).

This is accomplished by deriving the magnetic fluctuations in wavenumber and frequency space $\langle B^2 \rangle_{k\omega}/8\pi$ from the *fluctuation-dissipation theory*, then integrating over wavenumber. $\langle B^2 \rangle_\omega/8\pi$ is nearly a *black-body spectrum* at high frequencies, but, when plasma collisionality is taken into account, it has a high, narrow peak at frequency $\omega = 0$.

We look at waves in a homogeneous isotropic equilibrium plasma. However, since we are interested in spontaneous generation of magnetic fields, we consider a nonmagnetized plasma here. To start with, we assume a wavevector $\mathbf{k} = k\hat{x}$. The strength of electric field fluctuations may be found in Sitenko (1967):

$$\frac{1}{8\pi} \langle E_i E_j \rangle_{k\omega} = \frac{i}{2} \frac{\hbar}{e^{\hbar\omega/T} - 1} \left\{ \Lambda_{ij}^{-1} - \Lambda_{ij}^{-1*} \right\}, \tag{5.3.5}$$

where

$$\Lambda_{ij}(\omega, \mathbf{k}) = \frac{c^2 k^2}{\omega^2} \left(\frac{k_i k_j}{k^2} - \delta_{ij} \right) + \epsilon_{ij}(\omega, \mathbf{k}), \tag{5.3.6}$$

where $\epsilon_{ij}(\omega, \mathbf{k})$ being the dielectric tensor of the plasma. Since Faraday's law is $\mathbf{B} = \frac{ck}{\omega} \times \mathbf{E}$, and we have set $\mathbf{k} = k\hat{x}$, we find

$$\frac{\langle B_2^2 \rangle_{\mathbf{k}\omega}}{8\pi} = \frac{i}{2} \frac{\hbar}{e^{\hbar\omega/T} - 1} \frac{c^2 k^2}{\omega^2} \left\{ \Lambda_{33}^{-1} - \Lambda_{33}^{-1*} \right\}, \tag{5.3.7}$$

and

$$\frac{\langle B_3^2 \rangle_{\mathbf{k}\omega}}{8\pi} = \frac{i}{2} \frac{\hbar}{e^{\hbar\omega/T} - 1} \frac{c^2 k^2}{\omega^2} \left\{ \Lambda_{22}^{-1} - \Lambda_{22}^{-1*} \right\}, \tag{5.3.8}$$

where the subscript 1,2, and 3 refer to x, y, and z. We then have the total magnetic fluctuations as

$$\frac{\langle B_{tot}^2 \rangle_{\mathbf{k}\omega}}{8\pi} = \frac{i}{2} \frac{\hbar}{e^{\hbar\omega/T} - 1} \frac{c^2 k^2}{\omega^2} \left\{ \Lambda_{22}^{-1} + \Lambda_{33}^{-1} - \Lambda_{22}^{-1*} - \Lambda_{33}^{-1*} \right\}, \tag{5.3.9}$$

where c.c. refers to the complex conjugate.

In order to establish $\Lambda_{ij}(\omega, \mathbf{k})$, from the equation of motion of a plasma, here we introduce a multi-fluid model of a plasma. As a simple and analytically tractable model consider the case with finite and constant collisionality:

$$m_\alpha \frac{d\mathbf{v}_\alpha}{dt} = e_\alpha \mathbf{E} - \eta_\alpha m_\alpha \mathbf{v}_\alpha, \tag{5.3.10}$$

where α is a particle species label and η_α is the collisional frequency but can include the viscosity effect. We can show that a description of electron dynamics more accurate than Eq. (5.3.10) such as kinetic treatments leads to natural convergence. We note that η should tend to zero for very short wavelength EM waves because for large wavenumbers $\mathbf{k}$ the photon shifts momentum by a large amount so that interacting electron population with the thermal energy before and after the interaction becomes very small. Fourier transforming (5.3.10) gives

$$-i\omega m_\alpha \mathbf{v}_\alpha = e_\alpha \mathbf{E} - \eta_\alpha m_\alpha \mathbf{v}_\alpha, \tag{5.3.11}$$

which yields the current $\mathbf{j}_\alpha$

$$(-i\omega + \eta_\alpha) \mathbf{j}_\alpha = \frac{\omega_{p\alpha}^2}{4\pi} \mathbf{E}. \tag{5.3.12}$$

The susceptibility tensor $\chi_{\alpha ij}$ is defined to relate $\mathbf{j}_\alpha$ to $\mathbf{E}$ such that

$$j_{\alpha i} = -i\omega \chi_{\alpha ij}(\omega \mathbf{k}) E_j(\omega \mathbf{k}). \tag{5.3.13}$$

The dielectric tensor $\epsilon_{ij}(\omega \mathbf{k})$ is given by

$$\epsilon_{ij}(\omega \mathbf{k}) = \delta_{ij} + 4\pi \sum_\alpha \chi_{\alpha ij}, \tag{5.3.14}$$

so

$$4\pi \chi_{\alpha ij}^{(\omega, \mathbf{k})} = \frac{\omega_{p\alpha}^2}{\omega(\omega + i\eta_\alpha)} \delta_{ij}, \tag{5.3.15}$$

and

$$\epsilon_{ij}(\omega, \mathbf{k}) = \delta_{ij} - \sum_\alpha \frac{\omega_{p\alpha}^2}{\omega(\omega + i\eta_\alpha)} \delta_{ij}. \tag{5.3.16}$$

In an electron-positron plasma neglecting ions, we have $\omega_{pe+} = \omega_{pe-}$ and $\eta_{e+} = \eta_e = \eta$. So Eq. (5.3.16) becomes

$$\epsilon_{ij}(\omega, \mathbf{k}) = \delta_{ij} - \frac{\omega_p^2}{\omega(\omega + i\eta)} \delta_{ij}, \tag{5.3.17}$$

where $\omega_p^2 = \omega_{pe+}^2 + \omega_{pe-}^2$. We now obtain

$$\Lambda_{ij} = \begin{pmatrix} 1 - \dfrac{\omega_p^2}{\omega(\omega + i\eta)} & & \\ & 1 - \dfrac{c^2 k^2}{\omega^2} - \dfrac{\omega_p^2}{\omega(\omega + i\eta)} & \\ & & 1 - \dfrac{c^2 k^2}{\omega^2} - \dfrac{\omega^2}{\omega(\omega + i\eta)} \end{pmatrix}. \tag{5.3.18}$$

Combining Eqs. (5.3.9) and (5.3.19) after some algebra, we obtain

$$\frac{\langle B^2 \rangle_{\mathbf{k}\omega}}{8\pi} = \frac{2\hbar\omega}{e^{\hbar\omega/T} - 1} \eta \omega_p^2 \frac{k^2 c^2}{\omega^2} \frac{1}{\left[\omega^2 - k^2 c^2 - \omega_p^2\right]^2 + \eta^2 \left[\omega - k^2 c^2/\omega\right]^2}, \tag{5.3.19}$$

or

$$\frac{\langle B^2 \rangle_{\mathbf{k}\omega}}{8\pi} = \frac{2\hbar\omega}{e^{\hbar\omega/T} - 1} \eta \omega_p^2 \frac{k^2 c^2}{(\omega^2 + \eta^2)k^4 c^4 + 2\omega^2(\omega_p^2 - \omega^2 - \eta^2)k^2 c^2 + \left[(\omega^2 - \omega_p^2)^2 + \eta^2 \omega^2\right]\omega^2}. \tag{5.3.20}$$

Equation (5.3.20) shows the magnetic fluctuation spectrum in frequency–wavenumber space as a function of the temperature of the plasma.

5.3.2 Frequency Spectrum of Magnetic Fields*

Equation (5.3.20) has the well-known limit, as we take $\eta \to 0$ and $\omega_p \to 0$ (a plasmaless limit), which is the vacuum black-body radiation:

$$\frac{\langle B^2 \rangle_\omega}{8\pi} = \int d\mathbf{k} \, \frac{\langle B^2 \rangle_{\mathbf{k}\omega}}{8\pi} = \frac{\pi\hbar}{e^{\hbar\omega/T} - 1} \frac{1}{2} \frac{\omega^3}{c^3}. \tag{5.3.21}$$

When there is a plasma, its presence modifies the magnetic fluctuation. As we see in a moment, in this case the plasma effect dominates for small k (long wavelength) modes. In integrating $\langle B^2 \rangle_{\mathbf{k}\omega}$ over $\mathbf{k}$, we thus split the integral for the small wavenumber regime $(0 < k < k_c)$ where plasma effects dominate and for the large wavenumber regime $(k > k_c)$ where the spectrum is essentially black-body. A straightforward integration over dk diverges due to the lack of incorporation of quantum mechanical discrete $\mathbf{k}$ effects in high $\mathbf{k}$ regimes. An approximate treatment of this effect leads to an analytical expression of the frequency spectrum (Tajima *et al.*, 1992a)

$$\langle B^2 \rangle_\omega = \frac{1}{\pi^2} \frac{\hbar\omega}{e^{\hbar\omega/T} - 1} 2\eta \left(\frac{\omega_p}{c}\right)^3 \int_0^{k-c} dk \frac{k^4}{(\omega^2 + \eta^2)k^2 + \cdots}$$

$$+ \frac{\hbar(\omega^2 - \omega_p^2)^{3/2}}{2\pi(e^{\hbar\omega/T} - 1)} \left(\frac{\omega_p}{c}\right)^3 \Theta\left[\omega - \sqrt{c^2 k_c^2 + \omega_p^2}\right]. \qquad (5.3.22)$$

where Θ is the Heaviside step function and k_c is the cut-off frequency of integration and turns out to be $\sim k_p \equiv c/\omega_p$ and only the leading term is left in the denominator in the integral of the first term. This spectrum has two significant deviations from the *Rayleigh-Jeans-Planck distribution*. The first term did not exist in the absence of a plasma and the second term corresponds to the black-body radiation. The first deviation is the appearance of the cut-off frequency ω_p, below which no electromagnetic wave propagates, arising in the second term of Eq. (5.3.22). The second is the appearance of the first term in Eq. (5.3.22), the integral of which has a sharp peak at zero frequency $\omega = 0$, the zero frequency peak. The height of the peak is inversely proportional to the dissipation η and the width proportional to η.

On the other hand, integrating Eq. (5.3.20) over ω yields the $\mathbf{k}$-spectrum

$$\frac{\langle B^2 \rangle_\mathbf{k}}{8\pi} = \int d\omega \frac{\langle B^2 \rangle_{\mathbf{k}\omega}}{8\pi} = \frac{T}{2}\left[\frac{1}{1 + k^2 c^2/\omega_p^2} + \frac{\hbar}{e^{\hbar(\omega_p^2 + k^2 c^2)^{1/2}} - 1} \frac{k^2 c^2}{(\omega_p^2 + k^2 c^2)^{1/2}}\right], \qquad (5.3.23)$$

where the integral was carried out using the *Krommes-Kronig's relation*, the analytical property of causal media's responses (Tajima *et al.*, 1992). The first term in Eq. (5.3.23) arises only in the presence of a plasma and contributes importantly for small k-regime, while the second term corresponds to the black-body radiation modified by the plasma. When we let $\hbar \to 0$, Eq. (5.3.23) reduces to the well-known equipartition law

$$\frac{\langle B^2 \rangle_\mathbf{k}}{8\pi} = \frac{T}{2}. \qquad (5.3.24)$$

The zero-frequency peak arose from the small $\mathbf{k}$ mode (the first term).

The frequency spectrum Eq. (5.3.23) near zero frequency (which is due to the plasma current fluctuations) yields

$$\frac{\langle B^2 \rangle_{\omega=0}}{8\pi} = \frac{3}{2\pi^3}\sqrt{\frac{3}{\pi}} T \left(\frac{\omega_p}{c}\right)^3. \qquad (5.3.25)$$

This means that the zero frequency magnetic fluctuations are proportional to the temperature and the density to the 3/2 power ($n^{3/2}$) of the plasma. The earlier the plasma of the Universe, the higher the values of T and n and the greater the zero-frequency fluctuations are. Because of this scaling, the magnetic fluctuation Eq. (5.3.25) is negligible in nearly all plasmas (relatively low T and low n), except those in the primordial Universe. This may be

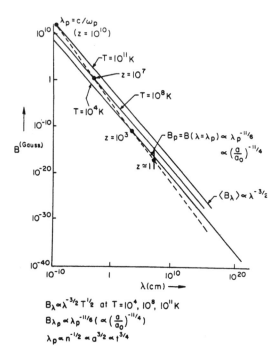

$$B_\lambda \propto \lambda^{-3/2} T^{1/2} \text{ at } T = 10^4, \ 10^8, \ 10^{11} \text{ K}$$
$$B_{\lambda_p} \propto \lambda_p^{-11/6} \left(\propto \left(\frac{a}{a_0} \right)^{-11/4} \right)$$
$$\lambda_p \propto n^{-1/2} \propto a^{3/2} \propto t^{3/4}$$

FIGURE 5.10 Magnetic fluctuations in primordial thermal plasmas.

one of the reasons why this effect has been overlooked, in addition to the divergent nature of the integral arriving at Eq. (5.3.22). The ratio of this zero frequency ($\omega = 0$) magnetic fields to the plasma pressure nT is proportional to $n^{1/2}$ and is greater as the epoch is earlier ($\propto a^{-3/2} \propto t^{-3/4}$). In Fig. 5.10 we show the characterization of the Universe between $t = 10^{-2}$ till the present. Typical magnetic fields generated spontaneously by thermal fluctuations in the early Universe and related physical quantities are shown in Table 5.1.

Since the magnetic fields couple with charged particles strongly, any magnetic fluctuations such as nonuniformity lead to plasma density fluctuations in such a way as to make the total of the magnetic and plasma pressure approximately constant. This means anti-correlation of the plasma density perturbation to the magnetic perturbation. Where strong magnetic fields (5.3.25) is created the plasma pressure is depressed and vice versa. As the magnetic fluctuations are nearly zero-frequency, the plasma response is nonadiabatic and isothermal. This is crucial. If the plasma perturbations incurred by the zero-frequency magnetic fluctuations were adiabatic, the perturbations would have influenced the photon pressure directly, which could be detectable if it were emitted at the time of recombination.

440

On the other hand, the *isothermal fluctuations* do not influence the photon pressure directly and thus could escape detection via the photon spectrum observation. The size of typical zero-frequency *magnetic bubbles* is c/ω_p, which is merely $\sim 10^{-10}$cm at the time $t = 10^{-2}$ second. Such a size is miniscule by itself. We had a large amount of plasma fluctuations $[\Delta n/n \sim \mathcal{O}(1)]$ but with miniscule scales ($\sim 10^{-10}$cm) at the dawn of the electron-positron epoch. Such a large amount of fluctuations may have influenced cosmological *nucleosynthesis* due to strong *inhomogeneity of baryons* (Kajino et al., 1990). However, the spatial size is too small for structure formation of the Universe as we know it. The question is can this large density fluctuation with small spatial scales become spatially large-scale structures in later epochs that have relations to the present day large-scale cosmological structures, including galaxies? We address this question in the next section.

We compare the size of the frequency $\omega = 0$ peak of $\langle B^2 \rangle_\omega/8\pi$ relative to the size of the black-body peak. The black-body spectrum has its maximum at frequency $\omega \approx 2.81\, T/\hbar$. So, the ratio of the zero frequency peak fluctuations to the black-body radiation in an electron-positron plasma is

$$\frac{\langle B^2 \rangle_{\omega=0}}{\langle B^2 \rangle_{b-b}} = \alpha_1\, \hbar^2\, \frac{\omega_{pe}^2\, ck_{cut}}{\eta T^2} = (2\pi)^3 \alpha_1 \delta \left(\frac{\lambda_B}{\lambda_{D_e}} \right)^3 \frac{T}{\hbar \eta}, \qquad (5.3.26)$$

where λ_B is the thermal *deBroglie wavelength* $2\pi\hbar/\sqrt{mT}$ and $\delta \sim 0.81$. For the electron-ion case α_1 in Eq. (5.3.26) becomes 0.47 for electron-ion plasma and 0.89 for $e^- - e^+$. Note, for example, that this ratio Eq. (5.3.26) can be as great as $0.1 - 1$ at $t = 10^{-2}$ sec (see Table 5.1). This is a surprisingly large value.

From Eq. (5.3.26) the plasma beta due to the magnetic field energy density associated with the zero frequency is evaluated to be

$$\beta = \frac{nT}{\langle B^2 \rangle^0/8\pi} = 2\pi^3\, n \left(\frac{c}{\omega_p} \right)^3. \qquad (5.3.27)$$

If we assume that at each instance of cosmic time the level of magnetic fluctuations is determined by the fluctuation-dissipation theorem, the plasma beta scales as $\beta \propto n^{-1/2} \propto a^{3/2}$ (see Table 5.1). The beta at $t = 10^{-2}$ sec is as small as $1-10$. Once again this is an impressive value. On the other hand, if the primordial magnetic fields are created at $t = t_d$ (at this moment we do not have sufficient knowledge to determine t_d) according to the fluctuation-dissipation theorem and the magnetic field evolution is detached from that of the plasma temperature or the photon temperature as the Universe continues to cool, the plasma beta may scale as $\beta \propto a^0$ (invariant), because $B \propto a^{-2}$ (the *flux conservation*). One, however, notes that most likely at some point of time the dynamo effect comes into play (see Sec. 5.2), which tends to amplify magnetic fields in competition of the cosmic expansion.

5.3.3   Collisional Effects*

In this section we examine the essential results of Sec. 5.3.2 and attach our physical interpretation to it. The basic intuitive physical picture of what the fluctuation-dissipation theorem says is: an individual mode (or field) decays by a certain dissipation, giving up energy to particles or other modes, while particles (or other modes) excite new modes and repeat the process and the amount of fluctuations is related to the dissipation. We find that the physical basis of the zero frequency peak is due to collisions (or other kinetic dissipation) or more precisely collision-induced quasi-modes. Imagine an individual charged particle, say, an electron propagates in a plasma, which itself is composed of an ensemble of such electrons (and other charged particles). With finite discreteness of the charge, the electron can contribute to a current fluctuation due to this ballistic motion of the electron over time until it encounters collisions (or other kinetic dissipation). Electromagnetic field fluctuations are induced from such a current fluctuation. However, the low frequency component of the field fluctuations cannot propagate, as it is evanescent, as distinct from the high frequency component which can propagate in a plasma as photon. Thus this gives rise to damping of the field fluctuations. Therefore, the lifetime of the fluctuations is related to the collision time (or other kinetic dissipation time).

From Maxwell's equations with all the terms on the right-hand side except the source term (the third term) written in terms of $\mathbf{E}$, we obtain the dispersion relation of the quasi-modes:

$$\omega^2 - k^2 c^2 - \frac{\omega_p^2}{1 + i\eta/\omega} = 0. \tag{5.3.28}$$

In the low frequency limit Eq. (5.3.28) yields the dispersion relation

$$\omega = i \frac{k^2 c^2}{\omega_p^2} \eta. \tag{5.3.29}$$

Or, equivalently, the spatial size λ of magnetic field fluctuations for a given lifetime $\tau_\ell (\equiv \omega^{-1})$ is

$$\lambda(\tau_\ell) = 2\pi \frac{c}{\omega_p} (\eta \, \tau_\ell)^{1/2}. \tag{5.3.30}$$

Equation (5.3.30) states that the lifetime τ_ℓ of magnetic fluctuations (or maybe called "magnetic bubble") of size λ is proportional to the size squared ($\tau_\ell \propto \lambda^2$); the larger the size of the bubble, the longer it lasts.

This entails a possible ramification. Suppose two magnetic bubbles touch or collide with each other and coalesce into one. The time for *coalescence* of magnetic bubbles involves reconnection of magnetic field lines. It is generally known (Parker, 1957; Bhattacharjee *et al.*, 1983; Tajima *et al.*, 1987) that this process (or related ones) is much faster than the diffusive time related to Eq. (5.3.30). (See Sec. 3.3). Thus the coalescence time is much shorter than the individual life time. Therefore, before bubbles die away, they can form a coalesced bubble when they collide with each other, as long as they collide frequently enough. Once a larger coalesced bubble is formed, its life time is substantially longer, as Eq. (5.3.30) shows the life time is proportional to the square of the size of the bubble. It is possible to imagine that once larger bubbles are formed, they become even more long-lived and may be able to encounter more opportunities to collide with other bubbles. In this way a preferential formation of larger bubbles may become possible. This process is not far different from that of polymerization.

442

5.3.4 Cosmological Implications

We discussed in the previous sections that electromagnetic waves in the primordial plasma fall into two categories: one with large wavelengths ($k \lesssim \omega_{pe}c$) and nearly zero frequency ($\omega \ll \omega_{pe}$) and one with small wavelengths ($k \gg \omega_{pe}/c$) and frequency greater than ω_{pe}. Those modes $k > \omega_p/c$ are not significantly modified by the presence of the plasma ('hard photon'), while those with $k < \omega_p/c$ significantly modified ('soft or plastic photon'). It is those 'plastic photons' or their magnetic fields that we are interested in, as they can have more 'magnetic' fields in nature and can leave possible structural imprints on the primordial plasma.

In Table 5.1 we summarized the results. Including the physical quantities we already discussed, we survey physical quantities of importance that characterize the radiation epoch (or the plasma epoch). The density scales as $n \propto a^{-3}$ where a is the cosmic scale factor, which increases as $a \propto t^{1/2}$ during the radiation epoch. The wavelength of photons also scales as $\lambda \propto a$ and thus the temperature of photons $T \propto a^{-1}$, as the frequency of the maximum intensity of the black-body radiation $\omega_{max} = 2.81\, T/\hbar$. It follows that the plasma frequency scales as $\omega_{pe} \propto n^{1/2} \propto a^{-3/2}$. The electron collision frequency goes like $\eta_e \propto n\,T^{-3/2} \propto a^{-3/2}$. The plasma parameter (and the collisionality) is therefore $g = (n\lambda_{De}^3)^{-1} \cong \eta_e/\omega_{pe} \propto a^0$ (independent of a) and thus invariant during the epoch in which the numbers of constituent particles are conserved; e.g. during $t = 10^{-2} - 10^0$ sec $g \sim 10^{-3}$ (invariant) and it changes around $t = 10^0$ sec as positrons annihilate with electrons to $g \sim 10^{-7}$ and stays invariant till the recombination. On the other hand, the collision frequency between electrons and photons may be given from the Thompson cross-section or in relativistic cases from the Klein-Nishina cross-section to be $\nu_{TH} \propto n\,T^{1/2} \propto a^{-7/2}$ and $\nu_{KN} \propto n\,T^{-1} \propto a^{-2}$, respectively. The Reynolds number Re is Lv/μ, where L, v, and μ are the typical sizes of the length, velocity, and viscosity. By taking v the thermal velocity, Lv/μ scales as $L\,a^{-1}$ and if we take L as the horizon size ct, Re $\propto t^{1/2}$ in the radiation epoch. The electron magnetic energy $\langle B^2 \rangle_\omega^{bb}$ contained in the black-body radiation is proportional to ω^2 and $\langle B^2 \rangle_\omega^{bb} \propto T^3 \propto a^{-3}$. On the other hand, the zero frequency magnetic fluctuation energy $\langle B^2 \rangle_{\omega \to 0} \propto T\omega_p^3/c^3 \propto a^{-11/2}$. Thus the ratio of the zero frequency fluctuations to the black-body energy is proportional to $a^{-5/2}$.

We review the numerical values of the magnetic fields sustained by the thermal equilibrium plasma. It should be emphasized that the strength of magnetic fields at early epoch of the radiation (or plasma) era is uncannily large (as large as 10^{16} Gauss at $t = 10^{-2}$ sec and 10^{13} Gauss at $t = 1$ sec though the spatial scale is small). These numbers are tabulated also in Table 5.1. It is not possible at this moment to accurately evaluate the magnetic field strength at later times such as at $t = 10^{13}$ sec. Some of the reasons for this difficulty will be explained in the following paragraphs. [Of course, after $t = 10^{13}$ most of the plasma is annihilated as a result of recombination and the influence of gravity cannot be ignored any more and thus the assumption of thermal plasma calculation is totally inapplicable]. Nevertheless, it may be of interest to put numbers for later epochs as reference numbers. For the moment let us exclude any possibility as such dynamo action, magnetic polymerization,

443

and the effects of noise re-entry to the horizon, to estimate the magnetic field strength solely based on the instantaneous conditions of the thermal plasma at each instance. Based on Eq. (5.3.23), we can define the strength of magnetic fields whose wavelength is larger than a certain size λ as $\langle B^2 \rangle_\lambda / 8\pi$. This becomes $\langle B^2 \rangle_\lambda / 8\pi = (T/2)(4\pi/3)\lambda^{-3}$. Based on this assumption that the magnetic fluctuations adjust to the plasma's instantaneous conditions at later epochs, the magnitude of magnetic fluctuations with wavelength $\lambda_p = 2\pi c/\omega_p$ is of the order of 10^{-12} Gauss at the dawn of recombination. [This value is already greater (though at a small spatial scale of λ_p) than 10^{-21} G assumed and needed in Kulsrud et al. (1997)]. An alternative estimate of magnitude of magnetic fields is based on the adiabatic expansion of magnetic fields in the early epoch (say at $t = 1\,\text{sec}$) for later epochs, which yields 10^3 Gauss at $\lambda = \lambda_p$ for the same recombination time and wavelengths. Both of these estimates for magnetic field evolution are likely to be extreme and simplistic, as the cosmological plasma is subject to evolution. A high β plasma such as the one under consideration is known to exhibit additional effects, including the tendency toward the accumulation of magnetic fields near the boundary layer region (between where B exists and where B does not), in which the plasma β may approach unity (Vekshtein, 1990).

It is theoretically important to note that our theory suggests the spontaneous generation of magnetic fields (spontaneous breakdown of symmetry) even in a perfectly isotropic and homogeneous thermal equilibrium plasma. Such generation provides a mechanism for seed magnetic fields for the Universe's later evolution. This is the first major consequence of the present theory on cosmology. There are, however, additional consequences. In the early epoch of the radiation era ($t \sim 10^{-2}\,\text{sec}$) the magnitude of magnetic fields is as large as 10^{16} gauss and associated density fluctuations as large as 100%. A possible consequence is the influence of the above induced density inhomogeneity on nucleosynthesis around this time. Some people (Kajino and Boyd, 1990) have considered inhomogeneous Big Bang nucleosynthesis due to the *quark-gluon phase transition* earlier than this epoch. Another possible impact of these magnetic fields at early times is the "polymerization" of magnetic fluctuations (these may be called magnetic bubbles) of spatially small scales. This is one of possible routes of magnetic evolution, without which evaluation of magnetic fields in later epochs seems incomplete. In the early epoch the fluctuations have a typical spatial scale of c/ω_p. In thermal equilibrium both the processes of coalescence and splitting of magnetic bubbles take place. However, the greater the spatial size of the fluctuation, the longer it lasts, as discussed in Sec. 5.3.3. This may provide the time arrow to "polymerize" magnetic bubbles toward larger and larger structures. Recently we (Tajima et al., 1992b) calculated the process and found the time scale t_* to "polymerize" the magnetic fluctuations from the typical size c/ω_p to the horizon size is $t_* \approx t_1(t/t_1)^\alpha$ with $\alpha \approx 2/3$, where t_1 is the beginning of polymerization (we take $\sim 10^{-2}\,\text{sec}$) and t is the cosmic time. (More detail is presented below.) This indicates, for example, that at $t = 10\,\text{sec}$ after the Big Bang the time to create the largest magnetic "polymer" of the size of the horizon at that time is approximately 1 sec, i.e. most likely largest scales have been reached. Similarly at $t = 10^{13}$ (about the dawn of recombination) the "polymerization" time is approximately 10^8 sec. Such a fast time scale (compared with the Hubble time of each epoch) has an impact on the observability of the

444

Sachs and Wolfe (1967) effect.

Two main scenarios (Rees, 1987; Silk, 1989) have been considered for primordial fluctuations, adiabatic fluctuations and isothermal fluctuations. The adiabatic (or isentropic) fluctuations are like those accompanied by ordinary sound waves and a cartoon illustration of this situation is displayed in Fig. 5.11(a). In such fluctuations the density of matter (electrons, positrons, and protons (and helium ions) for the case of the early radiation epoch) is accompanied by that of photons, as indicated in Fig. 5.11(a). Therefore, after electrons and positrons annihilate around $t = 1\,\mathrm{sec}$ or after electrons and ions recombine around $t = 10^{13}\,\mathrm{sec}$, the imprint of matter fluctuations would remain in photon fluctuations as a fossil of the primordial plasma structure. Thus the background microwave spectra would show a certain fluctuation or anisotropy/inhomogeneity on top of the black-body spectra. This would be a contradiction to the latest observations by COBE etc. (Mather, 1990; Gush *et al.*, 1990).

On the other hand, imagine that there exist static magnetic fields in the primordial plasma [Fig. 5.11(b)]. Charged particles in the early radiation epoch ($t \lesssim 1\,\mathrm{sec}$) or in the late radiation epoch ($t \lesssim 10^{13}\,\mathrm{sec}$) cam readily respond to magnetic fields. Where stronger magnetic fields exist, less matter of charged particles, be it electron-positron-proton or electrons-proton plasma, aggregates, in such a way to maintain the total matter pressure and magnetic pressure constant in space. Now on top of this, photons are present. Photons do couple strongly with charged particles but not as strongly as static magnetic fields do with charged particles. Furthermore, photons are less strongly coupled with magnetic fields. For example, we can look at the hierarchy of coupling strengths just before the time of recombination ($10^{13}\,\mathrm{sec}$). The "coupling length" (the mean free path) of an electron with the magnetic fields is $< 10^{10}\mathrm{cm}$, the "coupling length" of an electron with the blackbody photons is $\gtrsim 10^{11}\mathrm{cm}$. And the "coupling length" between the magnetic fields and photons is infinite. This should leave a landscape of fluctuations in such a way that the sum of the magnetic and charged particle pressure is constant in space, while the photon pressure remains nearly constant in space. Such fluctuations are similar to the second category of isothermal fluctuations (Rees, 1987), as they are nearly frequencyless.

Although there may remain a certain residual photon fluctuation incurred by the isothermal matter fluctuations sustained by the zero frequency magnetic fields, the level of photon fluctuations reflected in the bulk of microwave black-body radiation spectra is practically undetectably miniscule. The magnetic fluctuations are in the far end ("infra-red" side zero or much lower than the microwave) of the frequency of the electromagnetic fluctuations (see Fig. 5.12). Thus in microwave background frequency spectra it would be difficult to detect the imprint of magnetic and associated density fluctuations on its main bulk.

5.3.5   Structure Formation

We have seen that the plasma thermal energy can sustain large magnetic fluctuations ($\omega \sim 0$) at small wavelengths at the early Universe. These tiny 'magnetic bubbles' are created and decay spontaneously all the time. These small bubbles A and B are, however, capable

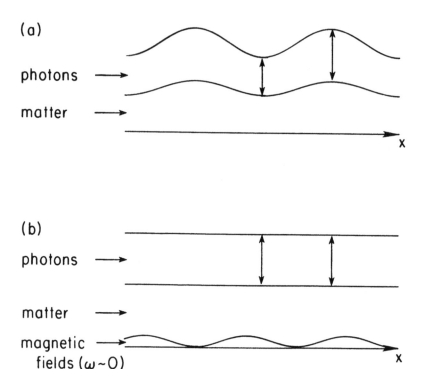

FIGURE 5.11 Imprints of fluctuations on photon distribution in the cosmological scale before recombination. (a) Adiabatic perturbation, which leaves imprint after the recombination. (b) Isothermal perturbation such as magnetic fluctuations, which leaves little imprint after the recombination.

of coalesce into large magnetic bubbles AB (see Sec. 2.4 and Sec. 2.5), a process of self-organization. At the same time larger magnetic bubbles AB can split into smaller bubbles A and B. In thermal equilibrium (steady state) the rate of coalescence and that of split are balanced. $A + B \overset{\longleftarrow}{\longrightarrow} ABC$.

These processes may be phenomenologically described by kinetic theory. A question is whether the chain of these processes is capable of making very large bubbles, starting from tiny bubbles under the condition of thermal equilibrium. At first inspection this looks impossible, as each step such as $AB + CA + B \overset{\longleftarrow}{\longrightarrow} ABC$ is in detailed balance and decay and coalesce are equally probable: Any growth of bubbles is followed equally by decay of larger ones. If this were the case for the formation of very large natural molecules, it would be a (near) miracle that such large molecules have been formed in (near) thermal equilibrium sea waters. In a closer scrutiny, it may be very large bubbles (or molecules) $ABC \cdots XY$ can coalesce into $ABC \cdots XYZ$ (and equally probably decay back into $ABC \cdots XY$ and Z,

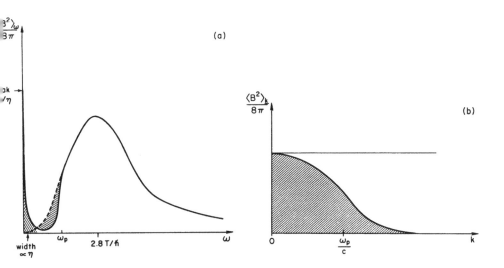

FIGURE 5.12 The energy spectrum of (electro)magnetic fields in the hot and dense primordial plasma. As opposed to the well-known vacuum black body radiation spectrum (shown by the broken line), the spectrum with the plasma shows a dip for $\omega < \omega_p$, while there appears a large zero frequency ($\omega \sim 0$) peak.

but there are very many such layers. Thus even though each reaction stage (coalesce and decay) has a (nearly) detailed balance, the very presence of the large number of stages in the chain makes the reactions into very large bubbles time-directed. This statement may be made slightly more quantitative as follows, though it remains as a hypothetical model:

Following the analogy of larger magnetic bubble formation from small ones to the 'polymer' formation from 'monomers,' let us call a bubble that is composed of i small bubbles as i-th (order) polymers. Here we assume crudely all small bubbles having the same size. Let n_i be the concentration of the i-th polymers. The time development of the concentration may be described by the following kinetic equations

$$\frac{\partial n_i}{\partial t} = -\frac{n_i}{\tau_0 i^2} + n_{i-1} \langle n_1 v \sigma \rangle (-3 h n_i),$$ (5.3.31)

where the first term on the right-hand side is the decay time of the bubble due to the

447

diffusion, the second term is the coalescence of the $i-1$-th bubble and the monomer into the i-th polymer and the last term in parenthesis is due to the cosmic expansion. We neglected split processes such as $i \rightarrow i - 1 + 1$, $i \rightarrow (i - 2) + 2$, as well as coalescence processes such as $(i - 2) + 2 \rightarrow i$, $(i - 3) + 3 \rightarrow i$. The v and σ are the typical collision velocity and cross-section of bubble coalescence and τ_0 is the unit decay tame for a monomer. However, the $i \rightarrow (i - 1) + 1$ split process may be thought to have been tucked into this diffusive decay time of the first term on the right-hand side. Equation (5.3.31) contains many crude approximations, as mentioned above. Mathematical results presented below, nonetheless, may be considered as a sign post of the possible trend for spontaneous macro bubbles. By defining the dimensionless parameter

$$k \equiv \tau_0 \langle n_1 v \sigma \rangle, \tag{5.3.32}$$

the ratio of the monomer decay time to the coalescence (collision) time. The stationary solution at $t \rightarrow \infty$ can be obtained by putting $\partial_t = 0$, yielding

$$n_i = k i^2 n_{i-1}, \tag{5.3.33}$$

and

$$n_i = \left(\frac{\sqrt{k} \, i}{e} \right)^{2i}. \tag{5.3.34}$$

From this we obtain the polymerization time

$$t_* = \tau_0 k^{-(1+1/\alpha)}, \tag{5.3.35}$$

where $\alpha = -\frac{1+b}{2} + \frac{27}{16} \sim \frac{2}{3}$, with b being related to the relaxation in the Universe. Equation (5.3.31) may be converted into a differential equation (Isichenko)

$$\frac{\partial n}{\partial t} = -\frac{n}{x^2} + k \left(n - \frac{\partial n}{\partial x} \right), \tag{5.3.36}$$

where $n_{i-1} \simeq n_i - \frac{\partial n}{\partial x}\big|_{x=i}$ is used. With the initial conditions $n(1,t) = 1$ and $n(x > 1, 0) = 0$, the solution to Eq. (5.3.36) is (Isichenko)

$$n(x,t) = \Theta \left[kt - (x - 1) \right] e^{x - 1 + 1/k(1/s - 1)}, \tag{5.3.37}$$

where Θ is the Heaviside step function. The corresponding solution to the expanding Universe is

$$n(x, h^{-1}) \sim \exp \left[x - 1 + \frac{h^{1+b}}{k} \left(\frac{1}{x} - 1 \right) \right], \tag{5.3.38}$$

where h is the Hubble parameter. These results imply that large (infinite or the horizon sized) polymers may be formed starting from tiny monomers in just about the Hubble time scale. If this crude kinetics indicates true (or approximately true) temporal evolution, one

448

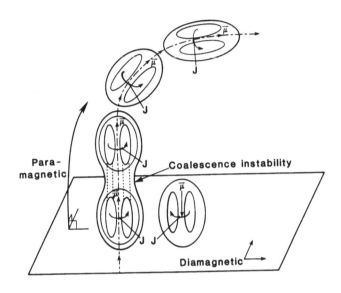

FIGURE 5.13 Magnetic bubbles and plasma filaments. Self-organization of a plasma.

should be able to construct statistical description of the final state in which large polymers do exist.

In order to investigate this, we introduce the filamentary construct of fluids and plasmas [see Sec. 2.4]. We mentioned in Chap. 2 that 2D hydrodynamic simulations with very large Reynolds numbers show filamentary structures and enstrophy conservation (McWilliams *et al.*, 1983). Theoretically it is known that 2D hydrodynamic statistical distribution of k-modes with enstrophy constant is (roughly) equivalent to the filamentary construct. A similar result has been obtained for 2D MHD for *filamentary MHD construct* (Kinney and McWilliams, 1994a). These results are in contrast with more conventional law of equipartition based on the Fourier mode statistics (see Sec. 2.3). Although this discrepancy is very surprising (Onsager, 1949) and subtle, the overall implication is that in very *high Reynolds number* hydrodynamic or magnetohydrodynamic fluids more highly *filamentary structures* tend to grow in time. In 3D MHD a similar tendency is found. We need a lot more work here and it remains to be seen, however, if these models do apply in some specific phases of the evolution of the cosmological plasma.

5.4 Electroweak Plasmas*

5.4.1 Neutrino Fluid and Electron Fluid*

The primordial plasma in the epoch approximately between 10^{-4} sec to 1 sec after the Big Bang is believed to be made up primarily of electrons, positrons, neutrinos and anti-neutrinos, and photons (with baryonic matter). During this epoch (particularly its early part) electrons and positrons not only coupled strongly with photons through electromagnetic interaction, but also coupled with neutrinos (and anti-neutrinos) through the weak interaction. As we have done in Sec. 5.3, we are interested in collective behaviors of this primordial matter. The properties arising simply from individual particles' nature have been well studied and surveyed (e.g. Weinberg, 1972). After the temperature of the plasma dropped below the rest mass of $W^{\pm}$ and Z bosons, by integrating out the boson propagators and using the *Fierz transformation*, we have an *effective Lagrangian* for the $e - \nu$ interaction. Then, the effective Lagrangian for the $e - \nu$ parts is:

$$\mathcal{L}_{\text{int}}^W = -\frac{\sqrt{2}G}{c^2} \left(\bar{\nu}\gamma_\mu \frac{1+\gamma_5}{2} \nu \right) \left[\bar{e}\gamma_\mu \left(\frac{1+\gamma_5}{2} + 2\sin^2\theta_w \right) e \right], \tag{5.4.1}$$

and the electromagnetic interaction Lagrangian is $\mathcal{L}_{\text{int}}^{\text{EM}} = \frac{q}{c}(i\bar{e}\gamma_\mu e)A_\mu$, where G is the Fermi constant and θ_w is the Weinberg angle in the *Weinberg-Salam theory* ($\sin^2\theta_w \approx 0.25$). Since we are interested in collective interaction, the present approach is to start from Eq. (5.4.1) in a semiclassical way to systematically derive classical hydrodynamic Lagrangian of the electroweak plasma.

The classicalization is done by suppressing the handedness and by reducing $i\bar{e}_L\gamma_\mu e_L$ to $\frac{1}{2}j_\mu^e = \frac{1}{2}(n_e \mathbf{v}_e, icn_e)$ and $i\bar{\nu}_L\gamma_\mu \nu_L$ to $\frac{1}{2}j_\mu^{\nu L} = \frac{1}{2}(n_\nu \mathbf{v}_\nu, icn_\nu)$. We then obtain the interaction Lagrangian as

$$L_{\text{int}} = \int \mathcal{L}_{\text{int}} d^3x = \frac{q}{c}\mathbf{v}_e \cdot \mathbf{A}(\mathbf{x}_e, t) - q\phi(\mathbf{x}_e, t)$$

$$+ \frac{\sqrt{2}G}{c^2} \left(n_\nu(\mathbf{x}_e, t)\mathbf{v}_e \cdot \mathbf{v}_\nu(\mathbf{x}_e, t) - c^2 n_\nu(\mathbf{x}_e, t) \right), \tag{5.4.2}$$

for a single electron and a similar one for a single neutrino, where $n_\sigma(\mathbf{x}) = \delta(\mathbf{x} - \mathbf{x}_\sigma)$ ($\sigma = e$, ν). The equations of motion for electrons (e) and neutrinos (ν) are obtained by the Euler-Lagrange equation.

Deduced from Weinberg-Salam electroweak theory in the above, a *Klimontovich equation*, then a *Boltzmann equation* and subsequent fluid equations are now derived for the primordial electron-positron-neutrino-photon plasma. Such a theoretical tool may be utilized to study collective modes in this electroweak plasma. Some new physics including a collective process that tends to separate the phases of electrons (and positrons) and neutrinos (and anti-neutrinos) are discussed.

The *Euler-Lagrange equation* thus derived defines the characteristics of each species of particles and thus allow us to construct the Boltzmann equation: $\frac{\partial f_\sigma}{\partial t} + \mathbf{v}_\sigma \cdot \nabla_{\mathbf{x}} f_\sigma + \mathbf{F}_\sigma \cdot \nabla_{\mathbf{p}} f_\sigma =$

C_σ, where C_σ is the appropriate collision operator including the annihilation and creation of particles. By making the velocity moments and summing over many particles, we finally arrive at the classical hydrodynamical equations (Lai and Tajima, 1997)

$$n_e \frac{d\mathbf{p}_e}{dt} = n_e q \left(\mathbf{E} + \frac{\mathbf{v}_e \times \mathbf{B}}{c} \right) - \nabla P_e - \nabla P_\gamma + \frac{\sqrt{2}Gn_e}{c^2} \left[\nabla \cdot (n_\nu \mathbf{v}_\nu)\mathbf{v}_e \right.$$

$$\left. + \nabla(n_\nu \mathbf{v}_\nu \cdot \mathbf{v}_e) - c^2\nabla n_\nu - \frac{d}{dt}(n_\nu \mathbf{v}_\nu) \right] - \eta_e n_e m_e \mathbf{v}_e, \tag{5.4.3}$$

and

$$n_\nu \frac{d\mathbf{p}_\nu}{dt} = \frac{\sqrt{2}Gn_\nu}{c^2} \left[\nabla \cdot (n_e \mathbf{v}_e)\mathbf{v}_\nu + \nabla(n_e \mathbf{v}_e \cdot \mathbf{v}_\nu) - c^2\nabla n_e - \frac{d}{dt}(n_e \mathbf{v}_e) \right]$$

$$-\eta_\nu n_\nu m_\nu \mathbf{v}_\nu, \tag{5.4.4}$$

where $\frac{d}{dt}$ is the Lagrangian time derivative, n_σ is now the density, η_σ is the collisional frequency of fluid species σ with photons or other fluid components, and P_e and P_γ are electron and photon pressures. Other equations are the continuity equations for each species: $\frac{\partial n_\sigma}{\partial t} + \nabla \cdot (n_\sigma \mathbf{v}_\sigma) = 0$. We have neglected the cosmological background expansion, as the Hubble expansion time proves to be much greater than the growth time as we shall see.

5.4.2 Primordial Instability*

Since the plasma is highly relativistic and opaque, we can safely assume $P_\gamma = P_e = n_e T_e = \frac{1}{\sigma_s^{1/3}}n_e^{4/3}$, where $\sigma_s = \frac{\pi^2}{45\hbar^3 c^3}$. We assume that the equilibrium of the plasma is uniform: $n_e = \bar{n}_e$ and $n_\nu = \bar{n}_\nu$ and no large scale EM fields and flows $\mathbf{E} = \mathbf{B} = \mathbf{v}_e = \mathbf{v}_\nu = 0$. We then solve Eqs. (5.4.3) and (5.4.4) by linearizing electron and neutrino equations of motion about the equilibrium and transforming these into Fourier space:

$$(-i\omega)m_e\gamma_e\bar{n}_e\delta\mathbf{v}_e = \bar{n}_e q_e\delta\mathbf{E} - \sqrt{2}G\bar{n}_e(i\mathbf{k})\delta n_\nu - \frac{\sqrt{2}G}{c^2}\bar{n}_e\bar{n}_\nu(-i\omega)\delta\mathbf{v}_\nu$$

$$- \eta_e\bar{n}_e m_e\delta\mathbf{v}_e - \frac{8}{3\sigma^{1/3}}\bar{n}_e^{1/3}(i\mathbf{k})\delta n_e, \tag{5.4.5}$$

$$(-i\omega)m_\nu\gamma_\nu\bar{n}_\nu\delta\mathbf{v}_\nu = -\sqrt{2}G\bar{n}_\nu(i\mathbf{k})\delta n_e$$

$$- \frac{\sqrt{2}G}{c^2}\bar{n}_e\bar{n}_\nu(-i\omega)\delta\mathbf{v}_e - \eta_\nu\bar{n}_\nu m_\nu\delta\mathbf{v}_\nu, \tag{5.4.6}$$

and similarly, for the continuity equations, where γ_σ is the relativistic factor for the species σ. These equations are coupled with Maxwell's equations for electric and magnetic fields.

451

As a standard procedure of derivation of the linear dispersion relation, the determinant of the matrix of these equations results in an equation of a type $f \cdot g^2 = 0$, which leads to two independent dispersion relations $f = 0$ or $g = 0$, where f and g are given by

$$f = (1-a)\omega^4 + i(\eta_e + \eta_\nu)\omega^3 - (bc^2 k^2 - 2ac^2 k^2 + \eta_e\eta_\nu + \omega_p^2)\omega^2$$

$$-i\eta_\nu(\omega_p^2 + bc^2 k^2)\omega - ac^4 k^4 = 0, \tag{5.4.7}$$

or

$$g = A - \eta_e\eta_\nu\omega^2(\omega^2 - c^2 k^2) + i(\eta_e + \eta_\nu)(\omega^5 - c^2 k^2\omega^3)$$

$$+i\eta_\nu(bc^4 k^4\omega - bc^2 k^2\omega^3 - \omega_p^2\omega^3) = 0, \tag{5.4.8}$$

and

$$A = (1-a)\omega^6 + (3ac^2 k^2 - bc^2 k^2 - c^2 k^2 - \omega_p^2)\omega^4 + (b - 3a)c^4 k^4\omega^2 + ac^6 k^6.$$

In these expressions, we include the viscosity effect μ by lumping $\eta + \mu k^2$ in a simple η, where we can approximately evaluate as $\eta = c\bar{n}_e\sigma_{KN}$ and $\mu = \frac{c}{3\bar{n}\sigma_{KN}}$, with the Klein-Nishina cross-section $\sigma_{KN} = \frac{\pi e^4}{m_e c^2 \hbar\omega}$. A dimensionless weak coupling coefficient a between electrons and neutrinos is defined as

$$a = \frac{2G^2\bar{n}_e\bar{n}_\nu}{c^4 m_e m_\nu},$$

and a dimensionless (stronger EM) coupling coefficient b between electrons and photons is

$$b = \frac{8}{3}\left(\frac{\bar{n}_e}{\sigma_s}\right)^{1/3}\frac{1}{m_e c^2}.$$

The dispersion relation $f = 0$ gives rise from the longitudinal mode (i.e. the electric polarization $\mathbf{E}$ is parallel to $\mathbf{k}$), while the dispersion relation $g^2 = 0$ from the transverse ($\mathbf{E} \perp \mathbf{k}$) ($g^2 = 0$ since two polarizations). Each relation reduces to the familiar form of $\omega^2 = \omega_p^2 + bc^2 k^2$ (plasmons) and $\omega^2 = \omega_p^2 + k^2 c^2$ (polaritons) respectively, when the electron-neutrino coupling $a \to 0$ and collisions $\eta \to 0$. Note that the coefficient a is mass (m_ν) dependent (we took the form for $m_\nu \neq 0$; for $m_\nu = 0$ case appropriate modifications to nonsingularize the present equations can be carried out).

Analytical solutions for Eqs. (5.4.7) and (5.4.8) can be sought. Let $\Gamma = -i\frac{\omega}{ck}$ (normalized growth rate). The dispersion relation becomes respectively from (5.4.7)

$$f(\Gamma) = (1-a)\Gamma^4 + \frac{\eta_e + \eta_\nu}{ck}\Gamma^3 + \left(b - 2a + \frac{\eta_e\eta_\nu}{c^2 k^2} + \left(\frac{\omega_p}{ck}\right)^2\right)\Gamma^2$$

$$+ \frac{\eta_\nu}{ck}\left(\frac{\omega_p}{ck}^2 + b\right)\Gamma - a = 0, \tag{5.4.9}$$

and from (5.4.8)

$$g(\Gamma) = (1-a)\Gamma^6 + \left(b + 1 + \left(\frac{\omega_p}{ck}\right)^2 - 3a\right)\Gamma^4 + (b - 3a)\Gamma^2 - a + \frac{\eta_e\eta_\nu}{c^2 k^2}\Gamma^2(\Gamma^2 + 1)$$

$$+\frac{\eta_e + \eta_\nu}{ck}\Gamma^3(\Gamma^2 + 1) + \frac{\eta_\nu}{ck}\Gamma\left(b + b\Gamma^2 + \left(\frac{\omega_p}{ck}\right)^2\Gamma^2\right) = 0. \tag{5.4.10}$$

In our plasma, the coupling coefficient a is a small parameter ($a \ll 1$) and the parameter $b \gg a$. We already know the high frequency behavior of these equilibrium plasmons and polaritons. We thus focus on low frequency behaviors of collective modes which are influenced by the presence of electroweak coupling: $|\Gamma| \ll 1$.

From $f(0) = -a < 0$, $f(\infty) > 0$ and $g(0) = -a < 0$, $g(\infty) > 0$ we know that there exists at least a solution with $\Gamma > 0$. Recalling the definition of Γ, we immediately realize that this is an unstable (exponentially temporary growing) mode. We can solve for small Γ by neglecting higher order terms like Γ^6, Γ^5 and Γ^4. For the transverse modes $g(\Gamma)$ we have

$$\left(b + \frac{\eta_e \eta_\nu}{c^2 k^2}\right)\Gamma^2 + \frac{\eta_\nu}{ck}b\Gamma - a \approx 0. \tag{5.4.11}$$

Solve Γ as $\Gamma(k)$. Find the wavenumber value for maximum growth $k_{mg} \approx \sqrt{\frac{\eta}{\mu}}$ such that $\frac{d\Gamma}{dk}(k_{mg}) = 0$, where we now express η and μ explicitly. And the maximum Γ value $\Gamma_{\max} = \Gamma(k_{mg}) \approx \frac{ac}{b\sqrt{\eta\mu}}$. Then the maximum growth rate is given by

$$\gamma_{\max} = ck_{mg}\Gamma_{\max} = \frac{ac^2}{b\mu}. \tag{5.4.12}$$

For the longitudinal modes $f(\Gamma)$ we have similar results.

From Eqs. (5.4.5), (5.4.6), and (5.4.11), along with the continuity equations, it is evident that the Fourier components of density perturbations δn_e and δn_ν are 180° out of phase (optical phonon-like). For example, the negative gradient of the *neutrino fluid* exerts a weak force to reinforce the positive density gradient of the electron fluid and vice versa. This mutual reinforcing force leads to the instability. The reason why both the transverse and longitudinal dispersion relations $f = 0$ and $g = 0$ produce the same instability is now clear. That is, the origin of this instability is not related to photon coupling but to the weak coupling (or *weak collisions*). The latter is blind to the optical polarization and comes in equally in transverse and longitudinal equations. On the other hand, the mode that moves the neutrino and electron fluids together (acoustic phonon-like) is stable, as expected due to the photon viscosity. The instability tries to separate the neutrinos (and anti-neutrinos) from the electrons (and positrons) and each species into small "bubbles." However, the unstable bubble size is limited to the neutrino penetration length (see Table 5.2). The growth rate as a function of the wavenumber is shown in Fig. 5.14(b). Because of the large neutrino mean free path (the typical unstable wavelength is the geometrical means of neutrino and electron scale lengths). This instability is thus viable only in the linear phase. Note that the present fluid theory fails to apply to quantum mechanical wavenumber regime $k > n_e^{-1/3}$ (Cable and Tajima, 1992). Perhaps the most important effect of this instability is not the bubbles of phase separation, but the phenomenon called the critical opalescence in a (near) unstable plasma, where the plasma fluctuations are enhanced.

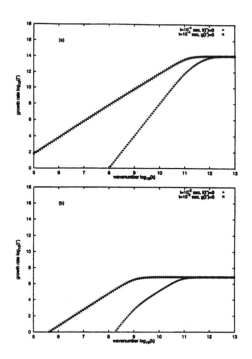

FIGURE 5.14 Linear growth rate of the phase separation at early cosmological epoch of electron-neutrino plasma. (a) a log-log plot of growth rate $\gamma(\mathrm{sec}^{-1})$ at $t = 10^{-4}$ sec after the Big Bang as a function of wavenumber $k(\mathrm{cm}^{-1})$. (b) a log-log plot of growth rate $\gamma(\mathrm{sec}^{-1})$ at $t = 10^{-2}$ sec after the Big Bang as a function of wavenumber $k(\mathrm{cm}^{-1})$.

5.5   Extended Yang-Mills Structures*

Be it the electroweak plasma as discussed in the previous section, or be it the *quark-gluon plasma* for the baryonic component of matter in the primordial Universe prior to $10^{-5\sim-3}$ sec after the Big Bang, the fields other than the more familiar electromagnetic ones strongly couple with the plasma. As Salam started out from Maxwell's equations to introduce the Weinberg-Salam electroweak theory, we are going to take a journey backward to gain more intuitive insight into these exotic field dynamics and their cosmological consequences. The best platform of doing so is to start from the *Yang-Mills field* equations. Usually the SU(3) Yang-Mills fields are related to the *strong interaction* (quark-gluon fields), while the SU(2) Yang-Mills fields to the electroweak interaction. In the very early Universe, either of these fields were strongly coupled. In order to understand the nature of strong coupling of these

454

Time (s)	$t = 10^{-4}$	$t = 10^{-2}$	$t = 1$
Temperature (eV)	$T = 10^8$	$T = 10^7$	$T = 10^6$
Density (cm^{-3})	$n = 6 \times 10^{37}$	$n = 5 \times 10^{34}$	$n = 4 \times 10^{31}$
a	1.132×10^{-5}	7.859×10^{-12}	5.03×10^{-18}
b	668.45	62.9	5.839
$\sigma_{KN}(\text{cm}^2)$	1.275×10^{-27}	1.275×10^{-26}	1.275×10^{-25}
$\mu(\text{cm}^2/s)$	0.131	15.682	1960.21
$\eta(\text{s}^{-1})$	2.292×10^{21}	1.91×10^{19}	1.53×10^{17}
$\omega_p(\text{s}^{-1})$	6.183×10^{23}	1.785×10^{22}	5.049×10^{20}
$f(\Gamma) = 0, \gamma_{\max}(\text{s}^{-1})$	1.164×10^{14}	7.16×10^6	0.3949
$f(\Gamma) = 0, k_{mg}(\text{cm}^{-1})$	6.872×10^{11}	8.59×10^{10}	5.369×10^9
$g(\Gamma) = 0, \gamma_{\max}(\text{s}^{-1})$	1.164×10^{14}	7.16×10^6	0.3949
$g(\Gamma) = 0, k_{mg}(\text{cm}^{-1})$	1.323×10^{11}	1.104×10^9	8.835×10^6

TABLE 5.2   Growth Rates of the Instability in a Primordial Electroweak Plasma

fields, we take up an example of SU(2) fields, without specifically restricting ourselves to the electroweak theory.

Although a large volume of literature on exact (or semi-exact) solutions on Yang-Mills fields equilibrium (i.e. time-independent solutions), it is our interest to look for cosmologically relevant solutions. And again we are interested in collective properties as opposed to single particle properties. It is of interest, therefore, to investigate what we call global (or nonlocal) equilibrium solutions of Yang-Mills equations. If there are nontrivial equilibrium solutions, these could be vacua with finite energy density. Such solutions might be related to dark matter.

The nonabelian (SU(2)) generalizations of the Maxwell equations are radically different from the abelian (U(1)) case, due to the non-linear terms defining the 'electric' and 'magnetic' fields. In nonabelian theories, it is not possible to re-write the equations of motion and the *Bianchi identities* (which in the abelian case precisely recover Maxwell equations) purely in terms of the fields $\mathbf{E}^a$ and $\mathbf{B}^a$, since these fields are not fundamental (the gauge potential 4-vectors A^a are). Related to this is the fact that in the electromagnetism the fields $\mathbf{E}$ and

B are *gauge invariant*, as they are the components of the gauge invariant field strength $F_{\mu\nu}$. In the *nonabelian* case $F^a_{\mu\nu}$ is no longer gauge invariant, and neither are therefore $\mathbf{E}^a$ or $\mathbf{B}^a$. We chose our $\mathbf{A}^a$ fields to satisfy the Coulomb gauge and generalized Maxwell equations. Our background calculations deal with divergence free and time independent fields only, that yield nontrivial global vacua, which might correspond to some primordial structures. Time-dependent perturbations lead to elementary excitations in such nontrivial vacua.

5.5.1 Introduction to Yang-Mills Vacua*

The Abelian Case

The basic quantity in (abelian) Yang-Mills theory is the 4-vector potential, $A_\mu(x)$, where x denotes the position 4-vector and has components $x = (x^0, x^1, x^2, x^3)$, with $x^0 = t$. [In the following $c = 1$.] If x and x' are distinct points in the spacetime, the 4-vectors $A_\mu(x)$ and $A_\mu(x')$ are in principle independent—up to the dynamics governed by the action. Thus, per point of spacetime, $A_\mu(x)$ associates a 4-vector of degrees of freedom.

The derived quantity

$$F_{\mu\nu}(x) = \partial_\mu A_\nu(x) - \partial_\nu A_\mu(x),\tag{5.5.1}$$

is called the field strength, where

$$\partial_0 = \frac{\partial}{\partial t}, \qquad \sum_{\mu=1}^{3} \hat{e}_\mu \partial_\mu = \boldsymbol{\nabla}.\tag{5.5.2}$$

Note that $F_{\mu\nu}(x) = -F_{\nu\mu}(x)$; $F_{\mu\nu}(x)$ is antisymmetric. In the 4×4 matrix $F_{\mu\nu}(x)$, this leaves 6 elements independent—per point in spacetime. Conventionally, these are assembled into two 3-vectors, defined as follows:

$$\mathbf{E} = \sum_{\mu=1}^{3} \hat{e}_\mu E_\mu \ \ E_\mu = F_{0\mu},\tag{5.5.3}$$

$$\mathbf{B} = \sum_{\mu=1}^{3} \hat{e}_\mu B_\mu \ \ B_\mu = \frac{1}{2} \sum_{\nu,\rho=1}^{3} \epsilon_{\mu\nu\rho} F_{\nu\rho}.\tag{5.5.4}$$

Here $\epsilon_{\mu\nu\rho}$ is the Levi-Civita alternating symbol. Thus, the electric field $\mathbf{E}$ and the magnetic field $\mathbf{B}$ are not fundamental quantities, but derived; they are merely components of the field-strength $F_{\mu\nu}$, defined in Eq. (5.5.1).

In non-covariant notation, splitting up spacetime into space and time, we write $\phi = A^0$ for the 'scalar' potential and $\mathbf{A} = \sum_{\mu=1}^{3} \hat{e}_\mu A_\mu$ for the 3-vector potential. Then,

$$\mathbf{E} = \sum_{\mu=1}^{3} \hat{e}_\mu (\partial_0 A_\mu - \partial_\mu A_0) = \frac{\partial}{\partial t} \mathbf{A} - \boldsymbol{\nabla}\phi,\tag{5.5.5}$$

and

$$\mathbf{B} = \sum_{\mu=1}^{3} \widehat{e}_\mu \left(\frac{1}{2} \sum_{\nu,\rho=1}^{3} \epsilon_{\mu\nu\rho}(\partial_\mu A_\nu(x) - \partial_\nu A_\mu) \right) = \nabla \times \mathbf{A}. \tag{5.5.6}$$

The next step is to construct a suitable action functional, such that Euler-Lagrange equations of motion will be the second order partial differential equations for the 4-vector potential A_μ. Any action functional that depends on A_μ only through $F_{\mu\nu}$ will automatically be gauge invariant, since $F_{\mu\nu}$ itself is gauge invariant

$$S[A_\mu] = -\int d^4x \|F\|^2 = \frac{1}{2}\int d^4x A^\nu(\partial^\mu F_{\mu\nu}). \tag{5.5.7}$$

As desired then, the equations of motion for A_μ will read:

$$0 = \partial^\mu F_{\mu\nu} = (\partial_\nu^\mu \partial^2 - \partial^m \partial_n)A_\mu. \tag{5.5.8}$$

Another equation is (trivially) obeyed by $F_{\mu\nu}$:

$$\partial_{[\mu} F_{\nu\rho]} = \frac{1}{3}[\partial_\mu F_{\nu\rho} + \partial_\nu F_{\rho\mu} + \partial_\rho F_{\mu\nu}] \equiv 0. \tag{5.5.9}$$

This equation is easily verified simply by substituting Eq. (5.5.1) and is often called the Bianchi identity.

The equations (5.5.7) and (5.5.8) easily produce the usual Maxwell equations when the above definitions for $\mathbf{E}$ and $\mathbf{B}$ are substituted. For example, Eq. (5.5.8) involves the rank-3 antisymmetric tensor $\partial_{[\mu} F_{\nu\rho]}$. Should any two indices be equal, $\partial_{[\mu} F_{\nu\rho]}$ vanishes merely by antisymmetry and holds no information. Thus, the time ($= 0$) index may occur either once or never. In the latter case, we can contract the whole equation with the rank-3 Levi-Civita symbol:

$$0 = \sum_{\mu,\nu,\rho=1}^{3} \epsilon_{\mu\nu\rho}\partial_\mu F_{\nu\rho} = 2\sum_{\mu=1}^{3} \partial_\mu B_\mu, \qquad \nabla \cdot \mathbf{B} = 0, \tag{5.5.10}$$

which is one of the conventional Maxwell equations.

If we set $\mu = 0$ and $\nu, \rho = 1, 2, 3$ in Eq. (5.5.8), we obtain:

$$\partial_0 F_{\nu\rho} + \partial_\nu F_{\rho 0} + \partial_\rho F_{0\nu} = 0. \tag{5.5.11}$$

Contracting again with $\frac{1}{2}\epsilon_{\nu\rho\sigma}$, where $\nu, \rho, \sigma = 1, 2, 3$, we have

$$\frac{\partial}{\partial t}\mathbf{B} + \nabla \times \mathbf{E} = 0. \tag{5.5.12}$$

Taking $\nu = 0$ and $\mu = 1, 2, 3$ in Eq. (5.5.7) leads to

$$\nabla \cdot \mathbf{E} = 0, \tag{5.5.13}$$

while letting $\nu = 1, 2, 3$ and $\mu = 0, 1, 2, 3$ in Eq. (5.5.7) produces

$$0 = -\frac{\partial}{\partial t}\mathbf{E} + \nabla \times \mathbf{B}. \tag{5.5.14}$$

457

This recovers the usual Maxwell equations.

Note that the equations (5.5.7) and (5.5.8) are linear partial differential equations in the 4-vector photon field A_μ, in the field strength $F_{\mu\nu}$ and its component fields $\mathbf{E}$ and $\mathbf{B}$. These equations are second order PDE in A_μ and first order in $F_{\mu\nu}$. The formalism in which $F_{\mu\nu}$ (and so also $\mathbf{E}$ and $\mathbf{B}$) is regarded as fundamental is therefore called the first order formalism. Alternatively, we take the point of view that A_μ is fundamental and work in the second order formalism.

For the abelian case, the distinction seems purely conventional; by contrast in the non-abelian case, it is not possible to eliminate A_μ entirely the first-order formalism and (the appropriate generalization of) $F_{\mu\nu}$ cannot be regarded as fundamental. Also, even in electromagnetism, $F_{\mu\nu}$ can always be defined in terms of A_μ, but not the other way around.

Finally, note that fundamental does not mean observable. As it should be known, there is a gauge-ambiguity in defining A_μ. For example, the value of the potential $\phi = cA_0$ is not an observable, but the relative potential between some two points is measurable and makes physical sense as an observable.

The Nonabelian Case

Again, the physical system is described by the 4-vectors $A_\mu^a(x)$, the gauge field. These are defined only up to the addition of the covariant 4-gradient of an arbitrary function. Contrary to the abelian case, $F_{\mu\nu}^a$ is no longer invariant under the gauge transformations. Instead, it transforms covariantly.

$$F_{\mu\nu}^a = (\partial_\mu A_\nu^a) - (\partial_\nu A_\mu^b) - g f^a{}_{bc} A_\mu^b A_\nu^c, \tag{5.5.15}$$

where a is a group index ($a = 1, 2, 3$ for SU(2) and $1, ..., 8$ for SU(3)), $f^a{}_{bc}$ is a structure constant, while g is a coupling constant. The equation above is, of course, very similar to Eq. (5.5.1) and the generalization shows up in the nonlinear term $-i f^a{}_{bc} A_\mu^b A_\nu^c$. The structure constant $f^a{}_{bc}$ is antisymmetric, $f^a{}_{bc} = -f^a{}_{cb}$, so that $f^a{}_{bc} = 0$ if the indices are allowed to range over a single value which is the case of U(1).

Next, we can again define "electric" and 'magnetic" fields as in the abelian case, $E_\mu^a = F_{0\mu}^a$, so that

$$\mathbf{E}^a = -\frac{\partial}{\partial t} \mathbf{A}^a - \nabla \phi^a + g f^a{}_{bc} \phi^b \mathbf{A}^c. \tag{5.5.16}$$

Also, $\mathbf{B}_\mu^a = \frac{1}{2} \epsilon_\mu{}^{\nu\rho} F_{\nu\rho}^a$ produces

$$\mathbf{B}^a = \nabla \times \mathbf{A}^a - \frac{1}{2} g f^a{}_{bc} \mathbf{A}^b \times \mathbf{A}^c = (\mathcal{D} \times \mathbf{A})^a. \tag{5.5.17}$$

Note that the nonlinear terms explicitly depending on ϕ^a and $\mathbf{A}^a$.

Finally, the nonabelian generalization of the Maxwell equations are radically different from the abelian case, owing to the nonlinear terms in the definition of the "electric" and "magnetic" fields. In sharp contrast to the abelian case, it is not possible to re-write the equations of motion and the Bianchi identities (which in the abelian case precisely recover

458

Maxwell's equations) purely in terms of the "electric" and "magnetic" fields $\mathbf{E}^a$ and $\mathbf{B}^a$. This reinforces our earlier observation that these fields are not fundamental, whereas the gauge potential 4-vectors A_μ^a are.

Related to this is the fact that in electromagnetism, the fields $\mathbf{E}$ and $\mathbf{B}$ are gauge-invariant, as they are components of the gauge-invariant field strength $F_{\mu\nu}$. In the nonabelian case, $F_{\mu\nu}^a$ is no longer gauge-invariant, and neither are therefore $\mathbf{E}^a$ and $\mathbf{B}^a$.

Still, the action

$$S[A_\mu] = -\int d^4x \|F\|^2 = -\tfrac{1}{4}\int d^4x (\partial^{\mu\rho}\partial^{\nu\sigma}\partial_{ab}F_{\mu\nu}^a F_{\rho\sigma}^b) \qquad (5.5.18)$$

is invariant under gauge-transformations and is a valid starting point for discussing the dynamics of nonabelian gauge-fields. The variation of the action with respect to A_μ^a yields the Euler-Lagrange equations of motion:

$$(\mathcal{D}^\mu F_{\mu\nu})^a = 0. \qquad (5.5.19)$$

In addition to these, the Bianchi identities become:

$$(\mathcal{D}_\mu F_{\nu\rho})^a + (\mathcal{D}_\nu F_{\rho\mu})^a + (\mathcal{D}_\rho F_{\mu\nu})^a = 0. \qquad (5.5.20)$$

In all of the equations above, $\mathcal{D}^\mu$ stands for covariant derivative:

$$\mathcal{D}^\mu{}_{a,b} = \partial_{a,b}\partial^\mu + gf^a{}_{bc}A^{c\mu}. \qquad (5.5.21)$$

The equations of motion and Bianchi identities can now be re-written in terms of the non-covariant component fields $\mathbf{E}^a$ and $\mathbf{B}^a$—and $\mathbf{A}^a$. Then we can recover the generalized Maxwell equations using the abelian case procedure.

First, we will look at the equations of motion. Taking $\nu = 0$ and $\mu = 1, 2, 3$ one gets the following expression:

$$\nabla \cdot \mathbf{E}^a - gf_a{}^{bc}\mathbf{A}_b \cdot \mathbf{E}_c = 0. \qquad (5.5.22)$$

If $\nu = 1, 2, 3$ and $\mu = 0, 1, 2, 3$ one can obtain another generalized equation:

$$-\frac{\partial}{\partial t}\mathbf{E}^a + \nabla \times \mathbf{B}^a - gf^a{}_{bc}\mathbf{A}^b \times \mathbf{B}^c - gf^a{}_{bc}\phi^a\mathbf{E}^b = 0. \qquad (5.5.23)$$

From Bianchi Identities, if none of the indices take the value 0, one gets:

$$\nabla \cdot \mathbf{B}^a - gf^a{}_{bc}\mathbf{A}^b \cdot \mathbf{B}^c = 0. \qquad (5.5.24)$$

The last equation is obtained by taking $\nu, \rho = 1, 2, 3$ and $\mu = 0$:

$$\frac{\partial}{\partial t}\mathbf{B}^a + \nabla \times \mathbf{E}^a - gf^a{}_{bc}\mathbf{E}^b \times \mathbf{A}^c - gf^a{}_{bc}\mathbf{B}^b \times \phi^c = 0. \qquad (5.5.25)$$

Notice that these equations are very general, i.e. they are valid for any SU(n) case, since we did not specify what the structure constant is. Also, no gauge condition has been imposed yet.

459

After introducing the Coulomb gauge, equations above will reduce to:

$$\nabla \cdot \mathbf{E}^a - g f^a{}_{bc} \mathbf{A}^b \cdot \mathbf{E}^c = 0. \tag{5.5.26}$$

$$-\frac{\partial}{\partial t} \mathbf{E}^a + \nabla \times \mathbf{B}^a - g f^a{}_{bc} \mathbf{A}^b \times \mathbf{B}^c = 0. \tag{5.5.27}$$

$$\nabla \cdot \mathbf{B}^a - g f^a{}_{bc} \mathbf{A}^b \cdot \mathbf{B}^c = 0. \tag{5.5.28}$$

$$\frac{\partial}{\partial t} \mathbf{B}^a + \nabla \times \mathbf{E}^a - g f^a{}_{bc} \mathbf{E}^b \times \mathbf{A}^c = 0. \tag{5.5.29}$$

For static, divergence free case (in vacuum), there are only two equations to deal with:

$$g f^a{}_{bc} \mathbf{A}^b \cdot \mathbf{B}^c = 0. \tag{5.5.30}$$

$$\nabla \times \mathbf{B}^a - g f^a{}_{bc} \mathbf{A}^b \times \mathbf{B}^c = 0. \tag{5.5.31}$$

In (the adjoint representation of) SU(2), the structure constant $f^a{}_{b,c}$ is just $\epsilon_{\mu\nu\rho}$, the Levi-Civita symbol.

5.5.2 Nontrivial (Non-zero Energy) Vacua*

Choices for A^a_μ and the Background Equilibrium Solutions

Let us see if, as a starting point, it will be possible to find stationary nontrivial equilibrium solutions for A^a_μ that the nonlinearities in the equations (5.5.30) and (5.5.31) will cancel. Solutions would than be reduced to U(1) case. We list here some of the background solutions we deals with (Tajima et al., 1994).

i. The simplest case—constant $\mathbf{B}$ field. The following choice of $\mathbf{A}^a$:

$$A^a_x = 0 \quad \text{for } a = 1, 2, 3,$$

$$A^a_y = 0 \quad \text{for } a = 1, 2, 3, \tag{5.5.32}$$

$$A^{1,2}_z = 0, \quad A^3_z = Ky,$$

will yield the following $\mathbf{B}^a$ fields:

$$B^{1,2}_x = 0, \quad B^3_z = K,$$

$$B^a_y = 0 \quad \text{for } a = 1, 2, 3, \tag{5.5.33}$$

$$B^a_z = 0 \quad \text{for } a = 1, 2, 3,$$

where K is an arbitrary constant. This solution is a "constant magnetic field" solution.

ii. The non-trivial case—quadrupole **B** field. For this case, the values for $\mathbf{A}^a$ were chosen to be:

$$A_x^a = f(z)g(y) \quad \text{for } a = 1, 2, 3,$$
$$A_y^a = f(z)g(x) \quad \text{for } a = 1, 2, 3, \qquad (5.5.34)$$
$$A_z^a = p(x)q(y) \quad \text{for } a = 1, 2, 3,$$

This choice gives the following $\mathbf{B}^a$ fields:

$$B_x^a = p(x)q'(y) - f'(z)g(x) \quad \text{for } a = 1, 2, 3$$
$$B_y^a = f'(z)g(y) - p'(x)q(y) \quad \text{for } a = 1, 2, 3, \qquad (5.5.35)$$
$$B_z^a = f(z)[g'(x) - g'(y)] \quad \text{for } a = 1, 2, 3,$$

where the derivative of the function is with the respect to the variable in the parenthesis. Plugging these expressions for $\mathbf{A}^a$s and $\mathbf{B}^a$s into Eq. (5.5.31) (Eq. (5.5.30) will automatically be satisfied), will yields some conditions on p, q, f and g functions.

$$\frac{f''(z)}{f(z)} = -\frac{g''(y)}{g(y)} = K,$$

$$\frac{f''(z)}{f(z)} = -\frac{g''(x)}{g(x)} = K, \qquad (5.5.36)$$

$$\frac{p''(x)}{p(x)} = -\frac{q''(y)}{q(y)} = C,$$

where K and C are some arbitrary constants.
Obviously, there are two possibilities:

a. trivial solution—take $K = C = 0$; then:

$$f(z) = az + b,$$
$$g(y) = cy + d,$$
$$g(x) = cx + d, \qquad (5.5.37)$$
$$p(x) = ex + f,$$
$$q(y) = my + n.$$

461

One can easily set $b = d = f = n = 0$. Than the $\mathbf{B}^a$ fields will be given by:

$$B_x^a = (me - ac)x = \alpha x,$$

$$B_y^a = (ac - me)y = -\alpha y, \qquad (5.5.38)$$

$$B_z^a = 0.$$

So, here we have a *quadrupole* "magnetic" field.

b. not-so-trivial solution-take $K \neq C \neq 0$; then:

$$f(z) = \cosh(az + b),$$

$$g(y) = \cos(cy + d),$$

$$g(x) = \cos(cx + d), \qquad (5.5.39)$$

$$p(x) = \cosh(ex + f),$$

$$q(y) = \cos(my + n).$$

These are sinusoidal in one direction, while exponential in another direction.

Solutions in Eqs. (5.5.38) and (5.5.39) diverge at $r \to \infty$. Therefore, by themselves these solutions cannot be realistic for physical solutions. Even Eq. (5.5.33) solution diverges in its energy over infinite extent. It is thus obvious that these solutions, nonlocal ones, are valid over some spatial extent over which these spatial characteristics manifest, but beyond which slowly varying additional spatial variation can take care of the infinite energy content. Such a theory is conveniently provided by the *WKB method*. In the standard technique of the WKB method, by dropping the square of the slowly varying (long wavelength) spatial derivatives that are relevant beyond the solutions spatial content, one usually arrives at a nonlinear Schrödinger equation for the slowly varying amplitude of these fields. These variations may come from the global constraints such as the general relativistic effect.

5.5.3   Perturbed Excitations*

We now consider elementary excitations of Yang-Mills "electromagnetic fields" around these nontrivial energy equilibria. We will also consider their stability and spectra. Tajima and Niu investigated this problem. A naive application of the perturbation theory to the equilibrium, e.g. Eq. (5.5.33), would lead to a gauge-dependent dispersion relation, i.e. the gauge boson's energy (or frequency) would depend on the gauge and thus spatial coordinate. This is clearly not acceptable, as the theory to start out was manifestly *gauge-independent*. In the simplest example of the "constant magnetic field" equilibrium, Tajima and Niu (1996) demonstrated, through the technique similar in the theory of quantum Hall effects, that a

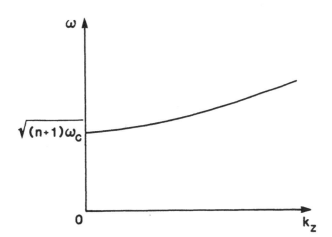

FIGURE 5.15 The dispersion relation of the boson perturbations around the nontrivial Yang-Mills vacuum. Because of the presence of "magnetic field"-like components in the Yang-Mills vacuum, originally massless bosons have now acquired a mass $\sqrt{(n+1)\omega_c}$.

proper treatment of the gauge and associated spatial operators can lead to gauge-independent dispersion relations such as

$$\omega^2 = k_z^2 c^2 + (n+1)\omega_c, \qquad (5.5.40)$$

where ω_c is the "cyclotron frequency" arising from the B_z^3, and n is an integer. See Fig. 5.15.

This result indicates that this boson now acquired a mass (similar to the *Higgs' mechanism*) due to the presence of the "magnetic field." Further, these perturbed solutions are all stable.

If the "magnetic field" energy density is not zero (and since the "magnetic field" is not the ordinary one), this could constitute a possible candidate for dark matter. Investigators, both theory and observation/experiment, are called for. The amount of such energy density at this time cannot be determined by theory.

Finally, we would like to briefly add our considerations on physical observables, possible implications and applications of the present theory. The *nontrivial vacuum* solutions found here are all extended throughout space and even diverge at infinity. They have infinite energies in an infinite volume, and therefore are unphysical by their own right. However, they may be regarded as local approximations of some finite (yet large scale) global solutions. To be more specific, one might have a finite global solution in the form $B^a(\mathbf{r}, \epsilon\mathbf{r})$, where $\epsilon \ll 1$, with $\epsilon\mathbf{r}$ describing the global behavior of the solution, and with the first $\mathbf{r}$ dependence describing the behavior on scales much smaller than the global scale. These solutions may be used to describe such relatively short scale behaviors.

Global equilibria of the Yang-Mills field have been considered because of potential applications to cosmology. While localized solutions have been sought as models of particles, extended global solutions obtained here and elsewhere (Tafel, 1993) may be relevant to cosmology, because such solutions represent nonzero energy density in the extended scale that may play the role of *dark matter*. Since there is not sufficient observational evidence to constraint the exact nature of the dark matter field, we have to be content with a hypothetical model represented by the Yang-Mills fields at this time. If the present Yang-Mills field corresponds to physical reality, a consequence of the theory is that elementary excitations in the nontrivial 'magnetic field' acquire masses, which may or may not correspond to particles in the standard theory of elementary particles. It remains to be seen, however, if this toy model (Tajima *et al.*, 1997) reflects nature.

Problem 5–1: Make perturbation $\mathbf{a}^a$ to $\mathbf{A}^a$ (i.e. $\mathbf{A}^a + \mathbf{a}^a$) in Eqs. (5.5.26)–(5.5.29) and linearize them with the background (zeroth order) "magnetic" field $B^3 = \Omega\hat{z}$ (and others are zero. Obtain for the $a = 3$ component

$$
\begin{cases}
\boldsymbol{\nabla} \cdot \mathbf{e}^3 = 0 \\
-\dfrac{\partial}{\partial t}\mathbf{e}^3 + \boldsymbol{\nabla} \times \mathbf{b}^3 = 0 \\
\boldsymbol{\nabla} \cdot \mathbf{b}^3 = 0 \\
\dfrac{\partial}{\partial t}\mathbf{b}^3 + \boldsymbol{\nabla} \times \mathbf{e}^3 = 0
\end{cases}
$$

and for $a = 1$ and 2 with $\mathbf{e}^\pm = \frac{1}{\sqrt{2}}(\mathbf{e}^2 \pm i\mathbf{e}^2)$ as

$$
\begin{cases}
\mathbf{D} \cdot \mathbf{e}^+ = 0 \\
-\dfrac{\partial}{\partial t}\mathbf{e}^+ + \mathbf{D} \times \mathbf{b}^+ + ig\mathbf{a}^+ \times \mathbf{B}^3 = 0 \\
\mathbf{D} \cdot \mathbf{b}^+ + ig\mathbf{a}^+ \cdot \mathbf{B} = 0 = 0 \\
\dfrac{\partial}{\partial t}\mathbf{b}^+ + \mathbf{D} \times \mathbf{e}^+ = 0,
\end{cases}
$$

where the so-called *covariant derivative* $\mathbf{D} \equiv \boldsymbol{\nabla} - ig\mathbf{A}^3$ is employed (Tajima and Niu, 1997).

Problem 5–2: Derive the dispersion relation Eq. (5.5.40) based on the results of Problem 5–1.

Problem 5–3: The background "magnetic" field in Sec. 5.5 could give rise to the vacuum energy density and thus "dark matter." Discuss possible implications on the cosmological equations such as the Einstein equation.

Problem 5–4: Discuss why the linear instability in Sec. 5.4 can enhance fluctuations in the neutrino-electron plasma, even though the instability is not expected to grow nonlinearly due to the long mean-free path of neutrinos.

Problem 5–5: In the model of magnetically constricted cosmological plasmas why does the so-called Compton y-parameter not become large?

Problem 5–6: In the vicinity of a black hole gravitational field the great gravitational accelerating causes a quantum electrodynamic effect of radiation, called the Hawking radiation, which emits a blackbody radiation with the Hawking temperature $T_H = \hbar g/2\pi c$. When T_H exceeds $2mc^2$ (the rest mass energy of electrons), the vacuum becomes an e^+e^- plasma. Estimate the radial distance away from the blackhole horizon (Schwarzschild radius) with the hole mass M.

Actor, A., Rev. Mod. Phys. **51**, 461 (1979).

Alcock, C., Fuller, G.M., and Mathews, G.J., Astrophys. J. **320**, 439 (1987).

Baierlein, R., Mon. Not. R. Astr. Soc. **184**, 843 (1978).

Bailin, D., and Love, A., *Introduction to Gauge Field Theory*, (Adam-Hilger, 1986).

Berera A., and Fang, L.Z., Phys. Rev. Lett. **72**, 458 (1994).

Bhattacharjee, A., Brunel, F., and Tajima, T., Phys. Fluids **26**, 3332 (1983).

Broadhurst, T.J., Ellis, R.S., Koo, D.C., and Szalay, A.S., Nature **343**, 1726 (1990).

Cable, S.B., and Tajima, T., Phys. Rev. A **46**, 3413 (1992); Also see Sitenko, A.G., *Electromagnetic Fluctuations in Plasma*, (Academic Press, New York, 1967).

Cable, S.B., and Tajima, T., Phys. Rev. A **46**, 3413 (1992).

Coles, P., Comments Astrophys. **16**, 45 (1992).

Dawson, J.M., in *Advances in Plasma Physics*, eds. Simon, A., and Thompson, W.B. (Academic, New York, 1968) Vol. 1, p. 1.

De Lapparent, V., Geller, M.J., and Huchra, J.P., Astrophys. J. **343**, 1 (1989).

Dombre, T., Frisch, U., Greene, J.M., Henon, H., Mehr, A., and Soward, A.M., J. Fluid Mech. **167**, 353 (1986).

Eguchi, T., Gilkey, P.B., and Hanson, A.J., Phys. Rep. **66**, 213 (1980).

Feingold, M., Kadanoff, L.P., and Piro, O., J. Stat. Phys. **50**, 529 (1988).

Field, G.B., and Perrenod, S.C., Astrophys. J. **215**, 717 (1977).

Finn, J.M., and Ott, E., Phys. Fluids **31**, 2992 (1988).

Fujimoto, M., Sofue, Y., and Kawabata, K., Prog. Theor. Phys. Suppl. **49**, 181 (1971).

Fujimoto, M., Publ. Astr. Soc. Jpn. **42**, L39 (1990).

Galloway, D., and Frisch, U., Geophys. Astrophys. Fluid Dyn. **36**, 53 (1986).

Giovannini, G., Kim, K-T., Kronberg, P.P., and Venturi, T., in *Galactic and Intergalactic Magnetic Fields*, ed. R. Wielebinski (Kluwer Academic, 1990) p. 492.

Guilbert, P.W., and Fabian, A.C., Mon. Not. R. Astr. Soc. **220**, 439 (1986).

Gunn, J.E., and Peterson, B.A., Astrophys. J. **142**, 1633 (1965).

Gush, H.P., Halpern, M., and Wishnow, E.H., Phys. Rev. Lett. **65**, 537 (1990).

Harrison, E.R., Phys. Rev. D **1**, 2726 (1970).

Harrison, E.R., Phys. Rev. Lett. **30**, 18 (1973).

Ichimaru, S., Rostoker, N., and Pines, D., Phys. Rev. Lett. **8**, 231 (1962).

Ichimaru, S. *Basic Principles of Plasma Physics, A Statistical Approach* (Benjamin, Reading, 1973).

Ichimaru, S., Astrophys. J. **202**, 528 (1975).

Isichenko, M., private communication (1994); an interesting thermodynamical argument for polymers has been given by C. Kittel, Am. J. Phys. **40**, 60 (1972).

Itzykson, C., *Quantum Field Theory*, (McGraw-Hill, Singapore, 1987); Weinberg, S., Phys. Rev. Lett. **19**, 1264 (1967).

Itzykson, C., and Zuber, J.B., *Quantum Field Theory*, (McGraw-Hill, 1980).

Jackson, J.D., *Classical Electrodynamics*, (John Wiley and Sons, 1975).

Kajantie, K., and Kurki-Suonio, T., Phys. Rev. D **34**, 1719 (1986).

Kajino, T., and Boyd, R.N., Astrophys. J. **359**, 267 (1990).

Kim, K-T., Kronberg, P.P., Giovannini, G., and Venturi, T., Nature **341**, 720 (1989).

Kim, K-T., Tribble, P.C., and Kronberg, P.P., Astrophys. J. **379**, 80 (1991).

Kinney, R.M., McWilliams, J.C., and Tajima, T., Phys. Plasmas **2**, 3623 (1995). to be published.

Kinney, R.M., Tajima, T., McWilliams, J.C., and Petviashvili, N., Phys. Plasmas **1**, 260 (1994).

Koyama, K. Makishima, K., Tanaka, Y., and Tsunei, H., Publ. Astr. Soc. Jpn. **38**, 121 (1986).

Krommes, J.A., Phys. Fluids **25**, 1393 (1982).

Kronberg, P.P., Rep. Prog. Phys. **57**, 325 (1994).

Kronberg, P.P., ed. M. Nagano, to be published in *Proc. International Symposium*, "Extremely high energy cosmic rays: Astrophysics and future observations," (Univ. Tokyo, Tokyo, 1997).

Kubo, R., J. Phys. Soc. Jpn. **12**, 570 (1957).

Kueny, C., and Tajima, T., notes (1996).

Kurki-Suonio, T., and Matzner, R., Phys. Rev. D **39**, 1046 (1989).

Kulsrud, R.M., Cen, R.Y., Ostriker, J.P., and Ryu, D.S., to appear in Astrophys. J. (May 10, 1997).

Kulsrud, R.M., and Anderson, S., Astrophys. J. **396**, 606 (1992).

Lahav, O., Loeb, A., and McKee, C.F., Astrophys. J. **329**, L9 (1990).

Lai, C.H., and Tajima, T., to be published (1997).

Le Bellac, M., *Quantum and Statistical Field Theory*, (Clarendon Press-Oxford, 1991).

Leboeuf, J-N., Tajima, T., and Dawson, J.M., Phys. Fluids **25**, 784 (1982).

Lynden-Bell, D., Mon. Not. R. Astr. Soc. **136**, 101 (1967).

Mahajan, S.M., and Valanju, P.M., Phys. Rev. **35**, 2543 (1987).

Mather, J.C., Cheng, E., Eplee, R., Isaacman, R., Meyer, S., Schafer, R., Weiss, R., Wright, E., Bennett, C., Boggess, N., Dwek, E., Gulkis, S., Hauser, M., Janssen, M., Kelsall, T., Lubin, P., Moseley, S., Murdock, T., Silverberg, R., Smoot, G.F., and Wilkinson, D., Astrophys. J. **354**, L37 (1990).

Matsumoto, R., Tajima, T., and Shibata, K. (1994), in preparation.

467

Meerson, B., Rev. Mod. Phys. **68**, 215 (1996).

McWilliams, J.C., *et al. Eddies and Marine Science*, ed. Robinson, A. (Springer-Verlag, Berlin, 1983).

McWilliams, J.C., J. Fluid Mech. **219**, 361 (1990) and reference therein.

Misner, C.W., Thorne, K.S., and Wheeler, J.A., *Gravitation*, (Freeman, W.H., San Francisco, 1970), p. 764.

Norman, C.A., in *Confrontation between Theories and Observations in Cosmology: Present Status and Future Programmes*, ed. J. Audouze and F. Melchiorri (North-Holland, 1990), p. 29.

Ono, Y., Morita, A., Katsurai, M., and Yamada, M., Phys. Fluids B **5**, 3691 (1993).

Ono, Y., Yamada, M., Tajima, T., and Matsumoto, R., Phys. Rev. Lett. **96**, 3328 (1996).

Onsager, L., Supplto. Nuovo Cim. **6**, 279 (1949).

Pacholczyk, A., *Radio Astrophysics* (San Francisco: Freeman, 1970).

Parker, E.N., J. Geophys. Res. **62**, 509 (1957).

Petschek, H.E., in *AAS-NASA Symposium on the Physics of Solar Flares*, ed. Hess, W.N. (NASA SP-50) (NASA, Washington, D.C., 1965), 425.

Pneuman, G.W., and Orrall, F.Q., in *Physics of the Sun*, eds. Sturrock, P.A., Holzer, T.E., Mihalas, D.M., and Ulrich, R.K., (Dordrecht: Reidel, 1986), p. 111.

Ratra, B., Phys. Rev. D **45**, 1913 (1992).

Rees, M., in *The Very Early Universe*, eds. Gibbons, G.W., Hawking, S.W., and Siklos, S.T.C. (Cambridge Univ. Press, 1987), p. 29.

Rostoker, N., Aamodt, R., and Eldridge, O., Ann. Phys. (NY) **31**, 243 (1965).

Ruzmaikin, A.A., and Sokoloff, D.D., Astrofizika **13**, 95 (1977).

Ryder, L.H., *Quantum Field Theory*, (Cambridge, 1985).

Sachs, R.K., and Wolfe, A.M., Astrophys. J. **147**, 73 (1967).

Sato, H., Matsuda, T., and Takeda, H., Prog. Theor. Phys. Suppl. **49**, 11 (1971).

Saunder, W., Frenk, C., Rowan-Robinson, M., Efstathiou, G., Lawrence, A., Kaiser, N., Ellis, R., Crawford, J., Xia, X., and Perry, I., Nature **349**, 32 (1991).

Shafranov, V.D., *Reviews of Plasma Physics* (Consultants Bureau, NY, 1970) Vol. 2, p. 103.

Silk, J., *The Big Bang*, (Freeman, W.H., New York, 1989).

Sitenko, A.G., *Electromagnetic Fluctuations in Plasmas*, (Academic Press, New York, 1967).

Smoot, G.F., *et al.*, Astrophys. J. **396**, L1 (1992).

Sunyaev, R.A., and Zeldovich, Ya.B., Astron. Astrophys. **20**, 189 (1972).

Sweet, P.A., in *Electromagnetic Phenomena in Cosmical Physics*, ed. Lehnert, B. (Cambridge, NY, 1953), p. 123.

Tajima, T., *Fusion Energy–1981* (International Centre for Theoretical Physics, Trieste, 1982), ed. B. McNamara, p. 403.

Tajima, T., Sakai, J-I., Nakajima, J-I., H., Kosugi, T., Brunel, F., and Kundu, M.R., Astrophys. J. **321**, 1031 (1987).

Tajima, T., *Computational Plasma Physics* (Addison-Wesley, Redwood City, CA, 1989).

Tajima, T., and Sakai, J.-I., Sov. J. Plasma Phys. **15**, 519 (1989).

Tajima, T., and Taniuti, T., Phys. Rev. A **42**, 3587 (1990).

Tajima, T., Shibata, K., Cable, S.B., and Kulsrud, R.M., Astrophys. J. **390**, 309 (1992).

Tajima, T., Cable, S.B., and Kulsrud, R.M., Phys. Fluids B **4**, 2338 (1992).

Tajima, T., Jancic, D., and Fisher, D. (notes) (1994).

Tajima, T., and Niu, Q., Inter. J. Mod. Phys. A (1997).

Tanuma, S., *et al.*, to be published (1997).

Taylor, J.B., Phys. Rev. Lett. **33**, 1139 (1978).

Taylor, J.B., Rev. Mod. Phys. **58**, 741 (1986).

Taylor, J.B., and Wright, E.L., Astrophys. J. **339**, 619 (1989).

Tribble, P.C., Mon. Not. R. Astr. Soc. **253**, 147 (1991).

Vekshtein, G.E., in *Reviews of Plasma Physics* ed. Kadomtsev, B.B. (Consultant Bureau, New York, 1990), Vol. 15, p. 1.

Weinberg, S., *Gravitation and Cosmology* (John Wiley, New York, 1972).

White, R.B., in *Handbook of Plasma Physics*, ed. by Rosenbluth, M.N., and Sagdeev, R.Z. (North-Holland, Amsterdam, 1983) vol. 1; White, R.B., Rev. Mod. Phys. **58**, 183 (1986). (North-Holland, Amsterdam, 1983) Vol. 1.

Yamada, M., *et al.*, Phys. Rev. Lett. **65**, 721 (1990).

Zaidman, E., and Tajima, T., Astrophys. J. **338**, 1139 (1989).

Zweibel, E.G., Astrophys. J. **329**, L1 (1988).

Epilogue

We have seen that the presence of magnetic fields and plasmas in the Universe (and astrophysical objects) plays an essential role in its properties and evolution. The inclusion of magnetic fields and plasma dynamics in astrophysics is thus fundamental in resolving many central questions. These effects immeasurably enrich the fabric of astrophysical phenomena as we have seen. In addition, more importantly, the presence of magnetic fields and plasmas in the Universe frequently alters the astrophysical processes in a fundamental way. The details have been described in Chapters 2-5. But broadly speaking, the magnetic and plasma presence tends: (i) to *initiate, enhance or maintain, the structure formation* of the Universe or astrophysical objects; (ii) to *provide efficient and sometimes violent paths to convert* gravitational *energies* into flows, radiation, or kinetic energies; (iii) to *facilitate or accelerate the evolution* of the Universe or astrophysical objects. Without the above elements concerning (i) *structures*, (ii) *properties*, and (iii) *evolution*, the correct grasp of astrophysical phenomena and our Universe is bleak. With deeper understanding of plasma astrophysics, we see the possibility that some of the outstanding problems of astrophysics may be unlocked, such as the structure of the cosmos, the violent nature of jets and flares, puzzles of ages of stars, galaxies and the Universe, etc.

List of References

Abramowitz, 312
Acheson, D.J., 138
Acton, L., 225
Adams, F.C., 331
Alfvén, H., 8, 36, 187
Anderson, S., 401
Anile, A.M., 98
Asseo, 20

Büchner, J., 242
Bachelor, 67
Baierlein, R., 399
Bak, P., 85
Balbus, S.A., 300–303, 305, 319, 320, 322, 326, 363, 364
Barnes, C.W., 197
Barnett, D.M., 242
Bartoe, J.D.-F., 202
Beck, R., 183
Belcher, S.A., 346
Benzi, R., 59
Berera, A., 429
Berk, H.L., 69, 80
Bernstein, I., 323
Bernstein, I.B., 35, 38
Bhattacharjee, A., 212, 394, 417
Biskamp, D., 53, 197, 208, 209, 221
Blaes, O., 368
Blandford, R.D., 352
Bloch, 69
Bogoliubov, 6
Bogoliubov, B., 34
Bohm, D., 82
Born, R., 180
Brachet, M., 53
Brandenburg, A., 135, 136, 185
Brants, J.J., 179
Broadhurst, T.J., 398
Brueckner, G., 202

Brunel, F., 212
Bruzek, A., 180, 199

Cable, S.B., 407, 428, 429
Cargil, P.J., 197
Carmichael, H., 197, 226
Carnevale, G.F., 59
Carreras, B., 68
Cattaneo, F., 131, 135
Chandrasekhar, S., 38, 45, 82, 130
Chatuvedi, P.K., 176
Chiba, M., 124
Chou, D., 179, 180, 186, 199
Cowling, T.G., 116
Culhane, L., 224

Daniel, J., 91, 95
Darwin, C., 31
Davidson, R.C., 177
Davies, R.M., 179
Dawson, J.M., 177
De Lapparent, V., 398
Defouw, R.J., 130
Devoucleurs, 27
Diamond, P.H., 70
Dixon, 98
Dombre, T., 402
Doxas, I., 244
Drake, J.F., 29, 69, 190, 191, 247

Eddy, J.A., 8, 124
Edwards, S.F., 52
Elsasser, 52
Elsasser, W.M., 54

Fabian, A.C., 389, 395, 398, 401
Fang, L.Z., 429
Feigenbaum, 40, 69
Feingold, M., 402

Subject Index

emerging flux region, 198
energy principle, 53
enstrophy, 66
epicyclic frequency, 331
equipartition, 60
Euler-Lagrange equation, 450
evolutionary, 7
explosion, 59
explosive instability, 391
exponential solution, 186

Fadeev equilibrium, 230
Faraday rotation, 413
fast dynamo, 142
fast mode MHD wave, 51
fast reconnection, 142
Fermi mechanism, 396
Ferraro's law of isorotation, 369, 370
fiducial observer, 107
Fierz transformation, 450
filamentary MHD construct, 71, 449
filamentary structure, 28, 61, 69, 449
first-order smoothing, 134
fluctuation-dissipation theorem, 47, 60, 96, 335, 433, 436
flux conservation, 441
flux tube constriction, 416
flux tube convective instability, 145, 146, 154
flux tube dynamo, 141
force-free solution, 186
fractal current sheet, 47, 241, 157, 200
Frobenius method, 332

fully-developed turbulence, 64

galactic cosmic rays, 395
galactic coronae, 25
galactic dynamo model, 139
galactic ridge brightening, 422
galactic ridge X-ray emission, 24
galaxy correlation, 413
gamma-ray bursts, 389
gauge invariant, 456
gauge-independent, 462
gaussian statistics, 92
general relativity, 106
generalized helicity, 82, 85
generalized vorticity, 84, 428
geo-dynamo, 144
glide-reflection-symmetric mode, 165
global current sheet, 241
global reach, 44
global spiral modes, 330
Grad-Shafranov equations, 91, 368, 372
Grad-Shafranov operator, 91, 116
grand unified model, 257
granular dark lane, 198
granules, 148
gravitational (or buoyancy) force, 161
gravitational drift, 53
gravitational epoch, 46
Greisen-Zatsepin cutoff, 396
guiding center, 51
gyroBohm, 96

Hartman number, 151
heat conductivity, 47, 100

heat energy, 170
heat flux, 98, 100
helioseismological, 140
hierarchical, 6, 7, 41
Higgs' mechanism, 463
high β plasma, 22, 144, 206
high Reynolds number, 54, 336, 449
hot cluster plasmas, 24, 31
HR diagram, 32
hydrostatic objects, 130

indicial equation, 332
inertial domain, 67
inflow, 249
inhomogeneity of baryons, 441
inside-out collapse, 315
interchange mode, 157
intergalactic magnetic, 412
intergalactic matter, 29
interleaved structure, 200
intermittent, 373
internal gravity wave, 147
ion Bernstein wave, 49
ion cyclotron mode, 49
isolated flux tubes, 22
isothermal perturbations, 92

Kelvin-Helmholtz instability, 316
Keplerian speed, 315
kinematic $\alpha\omega$ dynamo, 130, 131, 138
kinematic regime, 152
kink instability, 417
Kippenhahn-Schlüter model, 169
Klimontovich equation, 450
knot theory, 6

knotty, cellular, 69
Kolmogoroff power law, 428
Kolmogorov spectrum, 68
Krommes-Kronig's relation, 439

Landau damping, 53
Landau pole, 52
Landau-Ginzburg equation, 63, 103
lapse function, 107
large-scale arcade formations, 244
Larmor radius, 95
LDE flare, 248
linear dynamo,, 197, 131
Liouville's theorem, 77
longitudinal, 48
loop-loop interaction, 252
low β plasma, 21, 144, 206
lower hybrid wave, 49
Lundquist number, 55

magnetic braking, 375
magnetic bubbles, 441
magnetic collapse, 231, 393
magnetic flux concentration, 149
magnetic flux expulsion, 149
magnetic insulation, 95
magnetic islands, 55
magnetic pressure driven jet, 386
magnetic reconnection, 42, 236, 242, 248, 256
magnetic rotational instability, 106
magnetic shear, 99
magnetic tension force, 161
magnetic topology, 55
magnetic twist, 200
magnetic viscosity parameter, 336,

CREDITS

Figure 1.10 from A.M. Hillas, Ann. Rev. Astron. Astrophys. vol. 22, p. 425, Copyright © 1984.

Figure 1.12 from G.A. Dulk, *et al.*, Solar Phys. vol. 57, p. 279, Copyright © 1978.

Figures 1.13 and 1.14 (a) from G.S. Vaiana, *et al.*, Proc. IAU Symp. on Solar and Stellar Magnetic Fields, p. 165, Copyright © 1983. (b) from R. Rosner, L. Golub, and G.S. Vaiana, Ann. Rev. Astron. Astrophys. vol. 23, p. 413, Copyright © 1985.

Figure 1.15 from D. Mathewson and V.L. Ford, Mon. Not. R. Astr. Soc. vol. 74, p. 139, Copyright © 1970.

Figure 1.16 from Troland and C. Heiles, Astrophys. J. vol. 301, p. 339, Copyright © 1986.

Figure 1.17 from Y. Sofue, K. Wakamatsu, and D.F. Malin, Astron. J. vol. 108, p. 2102, Copyright © 1994.

Figure 1.18 from N.E.B. Killeen, G.V. Bicknell, and R.D. Ekers, Astrophys. J. vol. 302, p. 306, Copyright © 1986.

Figure 1.19 from F. Yusef-Zadeh, M. Morris, and O. Chance, Nature vol. 310, p. 557, Copyright © 1984.

Figure 1.21 from K. Makishima, Proc. New Horizon of X-ray Astronomy, eds. Makino, F., and Ohashi, T., p. 171, Copyright © 1995.

Figure 2.3 from M.E. Oakes, R.B. Michie, K.H. Tsui, and J.E. Copeland, Plasma Phys. vol. 21, p. 205, Copyright © 1979.

Figure 2.4 from T. Tajima and J.-N. Leboeuf, Phys. Fluids vol. 23, p. 884, Copyright © 1980.

Figure 2.7 from T. Stribling and W. H. Mattheus, Phys. Fluids B , vol. 2, No. 9, Copyright © 1990.

Figure 2.9 (a)–(f) from the Yohkoh, (b)–(c) H. Zirin, (d) from C. Parma,

Figure 3.10 from D.O. Gough, A.G. Kosovichev, T. Sekii, K.G. Libbrecht, and M.F. Woodard, in *Proc. Gong 1992: Seismic Investigation of the Sun and Stars*, ed. Brown, T.M., (ASP Conference Series, 1993) vol. 42, p. 213, Copyright © 1993.

Figure 3.11 from I. Kawaguchi, Solar Phys. vol. 65, p. 207, Copyright © 1980.

Figure 3.12 from R. Muller, in *Solar and Stellar Granulation*, eds. R.J. Rutten and G. Severino, (Kluwer Academic Pub., 1989), p. 101, Copyright © 1989.

Figure 3.13 from D.J. Galloway and N.O. Weiss, Astrophys. J. vol. 243, p. 945, Copyright © 1981.

Figure 3.15 from D.J. Galloway and D.R. Moore, Geophys. Astrophys. Fluid Dyn. vol. 12, p. 73, Copyright © 1979.

Figure 3.16 from H. Hanami and T. Tajima, Astrophys. J. vol. 377, p. 694, Copyright © 1991.

Figures 3.21 and 3.22 from T. Horiuchi, R. Matsumoto, T. Hanawa, and K. Shibata, Publ. Astr. Soc. Jpn. vol. 40, p. 147, Copyright © 1988.

Figure 3.23 from N. Nakai, N. Kuno, T. Hanada, and Y. Sofue, Publ. Astr. Soc. Jpn. vol. 46, p. 527, Copyright © 1994.

Figures 3.24, 3.25, 3.26, 3.27 and 3.28 from R. Matsumoto, R. Horiuchi, K. Shibata, and T. Hanawa, Publ. Astr. Soc. Jpn. vol. 40, p. 171, Copyright © 1988.

Figure 3.29 from R. Matasumoto, T. Horiuchi, T. Hanawa, and K. Shibata, Astrophys. J. vol. 356, p. 259, Copyright © 1990.

Figures 3.30, 3.31 and 3.32 from K. Shibata, T. Tajima, R.S. Steinolfson, and R. Matsumoto, Astrophys. J. vol. 345, p. 584), Copyright © 1989.

Figure 3.33 from K. Shibata, T. Tajima, and R. Matsumoto, Phys. Fluids B 2 , Copyright © 1989.

1987.

Figure 3.55 from K. Shibata, S. Nozawa, and R. Matsumoto, Publ. Astr. Soc. Jpn. vol. 44, p. 265, Copyright © 1992.

Figure 3.56 from T. Tajima, J-I. Sakai, H. Nakajima, T. Kosugi, F. Bruel, and R. Kundu, Astrophys. J. vol. 321, p. 1031, Copyright © 1987.

Figure 3.57 from K. Shibata and Y. Uchida, Solar Phys. vol. 103, p. 299, Copyright © 1986.

Figures 3.61 and 3.62 from R.S. Steinolfson and G. Van Hoven, Phys. Fluids vol. 27, p. 1207, Copyright © 1984.

Figure 3.63 from T. Tajima, J-I. Sakai, H. Nakajima, T. Kosugi, F. Brunel, and R. Kundu, Astrophys. J. vol. 321, p. 1031, Copyright © 1987.

Figures 3.64, 3.65 and 3.66 from T. Yokoyama and K. Shibata, Astrophys. J. Lett. vol. 436, p. L197, Copyright © 1994.

Figure 3.69 from (a) S. Tsuneta, et al., Publ. Astr. Soc. Jpn. vol. 44, p. L63, Copyright © 1992, (b) S. Masuda, et al., Nature vol. 371, p. 495, Copyright © 1994.

Figure 3.70 from A. McAllister, et al., J. Geophys. Res. in press, Copyright © 1995.

Figures 3.71 and 3.72 from K. Shibata, et al., Astrophys. J. Lett. vol. 451, p. L83, Copyright © 1995, and Figure 3.72 from Shibata, K., Adv. Sp. Res. vol. 17, p. 9, Copyright © 1996.

Figure 3.73 from T. Shimizu, Publ. Astr. Soc. Jpn. vol. 47, p. 251, Copyright © 1995.

Figure 3.74 from K. Shibata, et al., Publ. Astr. Soc. Jpn. vol. 44, p. L173, Copyright © 1992.

Figures 3.75 and 3.76 from T. Yokoyama and K. Shibata, Nature; Publ. Astr. Soc. Jpn. vol. 3754, p. 42, Copyright © 1995, and Publ. Astr. Soc. Jpn. vol. 48, p. 353, Copyright © 1996.

Figure 3.78 from K. Shibata, Adv. Sp. Res. vol. 17, p. 9, Copyright © 1996.

Figure 4.2 from C. Hayashi, IAU Symp. No 115, p. 403, Copyright © 1987.

Figure 4.4 from R.C. Davidson, *Methods in Nonlinear Plasma Theory* (Academic, New York, 1972) p. 39, Copyright © 1972.

Figures 4.5 and 4.6 from R.B. Larson, Mon. Not. R. Astr. Soc. vol. 145, p. 271, Copyright © 1969.

Figure 4.7 from R.H. Shu, Astrophys. J. vol. 214, p. 214, Copyright © 1977.

Figures 4.8, 4.9 and 4.10 from S.W. Stahler, F.H. Shu, and R.E. Taam, Astrophys. J. vol. 241, p. 637, Copyright © 1980.

Figures 4.11 and 4.12 from S. Narita, C. Hayashi, and S.M. Miyama, Prog. Theor. Phys. vol. 72, p. 1118, Copyright © 1984.

Figure 4.13 from C. Hayashi, S. Narita and S.M. Miyama, Prog. Theor. Phys. vol. 68, p. 1949, Copyright © 1982.

Figures 4.14 and 4.15 from S.M. Miyama, C. Hayashi, and S. Narita, Astrophys. J. vol. 279, p. 621, Copyright © 1984.

Figure 4.19 from T. Nakano, Fund. Cosmic Phys. vol. 9, p. 139, Copyright © 1984.

Figure 4.20 from K. Tomisaka, Publ. Astr. Soc. Jpn. vol. 48, p. 701, Copyright © 1996.

Figures 4.24 and 4.25 from J.F. Hawley and S.A. Balbus, Astrophys. J. vol. 376, p. 223, Copyright © 1991.

Figures 4.26, 4.27, 4.28, 4.29 and 4.30 from R. Matsumoto and T. Tajima, Astrophys. J. vol. 445, p. 767, Copyright © 1995.

Figures 4.31 and 4.32 from R. Matsumoto, *et al.*, in *Basic Physics of Accretion Disks*, eds. S. Kato and S. Inagaki, Copyright © 1996.

Figure 4.35 from S. Miyamoto in *Proc. IIAS Workshop on Mathematical Approach to Fluctuations II*, 2, ed. T. Hida (Inter. Inst. Adv. Stud., Kyoto,

1993), Copyright © 1993.

Figure 4.36 from S. Mineshige, M. Kusunose, and R. Matsumoto, Astrophys. J. Lett. vol. 445, p. L43, Copyright © 1995.

Figure 4.37 from A.H. Bridle and R.A. Perley, Ann. Rev. Astron. Astrophys. vol. 22, p. 319, Copyright © 1984.

Figure 4.38 from S.C. Unwin, in *Proc. Parsec-scale Radio Jets*, (Cambridge Univ. Press, 1990), p. 13, Copyright © 1990.

Figure 4.39 from Y. Sofue and T. Hanada, Nature vol. 310, p. 568, Copyright © 1984.

Figure 4.40 from R.L. Snell, *et al.*, Astrophys. J. vol. 290, p. 587, Copyright © 1985.

Figure 4.41 from N. Kaifu, S. Suzuki, T. Hasegawa, M. Morimoto, J. Inatani, J. Nagane, K. Miyazawa, Y. Chikada, T. Kanazawa, and K. Akabane, Astron. Astrophys. vol. 134, p. 7, Copyright © 1984.

Figure 4.43 from R.C. Vermeulen, *et al.*, Astron. Astrophys. vol. 270, p. 177, Copyright © 1993.

Figure 4.45 from E.J. Weber and L. Davis, Astrophys. J. vol. 148, p. 217, Copyright © 1967.

Figure 4.46 from T. Sakurai, Astron. Astrophys. vol. 152, p. 121, Copyright © 1985.

Figure 4.47 from T. Kudoh and K. Shibata, Astrophys. J. vol. 474, p. 362, Copyright © 1997.

Figures 4.48 and 4.49 from R.D. Blandford and D.G. Payne, Mon. Not. R. Astr. Soc. vol. 199, p. 883, Copyright © 1982.

Figure 4.50 from T. Sakurai, Astron. Astrophys. vol. 152, p. 121, Copyright © 1985.

Figure 4.51 from Y. Uchida and K. Shibata, Publ. Astr. Soc. Jpn. vol. 37, p. 515, Copyright © 1985.

Milton Keynes UK
Ingram Content Group UK Ltd.
UKHW040712141024
449569UK00012B/605